Single Variable Student's Solutions Manual

to accompany

CALCULUS

Early Transcendental Functions

Third Edition

Robert T. Smith
Millersville University of Pennsylvania

Roland B. Minton
Roanoke College

Prepared by
Rebecca Torrey
Brandeis University

with contributions by
Ji Li
Brandeis University

McGraw Hill Higher Education

Boston Burr Ridge, IL Dubuque, IA Madison, WI New York San Francisco St. Louis
Bangkok Bogotá Caracas Kuala Lumpur Lisbon London Madrid Mexico City
Milan Montreal New Delhi Santiago Seoul Singapore Sydney Taipei Toronto

The McGraw-Hill Companies

Single Variable Student's Solutions Manual to accompany
CALCULUS: EARLY TRANSCENDENTAL FUNCTIONS, THIRD EDITION
ROBERT T. SMITH AND ROLAND B. MINTON

Published by McGraw-Hill Higher Education, an imprint of The McGraw-Hill Companies, Inc., 1221 Avenue of the Americas, New York, NY 10020.

This book is printed on recycled, acid-free paper containing 10% postconsumer waste.

1 2 3 4 5 6 7 8 9 0 DCD / DCD 0 9 8 7 6

ISBN-13: 978-0-07-286969-9
ISBN-10: 0-07-286969-0

www.mhhe.com

Contents

Chapter 0

Preliminaries

0.1 Polynomials and Rational Functions

1. Yes. The slope of the line joining the points $(2, 1)$ and $(0, 2)$ is $-\frac{1}{2}$, which is also the slope of the line joining the points $(0, 2)$ and $(4, 0)$.

3. No. The slope of the line joining the points $(4, 1)$ and $(3, 2)$ is -1, while the slope of the line joining the points $(3, 2)$ and $(1, 3)$ is $-\frac{1}{2}$.

5. Slope is $\dfrac{6-2}{3-1} = \dfrac{4}{2} = 2$.

7. Slope is $\dfrac{-1-(-6)}{1-3} = \dfrac{5}{-2} = -\dfrac{5}{2}$.

9. Slope is

$$\frac{-0.4-(-1.4)}{-1.1-0.3} = \frac{1.0}{-1.4} = -\frac{5}{7}.$$

In exercises 11-15, the equation of the line is given along with the graph. Any point on the given line will suffice for a second point on the line.

11. $y = 2(x-1)+3 = 2x+1$

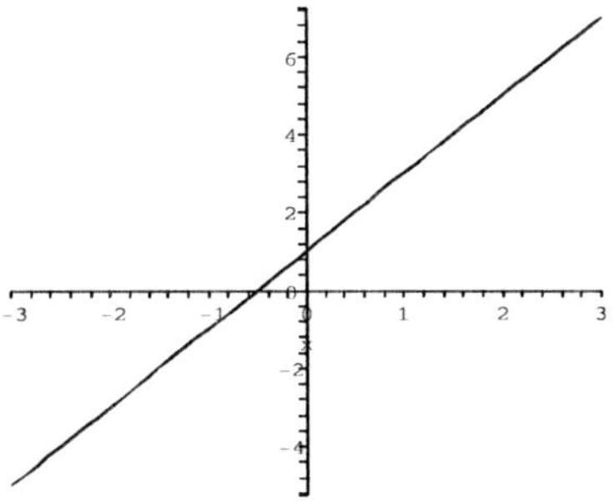

13. $y = 1$

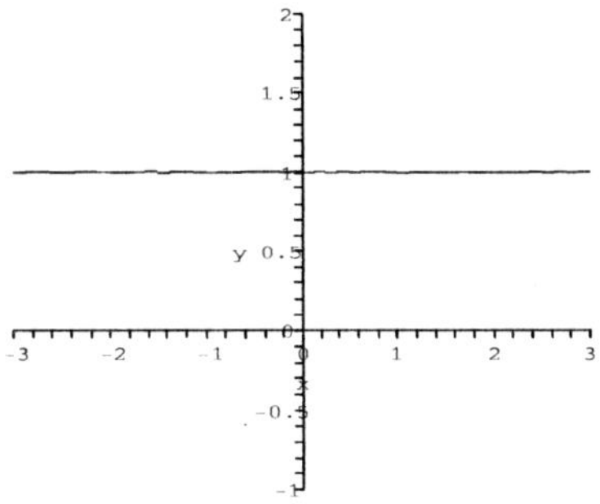

15. $y = 1.2(x-2.3)+1.1 = 1.2x - 1.66$

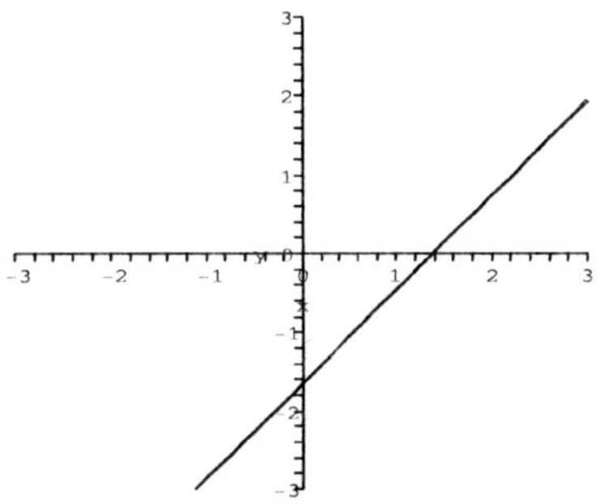

17. Parallel. Both have slope 3.

19. Perpendicular. Slopes are -2 and $\frac{1}{2}$.

21. Perpendicular. Slopes are 3 and $-\frac{1}{3}$.

23. (a) $y = 2(x-2)+1$

(b) $y = -\frac{1}{2}(x-2)+1$

25. (a) $y = 2(x-3)+1$

(b) $y = -\frac{1}{2}(x-3)+1$

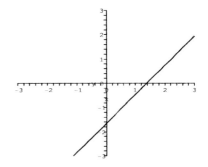

27. Slope $\frac{3-1}{2-1} = \frac{2}{1} = 2$ through the given points. One possibility:

$$y = 2(x-1)+1 = 2x-1.$$

When $x = 4$, $y = 7$.

29. Slope $\frac{3-4}{1-0.5} = \frac{-1}{0.5} = -2$ through the given points. One possibility:

$$y = -2(x-1)+3 = -2x+5.$$

When $x = 4$, $y = -3$.

31. Yes, passes vertical line test.

33. No. The vertical line $x = 0$ meets the curve twice; nearby vertical lines meet it three times.

35. Both: This is clearly a cubic polynomial, and also a rational function because it can be written as

$$f(x) = \frac{x^3 - 4x + 1}{1}.$$

(This shows that all polynomials are rational.)

37. Rational.

39. Neither: Contains square root.

41. We need the function under the square root to be non-negative. $x + 2 \geq 0$ when $x \geq -2$. The domain is $\{x \in \mathbb{R} | x \geq -2\} = [-2, \infty)$.

43. Negatives are permitted inside the cube root. There are no restrictions, so the domain is $(-\infty, \infty)$ or all real numbers.

45. The denominator cannot be zero. $x^2 - 1 = 0$ when $x = \pm 1$. The domain is $\{x \in \mathbb{R} | x \neq \pm 1\}$
$= (-\infty, -1) \cup (-1, 1) \cup (1, \infty)$

47.

$$f(0) = 0^2 - 0 - 1 = -1$$
$$f(2) = 2^2 - 2 - 1 = 1$$
$$f(-3) = (-3)^2 - (-3) - 1 = 11$$
$$f\left(\frac{1}{2}\right) = \left(\frac{1}{2}\right)^2 - \frac{1}{2} - 1 = -\frac{5}{4}$$

49.

$$f(0) = \sqrt{0+1} = 1$$
$$f(3) = \sqrt{3+1} = 2$$
$$f(-1) = \sqrt{-1+1} = 0$$
$$f\left(\frac{1}{2}\right) = \sqrt{\frac{1}{2}+1} = \sqrt{\frac{3}{2}} = \frac{\sqrt{6}}{2}$$

51. The only constraint we know is that the width should not be negative, so a reasonable domain would be $\{x | x > 0\}$.

53. Again, the only constraint we know for sure is that x should not be negative, i.e., a reasonable domain would be $\{x | x > 0\}$.

55. Answers vary. There may well be a positive correlation (more study hours = better grade), but not necessarily a functional relation.

57. Answers vary. While not denying a negative correlation (more exercise = less weight), there are too many other factors (metabolic rate, diet) to be able to quantify a person's weight as a function just of the amount of exercise.

59. A flat interval corresponds to an interval of constant speed; going up means that the speed is increasing while the graph going down means that the speed is decreasing. It is likely that the bicyclist is going uphill when the graph is going down and going downhill when the graph is going up.

61. The x-intercept occurs where $0 = x^2 - 2x - 8 = (x-4)(x+2)$, so $x = 4$ or $x = -2$; y-intercept at $y = 0^2 - 2(0) - 8 = -8$.

63. The x-intercept occurs where $0 = x^3 - 8 = (x-2)(x^2 + 2x + 4)$, so $x = 2$ (using the quadratic formula on the quadratic factor gives the solutions $x = -1 \pm \sqrt{-3}$, neither of which is real so neither contributes a solution); y-intercept at $y = 0^3 - 8 = -8$.

65. The x-intercept occurs where the numerator is zero, at $0 = x^2 - 4 = (x-2)(x+2)$, so $x = \pm 2$; y-intercept at
$$y = \frac{0^2 - 4}{0 + 1} = -4.$$

67. $x^2 - 4x + 3 = (x-3)(x-1)$, so the zeros are $x = 1$ and $x = 3$.

69. Quadratic formula gives
$$x = \frac{4 \pm \sqrt{16 - 8)}}{2} = 2 \pm \sqrt{2}$$

71.
$$\begin{aligned} x^3 - 3x^2 + 2x &= x(x^2 - 3x + 2) \\ &= x(x-2)(x-1), \end{aligned}$$
so the zeros are $x = 0$, 1, and 2.

73. With $t = x^3$, $x^6 + x^3 - 2$ becomes $t^2 + t - 2$ and factors as $(t+2)(t-1)$. The expression is zero only if one of the factors is zero, i.e., if $t = 1$ or $t = -2$. With $x = t^{1/3}$, the first occurs only if $x = (1)^{1/3} = 1$. The latter occurs only if $x = (-2)^{1/3}$, about -1.2599.

75. If $B(h) = -1.8h + 212$, then we can solve $B(h) = 98.6$ for h as follows:
$$\begin{aligned} 98.6 &= -1.8h + 212 \\ 1.8h &= 113.4 \\ h &= \frac{113.4}{1.8} = 63 \end{aligned}$$

This altitude (63,000 feet above sea-level, more than double the height of Mt. Everest) would be the elevation at which we humans boil alive in our skins. Of course the cold of space and the near-total lack of external pressure create additional complications which we shall not try to analyze.

77. This is a two-point line-fitting problem. If a point is interpreted as (x, y) =(temperature , chirp rate), then the two given points are $(79, 160)$ and $(64, 100)$. The slope being $\dfrac{160 - 100}{79 - 64} = \dfrac{60}{15} = 4$, we could write $y - 100 = 4(x - 64)$ or $y = 4x - 156$.

79. Her winning percentage is calculated by the formula $P = \dfrac{100w}{t}$, where P is the winning percentage, w is the number of games won and t is the total number of games. Plugging in $w = 415$ and $t = 415 + 120 = 535$, we find her winning percentage is approximately $P \approx 77.57$, so we see that the percentage displayed is rounded up from the actual percentage. Let x be the number of games won in a row. If she doesn't lose any games, her new winning percentage will be given by the formula $P = \dfrac{100(415 + x)}{535 + x}$. In order to have her winning percentage displayed as 80%, she only needs a winning percentage of 79.5 or greater. Thus, we must solve the inequality

$79.5 \leq \dfrac{100(415+x)}{535+x}$:

$$\begin{aligned} 79.5 &\leq \frac{100(415+x)}{535+x} \\ 79.5(535+x) &\leq 41500+100x \\ 42532.5+79.5x &\leq 41500+100x \\ 1032.5 &\leq 20.5x \\ 50.4 &\leq x \end{aligned}$$

(In the above, we are allowed to multiply both sides of the inequality by $535+x$ because we assume x (the number of wins in a row) is positive.) Thus she must win at least 50.4 times in a row to get her winning percentage to display as 80%. Since she can't win a fraction of a game, she must win at least 51 games in a row.

0.2 Graphing Calculators and Computer Algebra Systems

1. Intercepts $x = \pm 1$, $y = -1$. Minimum at $(0, -1)$. No asymptotes.

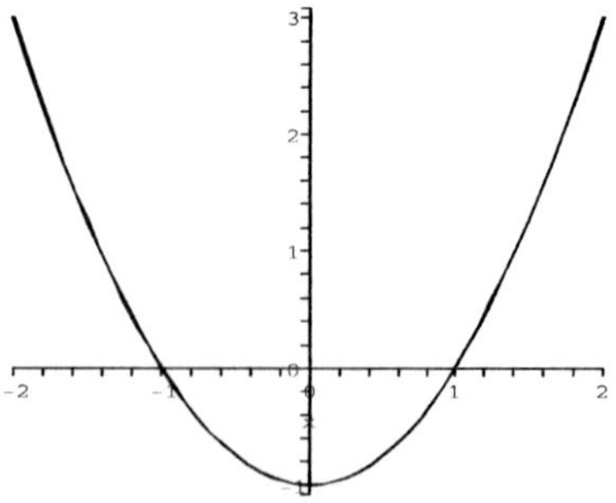

3. Intercepts $y = 8$ (no x-intercepts). Minimum at $(-1, 7)$. No asymptotes.

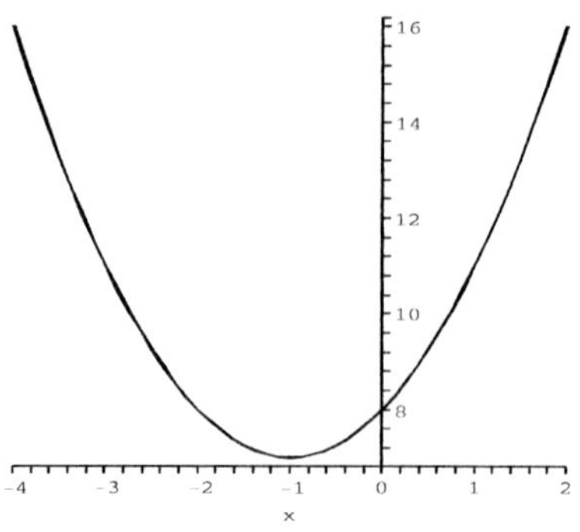

5. Intercepts $x = -1$, $y = 1$. No extrema or asymptotes.

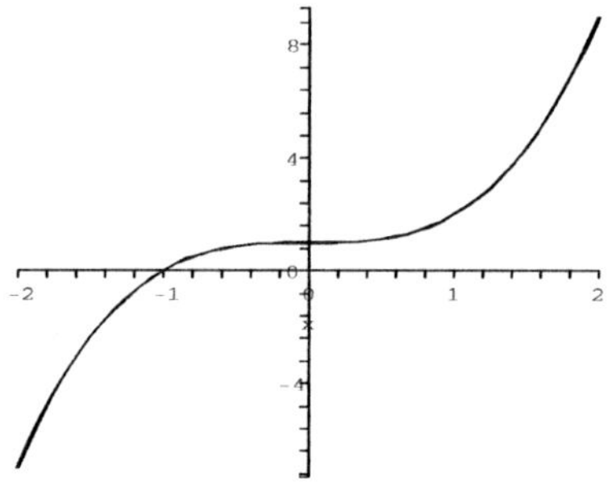

7. Intercepts $x \approx 0.453$, $y = -1$. No extrema or asymptotes.

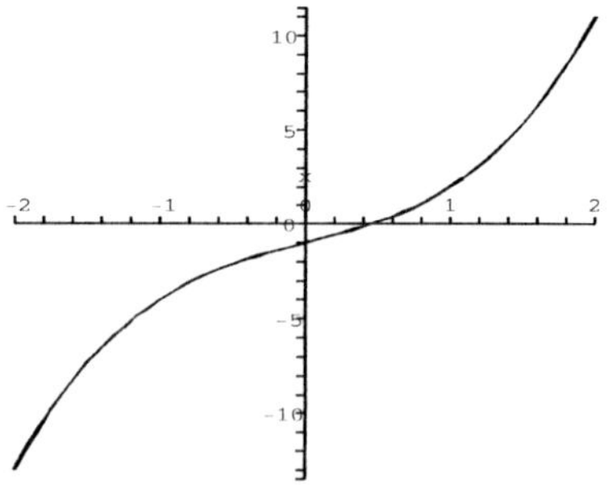

9. Intercepts $x \pm 1$, $y = -1$. Minimum at $(0, -1)$. No asymptotes.

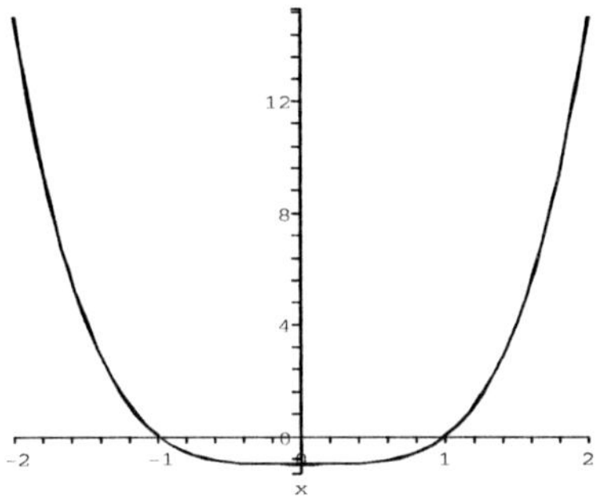

11. Intercepts $x \approx 0.475$, $x \approx -1.395$, $y = -1$. Minimum at (approximately) $(-1/\sqrt[3]{2}, -2.191)$. No asymptotes.

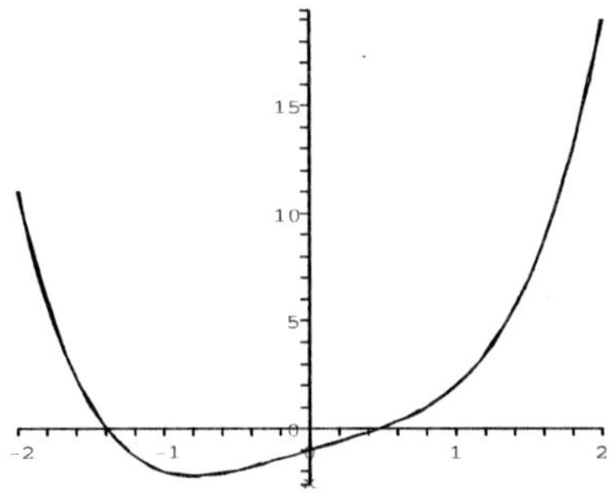

13. Intercepts $x \approx -1.149$, $y = 2$. No extrema or asymptotes.

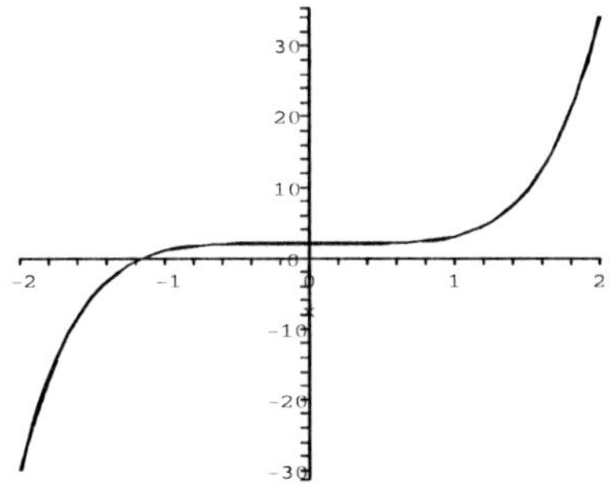

15. Intercepts $x \approx 0.050$, $y = -1$. The two local maxima occur at

$$x = \sqrt{\frac{24-\sqrt{176}}{10}} \text{ and } x = -\sqrt{\frac{24+\sqrt{176}}{10}},$$

while the two local minima occur at

$$x = \sqrt{\frac{24+\sqrt{176}}{10}} \text{ and } x = -\sqrt{\frac{24-\sqrt{176}}{10}}.$$

No asymptotes.

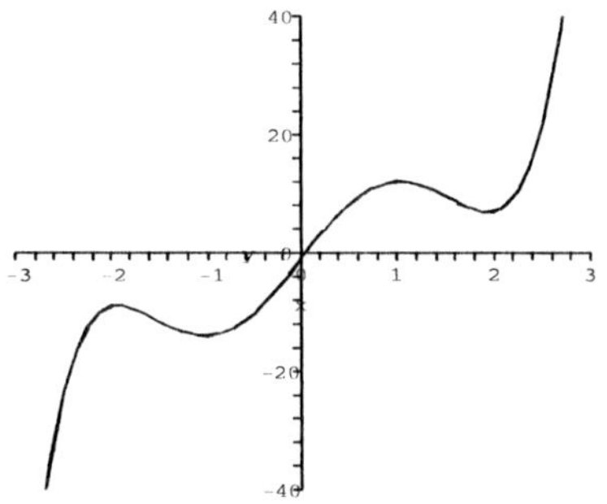

17. Intercepts $y = -3$ (no x-intercepts). No extrema. Horizontal asymptote $y = 0$. Vertical asymptote $x = 1$.

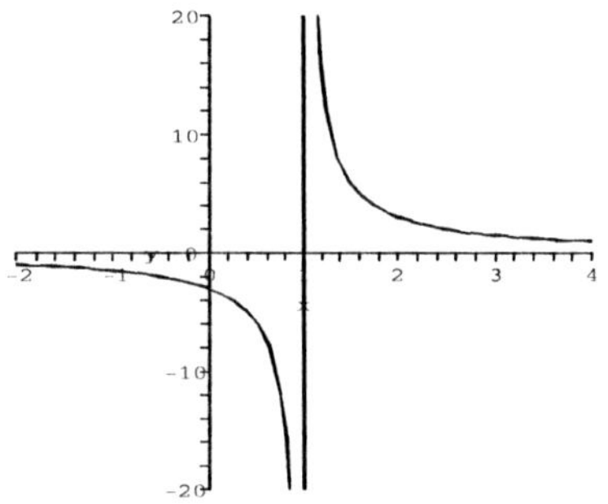

19. Intercepts $y = 0$ (and $x = 0$). No extrema. Horizontal asymptote $y = 3$. Vertical asymptote $x = 1$.

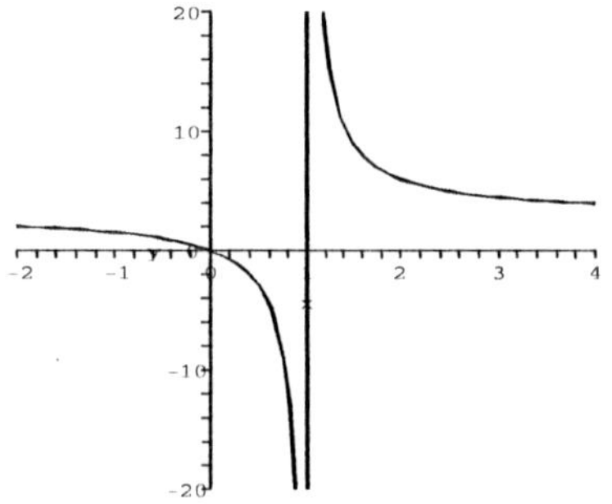

21. Intercepts $y = 0$ (and $x = 0$). Local maximum at $(0, 0)$. Local minimum at $(2, 12)$. Vertical asymptote $x = 1$. Slant asymptote $y = 3x + 3$.

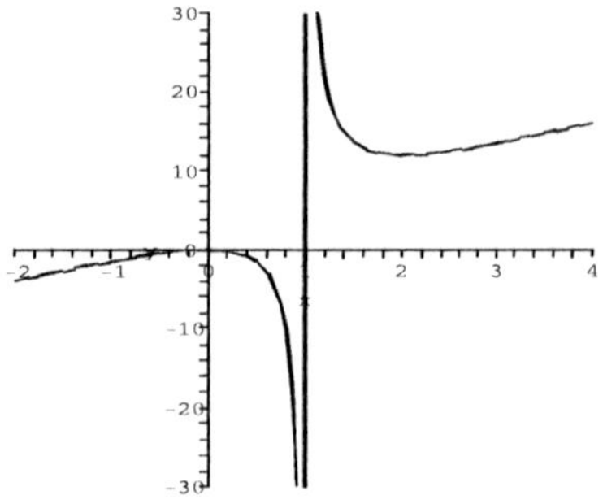

23. Intercepts $y = -1/2$ (no x-intercepts). Local maximum at $(0, -1/2)$. Vertical asymptotes $x = \pm 2$. Horizontal asymptote $y = 0$.

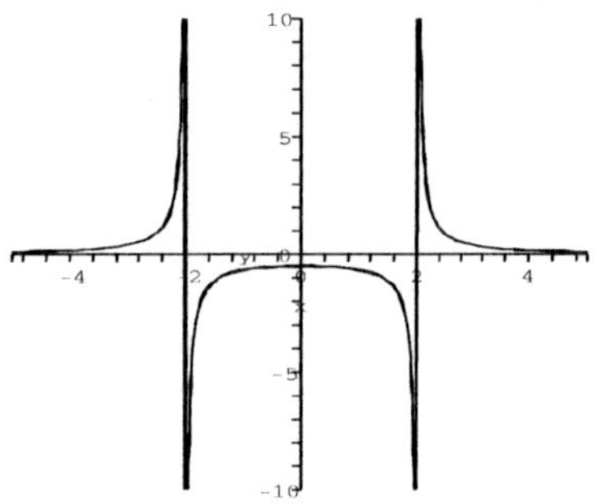

25. Intercepts $y = 3/4$ (no x-intercepts). Local maximum at $(0, 3/4)$. Horizontal asymptote $y = 0$.

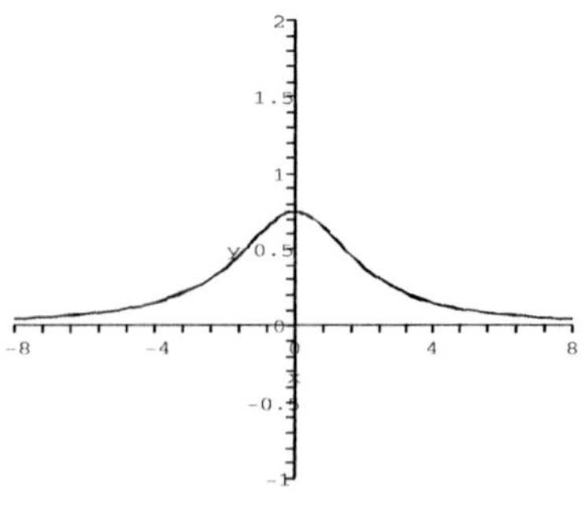

27. Intercepts $x = -2$, $y = -1/3$. No extrema. Horizontal asymptote $y = 0$. Vertical asymptotes at $x = -3$ and $x = 2$.

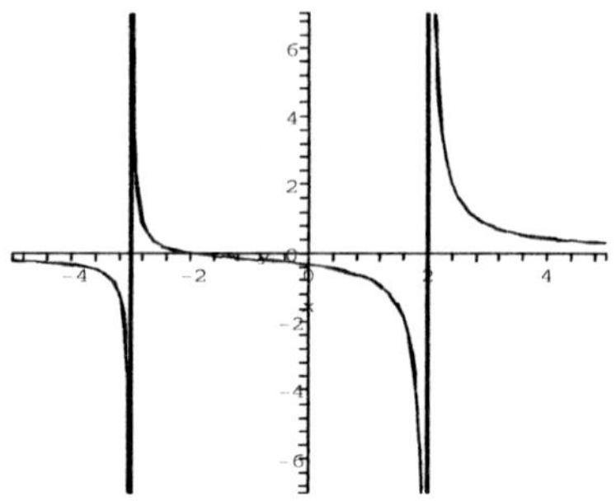

29. Intercepts $y = 0$ (and $x = 0$). No extrema. Horizontal asymptotes $y = \pm 3$.

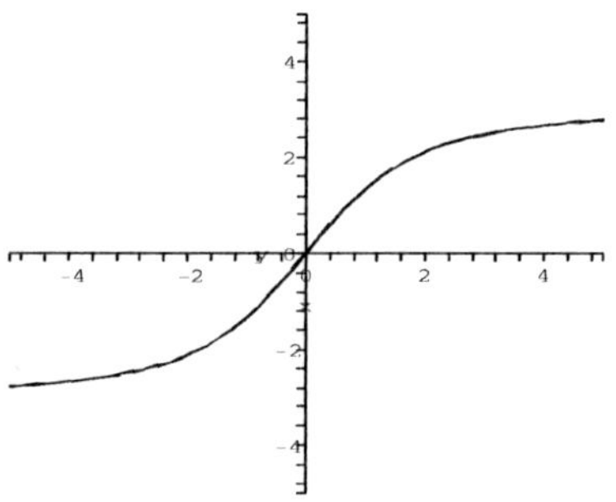

31. Vertical asymptotes where
$x^2 - 4 = 0 \Rightarrow x = \pm 2$.

33. Vertical asymptotes where
$x^2 + 3x - 10 = 0$
$\Rightarrow (x + 5)(x - 2) = 0$
$\Rightarrow x = -5$ or $x = 2$.

35. A vertical asymptote may occur when the denominator is zero. This denominator however is never zero, so there are no vertical asymptotes.

37. Vertical asymptotes where
$x^3 + 3x^2 + 2x = 0$
$\Rightarrow x(x^2 + 3x + 2) = 0$
$\Rightarrow x(x + 2)(x + 1) = 0$
$\Rightarrow x = 0$, $x = -2$, or $x = -1$.
Since none of these x values make the numerator zero, they are all vertical asymptotes.

39. A window with $-0.1 \le x \le 0.1$ and $-0.0001 \le y \le 0.0001$ shows all details.

41. A window with $-15 \le x \le 15$ and $-80 \le y \le 80$ shows all details.

43. From graph $x = 1$.

45. From graph $x = \pm 1$.

47. From graph $x = 0$.

49. The blow-up makes it appear that there are two intersection points. Solving algebraically, $\sqrt{x-1} = x^2 - 1$ (for $x \ge 1$) when
$x - 1 = (x^2-1)^2 = ((x-1)(x+1))^2$
$= (x-1)^2(x+1)^2$.
We see that $x = 1$ is one solution (obvious from the start), while for any other, we can cancel one factor of $x-1$ and find
$1 = (x-1)(x+1)^2 = (x^2-1)(x+1)$
$= x^3 + x^2 - x - 1$.
Hence $x^3 + x^2 - x - 2 = 0$.

By solver or spreadsheet, this equation has only the one solution $x \approx 1.206$.

51. The graph does not clearly show the number of intersection points. Solving algebraically,
$x^3 - 3x^2 = 1 - 3x$
$\Rightarrow x^3 - 3x^2 + 3x - 1 = 0$
$\Rightarrow (x-1)^3 = 0 \Rightarrow x = 1$.

So there is only one solution: $x = 1$.

53. After zooming out, the graph shows that there are two solutions: one near zero, and one around ten. Algebraically,
$(x^2-1)^{2/3} = 2x + 1$
$\Rightarrow (x^2-1)^2 = (2x+1)^3$
$\Rightarrow x^4 - 2x^2 + 1 = 8x^3 + 12x^2 + 6x + 1$
$\Rightarrow x^4 - 8x^3 - 14x^2 - 6x = 0$
$\Rightarrow x(x^3 - 8x^2 - 14x - 6) = 0$.
We thus confirm the obvious solution $x = 0$, and by solver or spreadsheet, find the second solution $x \approx 9.534$.

55. The graph shows that there are two solutions: $x \approx \pm 1.177$ by calculator or spreadsheet.

57. Calculator shows zeros at approximately -1.879, 0.347, and 1.532.

59. Calculator shows zeros at approximately .5637 and 3.0715.

61. Calculator shows zeros at approximately -5.248 and 10.006.

63. The graph of $y = x^2$ on the window $-10 \le x \le 10$, $-10 \le y \le 10$ appears identical (except for labels) to the graph of $y = 2(x-1)^2 + 3$ if the latter is drawn on a graphing window centered at the point $(1, 3)$ with

$$1 - 5\sqrt{2} \le x \le 1 + 5\sqrt{2}$$
$$-7 \le y \le 13.$$

65. $\sqrt{y^2}$ is the distance from (x, y) to the x-axis. $\sqrt{x^2 + (y-2)^2}$ is the distance from (x, y) to the point $(0, 2)$. If we require that these be the same, and we square both quantities, we have

$$\begin{aligned} y^2 &= x^2 + (y-2)^2 \\ y^2 &= x^2 + y^2 - 4y + 4 \\ 4y &= x^2 + 4 \\ y &= \frac{1}{4}x^2 + 1 \end{aligned}$$

In this relation, we see that y is a quadratic function of x. The graph is commonly known as a parabola.

0.3 Inverse Functions

1. $f(x) = x^5$ and $g(x) = x^{1/5}$

$$\begin{aligned} f(g(x)) &= f(x^{1/5}) \\ &= (x^{1/5})^5 = x^{(5/5)} = x \end{aligned}$$

$$\begin{aligned} g(f(x)) &= g(x^5) = (x^5)^{1/5} \\ &= x^{(5/5)} = x \end{aligned}$$

3.

$$\begin{aligned} f(g(x)) &= 2\left(\sqrt[3]{\frac{x-1}{2}}\right)^3 + 1 \\ &= 2\left(\frac{x-1}{2}\right) + 1 = x \end{aligned}$$

$$\begin{aligned} g(f(x)) &= \sqrt[3]{\frac{f(x)-1}{2}} \\ &= \sqrt[3]{\frac{2x^3+1-1}{2}} \\ &= \sqrt[3]{x^3} = x \end{aligned}$$

5. The function is one-to-one since $f(x) = x^3$ is one-to-one. To find the inverse function, write

$$\begin{aligned} y &= x^3 - 2 \\ y + 2 &= x^3 \\ \sqrt[3]{y+2} &= x \end{aligned}$$

So $f^{-1}(x) = \sqrt[3]{x+2}$.

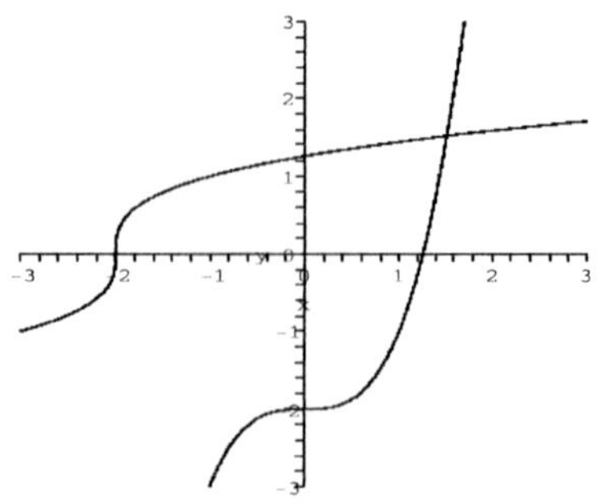

7. The graph of $y = x^5$ is one-to-one and hence so is $f(x) = x^5 - 1$. To find a formula for the inverse, write

$$\begin{aligned} y &= x^5 - 1 \\ y + 1 &= x^5 \\ \sqrt[5]{y+1} &= x \end{aligned}$$

So $f^{-1}(x) = \sqrt[5]{x+1}$.

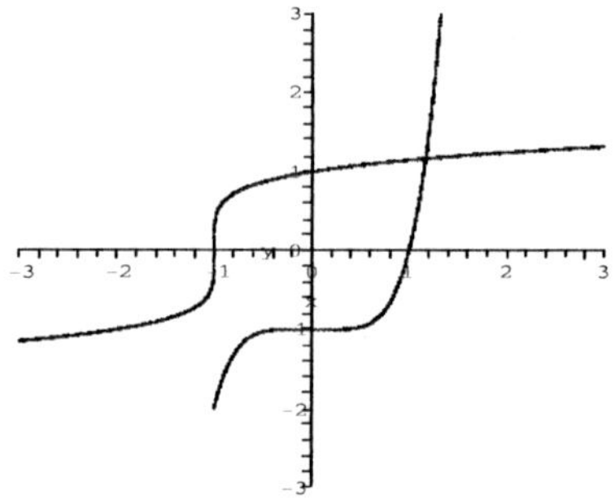

9. The function is not one-to-one since it is an even function ($f(-x) = f(x)$). In particular, $f(2) = 18 = f(-2)$.

11. Here, the natural domain requires that the radicand (the object inside the radical) be nonnegative. Hence $x \geq -1$ is required, while all function-values are nonnegative. Therefore the inverse, if defined at all, will be defined only for nonnegative numbers. Sometimes one can determine the existence of an inverse in the process of trying to find its formula. This is an example: Write

$$\begin{aligned} y &= \sqrt{x^3+1} \\ y^2 &= x^3 + 1 \\ y^2 - 1 &= x^3 \\ \sqrt[3]{y^2-1} &= x \end{aligned}$$

The left side is a formula for $f^{-1}(y)$, good for $y \geq 0$. Therefore, $f^{-1}(x) = \sqrt[3]{x^2-1}$ whenever $x \geq 0$.

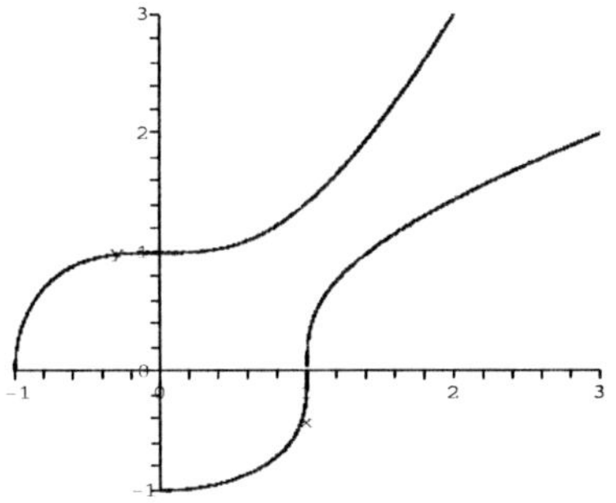

13. (a) Since $f(0) = -1$, we know $f^{-1}(-1) = 0$.

(b) Since $f(1) = 4$, we know $f^{-1}(4) = 1$.

15. (a) Since $f(-1) = -5$, we know $f^{-1}(-5) = -1$.

(b) Since $f(1) = 5$, we know $f^{-1}(5) = 1$.

17. (a) Since $f(2) = 4$, we know $f^{-1}(4) = 2$.

(b) Since $f(0) = 2$, we know $f^{-1}(2) = 0$.

19. Reflect the graph across the line $y = x$.

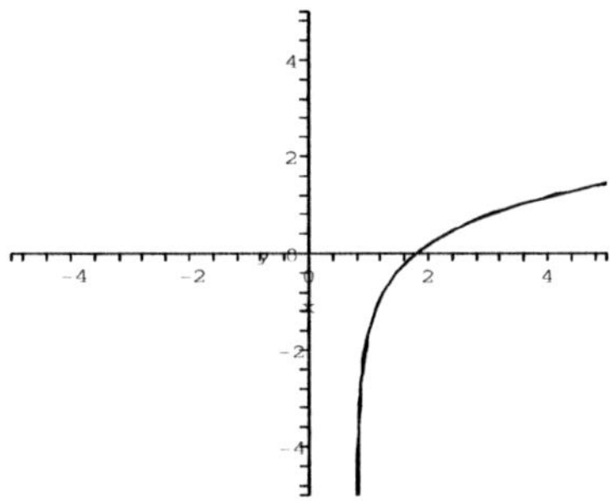

21. Reflect the graph across the line $y = x$.

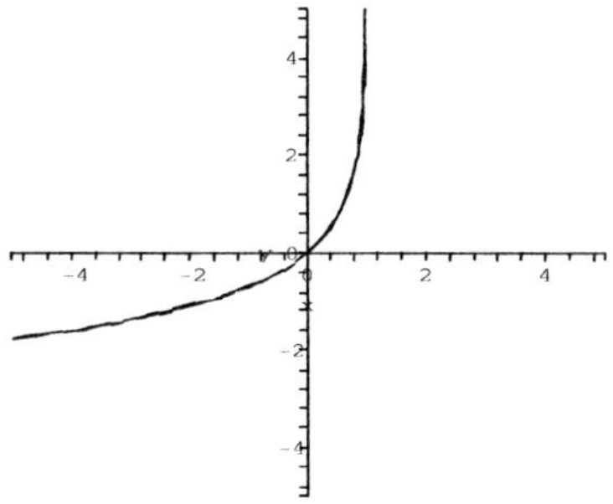

23. Since 23 is halfway between 20 ($= f(2)$) and 26 ($= f(3)$), the x-value for $y = 23$ should be halfway between 2 and 3, i.e., $f^{-1}(23)$ is estimated linearly by 2.5.

Since the lines between the points fall to the right of the apparent true curve of the graph, this estimate is too high.

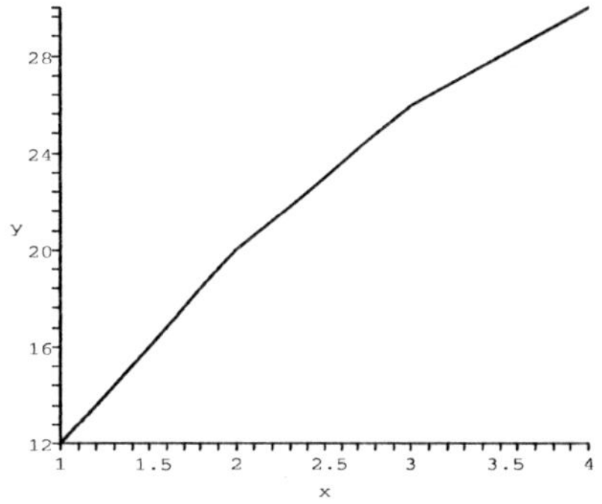

25. Since 5 is three-quarters of the way from 2 ($= f(3)$) to 6 ($= f(2)$), the x-value should be three-quarters of the way from 3 to 2, i.e., $f^{-1}(5)$ is estimated linearly by 2.25.

Since the lines between the points fall to the right of the apparent true curve of the graph, this estimate is too high.

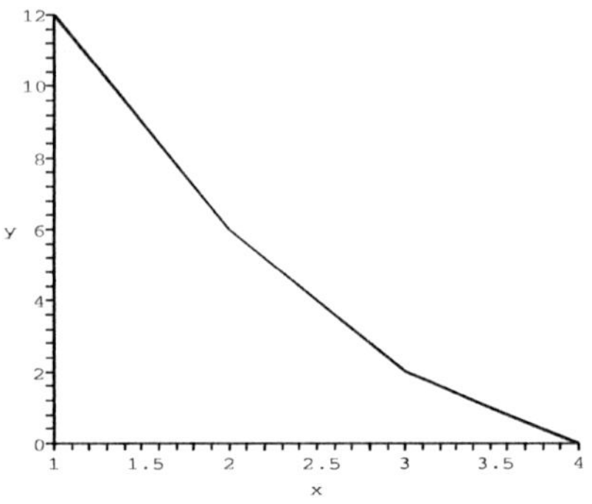

27. If $f(x) = x^3 - 5$, then the horizontal line test is passed, so $f(x)$ is one-to-one.

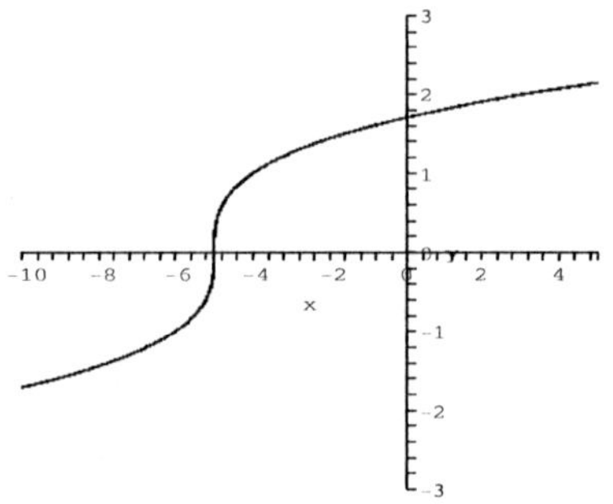

29. The function $f(x) = x^3 + 2x - 1$ easily passes the horizontal line test and is invertible.

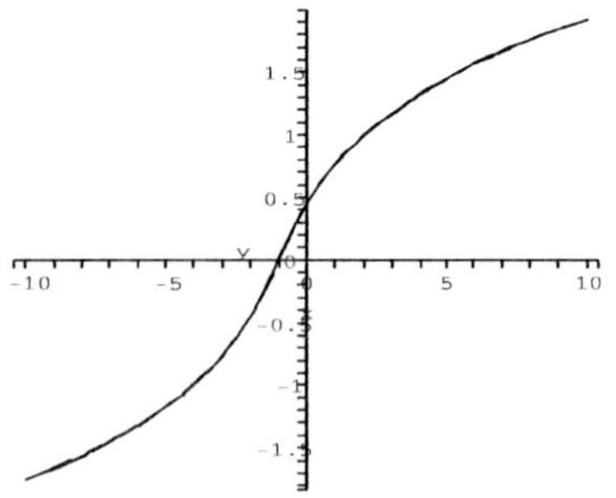

31. Not one-to-one. Fails horizontal line test.

33. If $f(x) = \dfrac{1}{x+1}$, then the horizontal line test is passed, so $f(x)$ is one-to-one.

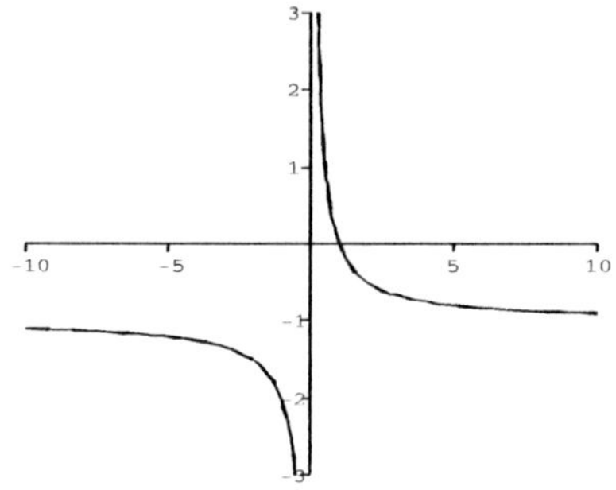

35. If $f(x) = \dfrac{x}{x+4}$, then the horizontal line test is passed so $f(x)$ is one-to-one.

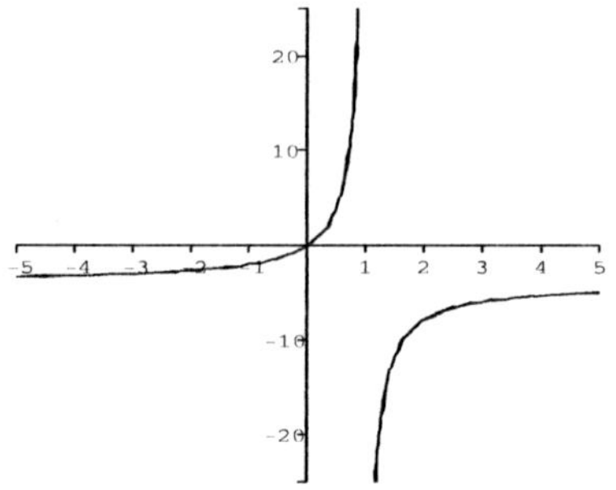

37. $f(g(x)) = (g(x))^2 = (\sqrt{x})^2 = x$
$g(f(x)) = \sqrt{f(x)} = \sqrt{x^2} = |x|$.
Because $x \geq 0$, the absolute value is the same as x. Thus these functions (both defined only when $x \geq 0$) are inverses.

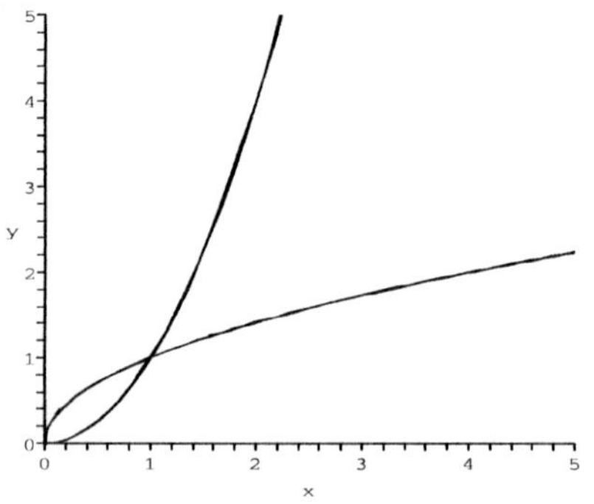

39. With $f(x) = x^2$ defined only for $x \leq 0$, (shown below as dotted) the horizontal line test is easily passed. The formula for the inverse function g is $g(x) = -\sqrt{x}$ shown below as solid and defined only for $x \geq 0$.

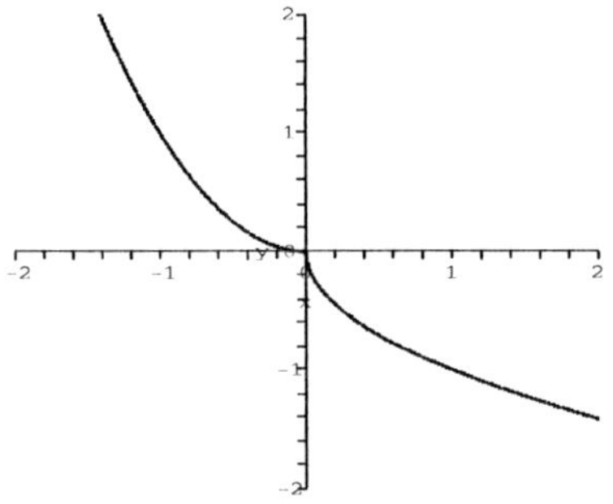

41. The graph of $y = (x-2)^2$ is a simple parabola with vertex at $(2,0)$. If we take only the right half $\{x \geq 2\}$ (shown below as the lower right graph) the horizontal line test is easily passed, and the formula for the inverse function g is $g(x) = 2 + \sqrt{x}$ defined only for $x \geq 0$ and shown below as the upper left graph.

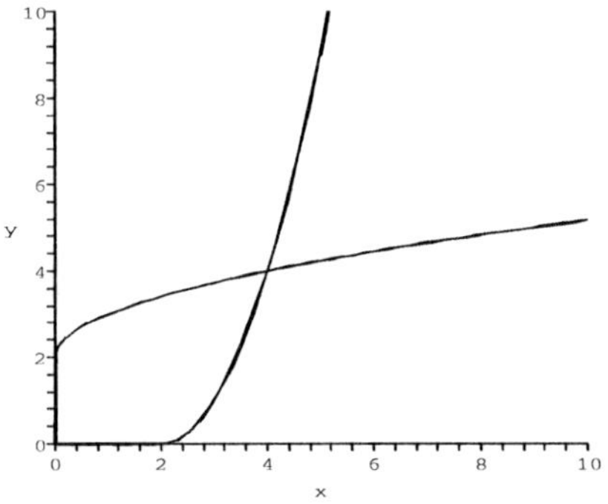

43. In the first place, for $f(x)$ to be defined, the radicand must be nonnegative, i.e., $0 \leq x^2 - 2x = x(x-2)$ which entails either $x \leq 0$ or $x \geq 2$. One can restrict the domain to either of these intervals and have an invertible function. Taking the latter for convenience, the inverse will be found as follows:

$$\begin{aligned} y &= \sqrt{x^2 - 2x} \\ y^2 &= x^2 - 2x = x^2 - 2x + 1 - 1 \\ &= (x-1)^2 - 1 \\ y^2 + 1 &= (x-1)^2 \\ \sqrt{y^2+1} &= \pm(x-1) \end{aligned}$$

With $x \geq 2$ and the left side nonnegative, we must choose the plus sign. We can then write $x = 1 + \sqrt{y^2+1}$.

The right side is now a formula for $f^{-1}(y)$, seemingly good for any y, but we recall from the original formula (as a radical) that y must be nonnegative. We summarize the conclusion:

$$f^{-1}(x) = 1 + \sqrt{x^2+1}, \ (x \geq 0).$$

This is the dotted graph below. The solid graph is the original $f(x) = \sqrt{x^2 - 2x}$.

Had we chosen $\{x \leq 0\}$, the "other half of the domain," and called the new function h, (same formula as f but a different domain, not shown) we would have come by choosing the minus sign, to the formula

$$h^{-1}(x) = 1 - \sqrt{x^2+1}, (x \geq 0).$$

The two inverse formulae, if graphed together, fill out the right half of the hyperbola $-x^2 + (y-1)^2 = 1$

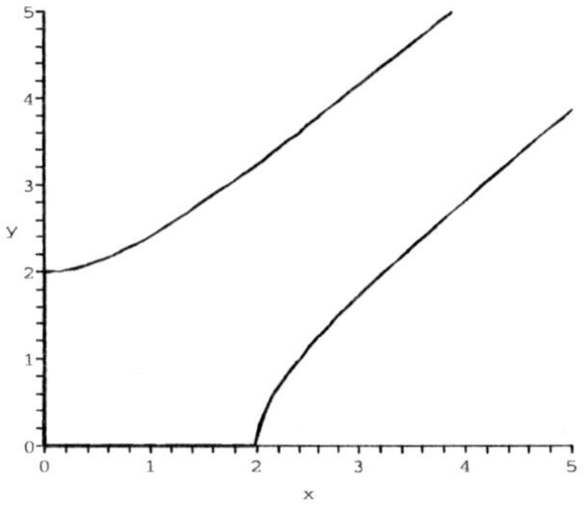

45. The function $\sin(x)$ (solid below) is increasing and one-to-one on the interval

$$-\frac{\pi}{2} \leq x \leq \frac{\pi}{2}.$$

One does not "find" the inverse in the sense of solving the equation $y = \sin(x)$ and obtaining a formula. It is done only in theory or as a graph. The name of the inverse is the "arcsin" function ($y = \arcsin(x)$ shown dotted), and some of its properties are developed in the next section.

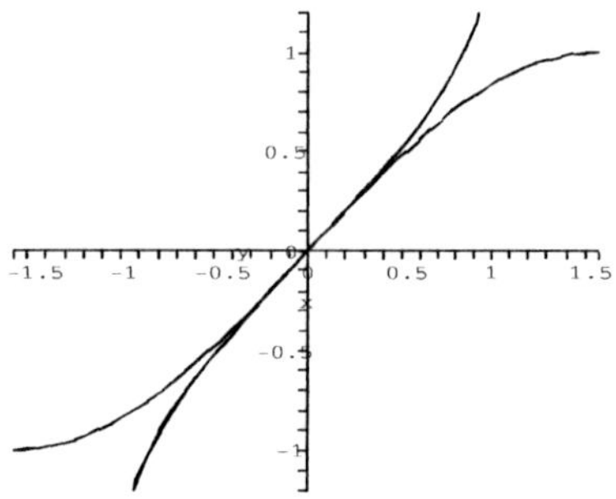

47. A company's income is not in fact a function of time, but a function of a time *interval* (income is defined as the change in *net worth*). When income *is* viewed as a function of time, it is usually after picking a fixed time interval (week, month, quarter, or year) and assigning the income for the period in a consistent manner to either the beginning or the ending date as in "... income for the quarter beginning... ." This much said, income more often than not rises and falls over time, so the function is unlikely to be one-to-one. In short, income functions *usually* do not have inverses.

49. During an interval of free fall following a drop, the height is decreasing with time and (barring a powerful updraft, as with hail) an inverse exists. After impact, if there is a bounce then some of the heights are repeated and the function is no longer one-to-one on the expanded time interval.

51. Two three-dimensional shapes with congruent profiles will cast identical shadows if the congruent profiles face the light source. Such objects need not be fully identical in shape. (For an example, think of a sphere and a hemisphere with the flat side of the latter facing the light.) The shadow as a function of shape is not one-to-one and does not have an inverse.

53. The usual meaning of a "ten percent cut in salary" is that the new salary is 90% of the old. Thus after a ten percent raise the salary is 1.1 times the original, and after a subsequent ten percent cut, the salary is 90% of the raised salary, or .9 times 1.1 times the original salary. The combined effect is 99% of the original, and therefore the ten percent raise and the ten percent cut are not inverse operations.

The 10%-raise function is $y = f(x) = (1.1)x$, and the inverse relation is $x = y/(1.1) = (0.90909\ldots)y$. Thus $f^{-1}(x) = (0.90909)x$ and in the language of cuts, this is a pay cut of fractional value $1 - 0.90909\ldots = 0.090909$ ... or $9.0909\ldots$ percent.

0.4 Trigonometric and Inverse Trigonometric Functions

1. (a) $\left(\frac{\pi}{4}\right)\left(\frac{180^\circ}{\pi}\right) = 45^\circ$

(b) $\left(\frac{\pi}{3}\right)\left(\frac{180^\circ}{\pi}\right) = 60^\circ$

(c) $\left(\frac{\pi}{6}\right)\left(\frac{180°}{\pi}\right) = 30°$

(d) $\left(\frac{4\pi}{3}\right)\left(\frac{180°}{\pi}\right) = 240°$

3. (a) $(180°)\left(\frac{\pi}{180°}\right) = \pi$

(b) $(270°)\left(\frac{\pi}{180°}\right) = \frac{3\pi}{2}$

(c) $(120°)\left(\frac{\pi}{180°}\right) = \frac{2\pi}{3}$

(d) $(30°)\left(\frac{\pi}{180°}\right) = \frac{\pi}{6}$

5. $2\cos(x) - 1 = 0$ when $\cos(x) = 1/2$. This occurs whenever $x = \frac{\pi}{3} + 2k\pi$ or $x = -\frac{\pi}{3} + 2k\pi$ for any integer k.

7. $\sqrt{2}\cos(x) - 1 = 0$ when $\cos(x) = 1/\sqrt{2}$. This occurs whenever $x = \frac{\pi}{4} + 2k\pi$ or $x = -\frac{\pi}{4} + 2k\pi$ for any integer k.

9. $\sin^2 x - 4\sin x + 3 = (\sin x - 1)(\sin x - 3) = 0$ when $\sin x = 1$ ($\sin x \neq 3$ for any x). This occurs whenever $x = \frac{\pi}{2} + 2k\pi$ for any integer k.

11. $\sin^2 x + \cos x - 1 = (1 - \cos^2 x) + \cos x - 1 = (\cos x)(\cos x - 1) = 0$ when $\cos x = 0$ or $\cos x = 1$. This occurs whenever $x = \frac{\pi}{2} + k\pi$ or $x = 2k\pi$ for any integer k.

13. $\cos^2 x + \cos x = (\cos x)(\cos x + 1) = 0$ when $\cos x = 0$ or $\cos x = -1$. This occurs whenever $x = \frac{\pi}{2} + k\pi$ or $x = \pi + 2k\pi$ for any integer k.

15. The graph of $f(x) = \sin 2x$.

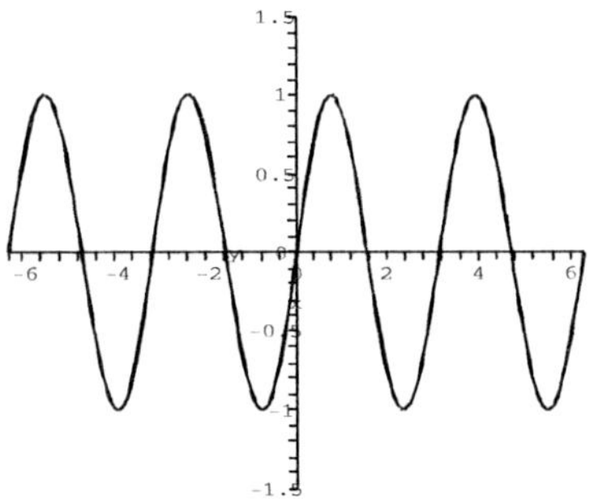

17. The graph of $f(x) = \tan 2x$.

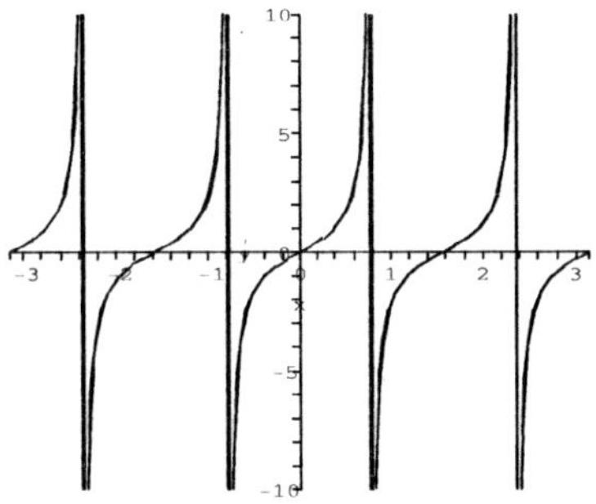

19. The graph of $f(x) = 3\cos(x - \pi/2)$.

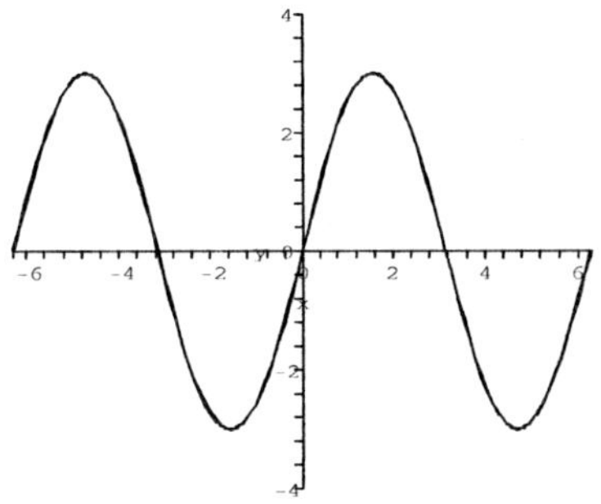

21. The graph of $f(x) = \sin 2x - 2\cos 2x$.

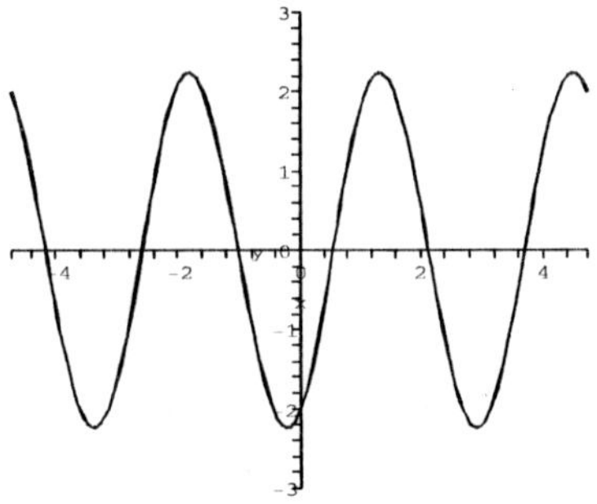

23. The graph of $f(x) = \sin x \sin 12x$.

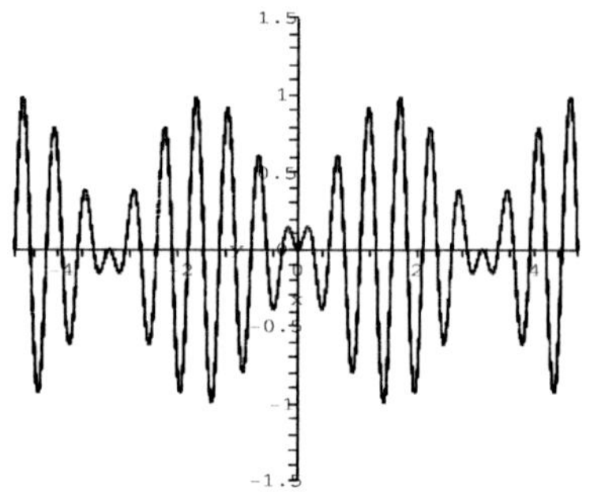

25. Amplitude is 3, period is $\frac{2\pi}{2} = \pi$, frequency is $\frac{1}{\pi}$.

27. Amplitude is 5, period is $\frac{2\pi}{3}$, frequency is $\frac{3}{2\pi}$.

29. Amplitude is 3, period is $\frac{2\pi}{2} = \pi$, frequency is $\frac{1}{\pi}$.

We are completely ignoring the presence of $-\pi/2$. This has an influence on the so-called "phase shift" which will be studied in Chapter 6.

31. Amplitude is 4 (the graph oscillates between -4 and 4, so we may ignore the minus sign), period is 2π, frequency is $\frac{1}{2\pi}$.

33. $\sin(\alpha - \beta) = \sin(\alpha + (-\beta))$
$= \sin\alpha\cos(-\beta) + \sin(-\beta)\cos\alpha$
$= \sin\alpha\cos\beta - \sin\beta\cos\alpha$

35. (a) $\cos(2\theta) = \cos(\theta + \theta)$
$= \cos(\theta)\cos(\theta) - \sin(\theta)\sin(\theta)$
$\cos^2\theta - \sin^2\theta = \cos^2\theta - (1 - \cos^2\theta)$
$= 2\cos^2\theta - 1$

(b) Just continue on, writing
$\cos(2\theta) = 2\cos^2\theta - 1$
$= 2(1 - \sin^2\theta) - 1 = 1 - 2\sin^2\theta$

37. From unit circle $\cos^{-1} 0 = \frac{\pi}{2}$.

39. From unit circle $\sin^{-1}(-1) = -\frac{\pi}{2}$.

41. From unit circle $\sec^{-1} 1 = 0$.

43. From unit circle $\sec^{-1} 2 = \frac{\pi}{3}$.

45. From unit circle $\cot^{-1}(1) = \frac{\pi}{4}$.

47. Use the formula
$\cos(x + \beta) = \cos x\cos\beta - \sin\beta\sin x$.
Now we see that $\cos\beta$ must equal $4/5$ and $\sin\beta$ must equal $3/5$. Since $(4/5)^2 + (3/5)^2 = 1$, this is possible. We see that $\beta = \sin^{-1}(3/5) \approx 0.6435$ radians, or $-36.87°$.

49. $\cos(2x)$ has period $\frac{2\pi}{2} = \pi$ and $\sin(\pi x)$ has period $\frac{2\pi}{\pi} = 2$. There are no common integer multiples of the periods, so the function $f(x) = \cos(2x) + 3\sin(\pi x)$ is not periodic.

51. $\sin(2x)$ has period $\frac{2\pi}{2} = \pi$ and $\cos(5x)$ has period $\frac{2\pi}{5}$. The smallest integer multiple of both of these is the fundamental period, and it is 2π.

53. $\cos^2\theta = 1 - \sin^2\theta$
$= 1 - \left(\frac{1}{3}\right)^2 = 1 - \frac{1}{9} = \frac{8}{9}$.

Because θ is in the first quadrant, its cosine is nonnegative. Hence

$\cos\theta = \sqrt{\frac{8}{9}} = \frac{2\sqrt{2}}{3} = 0.9428$.

55. Second quadrant, 1-$\sqrt{3}$-2 right triangle, so $\cos\theta = -\frac{\sqrt{3}}{2}$.

57. Assume $0 < x < 1$ and give the temporary name θ to $\sin^{-1}(x)$. In a right triangle with hypotenuse 1 and one leg of length x, the angle θ will show up opposite the x-side, and the adjacent side will have length $\sqrt{1 - x^2}$. Write

$$\cos(\sin^{-1}(x)) = \cos(\theta) = \frac{\sqrt{1 - x^2}}{1} = \sqrt{1 - x^2}.$$

The formula is numerically correct in the cases $x = 0$ and $x = 1$, and both sides are even functions of x, i.e. $f(-x) = f(x)$, so the formula is good for $-1 \le x \le 1$.

59. Assume $1 < x$ and give the temporary name θ to $\sec^{-1}(x)$. In a right triangle with hypotenuse x and one leg of length 1, the angle θ will show up adjacent to the side of length 1, and the opposite side will have length $\sqrt{x^2-1}$. Write

$$\tan(\sec^{-1}(x)) = \tan(\theta) = \frac{\sqrt{x^2-1}}{1} = \sqrt{x^2-1}.$$

The formula is numerically correct in the case $x \ge 1$.

Dealing with negative x is trickier: assume $x > 1$ for the moment. The key identity is $\sec^{-1}(-x) = \pi - \sec^{-1}(x)$. Taking tangents on both sides and applying the identity

$$\tan(a-b) = \frac{\tan(a) - \tan(b)}{1 + \tan(a)\tan(b)}$$

with $a = \pi$, $\tan(a) = 0$, $b = \sec^{-1}(x)$, we find

$$\tan(\sec^{-1}(-x)) = \frac{0 - \tan(\sec^{-1}(x))}{1+0} = -\sqrt{x^2-1} = -\sqrt{(-x)^2-1}$$

In this identity, $-x$ (on both sides) plays the role of an arbitrary number < -1. Consequently, the final formula is $\tan(\sec^{-1}(x)) = -\sqrt{x^2-1}$ whenever $x \le -1$.

61. One can use the formula $\sin(\cos^{-1} x) = \sqrt{1-x^2}$ derived in the text:

$$\sin(\cos^{-1}(\tfrac{1}{2})) = \sqrt{1 - \left(\tfrac{1}{2}\right)^2} = \tfrac{\sqrt{3}}{2}.$$

63. $\cos^{-1}(\frac{3}{5})$ relates to a triangle in the first quadrant with adjacent side 3 and hypotenuse 5, so the opposite side must be 4 and then

$$\tan\left(\cos^{-1}\left(\frac{3}{5}\right)\right) = \frac{4}{3}.$$

65. From graph the three solutions are 0, 1.109, and 3.698.

67. From graph the two solutions are ± 1.455.

69. Let h be the height of the rocket. Then $\frac{h}{2} = \tan 20°$
$h = 2\tan 20° \approx 0.73$ (miles)

71. Let h be the height of the steeple. Then $\dfrac{h}{80+20} = \tan 50°$
$h = 100\tan 50° \approx 119.2$ (feet).

73. Using feet as the measuring standard, we find

$$\tan A = \frac{20/12}{x} = \frac{5}{3x}$$
$$A(x) = \tan^{-1}\left(\frac{5}{3x}\right)$$

The graph of $y = A(x)$ (of course, one has to choose an appropriate range to make this a function):

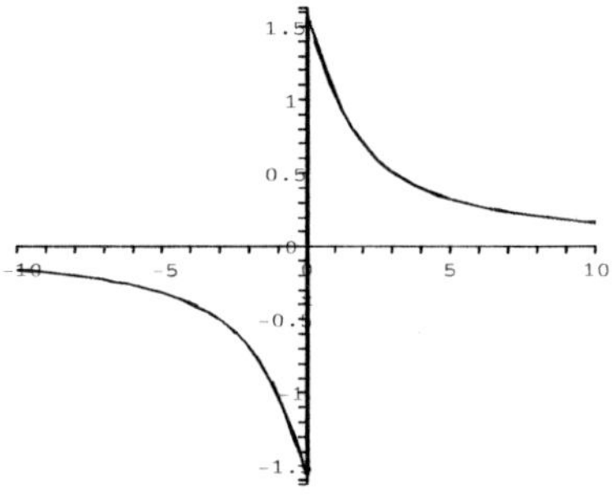

75. Presumably, the given amplitude (170) is the same as the "peak voltage" (v_p). Recalling an earlier discussion (#25 this section): the role of ω there is played by $2\pi f$ here, the frequency in cycles per second (Hz) was $\omega/2\pi$, which is now the f-parameter ($2\pi f/2\pi$). The period was $2\pi/\omega$ (which is now $1/f$), given in this case to be $\pi/30$ (seconds). So, apparently, the frequency is $f = 30/\pi$ (cycles per second) and the meter voltage is $\frac{170}{\sqrt{2}} \approx 120.2$.

77. There seems to be a certain slowly increasing base for sales ($110 + 2t$), and given that the sine function has period $\frac{2\pi}{\pi/6} = 12$ months, the sine term apparently represents some sort of seasonally cyclic pattern. If we assume that travel peaks at Thanksgiving, the effect is that time zero would correspond to a time one quarter-period (3 months) prior to Thanksgiving, or very late August.

The annual increase for the year beginning at time t is given by $s(t + 12) - s(t)$ and automatically ignores both the seasonal factor and the basic 110, and indeed it is the constant $2 \times 12 = 24$ (in thousands of dollars per year and independent of the reference point t).

79. As luck would have it, the trig functions csc and cot, being reciprocals respectively of sine and tangent, have inverses almost exactly where the other two do, both on the interval $\left[-\frac{\pi}{2}, \frac{\pi}{2}\right]$ but excluding the origin where neither is defined, and excluding the lower endpoint in the case of the cotangent. The range for the sine is $[-1, 1]$, hence the range for the csc is $\{|x| \geq 1\}$ and this is the domain for $\csc^{-1}$. The tangent assumes all values, and so does the cot (zero included as a value by convention when $x = \pi/2$ or $-\pi/2$), so the domain for $\cot^{-1}$ is universal. Finally, we simply copy the language of the others:

$y = \csc^{-1}(x)$ if $|x| \geq 1$,
y lies in $\left[-\frac{\pi}{2}, \frac{\pi}{2}\right]$ and $x = \csc(y)$.
$y = \cot^{-1}(x)$ if y lies in $\left(-\frac{\pi}{2}, \frac{\pi}{2}\right]$, and $x = \cot(y)$.

0.5 Exponential and Logarithmic Functions

1. $2^{-3} = \dfrac{1}{2^3} = \dfrac{1}{8}$

3. $3^{1/2} = \sqrt{3}$

5. $5^{2/3} = \sqrt[3]{5^2} = \sqrt[3]{25}$

7. $\dfrac{1}{x^2} = x^{-2}$

9. $\dfrac{2}{x^3} = 2x^{-3}$

11. $\dfrac{1}{2\sqrt{x}} = \dfrac{1}{2x^{1/2}} = \dfrac{1}{2}x^{-1/2}$

13. $4^{3/2} = \left(\sqrt{4}\right)^3 = 2^3 = 8$

15. $\dfrac{\sqrt{8}}{2^{1/2}} = \dfrac{\sqrt{8}}{\sqrt{2}} = \sqrt{4} = 2$

17. $2e^{-1/2} \approx 1.213$

19. $\dfrac{12}{e} \approx 4.415$

21. The graph of $f(x) = e^{2x}$:

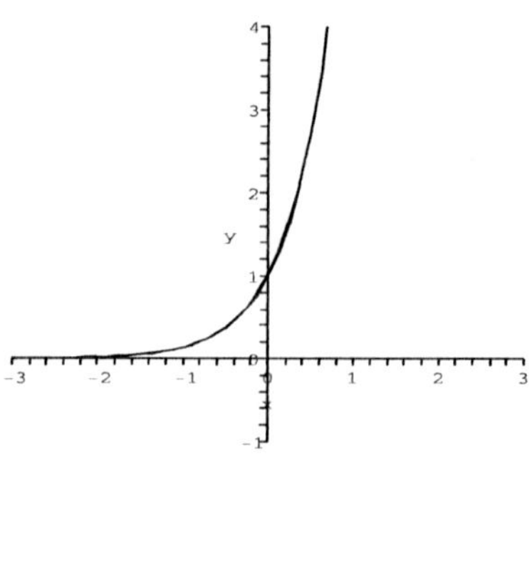

23. The graph of $f(x) = 2e^{x/4}$:

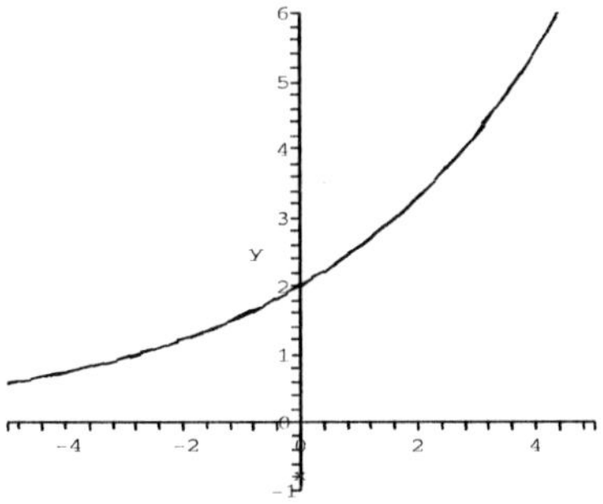

25. The graph of $f(x) = 3e^{-2x}$:

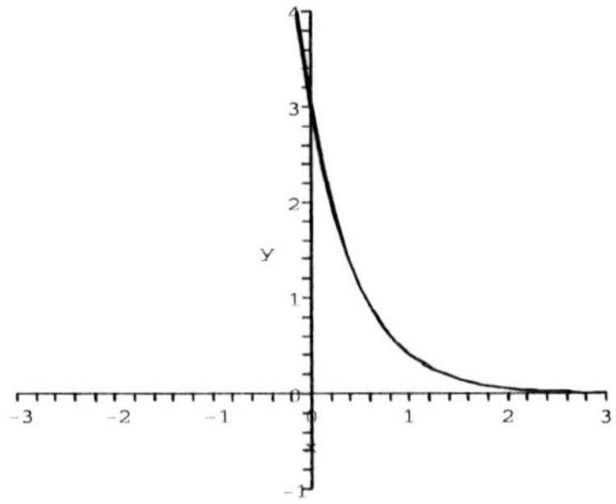

27. The graph of $f(x) = \ln 2x$:

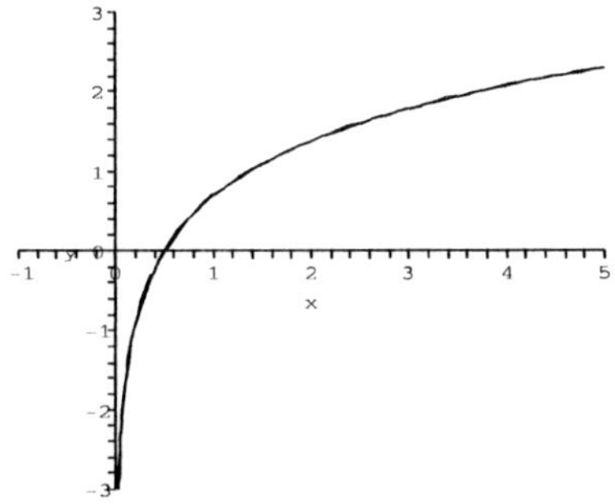

29. The graph of $f(x) = e^{2\ln x}$:

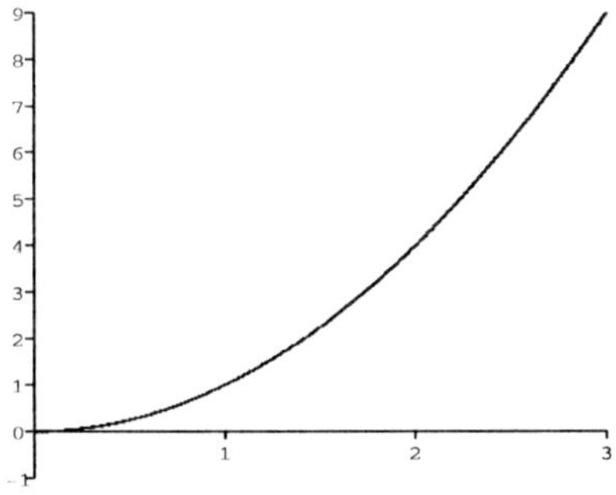

31. $e^{2x} = 2$
$\Rightarrow \ln e^{2x} = \ln 2$
$\Rightarrow 2x = \ln 2$
$\Rightarrow x = \frac{\ln 2}{2} \approx 0.3466$

33. $e^x(x^2 - 1) = 0$ implies either $x^2 - 1 = 0$ (hence $x = 1$ or $x = -1$), or $e^x = 0$ which has no solution.

35. $\ln 2x = 4$
$\Rightarrow 2x = e^4$
$\Rightarrow x = \frac{e^4}{2} \approx 27.299$

37. $4\ln x = -8$
$\Rightarrow \ln x = -2$
$\Rightarrow x = e^{-2} = \frac{1}{e^2} \approx 0.13533$

39. $e^{2\ln x} = 4$
$\Rightarrow 2\ln x = \ln 4$
$\Rightarrow \ln x^2 = \ln 4$
$\Rightarrow x^2 = 4$
$\Rightarrow x = \pm 2$,
but in the original equation we had the expression $e^{2\ln x}$ so $x \neq -2$ and thus the only solution is $x = 2$.

41. (a) $\log_3 9 = \log_3(3^2) = 2$

(b) $\log_4 64 = \log_4(4^3) = 3$

(c) $\log_3 \dfrac{1}{27} = \log_3(3^{-3}) = -3$

43. (a) $\log_3 7 = \dfrac{\ln 7}{\ln 3} \approx 1.771$

(b) $\log_4 60 = \dfrac{\ln 60}{\ln 4} \approx 2.953$

(c) $\log_3 \dfrac{1}{24} = \dfrac{\ln(1/24)}{\ln 3} \approx -2.893$

45. $\ln 3 - \ln 4 = \ln \frac{3}{4}$

47. $\frac{1}{2}\ln 4 - \ln 2 = \ln 4^{1/2} - \ln 2$
$= \ln 2 - \ln 2 = 0$

49. $\ln \frac{3}{4} + 4\ln 2 = \ln \frac{3}{2^2} + \ln 2^4$
$= \ln\left(\frac{3}{2^2} \cdot 2^4\right)$
$= \ln(3 \cdot 2^2) = \ln(12)$

51. $f(0) = 2 \Rightarrow a = 2$.
Then $f(2) = 6$ gives $2e^{2b} = 6$, so $2b = \ln 3$ and $b = \frac{1}{2}\ln 3$. So $f(x) = 2e^{(\frac{1}{2}\ln 3)x} = 2[e^{\ln(3)}]^{x/2} = 2 \cdot 3^{x/2}$.

53. $f(0) = 4 \Rightarrow a = 4$.
Then $f(2) = 2$ gives $4e^{2b} = 2$, so $2b = \ln\frac{1}{2}$ and $b = \frac{1}{2}\ln\frac{1}{2}$. So $f(x) = 4e^{(\frac{1}{2}\ln\frac{1}{2})x}$.

55. $1 - \left(\frac{9}{10}\right)^{10} \approx 0.651$

57. We take on faith, whatever it may mean, that

$$\lim_{n\to\infty}\left(1+\frac{1}{n}\right)^n = e$$

Just to take a sample starting with $n = 25$, the numbers are

$$\left(\frac{26}{25}\right)^{25}, \left(\frac{27}{26}\right)^{26}, \left(\frac{28}{27}\right)^{27},$$

and so on. If we were to try taking a similar look at the numbers in $\lim_{n\to\infty}\left(1-\frac{1}{n}\right)^n$, the numbers starting at $n = 26$ would be

$$\left(\frac{25}{26}\right)^{26}, \left(\frac{26}{27}\right)^{27}, \left(\frac{27}{28}\right)^{28},$$

and so on.

We could rewrite these as

$$\left[\left(\frac{25}{26}\right)^{25}\right]^{\frac{26}{25}}, \left[\left(\frac{26}{27}\right)^{26}\right]^{\frac{27}{26}}, \left[\left(\frac{27}{28}\right)^{27}\right]^{\frac{28}{27}}.$$

Here, the numbers inside the square brackets are the reciprocals of the numbers in the original list, which were all pretty close to e. Therefore these must all be pretty close to $1/e$. As to the external powers, they are all close to 1 and getting closer. This limit must be $1/e$. The expression in question must approach $1 - \frac{1}{e} \approx .632$

59.

$u = \ln x$	.78846	.87547	.95551
$v = \ln y$	2.6755	2.8495	3.0096

$u = \ln x$	1.0296	1.0986	1.1632
$v = \ln y$	3.1579	3.2958	3.4249

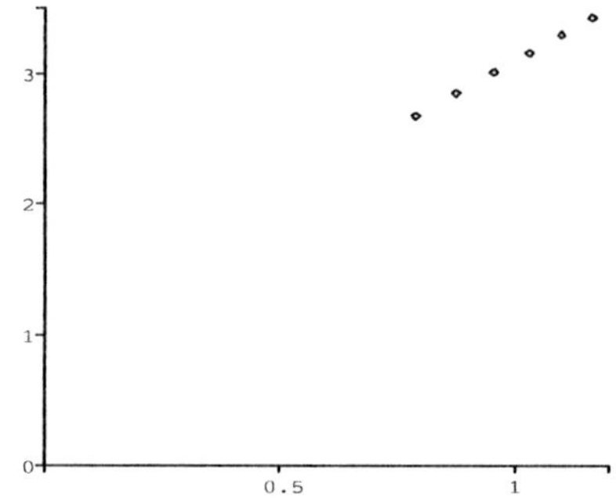

$m = \frac{3.4249-2.6775}{1.1632-.78846} \approx 2$.
Then we solve $2.6755 = 2.(.78846) + b$ to find $b \approx 1.099$. Now $b = \ln a$, so $a = e^b \approx 3.001$, and the function is $y = 3.001x^2$.

61. We compute $u = \ln x$ and $v = \ln y$ for x values in number of decades since 1780 and y values in millions.

$u = \ln x$	0	0.693	1.099	1.386
$v = \ln y$	1.36	1,668	1.974	2.262

$u = \ln x$	1.609	1.792	1.946	2.079
$v = \ln y$	2.549	2.839	3.14	3.447

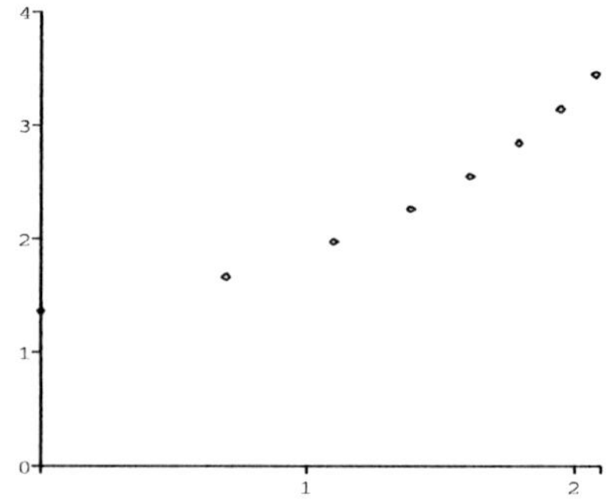

This plot does not look linear, which makes it clear that the population is *not* modeled by a power of x. The discussion in the Chapter has already strongly indicated that an exponential model is fairly good.

63. (a) $7 = -\log[H^+] \Rightarrow [H^+] = 10^{-7}$

(b) $[H^+] = 10^{-8}$

(c) $[H^+] = 10^{-9}$

For each increase in pH of one, $[H^+]$ is reduced to one tenth of its previous value.

65. (a) $\log E = 4.4 + 1.5(4) = 10.4 \Rightarrow E = 10^{10.4}$

(b) $\log E = 4.4 + 1.5(5) = 11.9 \Rightarrow E = 10^{11.9}$

(c) $\log E = 4.4 + 1.5(6) = 13.4 \Rightarrow E = 10^{13.4}$

For each increase in M of one, E is increased by a factor of $10^{1.5} \approx 31.6$.

67. (a) $80 = 10\log\left(\frac{I}{10^{-12}}\right) \Rightarrow$

$$8 = \log\left(\frac{I}{10^{-12}}\right) \Rightarrow$$

$$10^8 = \frac{I}{10^{-12}} \Rightarrow$$

$$I = 10^8 10^{-12} = 10^{-4}$$

(b) $I = 10^{-3}$

(c) $I = 10^{-2}$

For each increase in dB of ten, I increases by a factor of 10.

69. From the graphs, we estimate:
$y = xe^{-x}$ has a max value of $1/e$ at $x = 1$
$y = xe^{-2x}$ has a max value of $1/2e$ at $x = 1/2$
$y = xe^{-3x}$ has a max value of $1/3e$ at $x = 1/3$

So we guess that $y = xe^{-kx}$ has a max value of $1/ke$ at $x = 1/k$. If one believes the first of these, then the general case follows by writing

$$xe^{-kx} = \frac{kxe^{-kx}}{k} = \frac{ue^{-u}}{k}$$

if we let $u = kx$. The numerator has max value $1/e$ when $u = 1$, i.e., $x = 1/k$. Therefore the whole expression has max value $(1/k)(1/e)$ when $x = 1/k$.

71. We know that $\cosh x = \frac{e^x + e^{-x}}{2}$. To show that $\cosh x \geq 1$ for all x is the same as showing that $\cosh x - 1 \geq 0$ for all x. So we ask when is the expression

$$\cosh x - 1 = \frac{e^x + e^-x}{2} - 1$$

greater than or equal to 0? We have:

$\frac{e^x + e^{-x}}{2} - 1 \geq 0$ if and only if

$\frac{e^x + e^{-x} - 2}{2} \geq 0$ if and only if

$e^x + e^{-x} - 2 \geq 0$ if and only if

$e^{2x} + 1 - 2e^{-x} \geq 0$ if and only if *

$e^{2x} - 2e^{-x} + 1 \geq 0$ if and only if

$(e^x - 1)^2 \geq 0$

But $(e^x - 1)^2$ is always greater than or equal to 0 since it is squared. It is actually equal to 0 at $x = 0$ (i.e., $\cosh 0 = 1$), so the range of $y = \cosh x$ is $y \geq 1$.

* In the * step (above), we have multiplied on both sides by e^x, which we are allowed to do since $e^x > 0$ for all x.

To show that the range of the hyperbolic sine is all real numbers, let a be any real number and solve the equation $\sinh(x) = a$. Let $u = e^x$. Then

$\frac{u - \frac{1}{u}}{2} = a$ if and only if

$u^2 - 1 = 2au$ if and only if

$u^2 - 2au - 1 = 0$ if and only if

$$u = \frac{2a \pm \sqrt{4a^2 + 4}}{2} = a + \sqrt{a^2 + 1}.$$

We simplified and chose the positive square root because $u > 0$. Because we found a unique solution no matter what a we had started with, we have shown that the range of $y = \sinh x$ is the whole real line.

73. The issue is purely whether or not $y = 0$ when $x = 315$, i.e., whether or not $\cosh(315/127.7) = \cosh(2.4667\ldots) = 5.9343\ldots$ is the same as $(757.7)/(127.7) = 5.9334\ldots$ We see that it's pretty close, and these numbers would be considered equal according to the level of accuracy reported in the original measurements.

75. Since $\sinh^{-1}(0) = 0$, the equation is solved only by $x^2 - 1 = 0$, hence $x = 1$ or $x = -1$.

77. $f = f(x) = 220e^{x\ln(2)}$
$= 220e^{\ln(2^x)} = 220 \cdot 2^x$

0.6 Transformations of Functions

1. $(f \circ g)(x) = f(g(x))$
$= g(x) + 1 = \sqrt{x-3} + 1$
with domain $\{x | x \geq 3\}$.

$(g \circ f)(x) = g(f(x))$
$= \sqrt{f(x) - 1}$
$= \sqrt{(x+1) - 3} = \sqrt{x-2}$
with domain $\{x | x \geq 2\}$.

3. $(f \circ g)(x) = f(\ln x) = e^{\ln x} = x$
with domain $\{x | x > 0\}$

$(g \circ f)(x) = g(e^x) = \ln e^x = x$
with domain $(-\infty, \infty)$ or all real numbers.

5. $(f \circ g)(x) = f(\sin x) = \sin^2 x + 1$ with domain $(-\infty, \infty)$ or all real numbers.

$(g \circ f)(x) = g(x^2 + 1) = \sin(x^2 + 1)$ with domain $(-\infty, \infty)$ or all real numbers.

7. $\sqrt{x^4 + 1} = f(g(x))$ when $f(x) = \sqrt{x}$ and $g(x) = x^4 + 1$, for example.

9. $\dfrac{1}{x^2+1} = f(g(x))$ when $f(x) = 1/x$ and $g(x) = x^2 + 1$, for example.

11. $(4x+1)^2 + 3 = f(g(x))$ when $f(x) = x^2 + 3$ and $g(x) = 4x + 1$, for example.

13. $\sin^3 x = f(g(x))$ when $f(x) = x^3$ and $g(x) = \sin x$, for example.

15. $e^{x^2+1} = f(g(x))$ when $f(x) = e^x$ and $g(x) = x^2 + 1$, for example.

17. $\dfrac{3}{\sqrt{\sin x + 2}} = f(g(h(x)))$ when $f(x) = 3/x$, $g(x) = \sqrt{x}$, and $h(x) = \sin x + 2$, for example.

19. $\cos^3(4x - 2) = f(g(h(x)))$ when $f(x) = x^3$, $g(x) = \cos x$, and $h(x) = 4x - 2$, for example.

21. $4e^{x^2} - 5 = f(g(h(x)))$ when $f(x) = 4x - 5$, $g(x) = e^x$, and $h(x) = x^2$, for example.

23. Graph of $f(x) - 3$:

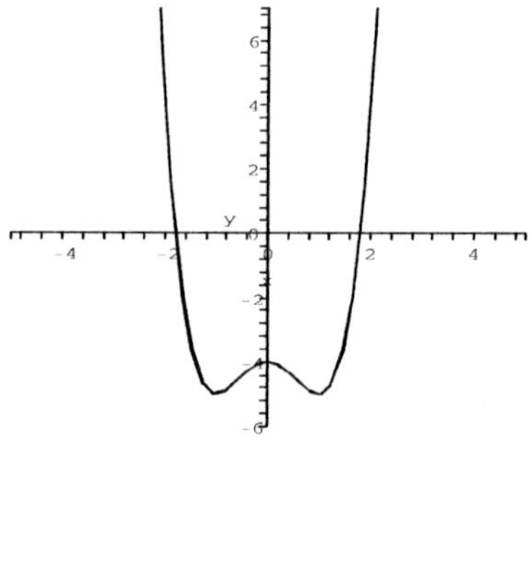

25. Graph of $f(x-3)$:

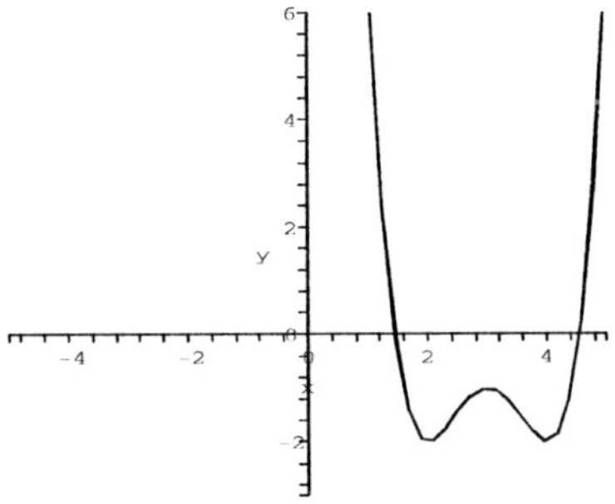

27. Graph of $f(2x)$:

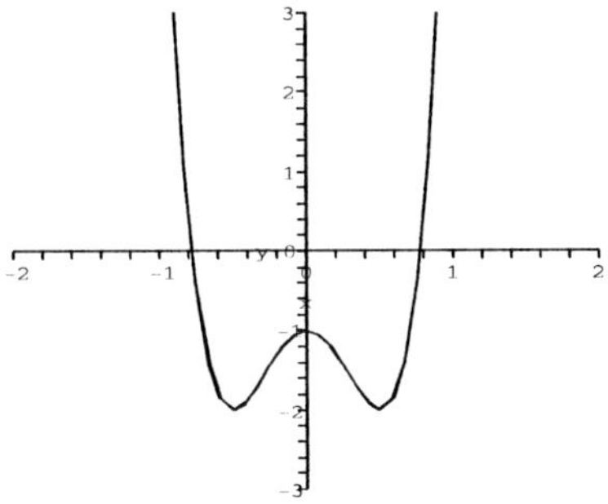

29. Graph of $4f(x)-1$:

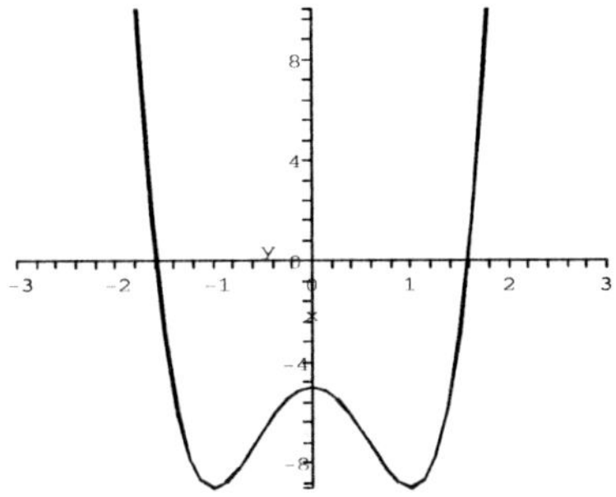

31. Graph of $f(x-4)$:

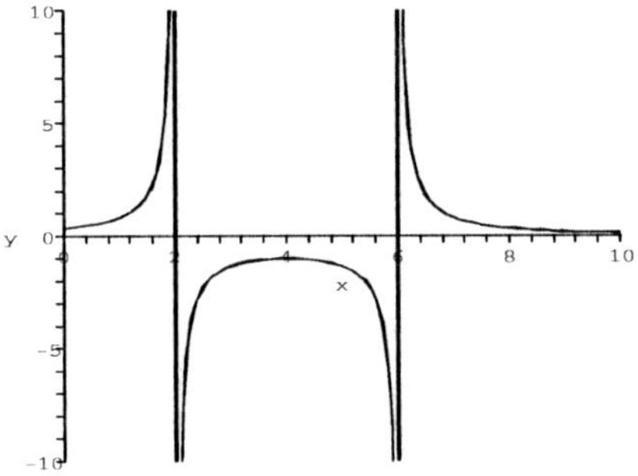

33. Graph of $f(2x)$:

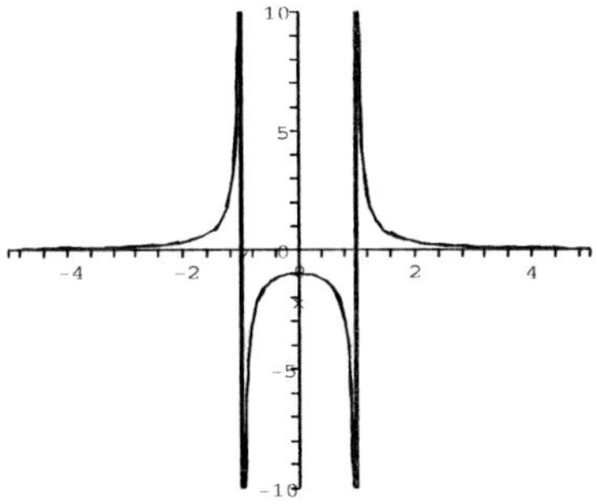

35. Graph of $f(3x+3)$:

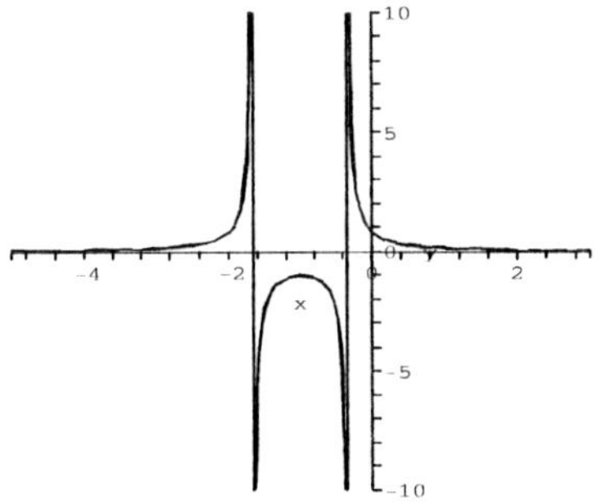

37. Graph of $2f(x)-4$:

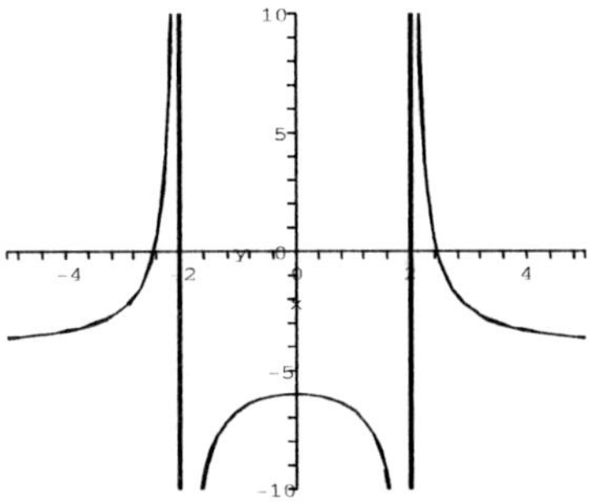

39. $f(x) = x^2 + 2x + 1 = (x+1)^2$.
Shift $y = x^2$ to the left 1 unit.

41. $f(x) = x^2+2x+4 = (x^2+2x+1)+4-1$
$= (x+1)^2 + 3$.
Shift $y = x^2$ to the left 1 unit and up 3 units.

43. $f(x) = 2x^2 + 4x + 4$
$= 2(x^2 + 2x + 1) + 4 - 2$
$= 2(x+1)^2 + 2$.
Shift $y = x^2$ to the left 1 unit, then

multiply the scale on the y-axis by 2, then shift up 2 units.

45. Graph is reflected across the x-axis and the scale on the y-axis is multiplied by 2.

47. Graph is reflected across the x-axis, the scale on the y-axis is multiplied by 3, and the graph is shifted up 2 units.

49. Graph is reflected across the y-axis.

51. Graph is reflected across the y-axis and shifted up 1 unit.

53. The graph is reflected across the x-axis and the scale on the y-axis is multiplied by $|c|$.

55. The graph of $y = |x|^3$ is identical to that of $y = x^3$ to the right of the y-axis because for $x > 0$ we have $|x|^3 = x^3$. For $y = |x|^3$ the graph to the left of the y-axis is the reflection through the y-axis of the graph to the right of the y-axis. In general to graph $y = f(|x|)$ based on the graph of $y = f(x)$, the procedure is to discard the part of the graph to the left of the y-axis, and replace it by a reflection in the y-axis of the part to the right of the y-axis.

57. The rest of the first 10 iterates of $f(x) = \cos x$ with $x_0 = 1$ are:

$$x_4 = \cos .65 \approx .796$$
$$x_5 = \cos .796 \approx .70$$
$$x_6 = \cos .70 \approx .765$$
$$x_7 = \cos .765 \approx .721$$
$$x_8 = \cos .721 \approx .751$$
$$x_9 = \cos .751 \approx .731$$
$$x_{10} = \cos .731 \approx .744$$

Continuing in this fashion and retaining more decimal places, one finds that x_{36} through x_{40} are all 0.739085. The same process is used with a different x_0.

59. They converge to 0. One of the problems in Chapter 2 asks the student to prove that $|\sin(x)| < |x|$ for all but $x = 0$. This would show that 0 is the only solution to the equation $\sin(x) = x$ and offers a partial explanation (see the comments for #61) of the phenomena which the student observes.

61. If the iterates of a function f (starting from some point x_0) are going to go toward (and remain arbitrarily close to) a certain number L, this number L must be a solution of the equation $f(x) = x$. For the list of iterates $x_0, x_1, x_2, x_3, \ldots$ is, apart from the first term, the same list as the list of numbers $f(x_0), f(x_1), f(x_2), f(x_3), \ldots$. (Remember that x_{n+1} is $f(x_n)$.) If any of the numbers in the first list are close to L, then the f-values (in the second list) are close to $f(L)$. But since the lists are *identical* (apart from the first term x_0 which is not in the second list), it must be true that L and $f(L)$ are the same number.

If conditions are right (and they are in the two cases $f(x) = \cos(x)$ (#57) and $f(x) = \sin(x)$ (#59)), this "convergence" will indeed occur, and since there is in these cases only *one* solution (x about 0.739085 in #57 and $x = 0$ in #59) it won't matter where you started.

Ch. 0 Review Exercises

1. $m = \dfrac{7-3}{0-2} = \dfrac{4}{-2} = -2$

3. These lines both have slope 3. They are parallel unless they are coincident. But the first line includes the point $(0, 1)$ which does not satisfy the equation of the second line. The lines are not coincident.

5. Let $P = (1, 2)$, $Q = (2, 4)$, $R = (0, 6)$.

Then PQ has slope $\dfrac{4-2}{2-1} = 2$

QR has slope $\dfrac{6-4}{0-2} = -1$

RP has slope $\dfrac{2-6}{1-0} = -4$

Since no two of these slopes are negative reciprocals, none of the angles are right angles. The triangle is not a right triangle.

7. The line apparently goes through $(1, 1)$ and $(3, 2)$. If so the slope would be $m = \frac{2-1}{3-1} = \frac{1}{2}$. The equation would be
$y = \frac{1}{2}(x-1) + 1$ or $y = \frac{1}{2}x + \frac{1}{2}$.
Using the equation with $x = 4$, we find $y = \frac{1}{2}(4) + \frac{1}{2} = \frac{5}{2}$.

9. Using the point-slope method, we find $y = -\frac{1}{3}(x+1) - 1$

11. The graph passes the vertical line test, so it is a function.

13. The radicand cannot be negative, hence we require $4 - x^2 \geq 0 \Rightarrow 4 \geq x^2$. Therefore the natural domain is $\{x | -2 \leq x \leq 2\}$ or, in "interval-language": $[-2, 2]$.

15. Intercepts at $x = -4$ and 2, and $y = -8$. Local minimum at $x = -1$. No asymptotes.

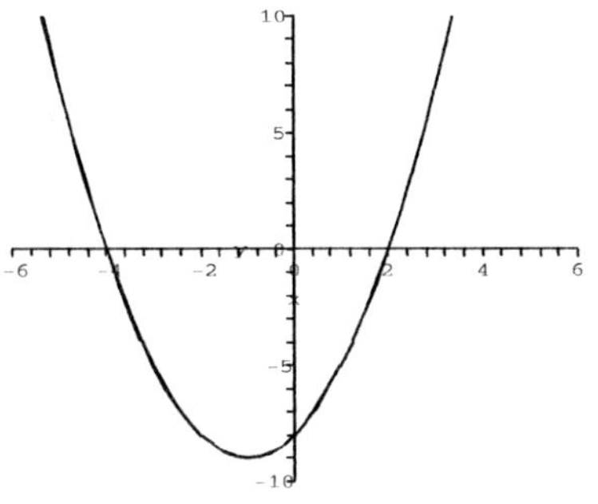

17. Intercepts at $x = -1$ and 1, and $y = 1$. Local minimum at $x = 1$ and at $x = -1$. Local maximum at $x = 0$. No asymptotes.

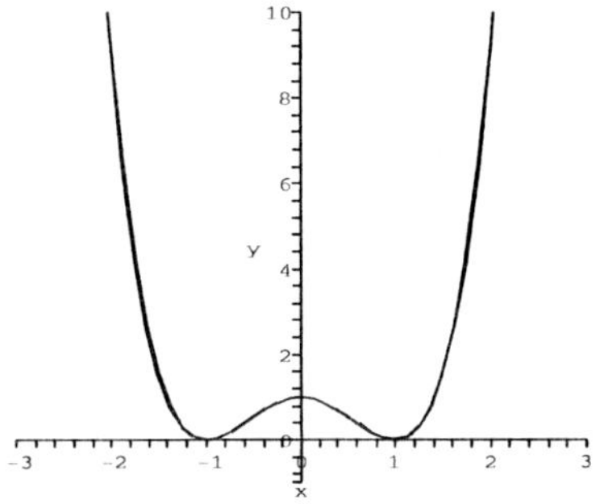

19. Intercept at $y = 0$ and at $x = 0$. No extrema. Horizontal asymptote $y = 4$. Vertical asymptote $x = -2$.

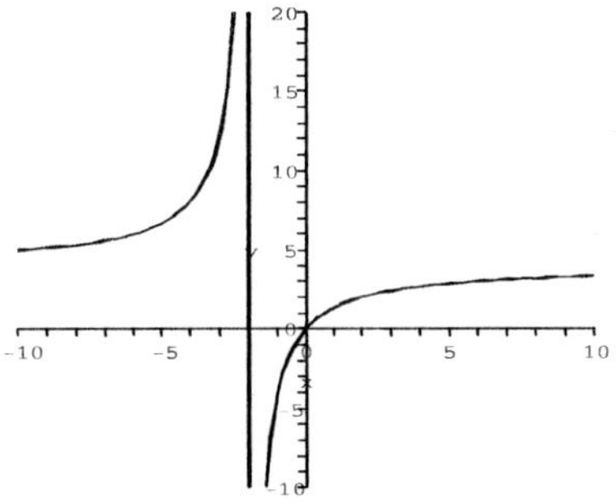

21. Intercept at $y = 0$ and $x = \frac{k\pi}{3}$ for integers k. Extrema: y takes maximum 1 and minimum -1 with great predictability and regularity. No asymptotes.

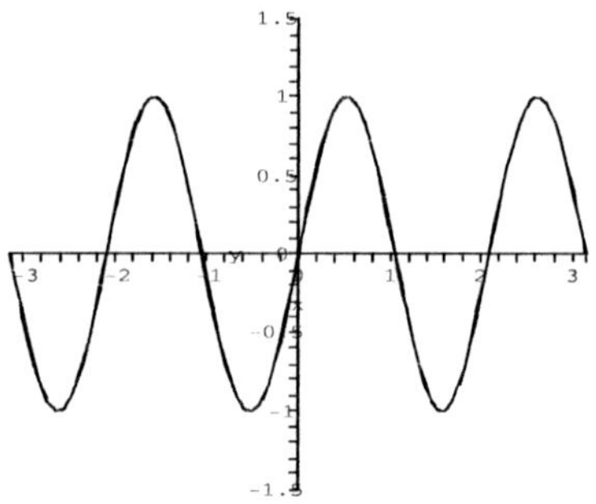

23. Intercept at $y = 2$ and from the amplitude/phase shift form $f(x) = \sqrt{5}\sin\left(x + \sin^{-1}(2/\sqrt{5})\right)$, we could write down all the intercepts only at considerable inconvenience. Extrema: y takes maximum $\sqrt{5}$ and minimum $-\sqrt{5}$ with great predictability and regularity. No asymptotes.

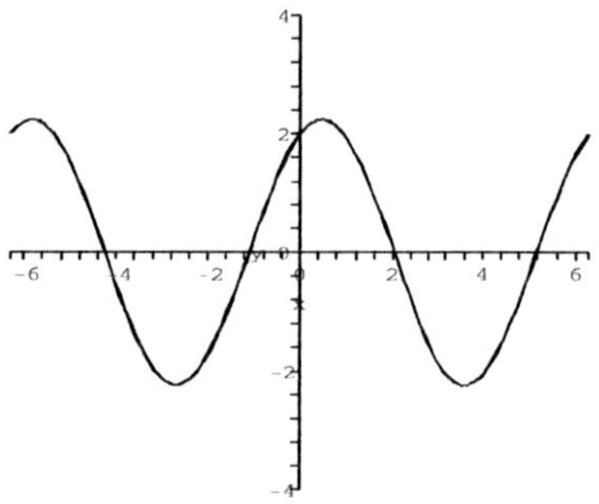

25. Intercept $y = 4$ (no x-intercepts). No extrema. Left horizontal asymptote $y = 0$.

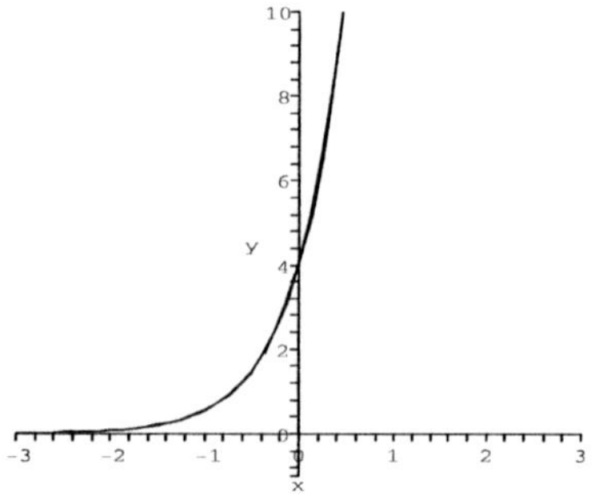

27. Intercept $x = 1/3$ (no y-intercepts). No extrema. Vertical asymptote $x = 0$.

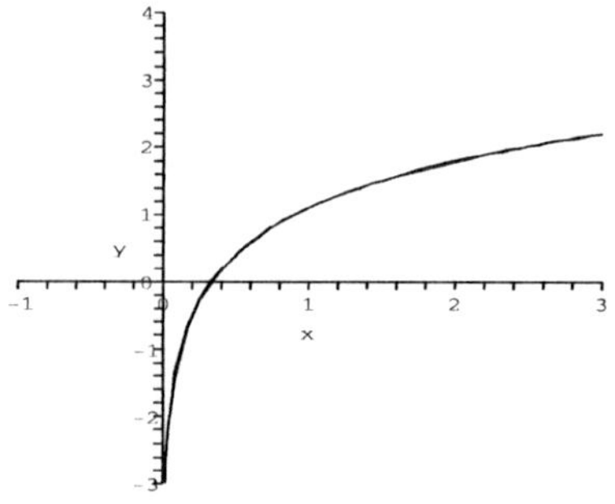

29. Intercepts at $x = -4$ and 2, and $y = -8$.

31. Vertical asymptote $x = -2$.

33. $x^2 - 3x - 10 = (x - 5)(x + 2)$. The zeros are when $x = 5$ and $x = -2$.

35. Guess a root: $x = 1$. Factor the left side: $(x - 1)(x^2 - 2x - 2)$. Solve the quadratic by formula:

$$x = \frac{2 \pm \sqrt{2^2 - 4(1)(-2)}}{2} = 1 \pm \sqrt{3}.$$

Complete list of three roots: $x = 1$, $x = 1 - \sqrt{3} \approx -.732$, $x = 1 + \sqrt{3} \approx 2.732$.

37. There are 3 solutions, one at $x = 0$ and the other two negatives of one another. The value in question is $.928632\ldots$, found using the function "Goal Seek" in Excel. The result can be checked, and a graphing calculator can find them by graphing $y = x^3$ and $y = \sin x$ on the same axes and finding the intersection points.

39. Let h be the height of the telephone pole. Then $\frac{h}{50} = \tan 34° \Rightarrow h = 50 \tan 34° \approx 33.7$ feet.

41. (a) $5^{-1/2} = \dfrac{1}{5^{1/2}} = \dfrac{1}{\sqrt{5}} = \dfrac{\sqrt{5}}{5}$

(b) $3^{-2} = \dfrac{1}{3^2} = \dfrac{1}{9}$

43. $\ln 8 - 2\ln 2 = \ln 8 - \ln 2^2$
$= \ln 8 - \ln 4 = \ln\left(\frac{8}{4}\right) = \ln 2$

45. $3e^{2x} = 8 \Rightarrow e^{2x} = \frac{8}{3}$
$\Rightarrow \ln e^{2x} = \ln\left(\frac{8}{3}\right)$
$\Rightarrow 2x = \ln\left(\frac{8}{3}\right)$
$\Rightarrow x = \frac{1}{2}\ln\frac{8}{3}$

47. The natural domain for f is the full real line. The natural domain for g is $\{x|1 \le x\}$. Because f has a universal domain, the natural domain for $f \circ g$ is the same as the domain for g, namely $\{x|1 \le x\}$. Because g requires its inputs be not less than 1, the domain for $g \circ f$ is the set of x for which $1 \le f(x)$, i.e., $\{x|1 \le x^2\} = \{x|1 \le |x|\}$, or in interval language $(-\infty, -1] \cup [1, \infty)$.

The formulae are easier:

$(f \circ g)(x) = f(\sqrt{x-1})$
$= (\sqrt{x-1})^2 = x - 1$

$(g \circ f)(x) = g(x^2) = \sqrt{x^2 - 1}$

Caution: the *formula* for $f \circ g$ is defined for any x, but the *domain* for $f \circ g$ is restricted as stated earlier. The formula must be viewed as irrelevant outside the domain.

49. $e^{3x^2+2} = f(g(x))$ for $f(x) = e^x$ and $g(x) = 3x^2 + 2$.

51. $x^2 - 4x + 1 = x^2 - 4x + 4 - 4 + 1$, so $f(x) = (x-2)^2 - 3$. The graph of $f(x)$ is the graph of x^2 shifted two units to the right and three units down.

53. Like x^3, the function $f(x) = x^3 - 1$ passes the horizontal line test and is one-to-one. To find a formula for the inverse, solve for x to find $(y+1)^{1/3} = x$ then switch x and y to get $f^{-1}(x) = (x+1)^{1/3}$ for all x.

55. The function is *even* $(f(-x) = f(x))$. Every horizontal line (except $y = 0$) which meets the curve at all automatically meets it at least twice. The function is not one-to-one. There is no inverse.

57. The inverse of $x^5 + 2x^3 - 1$:

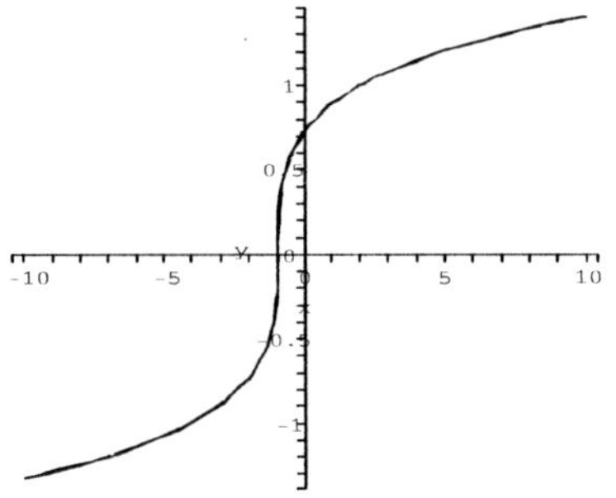

59. The inverse of $\sqrt{x^3 + 4x}$:

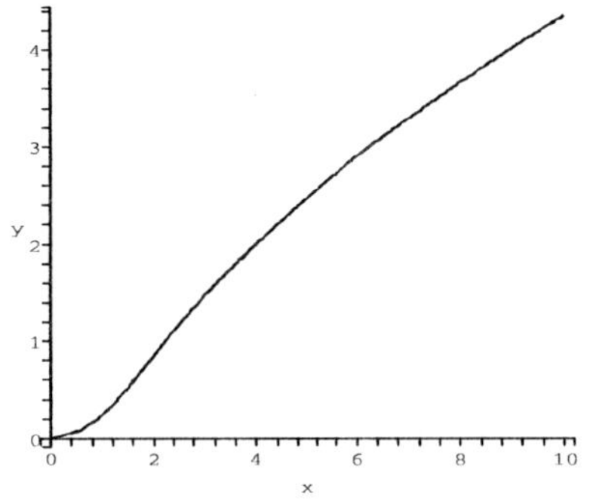

61. On the unit circle, $y = \sin\theta = 1$ when $\theta = \frac{\pi}{2}$. Hence, $\sin^{-1} 1 = \frac{\pi}{2}$.

63. Since $\tan\theta = \frac{\sin\theta}{\cos\theta}$ we want $y = \cos\theta$ to be equal to $-x = -\sin\theta$ on the unit circle. This happens when $\theta = -\pi/4$ and $\theta = 3\pi/4$. Hence, $\tan^{-1}(-1) = -\frac{\pi}{4}$ or $\tan^{-1}(-1) = \frac{3\pi}{4}$.

65. If an angle θ has $\sec(\theta) = 2$, then it has $\cos(\theta) = 1/2$. Its sine could be $\pm\frac{\sqrt{3}}{2}$. But if $\theta = \sec^{-1}(2)$, then in addition to all that has been stated, it is in the first quadrant, and the choice of sign (for its sine) is positive. In summary, $\sin(\sec^{-1} 2) = \sin\theta = \frac{\sqrt{3}}{2}$.

67. $\sin^{-1}\left(\sin\left(\frac{3\pi}{4}\right)\right) = \sin^{-1}\left(\frac{\sqrt{2}}{2}\right) = \frac{\pi}{4}$

69. $\sin 2x = 1 \Rightarrow$
$2x = \frac{\pi}{2} + 2k\pi$ for any integer k so
$x = \frac{\pi}{4} + k\pi$ for any integer k.

Chapter 1

Limits and Continuity

1.1 A Brief Preview of Calculus

1. The slope appears to be 2.

Second point	m_{sec}
$(2, 5)$	3
$(1.1, 2.21)$	2.1
$(1.01, 2.0201)$	2.01
$(0, 1)$	1
$(0.9, 1.81)$	1.9
$(0.99, 1.9801)$	1.99

3. The slope appears to be 0.

Second point	m_{sec}
(1, 0.5403)	-0.4597
(0.1, 0.995)	-0.05
(0.01, 0.99995)	-0.005
(-1, 0.5403)	0.4597
(-0.1, 0.995)	0.05
(-0.01, 0.99995)	0.005

5. The slope appears to be 3.

Second point	m_{sec}
$(2, 10)$	7
$(1.1, 3.331)$	3.31
$(1.01, 3.030301)$	3.0301
$(0, 2)$	1
$(0.9, 2.729)$	2.71
$(0.99, 2.970299)$	2.9701

7. The slope appears to be $\frac{1}{2}$.

Second point	m_{sec}
$(1, \sqrt{2})$	0.4142
(0.1, 1.0488)	0.488
(0.01, 1.004988)	0.4988
(-1, 0)	1
(-0.1, 0.9487)	0.513
(-0.01, 0.99499)	0.501

9. The slope appears to be 1.

Second point	m_{sec}
$(1, e)$	1.718282
(0.1, 1.1052)	1.051709
(0.01, 1.0101)	1.005017
(-1, 0.3679)	0.632121
(-0.1, 0.9048)	0.951626
(-0.01, 0.9901)	0.995017

11. The slope appears to be 1.

Second point	m_{sec}
(0.1, -2.3026)	2.5584
(0.9, -0.1054)	1.054
(0.99, -0.01005034)	1.005034
(2, 0.6931)	0.6931
(1.1, 0.09531)	0.9531
(1.01, 0.00995)	0.995

Note that we used 0.1 rather than 0 as an evaluation point because $\ln x$ is not defined at 0.

13. (a)

Left	Right	Length
(0, 1)	(0.5, 1.25)	0.559
(0.5, 1.25)	(1, 2)	0.901
(1, 2)	(1.5, 3.25)	1.346
(1.5, 3.25)	(2, 5)	1.820
	Total	4.6267

(b)

Left	Right	Length
(0, 1)	(0.25, 1.063)	0.258
(0.25, 1.063)	(0.5, 1.25)	0.313
(0.5, 1.25)	(0.75, 1.563)	0.400
(0.75, 1.563)	(1, 2)	0.504
(1, 2)	(1.25, 2.563)	0.616
(1.25, 2.563)	(1.5, 3.25)	0.732
(1.5, 3.25)	(1.75, 4.063)	0.850
(1.75, 4.063)	(2, 5)	0.970
	Total	4.6417

(c) Actual length approximately 4.6468.

15. (a) For the x-values of our points here we use (approximations of) 0, $\frac{\pi}{8}$, $\frac{\pi}{4}$, $\frac{3\pi}{8}$, and $\frac{\pi}{2}$.

Left	Right	Length
(0, 1)	(0.393, 0.92)	0.400
(0.393, 0.92)	(0.785, 0.71)	0.449
(0.785, 0.71)	(1.18, 0.383)	0.509
(1.18, 0.383)	(1.571, 0)	0.548
	Total	1.906

(b) For the x-values of our points here we use (approximations of) 0, $\frac{\pi}{16}$, $\frac{\pi}{8}$, $\frac{3\pi}{16}$, $\frac{\pi}{4}$, $\frac{5\pi}{16}$, $\frac{3\pi}{8}$, $\frac{7\pi}{16}$, and $\frac{\pi}{2}$.

Left	Right	Length
(0, 1)	(0.196, 0.98)	0.197
(0.196, 0.98)	(0.393, 0.92)	0.204
(0.393, 0.92)	(0.589, 0.83)	0.217
(0.589, 0.83)	(0.785, 0.71)	0.232
(0.785, 0.71)	(0.982, 0.56)	0.248
(0.982, 0.56)	(1.178, 0.38)	0.262
(1.178, 0.38)	(1.37, 0.195)	0.272
(1.37, 0.195)	(1.571, 0)	0.277
	Total	1.909

(c) Actual length approximately 1.9101.

17. (a)

Left	Right	Length
(0, 1)	(0.75, 1.323)	0.817
(0.75, 1.323)	(1.5, 1.581)	0.793
(1.5, 1.581)	(2.25, 1.803)	0.782
(2.25, 1.803)	(3, 2)	0.776
	Total	3.167

(b)

Left	Right	Length
(0, 1)	(0.375, 1.17)	0.413
(0.375, 1.17)	(0.75, 1.323)	0.404
(0.75, 1.323)	(1.125, 1.46)	0.399
(1.125, 1.46)	(1.5, 1.58)	0.395
(1.5, 1.58)	(1.88, 1.696)	0.392
(1.88, 1.696)	(2.25, 1.80)	0.390
(2.25, 1.80)	(2.63, 1.904)	0.388
(2.63, 1.904)	(3, 2)	0.387
	Total	3.168

(c) Actual length approximately 3.168.

19. (a)

Left	Right	Length
(-2, 5)	(-1, 2)	3.162
(-1, 2)	(0, 1)	1.414
(0, 1)	(1, 2)	1.414
(1, 2)	(2, 5)	3.162
	Total	9.153

(b)

Left	Right	Length
(-2, 5)	(-1.5, 3.25)	1.820
(-1.5, 3.25)	(-1, 2)	1.346
(-1, 2)	(-0.5, 1.25)	0.901
(-0.5, 1.25)	(0, 1)	0.559
(0, 1)	(0.5, 1.25)	0.559
(0.5, 1.25)	(1, 2)	0.901
(1, 2)	(1.5, 3.25)	1.346
(1.5, 3.25)	(2, 5)	1.820
	Total	9.253

(c) Actual length approximately 9.2936.

21. The sum of the areas of the rectangles is $11/8 = 1.375$.

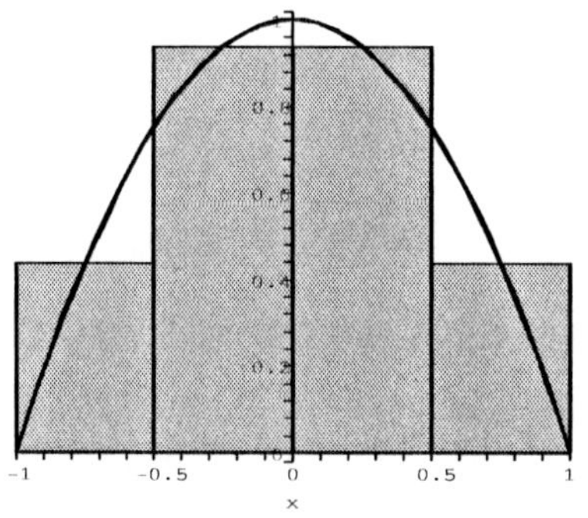

23. (a) The width of the entire region ($-1 \leq x \leq 1$) is 2, so the width of each rectangle is $2/16 = 0.125$. The left endpoints of the rectangles are
$-1, -1+\frac{2}{16}, \ldots, -1+\frac{28}{16}, -1+\frac{30}{16}$
so the midpoints of the rectangles are
$-1+\frac{1}{16}, -1+\frac{3}{16}, \ldots, -1+\frac{31}{16}.$
The heights of the rectangles are then given by the function $f(x) = 1 - x^2$ evaluated at those midpoints. We multiply each height by the width (0.125) and add them all to obtain the approximation 1.3359375 for the area.

(b) Using the same method as in (a), the width of the rectangles is now $2/32 = 0.0625$, and the midpoints are
$-1+\frac{1}{32}, -1+\frac{3}{32}, \ldots, -1+\frac{63}{32}.$
The approximation is 1.333984375.

(c) Using the same method as in (a), the width of the rectangles is now $2/64 = 0.03125$, and the midpoints are
$-1+\frac{1}{64}, -1+\frac{3}{64}, \ldots, -1+\frac{127}{64}.$
The approximation is 1.333496094.

The actual area is 4/3.

25. The following is a graph with 4 rectangles:

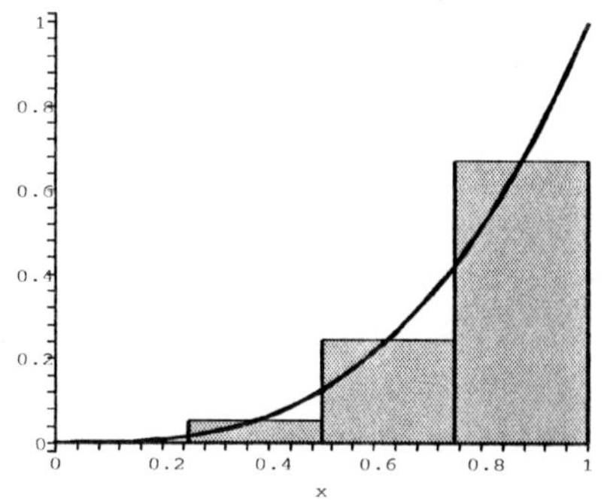

(a) Using the same method as in exercise 23, the width of the rectangles is 1/16, and the midpoints are
$\frac{1}{16}, \frac{3}{16}, \ldots, \frac{15}{16}.$
The approximation is 0.249511719.

(b) Using the same method as in exercise 23, the width of the rectangles is now 1/32, and the midpoints are
$\frac{1}{32}, \frac{3}{32}, \ldots, \frac{31}{32}.$
The approximation is 0.24987793.

(c) Using the same method as in exercise 23, the width of the rectangles is now 1/64, and the midpoints are
$\frac{1}{64}, \frac{3}{64}, \ldots, \frac{63}{64}.$
The approximation is 0.249969482.

The actual area is 1/4.

1.2 The Concept of Limit

1. (a) $\lim_{x\to 0^-} f(x) = -2$

(b) $\lim_{x\to 0^+} f(x) = 2$

(c) Does not exist.

(d) $\lim_{x\to 1^-} f(x) = 1$

(e) $\lim_{x\to -1} f(x) \approx 0.1$

(f) $\lim_{x\to 2^-} f(x) = -1$

(g) $\lim_{x\to 2^+} f(x) = 3$

(h) Does not exist.

(i) $\lim_{x\to -2} f(x) \approx 1.8$

(j) $\lim_{x\to 3} f(x) \approx 2.5$

3. (a) $\lim_{x\to 2^-} f(x) = \lim_{x\to 2^-} 2x = 4$

(b) $\lim_{x\to 2^+} f(x) = \lim_{x\to 2^+} x^2 = 4$

(c) $\lim_{x\to 2} f(x) = 4$

(d) $\lim_{x\to 1} f(x) = \lim_{x\to 1} 2x = 2$

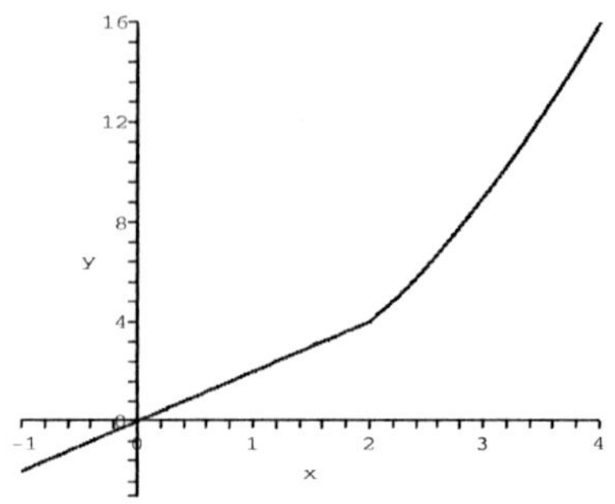

5. (a) $\lim_{x\to -1^-} f(x) = \lim_{x\to 1^-} x^2 + 1 = 2$

(b) $\lim_{x\to -1^+} f(x) = \lim_{x\to 1^+} 3x + 1 = -2$

(c) $\lim_{x\to -1} f(x)$ does not exist

(d) $\lim_{x\to 1} f(x) = \lim_{x\to 1} 3x + 1 = 4$

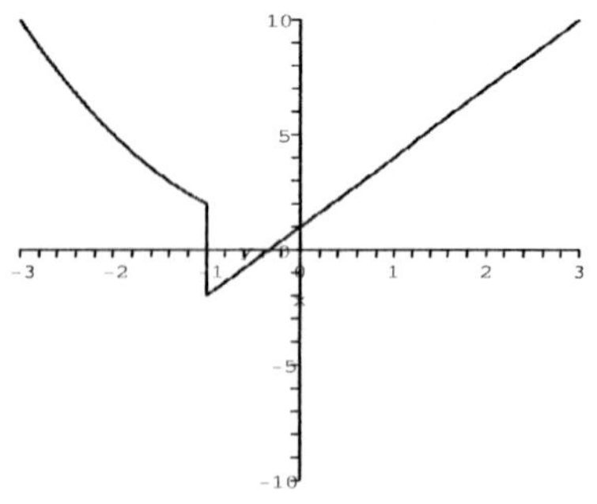

7. $f(1.5) = 2.22$, $f(1.1) = 2.05$, $f(1.01) = 2.01$, $f(1.001) = 2.00$.

The values of $f(x)$ seem to be approaching 2 as x approaches 1 from the right.

$f(0.5) = 1.71$, $f(0.9) = 1.95$, $f(0.99) = 1.99$, $f(0.999) = 2.00$.

The values of $f(x)$ seem to be approaching 2 as x approaches 1 from the left. Since the limits from the left and right exist and are the same, the limit exists.

9. By inspecting the graph, and using a sequence of values (as in exercises 7 and 8), we see that the limit is approximately 2.

11. By inspecting the graph, and using a sequence of values (as in exercises 7 and 8), we see that the limit is approximately 1.

13. By inspecting the graph, and using a sequence of values (as in exercises 7 and 8), we see that the limit is approximately 1.

15. The numerical evidence suggests that the function the function blows up at $x = 1$. From the graph we see that the function has a vertical asymptote at $x = 1$.

17. By inspecting the graph, and using a sequence of values (as in exercises 7 and 8), we see that the limit is approximately 3/2.

19. The limit does not exist because the graph oscillates wildly near $x = 0$.

21. The numerical evidence suggests that

$\lim_{x\to 2^-} \frac{x-2}{|x-2|} = -1$ while $\lim_{x\to 2^+} \frac{x-2}{|x-2|} = 1$

so $\lim_{x\to 2} \frac{x-2}{|x-2|}$ does not exist. There is a break in the graph at $x = 2$.

23. The function $\ln x$ is not defined for $x \leq 0$ so the limit does not exist. The numerical evidence suggests that the function blows up as x approaches 0 from the right. From the graph we see that the function has a one-sided vertical asymptote at $x = 0$.

25. The limit exists and equals 1.

27. Numerical and graphical evidence show that the limits

$$\lim_{x\to 1} \frac{x^2+1}{x-1} \text{ and } \lim_{x\to 2} \frac{x+1}{x^2-4}$$

do not exist (both have vertical asymptotes). Our conjecture is that if $g(a) = 0$ and $f(a) \neq 0$, $\lim_{x\to a} \frac{f(x)}{g(x)}$ does not exist.

29. One possibility:

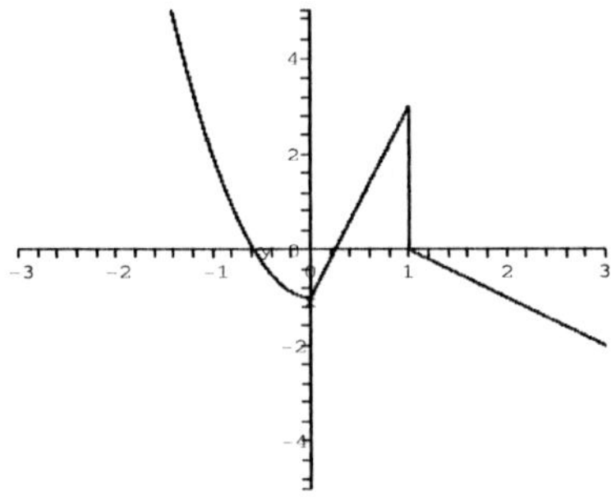

31. One possibility:

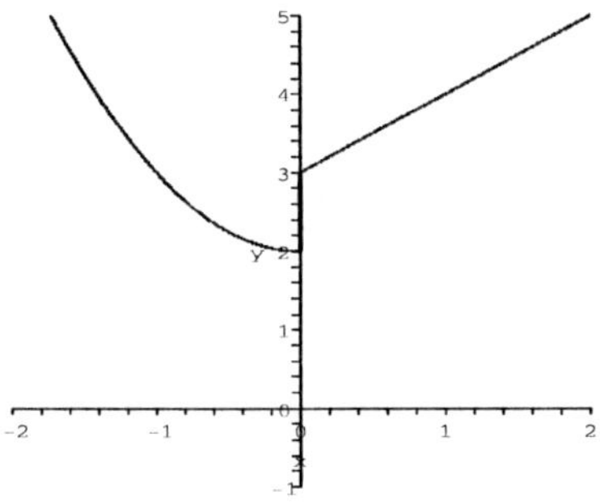

33. By inspecting the graph, and using a sequence of values (as in exercises 7 and 8), we see that the limit is approximately 1/2.

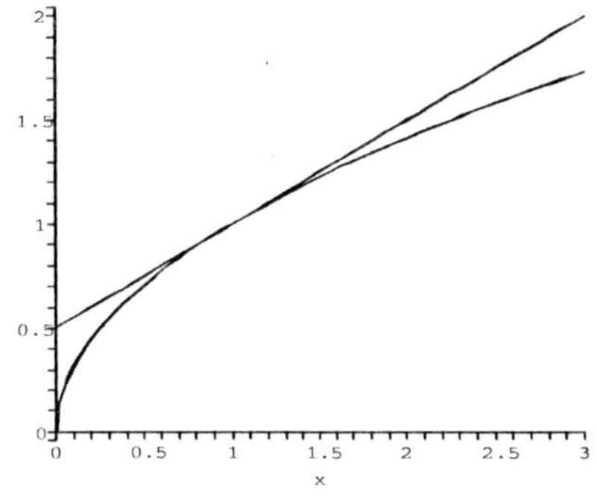

35. The first argument gives the correct value; the second argument is not valid because it looks only at certain values of x.

37.

x	$(1+x)^{\frac{1}{x}}$	x	$(1+x)^{\frac{1}{x}}$
0.1	2.59	−0.1	2.87
0.01	2.70	−0.01	2.73
0.001	2.7169	−0.001	2.7196

$\lim_{x\to 0}(1+x)^{1/x} \approx 2.7182818$

39.

x	$x^{\sec x}$
0.1	0.099
0.01	0.010
0.001	0.001

$\lim_{x\to 0^+} x^{\sec x} = 0$

For negative x the values of $x^{\sec x}$ are usually not real numbers, so $\lim_{x\to 0^-} x^{\sec x} = 0$ does not exist.

41. Possible answers:

$$f(x) = \frac{x^2}{x}$$

$$g(x) = \begin{cases} 1 & \text{if } x \leq 0 \\ -1 & \text{if } x > 0 \end{cases}$$

43. As x gets arbitrarily close to a, $f(x)$ gets arbitrarily close to L.

45. For $3 \le t \le 4$, $f(t) = 8$, so $\lim_{t\to 3.5} f(t) = 8$. Also $\lim_{t\to 4^-} f(t) = 8$. On the other hand, for $4 \le t \le 5$, $f(t) = 10$, so $\lim_{t\to 4^+} f(t) = 10$. Hence $\lim_{t\to 4} f(t)$ does not exist.

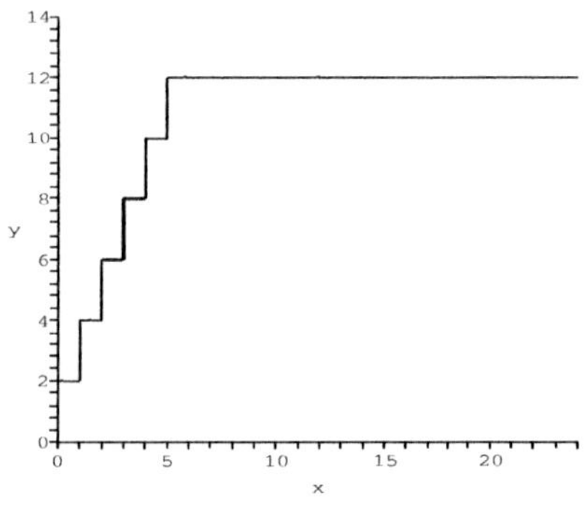

1.3 Computation of Limits

1. $\lim_{x\to 0}(x^2 - 3x + 1) = 0^2 - 3(0) + 1 = 1$

3. $\lim_{x\to 0}\cos^{-1}(x^2) = \cos^{-1} 0 = \dfrac{\pi}{2}$.

5. $\lim_{x\to 3}\dfrac{x^2 - x - 6}{x - 3}$
$= \lim_{x\to 3}\dfrac{(x-3)(x+2)}{x-3}$
$= \lim_{x\to 3}(x+2) = 3 + 2 = 5$

7. $\lim_{x\to 2}\dfrac{x^2 - x - 2}{x^2 - 4}$
$= \lim_{x\to 2}\dfrac{(x-2)(x+1)}{(x+2)(x-2)}$
$= \lim_{x\to 2}\dfrac{x+1}{x+2} = \dfrac{2+1}{2+2} = \dfrac{3}{4}$

9. $\lim_{x\to 0}\dfrac{\sin x}{\tan x} = \lim_{x\to 0}\dfrac{\sin x}{\frac{\sin x}{\cos x}}$
$= \lim_{x\to 0}\cos x = \cos 0 = 1$

11. $\lim_{x\to 0}\dfrac{xe^{-2x+1}}{x^2 + x}$
$= \lim_{x\to 0}\dfrac{x(e^{-2x+1})}{x(x+1)}$
$= \lim_{x\to 0}\dfrac{e^{-2x+1}}{x+1} = \dfrac{e^{-2(0)+1}}{0+1} = e$

13. $\lim_{x\to 0}\dfrac{\sqrt{x+4} - 2}{x}$
$= \lim_{x\to 0}\dfrac{\sqrt{x+4} - 2}{x}\left(\dfrac{\sqrt{x+4}+2}{\sqrt{x+4}+2}\right)$
$= \lim_{x\to 0}\dfrac{x + 4 - 4}{x(\sqrt{x+4}+2)}$
$= \lim_{x\to 0}\dfrac{x}{x(\sqrt{x+4}+2)}$
$= \lim_{x\to 0}\dfrac{1}{\sqrt{x+4}+2}$
$= \dfrac{1}{\sqrt{4}+2} = \dfrac{1}{2+2} = \dfrac{1}{4}$

15. $\lim_{x\to 1}\dfrac{x - 1}{\sqrt{x} - 1}$
$= \lim_{x\to 1}\dfrac{(\sqrt{x}+1)(\sqrt{x}-1)}{\sqrt{x}-1}$
$= \lim_{x\to 1}(\sqrt{x}+1) = \sqrt{1} + 1 = 2$

17. $\lim_{x\to 1}\left(\dfrac{1}{x-1} - \dfrac{2}{x^2-1}\right)$
$= \lim_{x\to 1}\left(\dfrac{1}{x-1} - \dfrac{2}{(x-1)(x+1)}\right)$
$= \lim_{x\to 1}\left(\dfrac{x+1}{(x-1)(x+1)} - \dfrac{2}{(x-1)(x+1)}\right)$
$= \lim_{x\to 1}\left(\dfrac{x-1}{(x-1)(x+1)}\right)$
$= \lim_{x\to 1}\left(\dfrac{1}{x+1}\right) = \dfrac{1}{2}$

19. $\lim_{x\to 0}\dfrac{1 - e^{2x}}{1 - e^x}$
$= \lim_{x\to 0}\dfrac{(1-e^x)(1+e^x)}{1-e^x}$
$= \lim_{x\to 0}(1 + e^x) = 2$

21. $\lim_{x\to 0^+}\dfrac{\sin(|x|)}{x} = \lim_{x\to 0^+}\dfrac{\sin(x)}{x} = 1$
$\lim_{x\to 0^-}\dfrac{\sin(|x|)}{x}$
$= \lim_{x\to 0^-}\dfrac{\sin(-x)}{x}$
$= \lim_{x\to 0^-}\dfrac{-\sin(x)}{x} = -1$

Since the limit from the left does not equal the limit from the right, we see that $\lim_{x\to 0} \frac{\sin(|x|)}{x}$ does not exist.

23. $\lim_{x\to 2^-} f(x) = \lim_{x\to 2^-} 2x = 2(2) = 4$
$\lim_{x\to 2^+} f(x) = \lim_{x\to 2^+} x^2 = 2^2 = 4$
$\lim_{x\to 2} f(x) = 4$

25. $\lim_{x\to 0} f(x) = \lim_{x\to 0}(3x+1) = 3(0)+1 = 1$

27. $\lim_{x\to -1^-} f(x) = \lim_{x\to -1^-} (2x+1) = 2(-1)+1 = -1$

$\lim_{x\to -1^+} f(x) = \lim_{x\to -1^+} 3 = 3$
Therefore $\lim_{x\to -1} f(x)$ does not exist.

29. $\lim_{h\to 0} \frac{(2+h)^2-4}{h}$
$= \lim_{h\to 0} \frac{(4+4h+h^2)-4}{h}$
$= \lim_{h\to 0} \frac{4h+h^2}{h} = \lim_{h\to 0} 4+h = 4$

31. $\lim_{h\to 0} \frac{h^2}{\sqrt{h^2+h+3}-\sqrt{h+3}}$
$= \lim_{h\to 0} \frac{h^2(\sqrt{h^2+h+3}+\sqrt{h+3})}{(h^2+h+3)-(h+3)}$
$= \lim_{h\to 0} \frac{h^2(\sqrt{h^2+h+3}+\sqrt{h+3})}{h^2}$
$= \lim_{h\to 0} \sqrt{h^2+h+3}+\sqrt{h+3} = 2\sqrt{3}$
To get from the first line to the second, we have multiplied by

$$\frac{\sqrt{h^2+h+3}+\sqrt{h+3}}{\sqrt{h^2+h+3}+\sqrt{h+3}}.$$

33. $\lim_{t\to -2} \frac{\frac{1}{2}+\frac{1}{t}}{2+t}$
$= \lim_{t\to -2} \frac{\frac{t+2}{2t}}{2+t}$
$= \lim_{t\to -2} \frac{1}{2t} = -\frac{1}{4}$

35.

x^2	$x^2 \sin(1/x)$
-0.1	0.0054
-0.01	5×10^{-5}
-0.001	-8×10^{-7}
0.1	-0.005
0.01	-5×10^{-5}
0.001	8×10^{-7}

Conjecture: $\lim_{x\to 0} x^2 \sin(1/x) = 0$.
Let $f(x) = -x^2, h(x) = x^2$. Then

$$f(x) \le x^2 \sin\left(\frac{1}{x}\right) \le h(x)$$

$$\lim_{x\to 0}(-x^2) = 0, \lim_{x\to 0}(x^2) = 0$$

Therefore, by the Squeeze Theorem,

$$\lim_{x\to 0} x^2 \sin\left(\frac{1}{x}\right) = 0.$$

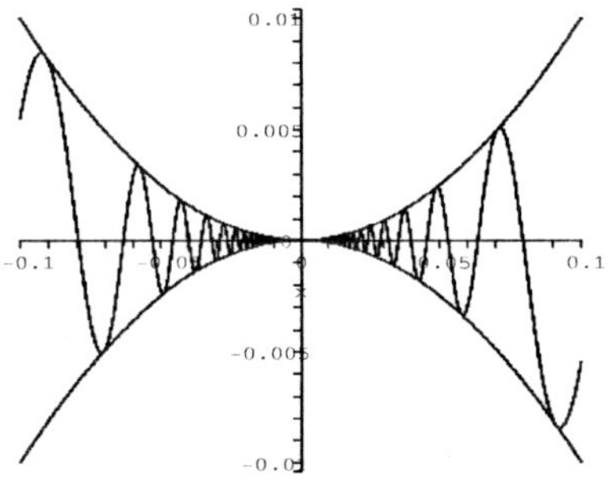

37. Let $f(x) = 0$, $h(x) = \sqrt{x}$. We see that

$$f(x) \le \sqrt{x}\cos^2(1/x) \le h(x),$$

$$\lim_{x\to 0^+} 0 = 0, \lim_{x\to 0^+} \sqrt{x} = 0$$

Therefore, by the Squeeze Theorem,

$$\lim_{x\to 0^+} \sqrt{x}\cos^2\left(\frac{1}{x}\right) = 0.$$

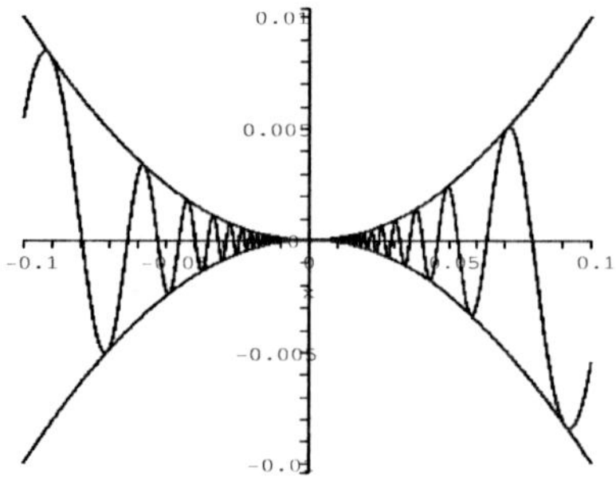

39. $\lim_{x \to 4^+} \sqrt{16 - x^2}$ does not exist because the domain of the function is $[-4, 4]$.

41. $\lim_{x \to -2^-} \sqrt{x^2 + 3x + 2} = 0$.

43. $\lim_{x \to 0^+} \frac{\sqrt{1 - \cos x}}{x} = \sqrt{\frac{1}{2}} = \frac{\sqrt{2}}{2}$

45. $\lim_{x \to a^-} f(x) = \lim_{x \to a^-} g(x) = g(a)$ because $g(x)$ is a polynomial. Similarly,

$$\lim_{x \to a^+} f(x) = \lim_{x \to a^+} h(x) = h(a).$$

47. (a) $\lim_{x \to 2}(x^2 - 3x + 1)$
$= 2^2 - 3(2) + 1$ (Theorem 3.2)
$= -1$

(b) $\lim_{x \to 0} \frac{x - 2}{x^2 + 1}$
$= \frac{\lim_{x \to 0}(x - 2)}{\lim_{x \to 0}(x^2 + 1)}$
(Theorem 3.1(iv))
$= \frac{\lim_{x \to 0} x - \lim_{x \to 0} 2}{\lim_{x \to 0} x^2 + \lim_{x \to 0} 1}$
(Theorem 3.1(ii))
$= \frac{0 - 2}{0 + 1}$
(Equations 3.1, 3.2, and 3.5)
$= -2$

49. Velocity is given by the limit

$$\lim_{h \to 0} \frac{f(2 + h) - f(2)}{h}$$
$$= \lim_{h \to 0} \frac{(2 + h)^2 + 2 - (2^2 + 2)}{h}$$
$$= \lim_{h \to 0} \frac{4h + h^2}{h}$$
$$= \lim_{h \to 0} 4 + h = 4.$$

51. Velocity is given by the limit

$$\lim_{h \to 0} \frac{f(0 + h) - f(0)}{h}$$
$$= \lim_{h \to 0} \frac{(0 + h)^3 - (0)^3}{h}$$
$$= \lim_{h \to 0} \frac{h^3}{h}$$
$$= \lim_{h \to 0} h^2 = 0.$$

53. $m = \lim_{h \to 0} \frac{\sqrt{1 + h} - 1}{h} \frac{\sqrt{1 + h} + 1}{\sqrt{1 + h} + 1}$

$$= \lim_{h \to 0} \frac{1 + h - 1}{h(\sqrt{1 + h} + 1)}$$
$$= \lim_{h \to 0} \frac{h}{h(\sqrt{1 + h} + 1)}$$
$$= \lim_{h \to 0} \frac{1}{\sqrt{1 + h} + 1}$$
$$= \frac{1}{\sqrt{1 + 0} + 1} = \frac{1}{2}.$$

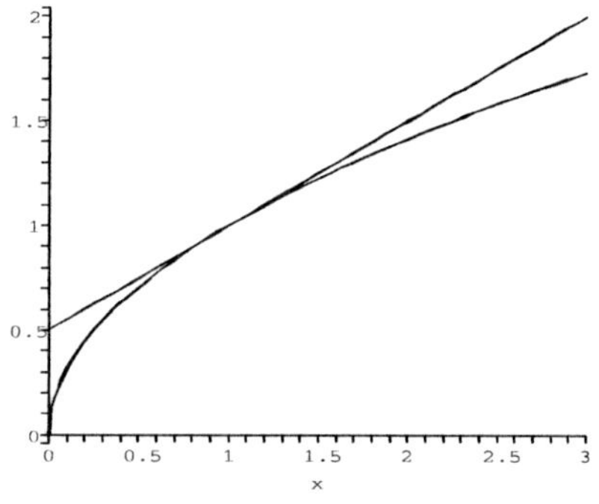

55. $\lim_{x \to 0^+} (1 + x)^{1/x} = e \approx 2.71828$

57. $\lim_{x \to 0^+} x^{-x^2} = 1$

59. As x gets close to 0, $1/x$ gets larger and larger in absolute value, so $\sin(1/x)$ oscillates more and more rapidly between 1 and -1, so the limit does not exist.

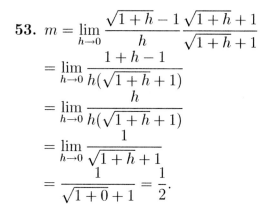

61. When x is small and positive, $1/x$ is large and positive, so $\tan^{-1}(1/x)$ approaches $\pi/2$. But when x is small and negative, $1/x$ is large and negative, so $\tan^{-1}(1/x)$ approaches $-\pi/2$. So the limit does not exist.

63. $\lim_{x\to a} [2f(x) - 3g(x)]$
$= 2\lim_{x\to a} f(x) - 3\lim_{x\to a} g(x)$
$= 2(2) - 3(-3) = 13$

65. $\lim_{x\to a} \left[\dfrac{f(x)+g(x)}{h(x)}\right]$
does not exist, because

$\lim_{x\to a} [f(x) + g(x)]$
$= \lim_{x\to a} f(x) + \lim_{x\to a} g(x)$
$= 2 - 3 = -1$

and $\lim_{x\to a} h(x) = 0$.

67. $\lim_{x\to a} [f(x)]^3$

$= \left[\lim_{x\to a} f(x)\right]\left[\lim_{x\to a} f(x)\right]\left[\lim_{x\to a} f(x)\right]$
$= L \cdot L \cdot L = L^3$
$\lim_{x\to a} [f(x)]^4 = \left[\lim_{x\to a} f(x)\right]\left[\lim_{x\to a} [f(x)]^3\right]$
$= L \cdot L^3 = L^4$

69. We can't split the limit of a product into a product of limits unless we know that both limits exist; the limit of the product of a term tending toward 0 and a term with an unknown limit is not necessarily 0 but instead is unknown.

71. One possibility is

$$f(x) = \frac{1}{x}, g(x) = -\frac{1}{x}.$$

73. Yes. If $\lim_{x\to a}[f(x) + g(x)]$ exists, then, it would also be true that

$$\lim_{x\to a}[f(x) + g(x)] - \lim_{x\to a} f(x)$$

exists. But by Theorem 3.1 (ii)

$\lim_{x\to a}[f(x) + g(x)] - \lim_{x\to a} f(x)$
$= \lim_{x\to a} [[f(x) + g(x)] - [f(x)]]$
$= \lim_{x\to a} g(x)$

so $\lim_{x\to a} g(x)$ would exist, but we are given that $\lim_{x\to a} g(x)$ does not exist.

75. $\lim_{x\to 0^+} T(x) = \lim_{x\to 0^+} (0.14x) = 0 = T(0)$.

$\lim_{x\to 10{,}000^-} T(x) = 0.14(10{,}000) = 1400$

$\lim_{x\to 10{,}000^+} T(x)$
$= 1500 + 0.21(10{,}000) = 3600$

Therefore $\lim_{x\to 10{,}000} T(x)$ does not exist. A small change in income should result in a small change in tax liability. This is true near $x = 0$ but is not true near $x = 10{,}000$. As your income grows past \$10,000 your tax liability jumps enormously.

77. $\lim_{x\to 3^-} [x] = 2$; $\lim_{x\to 3^+} [x] = 3$
Therefore $\lim_{x\to 3}[x]$ does not exist.

1.4 Continuity and its Consequences

1. Discontinuous at $x = -2$ (limit does not exist), and at $x = 2$ (function undefined).

3. Discontinuous at $x = -2$ (function undefined), at $x = 1$ (function undefined), and at $x = 4$ (limit does not exist).

5. Discontinuous at $x = -2$ (limit does not exist), at $x = 2$ (function undefined), and at $x = 4$ (limit does not exist).

7. $f(1)$ is not defined and $\lim_{x\to 1} f(x)$ does not exist.

9. $f(0)$ is not defined and $\lim_{x\to 0} f(x)$ does not exist.

11. $\lim_{x\to 2^-} f(x) = \lim_{x\to 2^-} (x^2) = 4$
$\lim_{x\to 2^+} f(x) = \lim_{x\to 2^+} (3x-2) = 4$
$\lim_{x\to 2} f(x) = 4; f(2) = 3$
$\lim_{x\to 2} f(x) \neq f(2)$

13. $f(x) = \dfrac{x-1}{(x+1)(x-1)}$ has a removable discontinuity at $x = 1$ and a non-removable discontinuity at $x = -1$; the removable discontinuity is removed by
$g(x) = \dfrac{1}{x+1}$.

15. No discontinuities.

17. $f(x) = \dfrac{x^2 \sin x}{\cos x}$ has non-removable discontinuities at $x = \frac{\pi}{2} + k\pi$ for any integer k.

19. By sketching the graph, or numerically, one can see that $\lim_{x\to 0} x \ln x^2 = 0$. Thus, one can remove the discontinuity at $x = 0$ by defining

$$g(x) = \begin{cases} x \ln x^2 & \text{if } x \neq 0 \\ 0 & \text{if } x = 0 \end{cases}$$

21. $f(x)$ has a non-removable discontinuity at $x = 1$.

23. $f(x)$ has a non-removable discontinuity at $x = 1$:
$\lim_{x\to -1^-} f(x) = \lim_{x\to -1^-} (3x-1) = -4$
$\lim_{x\to -1^+} f(x) = \lim_{x\to -1^+} (x^2+5x) = -4$
$\lim_{x\to 1^-} f(x) = \lim_{x\to 1^-} (x^2+5x) = 6$
$\lim_{x\to 1^+} f(x) = \lim_{x\to 1^+} (3x^3) = 3$

25. Continuous where $x + 3 > 0$, i.e. on $(-3, \infty)$

27. Continuous everywhere, i.e. on $(-\infty, \infty)$.

29. Continuous everywhere, i.e. on $(-\infty, \infty)$.

31. Continuous where $x + 1 > 0$, i.e. on $(-1, \infty)$.

33.

$$\lim_{x\to 0^-} f(x) = \lim_{x\to 0^-} 2\frac{\sin x}{x} = 2 \lim_{x\to 0^-} \frac{\sin x}{x} = 2$$

Hence a must equal 2 if f is continuous.

$$\lim_{x\to 0^-} f(x) = \lim_{x\to 0^-} b \cos x = b \lim_{x\to 0^-} \cos x = b,$$

so b and a must equal 2 if f is continuous.

35. First note that
$$\lim_{x\to 3^+} f(x) = \lim_{x\to 3^+} \ln(x-2) + x^2 = \ln(3-2) + 3^2 = 9.$$

Also $f(3) = 2e^{3b} + 1$, so if f is continuous, $2e^{3b} + 1$ must equal 9; that is $e^{3b} = 4$, so $b = \frac{\ln 4}{3}$. Then note that

$$f(0) = 2e^{(b)(0)} + 1 = 3.$$

Also,
$$\lim_{x\to 0^-} f(x) = \lim_{x\to 0^-} a(\tan^{-1} x + 2) = a(\tan^{-1} 0 + 2) = a(0+2) = 2a,$$

so a must equal 3/2 if f is continuous.

37. $$\lim_{x\to 10000^-} T(x) = \lim_{x\to 10000^-} 0.14x = 0.14(10{,}000) = 1400$$

$$\lim_{x\to 10000^+} T(x) = \lim_{x\to 10000^+} (c + 0.21x) = c + 0.21(10{,}000) = c + 2100$$

$c + 2100 = 1400$
$c = -700$

A small change in income should not result in a big change in tax, so the tax function should be continuous.

39. For $T(x)$ to be continuous at $x = 141{,}250$ we must have

$$\lim_{x\to 141{,}250^-} T(x) = \lim_{x\to 141{,}250^+} T(x).$$

Now

$$\lim_{x\to 141{,}250^-} T(x) = \lim_{x\to 141{,}250^-} (.30)(x)a$$
$$= (.30)(141{,}250) - 5685$$
$$= 36690.$$

On the other hand,

$$\lim_{x\to 141{,}250^+} T(x) = \lim_{x\to 141{,}250^+} (.35)(x) - b$$
$$= (.35)(141{,}250) - b$$
$$= 49437.50 - b.$$

Hence
$b = 49437.50 - 36690 = 12{,}747.50.$

For $T(x)$ to be continuous at $x = 307{,}050$ we must have

$$\lim_{x\to 307{,}050^-} T(x) = \lim_{x\to 307{,}050^+} T(x).$$

Now

$$\lim_{x\to 307{,}050^-} T(x)$$
$$= \lim_{x\to 307{,}050^-} (.35)(x) - b$$
$$= (.35)(307{,}050) - 12{,}747.5$$
$$= 94{,}720.$$

On the other hand,

$$\lim_{x\to 307{,}050^+} T(x)$$
$$= \lim_{x\to 307{,}050^+} (.386)(x) - c$$
$$= (.386)(307{,}050) - c$$
$$= 118521.3 - c.$$

Hence
$c = 118{,}521.3 - 94720 = 23801.3.$

41. The first two rows of the following table (together with the Intermediate Value Theorem) show that $f(x)$ has a root in $[2, 3]$. In the following rows, we use the midpoint of the previous interval as our new x. When $f(x)$ is positive, we use the left half, and when $f(x)$ is negative, we use the right half of the interval. (Because the function goes from negative to positive. If the function went from positive to negative, the intervals would be reversed.)

x	$f(x)$
2	−3
3	2
2.5	−0.75
2.75	0.5625
2.625	−0.109375
2.6875	0.223
2.65625	0.557

The zero is in the interval $[2.625, 2.65625]$.

43. The first two rows of the following table (together with the Intermediate Value Theorem) show that $f(x)$ has a root in $[2, 3]$. In the following rows, we use the midpoint of the previous interval as our new x. When $f(x)$ is positive, we use the right half, and when $f(x)$ is negative, we use the left half of the interval.

x	$f(x)$
−1	1
0	-2
−0.5	−0.125
−0.625	0.256
−0.5625	0.072
−0.53125	−0.025

The zero is in the interval $[-0.5625, -0.53125]$.

45. The first two rows of the following table (together with the Intermediate Value Theorem) show that $f(x)$ has a root in $[-2,-1]$. In the following rows, we use the midpoint of the previous interval as our new x. When $f(x)$ is positive, we use the right half, and when $f(x)$ is negative, we use the left half of the interval.

x	$f(x)$
0	1
1	−0.46
0.5	0.378
0.75	−0.018
0.625	0.186
0.6875	0.085
0.71875	0.034

The zero is in the interval $[0.71875, 0.75]$.

47. $\lim_{x\to 2^+} f(x) = \lim_{x\to 2^+} (3x-1) = 5$

$f(2) = 3(2)\text{-}1 = 5$

Thus $f(x)$ is continuous from the right at $x = 2$.

49. $\lim_{x\to 2^+} f(x) = \lim_{x\to 2^+} (3x-3) = 3$

$f(2) = 2^2 = 4$

Thus $f(x)$ is not continuous from the right at $x = 2$.

51. A function is continuous from the left at $x = a$ if $\lim_{x\to a^-} f(x) = f(a)$.

(a) $\lim_{x\to 2^-} f(x) = \lim_{x\to 2^-} x^2 = 4$
$f(2) = 5$
Thus $f(x)$ is not continuous from the left at $x = 2$.

(b) $\lim_{x\to 2^-} f(x) = \lim_{x\to 2^-} x^2 = 4$
$f(2) = 3$
Thus $f(x)$ is not continuous from the left at $x = 2$.

(c) $\lim_{x\to 2^-} f(x) = \lim_{x\to 2^-} x^2 = 4$
$f(2) = 4$
Thus $f(x)$ is continuous from the left at $x = 2$.

(d) $f(x)$ is not continuous from the left at $x = 2$ because $f(2)$ is undefined.

53. Need $g(30) = 100$ and $g(34) = 0$. We may take $g(T)$ to be linear.
$m = \dfrac{0-100}{34-30} = -25$
$y = -25(x-34)$
$g(T) = -25(T-34)$

55.

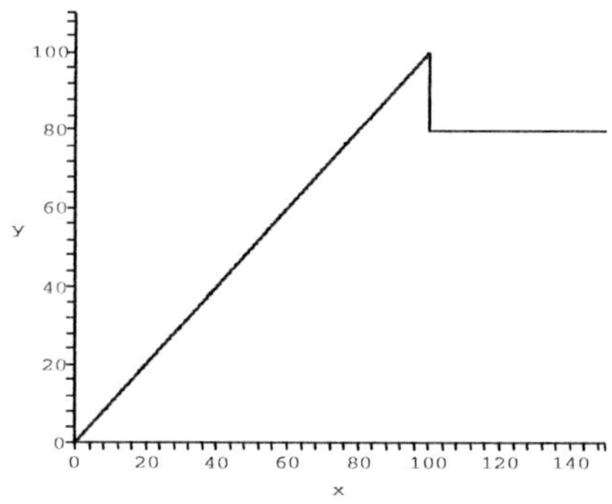

The graph is discontinuous at $x = 100$. This is when the box starts moving.

57. Let $f(t)$ be her distance from home as a function of time on Monday. Let $g(t)$ be her distance from home as a function of time on Tuesday. Let t be given in minutes, with $t = 0$ corresponding to 7:13 a.m. Then she leaves home at $t = 0$ and arrives at her destination at $t = 410$. Let $h(t) = f(t) - g(t)$. If $h(t) = 0$ for some t, then the saleswoman was at exactly the same place at the same time on both Monday and Tuesday. $h(0) = f(0) - g(0) = -g(0) < 0$ and $h(410) = f(410) - g(410) = f(410) >$

0. By the Intermediate Value Theorem, there is a t in the interval $[0, 410]$ such that $h(t) = 0$.

59.

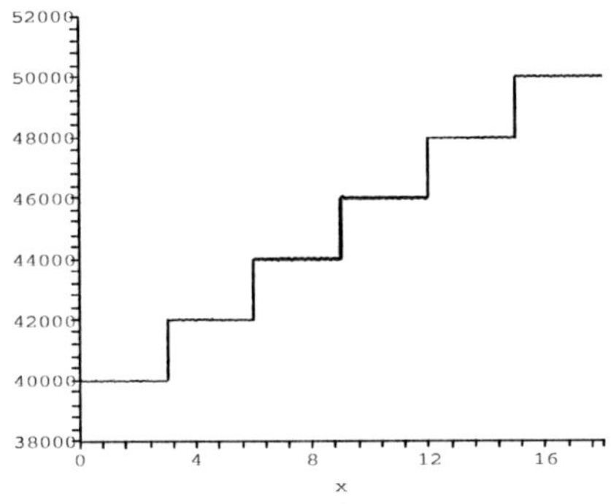

The function $s(t)$ has jump discontinuities every three months when the salary suddenly increases by \$2000. In the function $f(t)$, the \$2000 increase occurs gradually over the 3 month period, so $f(t)$ is continuous. It might be easier to do calculations with $f(t)$ because it is continuous and because it is given by a simpler formula.

61. We already know $f(x) \neq 0$ for $a < x < b$. Suppose $f(d) < 0$ for some d, $a < d < b$. Then by the Intermediate Value Theorem, there is an e in the interval $[c, d]$ such that $f(e) = 0$. But this e would also be between a and b, which is impossible. Thus, $f(x) > 0$ for all $a < x < b$.

63. $$\lim_{x\to 0} xf(x) = \lim_{x\to 0} x \lim_{x\to 0} f(x) = 0 f(0) = 0$$

65. $$\lim_{x\to a} g(x) = \lim_{x\to a} |f(x)| = \left|\lim_{x\to a} f(x)\right| = |f(a)| = g(a).$$

67. Let $b \geq a$. Then

$$\begin{aligned} \lim_{x\to b} h(x) &= \lim_{x\to b}\left(\max_{a\leq t\leq b} f(t)\right) \\ &= \max_{a\leq t\leq b}\left(\lim_{t\to b} f(t)\right) \\ &= h(b) \end{aligned}$$

since f is continuous. Thus, h is continuous for $x \geq a$.

No, the property would not be true if f were not assumed to be continuous. A counterexample is

$$f(x) = \begin{cases} 1 & \text{if } a \leq x < b \\ 2 & \text{if } b \leq x \end{cases}$$

Then $h(x) = 1$ for $a \leq x < b$, and $h(x) = 2$ for $x \geq b$. Thus, h is not continuous at $x = b$.

1.5 Limits Involving Infinity

1. (a) $\displaystyle \lim_{x\to 1^-} \frac{1-2x}{x^2-1} = \infty.$

(b) $\displaystyle \lim_{x\to 1^+} \frac{1-2x}{x^2-1} = -\infty.$

(c) Does not exist.

3. (a) $\displaystyle \lim_{x\to 2^-} \frac{x-4}{x^2-4x+4} = -\infty$

(b) $\displaystyle \lim_{x\to 2^+} \frac{x-4}{x^2-4x+4} = -\infty$

(c) $\displaystyle \lim_{x\to 2} \frac{x-4}{x^2-4x+4} = -\infty$

5. $\displaystyle \lim_{x\to 2^-} \frac{-x}{\sqrt{4-x^2}} = -\infty.$

As x approaches 2 from below, the numerator is near -2 and the denominator is small and positive, so the faction goes to $-\infty$.

7. $\lim_{x\to-\infty} \dfrac{-x}{\sqrt{4+x^2}}$
$= \lim_{x\to-\infty} \dfrac{-x}{-x\sqrt{\frac{4}{x^2}+1}}$
$= \lim_{x\to-\infty} \dfrac{1}{\sqrt{\frac{4}{x^2}+1}}$
$= \dfrac{1}{\sqrt{1}} = 1$

9. $\lim_{x\to\infty} \dfrac{x^3 - 2\cos x}{3x^2+4x-1}$
$= \lim_{x\to\infty} \dfrac{x^2\left(x - \frac{2\cos x}{x^2}\right)}{x^2\left(3+\frac{4}{x}-\frac{1}{x^2}\right)}$
$= \lim_{x\to\infty} \dfrac{\left(x - \frac{2\cos x}{x^2}\right)}{3+\frac{4}{x}-\frac{1}{x^2}} = \infty$

11. $\lim_{x\to\infty} \ln 2x = \infty$.

Note that $\ln 2x = \ln 2 + \ln x$, so it is enough to show that $\ln x$ goes to ∞ as x goes to ∞. This can be seen from the graph of the function $\ln x$ on page 51.

13. $\lim_{x\to 0^+} e^{-2/x} = 0$.

When x is small and positive, $-2/x$ is large and negative, and e raised to a large negative power is very small.

15. $\lim_{x\to\infty} \cot^{-1} x = 0$.

(Compare Example 5.8) We are looking for the angle that θ must approach as $\cot\theta$ goes to ∞. Look at the graph of $\cot\theta$. To define the inverse cotangent, you must pick one branch of this graph, and the standard choice is the branch immediately to the right of the y-axis. Then as $\cot\theta$ goes to ∞, the angle goes to 0.

17. $\lim_{x\to\infty} e^{2x-1} = \infty$.

As x gets large, $2x-1$ gets large, and e raised to a large positive power is large and positive.

19. $\lim_{x\to\infty} \sin 2x$ does not exist. As x gets larger and larger, the values of $\sin 2x$ oscillate between 1 and -1.

21. As x goes to ∞, both e^{3x} and e^x go to ∞ as well. Furthermore, as x goes to ∞, so does $\ln x$. Thus it looks like

$$\lim_{x\to\infty} \left(\frac{\ln(2+e^{3x})}{\ln(1+e^x)}\right) = \frac{\infty}{\infty}.$$

This is an indeterminate form, i.e., we don't know from this analysis what happens in this limit. Looking at numerical and/or graphing evidence, we guess that the limit is 3.

23. $\lim_{x\to\frac{\pi}{2}^-} e^{-\tan x} = \lim_{x\to\infty} e^{-x}$
$= \lim_{x\to-\infty} e^{x} = 0$, but
$\lim_{x\to\frac{\pi}{2}^+} e^{-\tan x} = \lim_{x\to-\infty} e^{-x}$
$= \lim_{x\to\infty} e^{x} = \infty$,

so the limit does not exist.

25. Since $4+x^2$ is never 0, there are no vertical asymptotes. We have
$\lim_{x\to\infty} \dfrac{x}{\sqrt{4+x^2}}$
$= \lim_{x\to\infty} \dfrac{x}{x\sqrt{\frac{4}{x^2}+1}}$
$= \lim_{x\to\infty} \dfrac{1}{\sqrt{\frac{4}{x^2}+1}}$
$= \dfrac{1}{\sqrt{1}} = 1$

and
$\lim_{x\to-\infty} \dfrac{x}{\sqrt{4+x^2}}$
$= \lim_{x\to-\infty} \dfrac{x}{-x\sqrt{\frac{4}{x^2}+1}}$
$= \lim_{x\to-\infty} \dfrac{-1}{\sqrt{\frac{4}{x^2}+1}}$
$= \dfrac{-1}{\sqrt{1}} = -1,$

so there are horizontal asymptotes at $y = 1$ and $y = -1$.

27. $4 - x^2 = 0 \Rightarrow 4 = x^2$ so we have vertical asymptotes at $x = \pm 2$. We have

$$\lim_{x\to\pm\infty} \frac{x}{4-x^2} = \lim_{x\to\pm\infty} \frac{x}{x^2\left(\frac{4}{x^2}-1\right)} = \lim_{x\to\pm\infty} \frac{1}{x\left(\frac{4}{x^2}-1\right)} = 0.$$

So there is a horizontal asymptote at $y = 0$.

29. The denominator factors: $x^2 - 2x - 3 = (x-3)(x+1)$. Since neither $x = 3$ nor $x = -1$ are zeros of the numerator, we see that $f(x)$ has vertical asymptotes at $x = 3$ and $x = -1$.

$f(x) \to -\infty$ as $x \to 3^-$,
$f(x) \to \infty$ as $x \to 3^+$,
$f(x) \to \infty$ as $x \to -1^-$, and
$f(x) \to -\infty$ as $x \to -1^+$.

We have

$$\lim_{x\to\pm\infty} \frac{3x^2+1}{x^2-2x-3}$$

$$\lim_{x\to\pm\infty} \frac{3+1/x^2}{1-2/x-3/x^2} = 3.$$

So there is a horizontal asymptote at $y = 3$.

31. The function $\ln x$ has a one-sided vertical asymptote at $x = 0$, so $f(x) = \ln(1-\cos x)$ will have a vertical asymptote whenever $1 - \cos x = 0$, i.e., whenever $\cos x = 1$. This happens when $x = 2k\pi$ for any integer k. Since $1 - \cos x \geq 0$ for all x, $f(x)$ is defined at all points except for these vertical asymptotes. Thus as $f(x)$ approaches any of these asymptotes (from either side), it behaves like $\ln x$ approaching 0 from the right, so $f(x) \to -\infty$ as x approaches any of these asymptotes from either side.

33. The function is continuous for all x, so no vertical asymptotes. We have

$$\begin{aligned}\lim_{x\to\infty} 4\tan^{-1} x - 1 &= 4(\lim_{x\to\infty} \tan^{-1} x) - 1\\ &= 4(\pi/2) - 1\\ &= 2\pi - 1\end{aligned}$$

and

$$\begin{aligned}&\lim_{x\to-\infty} 4\tan^{-1} x - 1\\ &= 4(\lim_{x\to-\infty} \tan^{-1} x) - 1\\ &= 4(-\pi/2) - 1\\ &= -2\pi - 1,\end{aligned}$$

so there are horizontal asymptotes at $y = 2\pi - 1$ and $y = -2\pi - 1$.

35. Vertical asymptotes at $x = \pm 2$. The slant asymptote is $y = -x$.

37. Vertical asymptotes at

$$x = \frac{-1 \pm \sqrt{17}}{2}.$$

The slant asymptote is $y = x - 1$.

39. When x is large, the value of the fraction is close to 0.

41. When x is large, the value of the fraction is very close to $\frac{1}{2}$.

43. $\displaystyle\lim_{x\to\infty} \frac{x^3+4x+5}{e^{x/2}} = 0.$

45. When x is close to -1, the value of the fraction is close to 1.

47. When x is close to 0, the value of the fraction is large and negative, so the limit appears to be $-\infty$.

49. We multiply by

$$\frac{\sqrt{4x^2-2x+1}+2x}{\sqrt{4x^2-2x+1}+2x}$$

to get:

$$\lim_{x\to\infty}(\sqrt{4x^2-2x+1}-2x)$$
$$=\lim_{x\to\infty}\frac{-2x+1}{\sqrt{4x^2-2x+1}+2x}\cdot\frac{1/x}{1/x}$$
$$=\lim_{x\to\infty}\frac{-2+1/x}{\sqrt{4-2/x+1/x^2}+2}$$
$$=\frac{-2}{\sqrt{4}+2}=-\frac{1}{2}.$$

51. $\lim_{x\to\infty}(\sqrt{5x^2+4x+7}-\sqrt{5x^2+x+3})$

If we multiply by
$$\frac{\sqrt{5x^2+4x+7}+\sqrt{5x^2+x+3}}{\sqrt{5x^2+4x+7}+\sqrt{5x^2+x+3}},$$
we get
$$\lim_{x\to\infty}\frac{(5x^2+4x+7)-(5x^2+x+3)}{\sqrt{5x^2+4x+7}+\sqrt{5x^2+x+3})}$$
$$=\lim_{x\to\infty}\frac{3x+4}{\sqrt{5x^2+4x+7}+\sqrt{5x^2+x+3}}$$
$$=\lim_{x\to\infty}\frac{3+\frac{4}{x}}{\sqrt{5+\frac{4}{x}+\frac{7}{x^2}}+\sqrt{5+\frac{1}{x}+\frac{3}{x^2}}}$$
$$=\frac{3}{2\sqrt{5}}=\frac{3\sqrt{5}}{10}$$

53.

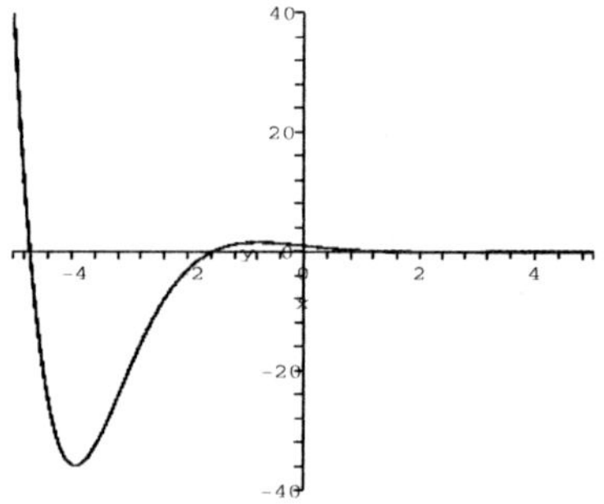

on $[-10,10]$ by $[-100,100]$

The horizontal asymptote is $y=0$ approached only as $x\to\infty$. The graph crosses the horizontal asymptote an infinite number of times.

55. $\lim_{x\to\infty}\left(1+\frac{1}{x}\right)^x=\lim_{x\to0^+}(1+x)^{1/x}$
$$=\lim_{x\to0^-}(1+x)^{1/x}=\lim_{x\to-\infty}\left(1+\frac{1}{x}\right)^x$$

57. $h(0)=\frac{300}{1+9(.8^0)}=\frac{300}{10}=30$ mm
$$\lim_{t\to\infty}\frac{300}{1+9(.8^t)}=300\text{ mm}$$

59. $\lim_{x\to0^+}\frac{80x^{-.3}+60}{2x^{-.3}+5}\left(\frac{x^{.3}}{x^{.3}}\right)$
$$=\lim_{x\to0^+}\frac{80+60x^{.3}}{2+5x^{.3}}$$
$$=\frac{80}{2}=40\text{ mm}$$
$$\lim_{x\to\infty}\frac{80x^{-.3}+60}{2x^{-.3}+5}=\frac{60}{5}=12\text{ mm}$$

61. $f(x)=\frac{80x^{-0.3}+60}{10x^{-0.3}+30}$

63. $\lim_{t\to\infty}v_N=\lim_{t\to\infty}\frac{Ft}{m}=\infty$
$$\lim_{t\to\infty}v_E=\lim_{t\to\infty}\frac{Fct}{\sqrt{m^2c^2+F^2t^2}}$$
$$=\lim_{t\to\infty}\frac{Fct}{t\sqrt{\frac{m^2c^2}{t^2}+F^2}}$$
$$=\lim_{t\to\infty}\frac{Fc}{\sqrt{\frac{m^2c^2}{t^2}+F^2}}$$
$$=\frac{Fc}{\sqrt{F^2}}=c$$

65. As in Example 5.10, the terminal velocity is $-\sqrt{\frac{32}{k}}$. When $k=0.00064$, the terminal velocity is $-\sqrt{\frac{32}{.00064}}\approx -224$. When $k=0.00128$, the terminal velocity is $-\sqrt{\frac{32}{.00128}}\approx-158$.

Solve $\sqrt{\frac{32}{ak}}=\frac{1}{2}\sqrt{\frac{32}{k}}$. Squaring both sides, $\frac{32}{ak}=\frac{1}{4}\cdot\frac{32}{k}$ so $a=4$.

67. We must restrict the domain to $v_0\geq0$ because the formula makes sense only if the rocket is launched upward. To find v_e, set $19.6R-v_0^2=0$. Using $R\approx$ 6,378,000 meters, we get $v_0=\sqrt{19.6R}\approx$ 11,180m/s. If the rocket is launched with initial velocity $\geq v_e$, it will never return to earth; hence v_e is called the escape velocity.

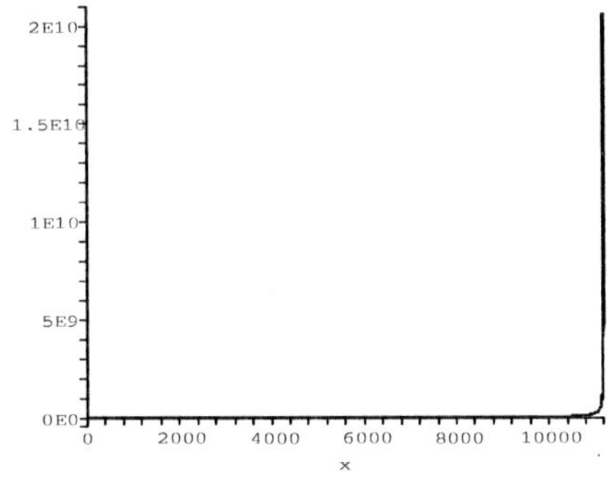

69. Suppose the degree of q is n. If we divide both $p(x)$ and $q(x)$ by x^n, then the new denominator will approach a constant while the new numerator tends to ∞, so there is no horizontal asymptote.

71. When we do long division, we get a remainder of $x+2$, so the degree of p is one greater than the degree of q.

73. The function $q(x) = -2(x-2)(x-3)$ satisfies the given conditions.

75. True.

77. False.

79. True.

81. Vertical asymptote at $x = 2$. Horizontal asymptotes at $y = 4$ and $y = 0$.

83. For any positive constant a, $e^{-at} \to 0$ as $t \to \infty$. Since $\sin t$ oscilates between -1 and 1, $e^{-at}\sin t \to 0$ as $t \to \infty$. In the following graph, we see that suspension system A damps out at about 5 seconds, while system B takes about 18 seconds to damp out.

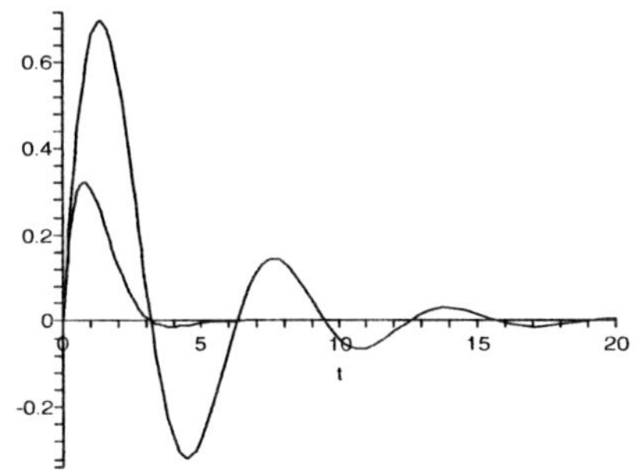

85. $g(x) = \sin x$, $h(x) = x$ at $a = 0$

87. $\lim\limits_{x\to 0^+} x^{1/(\ln x)} = e \approx 2.71828$

89. $\lim\limits_{x\to\infty} x^{1/x} = 1$

1.6 Formal Definition of the Limit

1. (a) From the graph, we determine that we can take $\delta = 0.316$, as shown below.

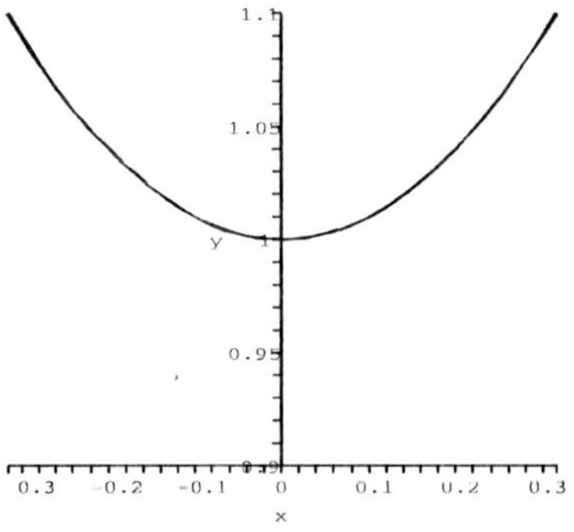

(b) From the graph, we determine that we can take $\delta = 0.223$, as shown below.

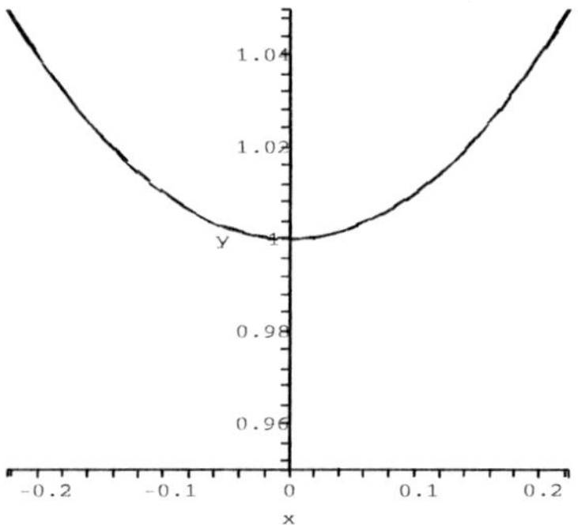

3. (a) From the graph, we determine that we can take $\delta = 0.45$, as shown below.

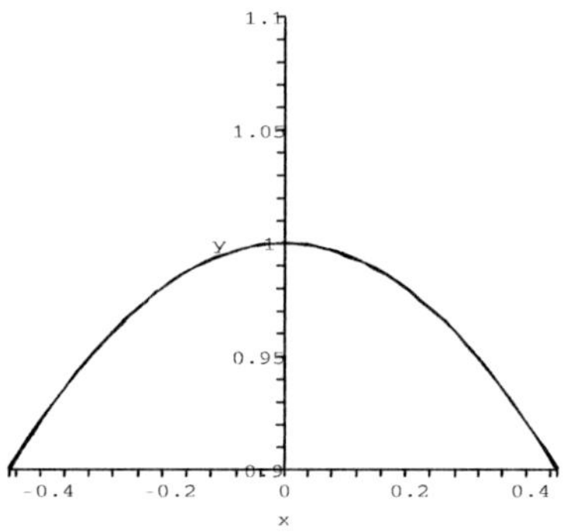

(b) From the graph, we determine that we can take $\delta = 0.315$, as shown below.

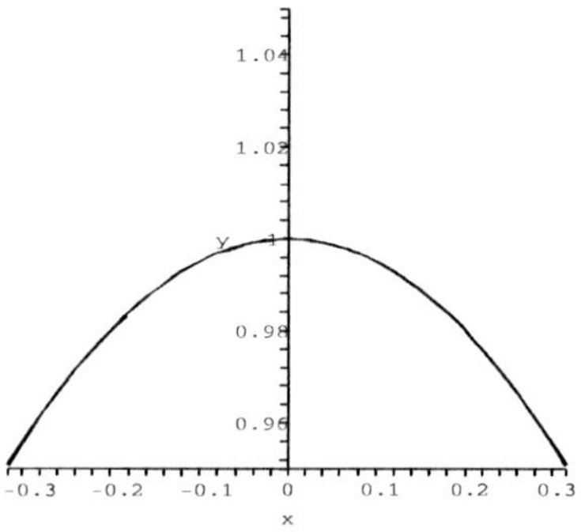

5. (a) From the graph, we determine that we can take $\delta = 0.38$, as shown below.

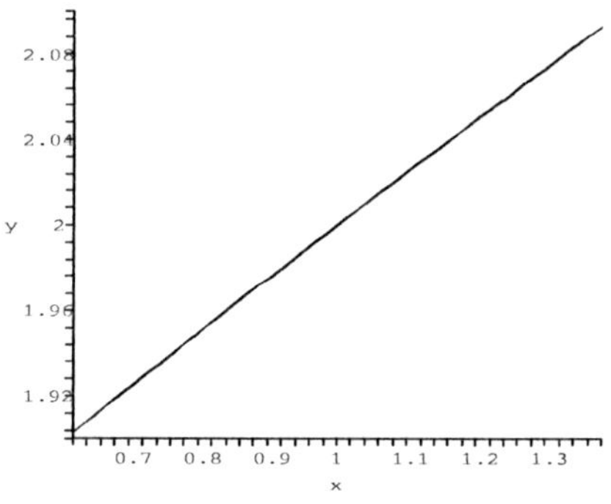

(b) From the graph, we determine that we can take $\delta = 0.2$, as shown below.

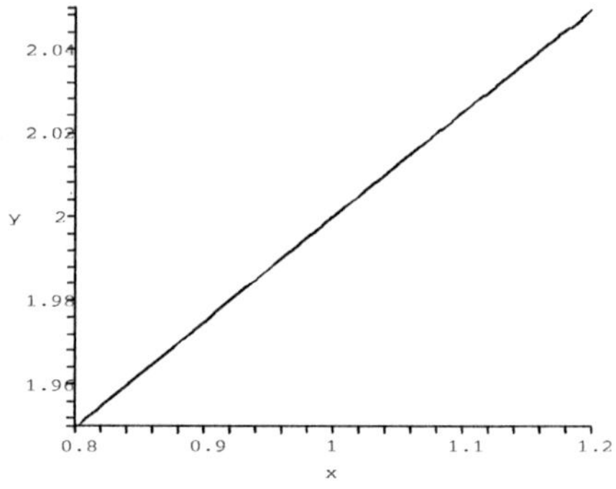

7. (a) From the graph, we determine that we can take $\delta = 0.02$, as shown below.

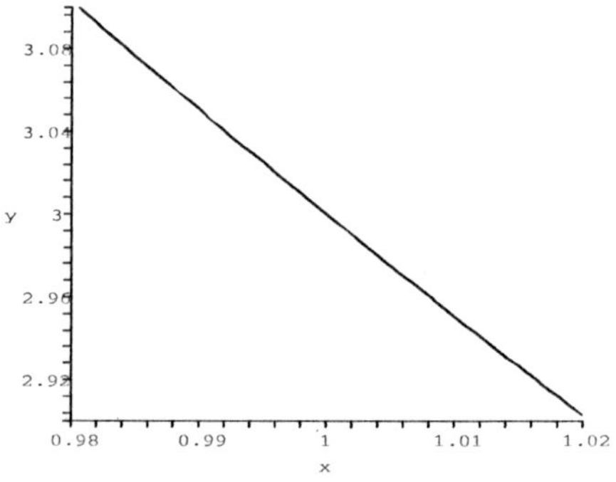

(b) From the graph, we determine that we can take $\delta = 0.01$, as shown below.

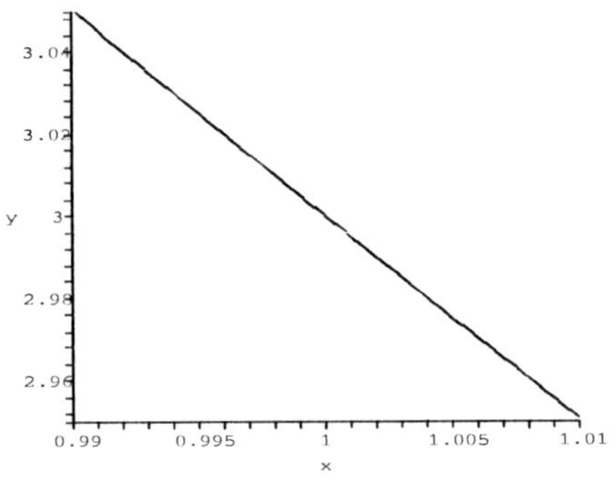

9. We want $|3x - 0| < \varepsilon$
   $\Leftrightarrow 3|x| < \varepsilon$
   $\Leftrightarrow |x| = |x - 0| < \varepsilon/3$
   Take $\delta = \varepsilon/3$.

11. We want $|3x + 2 - 8| < \varepsilon$
    $\Leftrightarrow |3x - 6| < \varepsilon$
    $\Leftrightarrow 3|x - 2| < \varepsilon$
    $\Leftrightarrow |x - 2| < \varepsilon/3$
    Take $\delta = \varepsilon/3$.

13. We want $|3 - 4x - (-1)| < \varepsilon$
    $\Leftrightarrow |-4x + 4| < \varepsilon$
    $\Leftrightarrow 4|-x + 1| < \varepsilon$
    $\Leftrightarrow 4|x - 1| < \varepsilon$
    $\Leftrightarrow |x - 1| < \varepsilon/4$
    Take $\delta = \varepsilon/4$.

15. We want $\left|\dfrac{x^2 + x - 2}{x - 1} - 3\right| < \varepsilon$.
    We have
    $$\left|\dfrac{x^2 + x - 2}{x - 1} - 3\right|$$
    $$= \left|\dfrac{(x + 2)(x - 1)}{x - 1} - 3\right|$$
    $$= |x + 2 - 3|$$
    $$= |x - 1|$$
    Take $\delta = \varepsilon$.

17. We want $|x^2 - 1 - 0| < \varepsilon$.
    We have $|x^2 - 1| = |x - 1||x + 1|$. We require that $\delta < 1$, i.e., $|x - 1| < 1$ so $0 < x < 2$ and $|x + 1| < 3$. Then $|x^2 - 1| = |x - 1||x + 1| < 3|x - 1|$.
    Requiring this to be less than ε gives $|x - 1| < \varepsilon/3$, so $\delta = \min\{1, \varepsilon/3\}$.

19. We want $|x^2 - 1 - 3| < \varepsilon$.
    We have $|x^2 - 4| = |x - 2||x + 2|$. We require that $\delta < 1$, i.e., $|x - 1| < 1$ so $1 < x < 3$ and $|x + 2| < 5$. Then $|x^2 - 4| = |x - 2||x + 2| < 5|x - 2|$.
    Requiring this to be less than ε gives $|x - 2| < \varepsilon/5$, so $\delta = \min\{1, \varepsilon/5\}$.

21. Let $f(x) = mx + b$. Since $f(x)$ is continous, we know that $\lim\limits_{x \to a} f(x) = ma + b$. So we want to find a δ which forces $|mx + b - (ma + b)| < \varepsilon$. But
    $|mx + b - (ma + b)| = |mx - ma|$
    $ = |m||x - a|$.
    So as long as $|x - a| < \delta = \varepsilon/|m|$, we will have $|f(x) - (ma + b)| < \varepsilon$. This δ clearly does not depend on a. This is due to the fact that $f(x)$ is a linear function, so the slope is constant, which means that the ratio of the change in y to the change in x is constant.

23. For a function $f(x)$ defined on some open interval (c, a) we say
    $$\lim_{x \to a^-} f(x) = L$$
    if, given any number $\varepsilon > 0$, there is another number $\delta > 0$ such that whenever $x \in (c, a)$ and $a - \delta < x < a$, we have $|f(x) - L| < \varepsilon$.

    For a function $f(x)$ defined on some open interval (a, c) we say
    $$\lim_{x \to a^+} f(x) = L$$
    if, given any number $\varepsilon > 0$, there is another number $\delta > 0$ such that whenever $x \in (a, c)$ and $a < x < a + \delta$, we have $|f(x) - L| < \varepsilon$.

25. As $x \to 1^+$, $x - 1 > 0$ so we compute

$$\begin{aligned} \frac{2}{x-1} &> 100 \\ 2 &> 100(x-1) \\ \frac{2}{100} &> x - 1 \end{aligned}$$

So take $\delta = 2/100$.

27. We look at the graph of $\cot x$ as $x \to 0^+$ and we find that we should take $\delta = 0.00794$.

29. As $x \to 2^-$, $4 - x^2 > 0$ so we compute

$$\begin{aligned} \frac{2}{\sqrt{4-x^2}} &> 100 \\ 2 &> 100\sqrt{4-x^2} \\ \frac{2}{100} &> \sqrt{4-x^2} \\ \frac{4}{10000} &> 4 - x^2 = (2-x)(2+x) \end{aligned}$$

Take $\delta < 1$ so that $1 < x < 3$ so we have $2 + x < 5$. Then $(2-x)(2+x) < (2-x)5$. Now, if we require $|x-2| < \frac{4}{50000}$ then $\frac{2}{\sqrt{4-x^2}} > 100$. So let $\delta = \frac{4}{50000}$.

31. We want M such that if $x > M$,

$$\left|\frac{x^2-2}{x^2+x+1} - 1\right| < 0.1$$

We have

$$\begin{aligned} &\left|\frac{x^2-2}{x^2+x+1} - 1\right| \\ &= \left|\frac{x^2 - 2 - (x^2+x+1)}{x^2+x+1}\right| \\ &= \left|\frac{-x-3}{x^2+x+1}\right| \\ &= \left|\frac{x+3}{x^2+x+1}\right| \end{aligned}$$

Now, as long as $x > 3$, we have

$$\begin{aligned} \left|\frac{x+3}{x^2+x+1}\right| &< \left|\frac{2x}{x^2+x}\right| \\ &= \left|\frac{2}{x+1}\right| \end{aligned}$$

We want $\left|\frac{2}{x+1}\right| < 0.1$. Since $x \to \infty$, we can take $x > 0$, so we solve $\frac{2}{x+1} < 0.1$ to get $x > 19$, i.e., $M = 19$.

33. We have

$$\begin{aligned} \left|\frac{x^2+3}{4x^2-4} - \frac{1}{4}\right| &= \left|\frac{x^2+3-(x^2-1)}{4x^2-4}\right| \\ &= \left|\frac{4}{4x^2-4}\right| \\ &= \left|\frac{1}{x^2-1}\right| \end{aligned}$$

Since $x \to -\infty$, we may take $x < -1$ so that $x^2 - 1 > 0$. We now need $\frac{1}{x^2-1} < 0.1$. Solving for x gives $|x| > \sqrt{11} \approx 3.3166$. So we can take $N = -4$.

35. We want $|e^{-2x}| < 0.1$. Since $e^{-2x} > 0$ for any x, this is the same as $e^{-2x} < 0.1$ so $-2x < \ln(0.1)$ and then $x > \frac{\ln(0.1)}{-2} \approx 1.15$. We may take $M = 2$.

37. Let $\varepsilon > 0$ be given and let $M = \sqrt[3]{2/\varepsilon}$. Then if $x > M$,

$$\left|\frac{2}{x^3}\right| < \left|\frac{2}{\left(\sqrt[3]{2/\varepsilon}\right)^3}\right| = \varepsilon$$

39. Let $\varepsilon > 0$ be given and let $M = \varepsilon^{-1/k}$. Then if $x > M$,

$$\left|\frac{1}{x^k}\right| < \left|\frac{1}{(\varepsilon^{-1/k})^k}\right| = \varepsilon$$

41. Let $\varepsilon > 0$ be given and assume $\varepsilon \le 1/2$. Let $N = -(\frac{1}{\varepsilon} - 2)^{1/2}$. Then if $x < N$,

$$\left|\frac{1}{x^2+2} - 3 - (-3)\right| = \left|\frac{1}{x^2+2}\right| < \left|\frac{1}{\left(-(\frac{1}{\varepsilon}-2)^{1/2}\right)^2 + 2}\right| = \varepsilon$$

43. Let $N < 0$ be given and let $\delta = \sqrt[4]{-2/N}$. Then for any x such that $|x+3| < \delta$,

$$\left|\frac{-2}{(x+3)^4}\right| > \left|\frac{-2}{(\sqrt[4]{-2/N})^4}\right| = |N|$$

45. Let $M > 0$ be given and let $\delta = \sqrt{4/M}$. Then for any x such that $|x-5| < \delta$,

$$\left|\frac{4}{(x-5)^2}\right| > \left|\frac{4}{(\sqrt{4/M})^2}\right| = |M|$$

47. We observe that $\lim_{x\to1^-} f(x) = 2$ and $\lim_{x\to1^+} f(x) = 4$. For any $x \in (1,2)$,

$$|f(x)-2| = |x^2+3-2| = |x^2+1| > 2.$$

So if $\varepsilon \le 2$, there is no $\delta > 0$ to satisfy the definition of limit.

49. We observe that $\lim_{x\to1^-} f(x) = 2$ and $\lim_{x\to1^+} f(x) = 4$. For any $x \in (1, \sqrt{2})$,

$$\begin{aligned}|f(x)-2| &= |5-x^2-2| \\ &= |3-x^2| > |3-(\sqrt{2})^2| = 1.\end{aligned}$$

So if $\varepsilon \le 1$, there is no $\delta > 0$ to satisfy the definition of limit.

51. We want to find, for any given $\varepsilon > 0$, a $\delta > 0$ such that whenever $0 < |r-2| < \delta$, we have $|2r^2-8| < \varepsilon$. We see that

$$|2r^2-8| = 2|r^2-4| = 2|r-2||r+2|.$$

Since we want a radius close to 2, we may take $|r-2| < 1$ which implies $|r+2| < 5$ and so

$$|2r^2-8| < 10|r-2|$$

whenever $|r-2| < 1$. If we then take $\delta = \min\{1, \varepsilon/10\}$, we see that whenever $0 < |r-2| < \delta$, we have

$$|2r^2-8| < 10\cdot\delta \le 10\cdot\frac{\varepsilon}{10} = \varepsilon.$$

53. Let $L = \lim_{x\to a} f(x)$. Given any $\varepsilon > 0$, we know there exists $\delta > 0$ such that whenever $0 < |x-a| < \delta$, we have

$$|f(x)-L| < \frac{\varepsilon}{|c|}.$$

Here, we can take $\varepsilon/|c|$ instead of ε because there is such a δ for *any* ε, including $\varepsilon/|c|$. But now we have

$$\begin{aligned}|c\cdot f(x) - c\cdot L| &= |c|\cdot|f(x)-L| \\ &< |c|\cdot\frac{\varepsilon}{|c|} = \varepsilon.\end{aligned}$$

Therefore, $\lim_{x\to a} c\cdot f(x) = c\cdot L$, as desired.

55. Let $\varepsilon > 0$ be given. Since $\lim_{x\to a} f(x) = L$, there exists $\delta_1 > 0$ such that whenever $0 < |x-a| < \delta_1$, we have

$$|f(x)-L| < \varepsilon.$$

In particular, we know that

$$L - \varepsilon < f(x).$$

Similarly, since $\lim_{x\to a} h(x) = L$, there exists $\delta_2 > 0$ such that whenever $0 < |x-a| < \delta_2$, we have

$$|h(x)-L| < \varepsilon.$$

In particular, we know that

$$h(x) < L + \varepsilon.$$

Let $\delta = \min\{\delta_1, \delta_2\}$. Then whenever $0 < |x - a| < \delta$, we have

$$L - \varepsilon < f(x) \le g(x) \le h(x) < L + \varepsilon.$$

Therefore

$$|g(x) - L| < \varepsilon$$

and so $\lim_{x \to a} g(x) = L$ as desired.

57. Let $\varepsilon > 0$ be given. If $\lim_{x \to a} f(x) = L$, then there exists $\delta > 0$ such that if $0 < |x - a| < \delta$, then $|f(x) - L| < \varepsilon$. Since $|f(x) - L - 0| = |f(x) - L|$, this is precisely what we need to see that $\lim_{x \to a}[f(x) - L] = 0$.

59. If $2 < x < \sqrt{4.1}$ then $4 < x^2 < 4.1$ so (for $x \in (2, \sqrt{4.1})$), $x^2 - 4 < 4.1 - 4 = 0.1$.

If $\sqrt{3.9} < x < 2$ then $3.9 < x^2 < 4$ so (for $x \in (\sqrt{3.9}, 2)$), $x^2 - 4 > 3.9 - 4 = -0.1$.

For the limit definition, we need to take $\delta = \min\{\delta_1, \delta_2\} = \delta_1$ to ensure that x^2 is within 0.1 of 4 on *both* sides of $x = 2$.

1.7 Limits and Loss-of-Significance Errors

1. The limit is $\frac{1}{4}$.

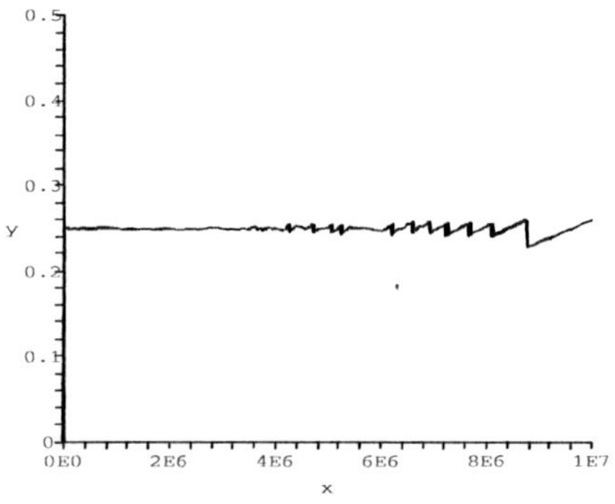

We can rewrite the function as

$$f(x) = x(\sqrt{4x^2+1} - 2x) \cdot \frac{\sqrt{4x^2+1} + 2x}{\sqrt{4x^2+1} + 2x}$$
$$= \frac{x(4x^2 + 1 - 4x^2)}{\sqrt{4x^2+1} + 2x}$$
$$= \frac{x}{\sqrt{4x^2+1} + 2x}$$

to avoid loss-of-significance errors.

In the table below, the middle column contains values calculated using $f(x) = x(\sqrt{4x^2+1} - 2x)$, while the third column contains values calculated using the rewritten $f(x)$.

x	old $f(x)$	new $f(x)$
1	0.236068	0.236068
10	0.249844	0.249844
100	0.249998	0.249998
1000	0.250000	0.250000
10000	0.250000	0.250000
100000	0.249999	0.250000
1000000	0.250060	0.250000
10000000	0.260770	0.250000
100000000	0.000000	0.250000
1000000000	0.000000	0.250000

3. The limit is 1.

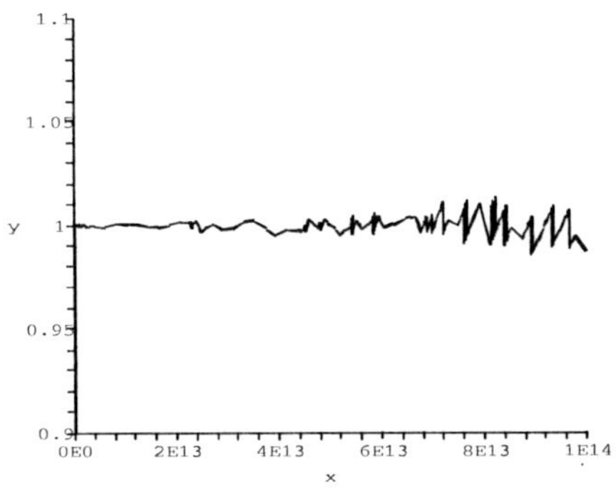

We can rewrite the function as

$$\sqrt{x}(\sqrt{x+4}-\sqrt{x+2})\cdot\frac{\sqrt{x+4}+\sqrt{x+2}}{\sqrt{x+4}+\sqrt{x+2}}$$
$$=\frac{\sqrt{x}[(x+4)-(x+2)]}{\sqrt{x+4}+\sqrt{x+2}}$$
$$=\frac{2\sqrt{x}}{\sqrt{x+4}+\sqrt{x+2}}$$

to avoid loss-of-significance errors.

In the table below, the middle column contains values calculated using $f(x)=\sqrt{x}\left(\sqrt{x+4}-\sqrt{x+2}\right)$, while the third column contains values calculated using the rewritten $f(x)$.

x	old $f(x)$	new $f(x)$
1	0.504017	0.504017
10	0.877708	0.877708
100	0.985341	0.985341
1000	0.998503	0.998503
10000	0.999850	0.999850
100000	0.999985	0.999985
1000000	0.999998	0.999999
10000000	1.000000	1.000000
100000000	1.000000	1.000000
1000000000	1.000000	1.000000
10000000000	1.000000	1.000000
1E+11	0.999990	1.000000
1E+12	1.000008	1.000000
1E+13	0.999862	1.000000
1E+14	0.987202	1.000000
1E+15	0.942432	1.000000
1E+16	0.000000	1.000000
1E+17	0.000000	1.000000

5. The limit is 1.

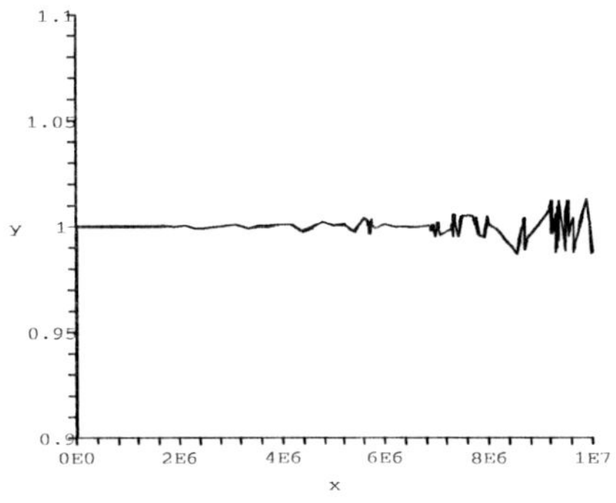

We can multiply $f(x)$ by

$$\frac{\sqrt{x^2+4}+\sqrt{x^2+2}}{\sqrt{x^2+4}+\sqrt{x^2+2}}$$

to rewrite the function as

$$\frac{x[x^2+4-(x^2+2)]}{\sqrt{x^2+4}+\sqrt{x^2+2}}$$
$$=\frac{2x}{\sqrt{x^2+4}+\sqrt{x^2+2}}$$

to avoid loss-of-significance errors.

In the table below, the middle column contains values calculated using $f(x)=\left(\sqrt{x^2+4}-\sqrt{x^2+2}\right)$, while the third column contains values calculated using the rewritten $f(x)$.

x	old $f(x)$	new $f(x)$
1	0.504017	0.504017
10	0.985341	0.985341
100	0.999850	0.999850
1000	0.999998	0.999999
10000	1.000000	1.000000
100000	1.000000	1.000000
1000000	1.000008	1.000000
10000000	0.987202	1.000000
100000000	0.000000	1.000000
1000000000	0.000000	1.000000

7. The limit is 1/6.

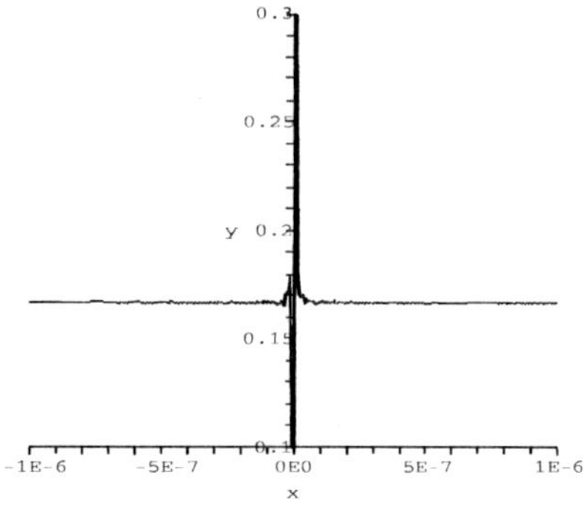

We can rewrite the function as

$$\frac{1-\cos 2x}{12x^2}\cdot\frac{1+\cos 2x}{1+\cos 2x} = \frac{\sin^2 2x}{12x^2(1+\cos 2x)}$$

to avoid loss-of-significance errors.

In the table below, the middle column contains values calculated using $f(x) = \dfrac{1-\cos 2x}{12x^2}$, while the third column contains values calculated using the rewritten $f(x)$. Note that $f(x) = f(-x)$ and so we get the same values when x is negative (which allows us to conjecture the two-sided limit as $x \to 0$).

x	old $f(x)$	new $f(x)$
1	0.118012	0.118012
0.1	0.166112	0.166112
0.01	0.166661	0.166661
0.001	0.166667	0.166667
0.0001	0.166667	0.166667
0.00001	0.166667	0.166667
0.000001	0.166663	0.166667
0.0000001	0.166533	0.166667
0.00000001	0.185037	0.166667
0.000000001	0	0.166667
1E-10	0	0.166667

9. The limit is $\frac{1}{2}$.

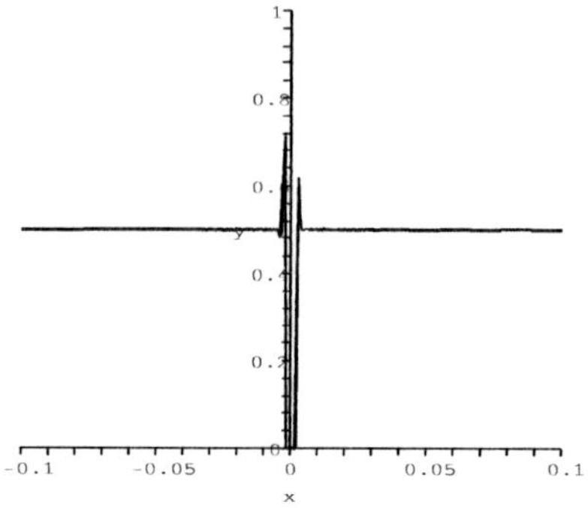

We can rewrite the function as

$$\frac{1-\cos x^3}{x^6}\frac{1+\cos x^3}{1+\cos x^3} = \frac{\sin^2(x^3)}{x^6(1+\cos x^3)}$$

to avoid loss-of-significance errors.

In the table below, the middle column contains values calculated using $f(x) = \frac{1-\cos x^3}{x^6}$, while the third column contains values calculated using the rewritten $f(x)$. Note that $f(x) = f(-x)$ and so we get the same values when x is negative (which allows us to conjecture the two-sided limit as $x \to 0$).

x	old $f(x)$	new $f(x)$
1	0.459698	0.459698
0.1	0.500000	0.500000
0.01	0.500044	0.500000
0.001	0.000000	0.500000
0.0001	0.000000	0.500000

11. The limit is 2/3.

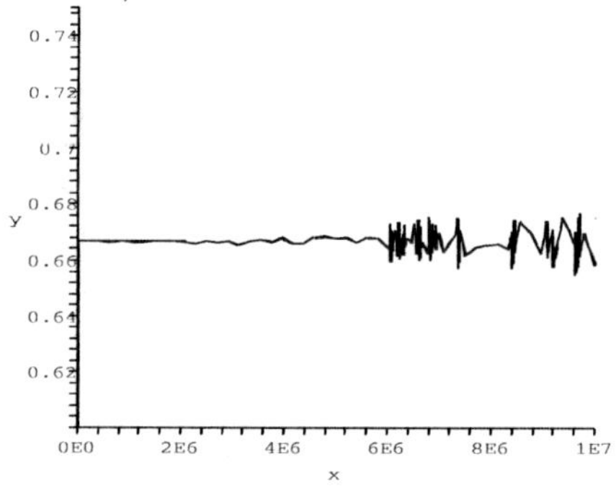

We can multiply $f(x)$ by

$$1 = \frac{g(x)}{g(x)}$$

where

$$g(x) = (x^2+1)^{\frac{2}{3}} + (x^2+1)^{\frac{1}{3}}(x^2-1)^{\frac{1}{3}} + (x^2-1)^{\frac{2}{3}}$$

to rewrite the function as

$$\frac{2x^{4/3}}{g(x)}$$

to avoid loss-of-significance errors.

In the table below, the middle column contains values calculated using $f(x) = x^{4/3}(\sqrt[3]{x^2+1} - \sqrt[3]{x^2-1})$, while the third column contains values calculated using the rewritten $f(x)$.

x	old $f(x)$	new $f(x)$
1	1.259921	1.259921
10	0.666679	0.666679
100	0.666667	0.666667
1000	0.666667	0.666667
10000	0.666668	0.666667
100000	0.666532	0.666667
1000000	0.63	0.666667
10000000	2.154435	0.666667
100000000	0.000000	0.666667
1000000000	0.000000	0.666667

13. $\lim_{x\to 1} \frac{x^2+x-2}{x-1}$
$= \lim_{x\to 1} \frac{(x+2)(x-1)}{x-1}$
$= \lim_{x\to 1}(x+2) = 3$

$\lim_{x\to 1} \frac{x^2+x-2.01}{x-1}$ does not exist, since when x is close to 1, the numerator is close to $-.01$ (a small but nonzero number) and the denominator is close to 0.

15. $f(1) = 0; g(1) = 0.00159265$
$f(10) = 0; g(10) = -0.0159259$
$f(100) = 0; g(100) = -0.158593$
$f(1000) = 0; g(1000) = -0.999761$

17. $(1.000003 - 1.000001) \times 10^7 = 20$
On a computer with a 6-digit mantissa, the calculation would be $(1.00000 - 1.00000) \times 10^7 = 0$.

Ch. 1 Review Exercises

1. The slope appears to be 2.

Second point	m_{sec}
$(3, 3)$	3
$(2.1, 0.21)$	2.1
$(2.01, 0.0201)$	2.01
$(1, -1)$	1
$(1.9, -0.19)$	1.9
$(1.99, -0.0199)$	1.99

3. (a) For the x-values of our points here we use (approximations of) 0, $\frac{\pi}{16}$, $\frac{\pi}{8}$, $\frac{3\pi}{16}$, and $\frac{\pi}{4}$.

Left	Right	Length
(0, 0)	(0.2, 0.2)	0.276
(0.2, 0.2)	(0.39, 0.38)	0.272
(0.39, 0.38)	(0.59, 0.56)	0.262
(0.59, 0.56)	(0.785, 0.71)	0.248
	Total	1.058

(b) For the x-values of our points here we use (approximations of) 0, $\frac{\pi}{32}$, $\frac{\pi}{16}$, $\frac{3\pi}{32}$, $\frac{\pi}{8}$, $\frac{5\pi}{32}$, $\frac{3\pi}{16}$, $\frac{7\pi}{32}$, and $\frac{\pi}{4}$.

Left	Right	Length
(0, 0)	(0.1, 0.1)	0.139
(0.1, 0.1)	(0.2, 0.2)	0.138
(0.2, 0.2)	(0.29, 0.29)	0.137
(0.29, 0.29)	(0.39, 0.38)	0.135
(0.39, 0.38)	(0.49, 0.47)	0.132
(0.49, 0.47)	(0.59, 0.56)	0.129
(0.59, 0.56)	(0.69, 0.63)	0.126
(0.69, 0.63)	(0.785, 0.71)	0.122
	Total	1.058

5. Let $f(x) = \dfrac{\tan^{-1} x^2}{x^2}$.

x	$f(x)$
0.1	0.999966669
0.01	0.999999997
0.001	1.000000000
0.0001	1.000000000
0.00001	1.000000000
0.000001	1.000000000

Note that $f(x) = f(-x)$, so the results for negative x will be the same as above. The limit appears to be 1.

7. Let $f(x) = \dfrac{x+2}{|x+2|}$.

x	$f(x)$
−1.9	1
−1.99	1
−1.999	1
−2.1	−1
−2.01	−1
−2.001	−1

$\lim_{x\to -2} \dfrac{x+2}{|x+2|}$ does not exist.

9. Let $f(x) = \left(1 + \dfrac{2}{x}\right)^x$.

x	$f(x)$
10	6.1917
100	7.2446
1000	7.3743
10,000	7.3876

$\lim_{x\to\infty} \left(1 + \dfrac{2}{x}\right)^x = e^2 \approx 7.4$

11. (a) $\lim_{x\to -1^-} f(x) = 1$.

(b) $\lim_{x\to -1^+} f(x) = -2$.

(c) $\lim_{x\to -1} f(x)$ does not exist.

(d) $\lim_{x\to 0} f(x) = 0$.

13. $x = -1, x = 1$

15. $\lim_{x\to 2} \dfrac{x^2 - x - 2}{x^2 - 4}$

$= \lim_{x\to 2} \dfrac{(x-2)(x+1)}{(x+2)(x-2)}$

$= \lim_{x\to 2} \dfrac{x+1}{x+2} = \dfrac{3}{4}$.

17. $\lim_{x\to 0^+} \dfrac{x^2 + x}{\sqrt{x^4 + 2x^2}}$

$= \lim_{x\to 0^+} \dfrac{x(x+1)}{x\sqrt{x^2+2}}$

$= \lim_{x\to 0^+} \dfrac{x+1}{\sqrt{x^2+2}}$

$= \dfrac{1}{\sqrt{2}}$

but

$\lim_{x\to 0^-} \dfrac{x^2 + x}{\sqrt{x^4 + 2x^2}}$

$= \lim_{x\to 0^-} \dfrac{x(x+1)}{(-x)\sqrt{x^2+2}}$

$= \lim_{x\to 0^-} -\dfrac{x+1}{\sqrt{x^2+2}}$

$= -\dfrac{1}{\sqrt{2}}$

Since the left and right limits are not equal, $\lim_{x\to 0} \dfrac{x^2 + x}{\sqrt{x^4 + 2x^2}}$ does not exist.

19. $\lim_{x\to 0} (2 + x)\sin(1/x)$

$= \lim_{x\to 0} 2\sin(1/x)$;

however, since $\lim_{x\to 0} \sin(1/x)$ does not exist, it follows that $\lim_{x\to 0}(2 + x)\sin(1/x)$ also does not exist.

21. $\lim_{x\to 2} f(x) = 5$.

23. Multiply the function by

$\dfrac{(1+2x)^{\frac{2}{3}} + (1+2x)^{\frac{1}{3}} + 1}{(1+2x)^{\frac{2}{3}} + (1+2x)^{\frac{1}{3}} + 1}$

to get

$\lim_{x\to 0} \dfrac{\sqrt[3]{1+2x} - 1}{x}$

$= \lim_{x\to 0} \dfrac{2}{(1+2x)^{\frac{2}{3}} + (1+2x)^{\frac{1}{3}} + 1} = \dfrac{2}{3}$

25. $\lim_{x\to 0}\cot(x^2) = \infty$

27. $\lim_{x\to\infty}\frac{x^2-4}{3x^2+x+1}$
$= \lim_{x\to\infty}\frac{x^2\left(1-\frac{4}{x^2}\right)}{x^2\left(3+\frac{1}{x}+\frac{1}{x^2}\right)}$
$= \lim_{x\to\infty}\frac{1-\frac{4}{x^2}}{3+\frac{1}{x}+\frac{1}{x^2}} = \frac{1}{3}$

29. Since $\lim_{x\to\pi/2}\tan^2 x = +\infty$, it follows that $\lim_{x\to\pi/2} e^{-\tan^2 x} = 0$.

31. $\lim_{x\to\infty}\ln 2x = \lim_{x\to\infty}(\ln 2 + \ln x)$
$= \ln 2 + \lim_{x\to\infty}\ln x = \infty$

33. $\lim_{x\to-\infty}\frac{2x}{x^2+3x-5}$
$= \lim_{x\to-\infty}\frac{2x}{x^2\left(1+\frac{3}{x}+\frac{5}{x^2}\right)}$
$= \lim_{x\to-\infty}\frac{2}{x\left(1+\frac{3}{x}+\frac{5}{x^2}\right)} = 0$

35. Let $u = -\frac{1}{3x}$, so that $\frac{2}{x} = -6u$. Then,

$\lim_{x\to 0^+}(1-3x)^{2/x}$
$= \lim_{u\to-\infty}\left(1+\frac{1}{u}\right)^{-6u}$
$= \left[\lim_{u\to-\infty}\left(1+\frac{1}{u}\right)^u\right]^{-6} = e^{-6}$

and

$\lim_{x\to 0^-}(1-3x)^{2/x}$
$= \lim_{u\to\infty}\left(1+\frac{1}{u}\right)^{-6u}$
$= \left[\lim_{u\to\infty}\left(1+\frac{1}{u}\right)^u\right]^{-6} = e^{-6}$

Thus, $\lim_{x\to 0}(1-3x)^{2/x} = e^{-6}$.

37. $0 \le \frac{x^2}{x^2+1} < 1$
$\Rightarrow -2|x| \le \frac{2x^3}{x^2+1} < 2|x|$

$\lim_{x\to 0} -2|x| = 0$; $\lim_{x\to 0} 2|x| = 0$

By the Squeeze Theorem,

$\lim_{x\to 0}\frac{2x^3}{x^2+1} = 0.$

39. $f(x) = \frac{x-1}{x^2+2x-3} = \frac{x-1}{(x+3)(x-1)}$

has a non-removable discontinuity at $x = -3$ and a removable discontinuity at $x = 1$.

41. $\lim_{x\to 0^-} f(x) = \lim_{x\to 0^-}\sin x = 0$
$\lim_{x\to 0^+} f(x) = \lim_{x\to 0^+} x^2 = 0$
$\lim_{x\to 2^-} f(x) = \lim_{x\to 2^-} x^2 = 4$
$\lim_{x\to 2^+} f(x) = \lim_{x\to 2^+}(4x-3) = 5$

f has a non-removable discontinuity at $x = 2$.

43. $f(x) = \frac{x+2}{x^2-x-6} = \frac{x+2}{(x-3)(x+2)}$

continuous on $(-\infty,-2)$, $(-2,3)$ and $(3,\infty)$.

45. $f(x) = \sin(1+e^x)$ is continuous on the interval $(-\infty,\infty)$.

47. $f(x) = \frac{x+1}{(x-2)(x-1)}$ has vertical asymptotes at $x = 1$ and $x = 2$.

$\lim_{x\to\pm\infty}\frac{x+1}{x^2-3x+2}$
$= \lim_{x\to\pm\infty}\frac{x\left(1+\frac{1}{x}\right)}{x^2\left(1-\frac{3}{x}+\frac{2}{x^2}\right)}$
$= \lim_{x\to\pm\infty}\frac{1+\frac{1}{x}}{x\left(1-\frac{3}{x}+\frac{2}{x^2}\right)} = 0$

So $f(x)$ has a horizontal asymptote at $y = 0$.

49. $f(x) = \frac{x^2}{x^2-1} = \frac{x^2}{(x+1)(x-1)}$

has vertical asymptotes at $x = -1$ and $x = 1$.

$$\lim_{x\to\pm\infty} \frac{x^2}{x^2-1}$$
$$= \lim_{x\to\pm\infty} \frac{x^2}{x^2\left(1-\frac{1}{x^2}\right)}$$
$$= \lim_{x\to\pm\infty} \frac{1}{1-\frac{1}{x^2}}$$
$$= \frac{1}{1} = 1$$

So $f(x)$ has a horizontal asymptote at $y = 1$.

51. $\lim_{x\to 0^+} 2e^{1/x} = \infty$, so $x = 0$ is a vertical asymptote.

$$\lim_{x\to\infty} 2e^{1/x} = 2, \quad \lim_{x\to-\infty} 2e^{1/x} = 2,$$

so $y = 2$ is a horizontal asymptote.

53. $f(x)$ has a vertical asymptote when $e^x = 2$, that is, $x = \ln 2$.

$$\lim_{x\to\infty} \frac{3}{e^x-2} = 0$$
$$\lim_{x\to-\infty} \frac{3}{e^x-2} = -\frac{3}{2}$$

so $y = 0$ and $y = -3/2$ are horizontal asymptotes.

55. The limit is $\frac{1}{4}$.

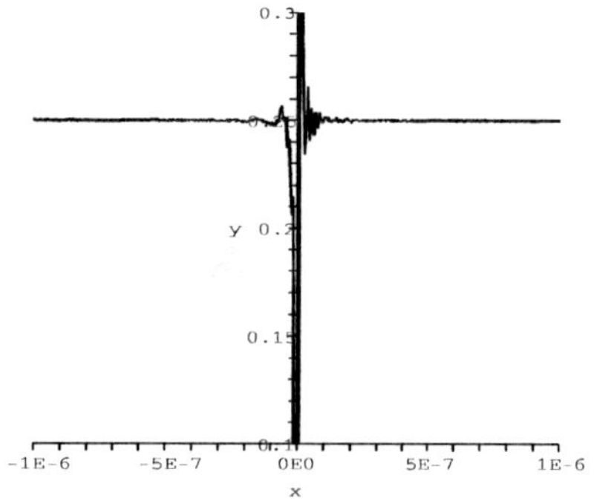

We can rewrite the function as

$$\frac{1-\cos x}{2x^2} = \left(\frac{1-\cos x}{2x^2}\right)\left(\frac{1+\cos x}{1+\cos x}\right)$$
$$= \frac{1-\cos^2 x}{2x^2(1+\cos x)} = \frac{\sin^2 x}{2x^2(1+\cos x)}$$

to avoid loss-of-significance errors.

In the table below, the middle column contains values calculated using $f(x) = \frac{1-\cos x}{2x^2}$, while the third column contains values calculated using the rewritten $f(x)$. Note that $f(x) = f(-x)$ and so we get the same values when x is negative (which allows us to conjecture the two-sided limit as $x \to 0$).

x	old $f(x)$	new $f(x)$
1	0.229849	0.229849
0.1	0.249792	0.249792
0.01	0.249998	0.249998
0.001	0.250000	0.250000
0.0001	0.250000	0.250000
0.00001	0.250000	0.250000
0.000001	0.250022	0.250000
0.0000001	0.249800	0.250000
0.00000001	0.000000	0.250000
0.000000001	0.000000	0.250000

57. The limit of θ' as x approaches 0 is 66 radians per second, far faster than the player can maintain focus. From about 9 feet on in to the plate the player can't keep her eye on the ball.

Chapter 2

Differentiation

2.1 Tangent Lines and Velocity

1.

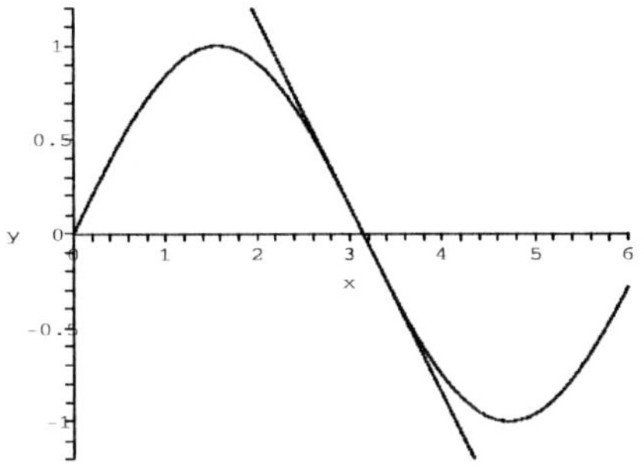

3. The tangent line is vertical and coincides with the y-axis:

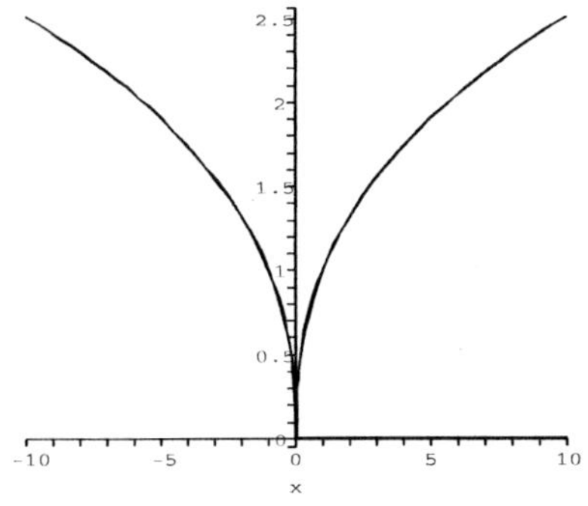

5. At $x = 1$ the slope of the tangent line appears to be about -1.

7. C, B, A, D. At the point labeled C, the slope is very steep and negative. At point B, the slope is zero and at point A, the slope is just more than zero. The slope of the line tangent to point D is large and positive.

9. (a) Points $(1, 0)$ and $(2, 6)$.
Slope is $\frac{6-0}{1} = 6$.

(b) Points $(2, 6)$ and $(3, 24)$.
Slope is $\frac{24-6}{1} = 18$.

(c) Points $(1.5, 1.875)$ and $(2, 6)$.
Slope is $\frac{6-1.875}{.5} = 8.25$.

(d) Points $(2, 6)$ and $(2.5, 13.125)$.
Slope is $\frac{13.125-6}{.5} = 14.25$.

(e) Points $(1.9, 4.959)$ and $(2, 6)$.
Slope is $\frac{6-4.959}{.1} = 10.41$.

(f) Points $(2, 6)$ and $(2.1, 7.161)$.
Slope is $\frac{7.161-6}{.1} = 11.61$.

(g) Slope seems to be approximately 11.

11. (a) Points $(1, .54)$ and $(2, -.65)$.
Slope is $\frac{-.65-.54}{1} = -1.19$.

(b) Points $(2, -.65)$ and $(3, -.91)$.
Slope is $\frac{-.91-(-.65)}{1} = -.26$.

(c) Points $(1.5, -.628)$ and $(2, -.654)$.
Slope is $\frac{-.654-(-.628)}{.5} = -.05$.

(d) Points $(2, -.65)$ and $(2.5, 1.00)$.
Slope is $\frac{1.00-(-.65)}{.5} = 3.3$.

(e) Points $(1.9, -.89)$ and $(2, -.65)$.
Slope is $\frac{-.65-(-.89)}{.1} = 2.4$.

(f) Points $(2, -.654)$ and $(2.1, -.298)$.
Slope is $\frac{-.298-(-.654)}{.1} = 3.56$.

(g) Slope seems to be approximately 3.

13.

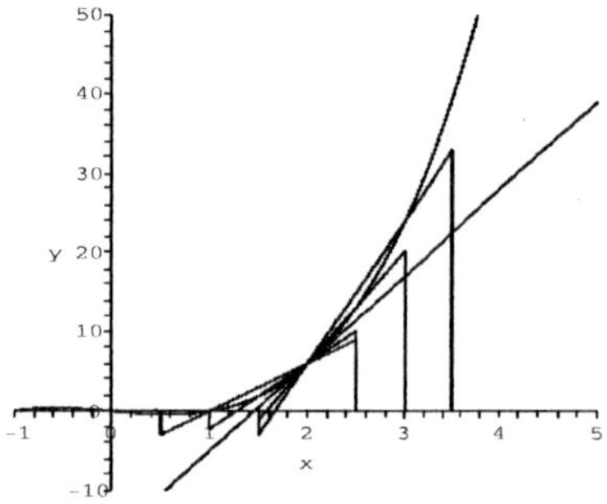

15. The sequence of graphs should look like:

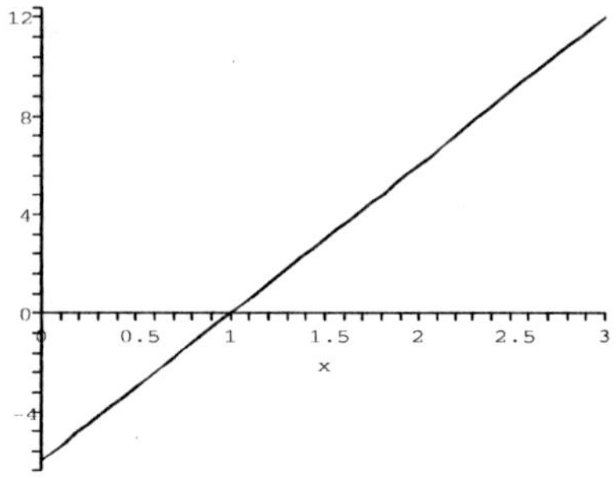

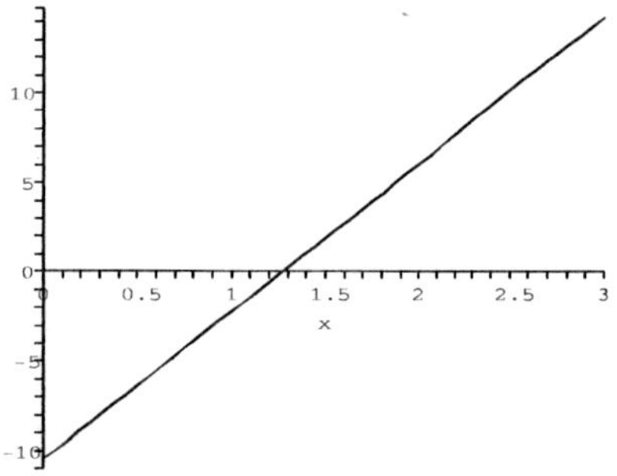

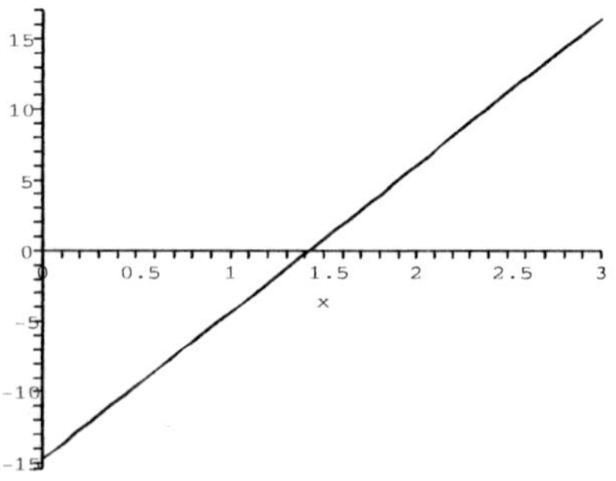

The third secant line is indistinguishable from the tangent line.

17. Slope is

$$\lim_{h\to 0} \frac{f(1+h)-f(1)}{h}$$
$$= \lim_{h\to 0} \frac{(1+h)^2-2-(-1)}{h}$$
$$= \lim_{h\to 0} \frac{h^2+2h}{h} = \lim_{h\to 0} h+2 = 2.$$

Tangent line is $y-(-1) = 2(x-1)$ or $y = 2x-3$.

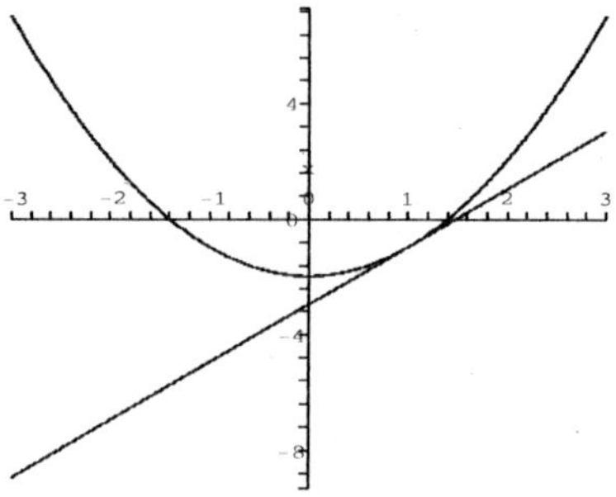

19. Slope is

$$\lim_{h\to 0} \frac{f(-2+h)-f(-2)}{h}$$
$$= \lim_{h\to 0} \frac{(-2+h)^2-3(-2+h)-(10)}{h}$$
$$= \lim_{h\to 0} \frac{4-4h+h^2+6-3h-10}{h}$$
$$= \lim_{h\to 0} \frac{-7h+h^2}{h} = \lim_{h\to 0} -7+h = -7.$$

Tangent line is $y - 10 = -7(x+2)$ or $y = -7x-4$.

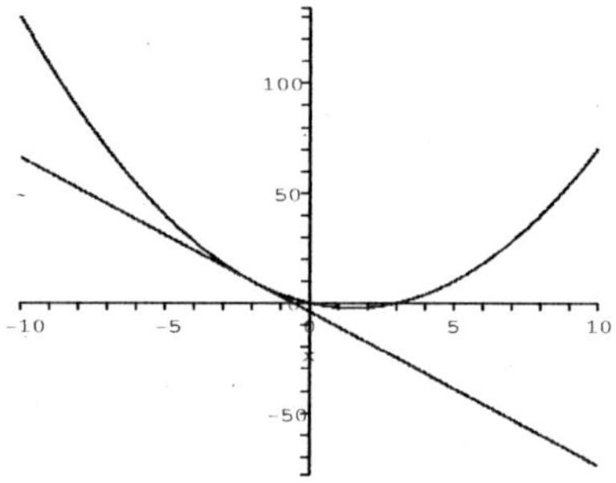

21. Slope is

$$\lim_{h\to 0} \frac{f(1+h)-f(1)}{h}$$
$$= \lim_{h\to 0} \frac{\frac{2}{(1+h)+1} - \frac{2}{1+1}}{h}$$
$$= \lim_{h\to 0} \frac{\frac{2}{2+h} - 1}{h}$$

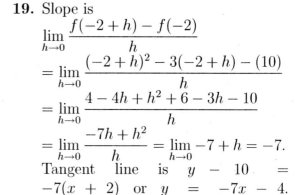

$$= \lim_{h\to 0} \frac{\left(\frac{2-(2+h)}{2+h}\right)}{h}$$
$$= \lim_{h\to 0} \frac{\left(\frac{-h}{2+h}\right)}{h}$$
$$= \lim_{h\to 0} \frac{-1}{2+h} = \frac{-1}{2}$$

Tangent line is $y - 1 = -\frac{1}{2}(x-1)$ or $y = -\frac{x}{2} + \frac{3}{2}$.

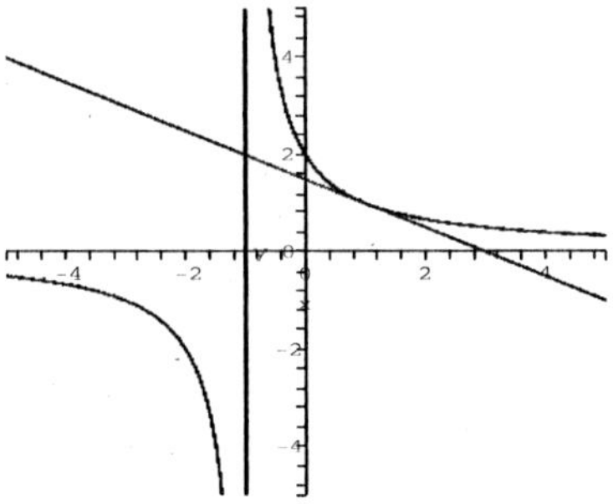

23. Slope is

$$\lim_{h\to 0} \frac{f(-2+h) - f(-2)}{h}$$
$$= \lim_{h\to 0} \frac{\sqrt{(-2+h)+3} - 1}{h}$$
$$= \lim_{h\to 0} \frac{\sqrt{h+1} - 1}{h}$$
$$= \lim_{h\to 0} \frac{\sqrt{h+1} - 1}{h} \cdot \frac{\sqrt{h+1}+1}{\sqrt{h+1}+1}$$
$$= \lim_{h\to 0} \frac{(h+1) - 1}{h(\sqrt{h+1}+1)}$$
$$= \frac{1}{\sqrt{h+1}+1} = \frac{1}{2}$$

Tangent line is $y - 1 = \frac{1}{2}(x+2)$ or $y = \frac{1}{2}x + 2$.

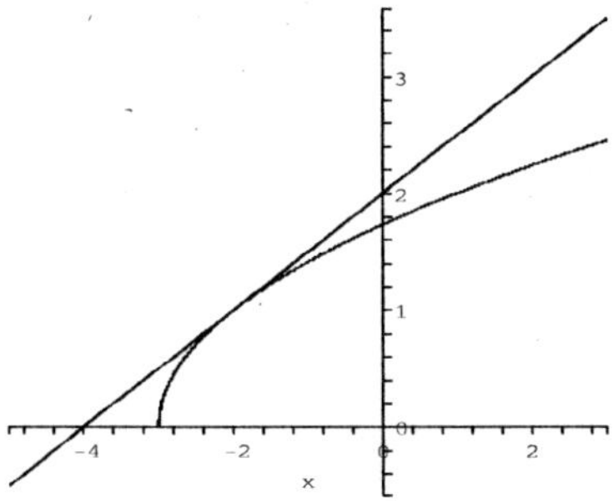

25. Numerical evidence suggests that

$$\lim_{h\to 0^+} \frac{f(1+h) - f(1)}{h} = 1$$

while

$$\lim_{h\to 0^-} \frac{f(1+h) - f(1)}{h} = -1.$$

Since these are not equal, there is no tangent line. A graph makes it apparent that this function has a "corner" at $x = 1$.

27. Numerical evidence suggests that

$$\lim_{h\to 0^+} \frac{f(0+h) - f(0)}{h}$$
$$= \lim_{h\to 0^-} \frac{f(0+h) - f(0)}{h}$$
$$= 0$$

Since the slope of the tangent line from the left equals that from the right and the function appears to be continuous in the graph, we conjecture that the tangent line exists and has slope 0.

29. Looking at the graph, we see that there is a jump discontinuity at $a = 0$. Thus there cannot be a tangent line, as the tangent line from the left would be different from the tangent line from the right.

31. (a) Points $(0, 10)$ and $(2, 74)$. Average velocity is $\frac{64-0}{2} = 32$.

(b) Second point $(1, 26)$. Average velocity is $\frac{64-26}{1} = 48$.

(c) Second point $(1.9, 67.76)$. Average velocity is $\frac{74-67.76}{.1} = 62.4$.

(d) Second point $(1.99, 73.3616)$. Average velocity is $\frac{74-73.3616}{.01} = 63.84$.

(e) The instantaneous velocity seems to be approaching 64.

33. (a) Points $(0, 0)$ and $(2, \sqrt{20})$. Average velocity is $\frac{\sqrt{20}-0}{2-0} = 2.236068$.

(b) Second point $(1, 3)$. Average velocity is $\frac{\sqrt{20}-3}{2-1} = 1.472136$.

(c) Second point $(1.9, \sqrt{18.81})$. Average velocity is $\frac{\sqrt{20}-\sqrt{18.81}}{2-1.9} = 1.3508627$.

(d) Second point $(1.99, \sqrt{19.8801})$. Average velocity is $\frac{\sqrt{20}-\sqrt{19.8801}}{2-1.99} = 1.3425375$.

(e) One might conjecture that these numbers are approaching 1.34. The exact limit is $\frac{6}{\sqrt{20}} \approx 1.341641$.

35. (a) Velocity at time $t = 1$ is

$$\lim_{h\to 0} \frac{f(1+h)-f(1)}{h}$$
$$= \lim_{h\to 0} \frac{-16(1+h)^2+5-(-11)}{h}$$
$$= \lim_{h\to 0} \frac{-16-32h-16h^2+5+11}{h}$$
$$= \lim_{h\to 0} \frac{-32h-16h^2}{h}$$
$$= \lim_{h\to 0} -32-16h = -32.$$

(b) Velocity at time $t = 2$ is

$$\lim_{h\to 0} \frac{f(2+h)-f(2)}{h}$$
$$= \lim_{h\to 0} \frac{-16(4+4h+h^2)+5+59}{h}$$
$$= \lim_{h\to 0} \frac{-64-64h-16h^2+64}{h}$$
$$= \lim_{h\to 0} -64-16h = -64.$$

37. The slope of the tangent line at $p = 1$ is approximately

$$\frac{-20-0}{2-0} = -10$$

which means that at $p = 1$, the freezing temperature of water decreases by 10 degrees Celsius per 1 atm increase in pressure. The slope of the tangent line at $p = 3$ is approximately

$$\frac{-11-(-20)}{4-2} = 4.5$$

which means that at $p = 3$, the freezing temperature of water increases by 4.5 degrees Celsius per 1 atm increase in pressure.

39. The hiker reached the top at the highest point on the graph (about 1.75 hours). The hiker was going the fastest on the way up at this point. The hiker was going the fastest on the way down at the point where the tangent line has the least (i.e. most negative) slope, at about 3 hours, at the end of the hike. Where the graph is level, the hiker was either resting, or walking on flat ground.

41. A possible graph of the temperature with respect to time:

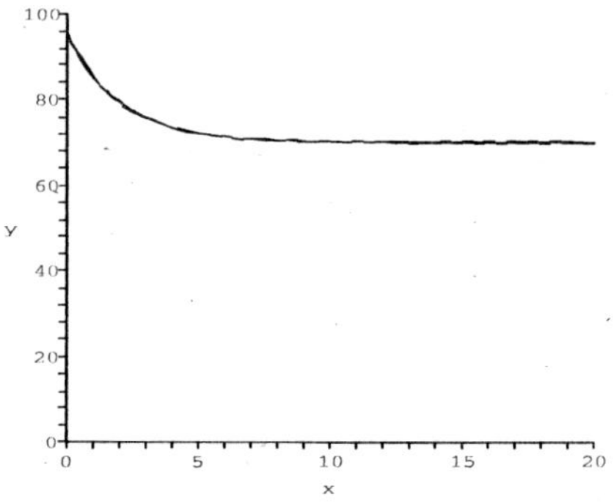

Graph of the rate of change of the temperature:

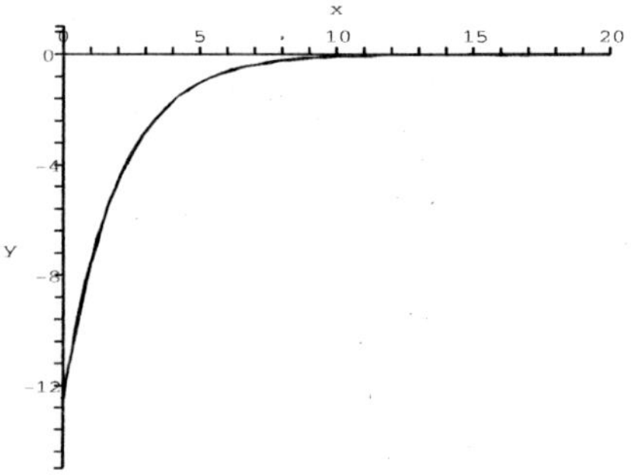

43. (a) To say that

$$\frac{f(4)-f(2)}{2} = 21{,}034$$

per year is to say that the average rate of change in the bank balance between Jan. 1, 2002 and Jan. 1, 2004 was 21,034 ($ per year).

(b) To say that

$$2[f(4) - f(3.5)] = 25{,}036$$

(note that $2[f(4) - f(3.5)] = \frac{f(4)-f(3.5)}{1/2}$) per year is to say that the average rate of change between July 1, 2003 and Jan. 1, 2004 was 25,036 ($ per year).

(c) To say that

$$\lim_{h\to 0} \frac{f(4+h) - f(4)}{h} = \$30{,}000$$

is to say that the instantaneous rate of change in the balance on Jan. 1, 2004 was 30,000 ($ per year).

45. We are given $\theta(t) = 0.4t^2$. We are advised that θ is measured in radians, and that t is time. Let us assume that t is measured in seconds.

Three rotations corresponds to $\theta = 6\pi$. Proceeding, if $\theta(t) = 6\pi$ then $0.4t^2 = 6\pi$ and solving for t yields $t = \sqrt{15\pi} \approx 6.865$ (seconds).

At that exact moment of time (call it a) , the exact angular velocity is

$$\lim_{h\to 0} \frac{\theta(a+h) - \theta(a)}{h}$$
$$= \lim_{h\to 0} \frac{.4(\sqrt{15\pi} + h)^2 - 6\pi}{h}$$
$$= \lim_{h\to 0} \frac{.4(15\pi + 2h\sqrt{15\pi} + h^2) - 6\pi)}{h}$$
$$= \lim_{h\to 0} \frac{.8h\sqrt{15\pi} + .4h^2}{h}$$
$$= \lim_{h\to 0} .8\sqrt{15\pi} + .4h = .8\sqrt{15\pi} \approx 5.492$$

and the units would be *radians per second.*

47. $$v_{avg} = \frac{f(s) - f(r)}{s - r}$$
$$= \frac{as^2 + bs + c - (ar^2 + br + c)}{s - r}$$
$$= \frac{a(s^2 - r^2) + b(s - r)}{s - r}$$
$$= \frac{a(s + r)(s - r) + b(s - r)}{s - r}$$
$$= a(s + r) + b$$

Let $v(r)$ be the velocity at $t = r$. We have

$$v(r) = \lim_{h\to 0} \frac{f(r+h) - f(r)}{h}$$
$$= \lim_{h\to 0} \frac{a(r^2 + 2rh + h^2) + bh - ar^2}{h}$$
$$= \lim_{h\to 0} \frac{h(2ar + ah + b)}{h}$$
$$= \lim_{h\to 0} 2ar + ah + b = \text{2ar + b.}$$

So $v(r) = 2ar + b$. The same argument shows that $v(s) = 2as + b$.

Finally,

$$\frac{v(r) + v(s)}{2} = \frac{(2ar + b) + (2as + b)}{2}$$
$$\frac{2a(s + r) + 2b}{2} = a(s + r) + b = v_{avg}$$

49. Let $x = h + a$. Then $h = x - a$, and clearly

$$\frac{f(a+h) - f(a)}{h} = \frac{f(x) - f(a)}{x - a}.$$

It is also clear that $x \to a$ if and only if $h \to 0$. Therefore if one of the two limits exists, then so does the other and

$$\lim_{h\to 0} \frac{f(a+h) - f(a)}{h} = \lim_{x\to a} \frac{f(x) - f(a)}{x - a}.$$

51. First, compute the slope of the tangent line. Using the result of #49, it is convenient to assume x is near but not exactly $1/2$, and write

$$\lim_{x\to 1/2}\frac{f(x)-f(1/2)}{x-(1/2)}=\frac{x^2-(1/4)}{x-(1/2)}$$
$$=\lim_{x\to 1/2}\frac{(x-(1/2))(x+(1/2))}{x-(1/2)}$$
$$=\lim_{x\to 1/2}x+(1/2)=1$$

Next, we quickly write the equation of the tangent line in point-slope form:

$y-(1/4)=1(x-(1/2))$ or $y=x-(1/4)$.

The location of the tree is the point $(x,y)=(1,3/4)$ and this point is indeed on the tangent line. The tree will be hit if the car gets that far (that being something we have no way of knowing).

2.2 The Derivative

1. Using (2.1):

$$f'(1)=\lim_{h\to 0}\frac{f(1+h)-f(1)}{h}$$
$$=\lim_{h\to 0}\frac{3(1+h)+1-(4)}{h}$$
$$=\lim_{h\to 0}\frac{3h}{h}=\lim_{h\to 0}3=3$$

Using (2.2):

$$\lim_{b\to 1}\frac{f(b)-f(1)}{b-1}$$
$$=\lim_{b\to 1}\frac{3b+1-(3+1)}{b-1}$$
$$=\lim_{b\to 1}\frac{3b-3}{b-1}$$
$$=\lim_{b\to 1}\frac{3(b-1)}{b-1}=\lim_{b\to 1}3=3$$

3. Using (2.1): Since

$$\frac{f(1+h)-f(1)}{h}=\frac{\sqrt{3(1+h)+1}-2}{h}$$
$$=\frac{\sqrt{4+3h}-2}{h}\cdot\frac{\sqrt{4+3h}+2}{\sqrt{4+3h}+2}$$
$$=\frac{4+3h-4}{h(\sqrt{4+3h}+2)}=\frac{3h}{h(\sqrt{4+3h}+2)}$$
$$=\frac{3}{\sqrt{4+3h}+2},\text{ we have:}$$
$$f'(1)=\lim_{h\to 0}\frac{f(1+h)-f(1)}{h}$$
$$=\lim_{h\to 0}\frac{3}{\sqrt{4+3h}+2}$$
$$=\frac{3}{\sqrt{4+3(0)}+2}=\frac{3}{4}.$$

Using (2.2): Since

$$\frac{f(b)-f(1)}{b-1}$$
$$=\frac{\sqrt{3b+1}-2}{b-1}$$
$$=\frac{(\sqrt{3b+1}-2)(\sqrt{3b+1}+2)}{(b-1)(\sqrt{3b+1}+2)}$$
$$=\frac{(3b+1)-4}{(b-1)\sqrt{3b+1}+2}$$
$$=\frac{3(b-1)}{(b-1)\sqrt{3b+1}+2}$$
$$=\frac{3}{\sqrt{3b+1}+2},\text{ we have:}$$
$$f'(1)=\lim_{b\to 1}\frac{f(b)-f(1)}{b-1}$$
$$=\lim_{b\to 1}\frac{3}{\sqrt{3b+1}+2}$$
$$=\frac{3}{\sqrt{4}+2}=\frac{3}{4}.$$

5. $$\lim_{h\to 0}\frac{f(x+h)-f(x)}{h}$$
$$=\lim_{h\to 0}\frac{3(x+h)^2+1-(3(x)^2+1)}{h}$$
$$=\lim_{h\to 0}\frac{3x^2+6xh+3h^2+1-(3x^2+1)}{h}$$
$$=\lim_{h\to 0}\frac{6xh+3h^2}{h}=\lim_{h\to 0}6x+3h=6x$$

7. $$\lim_{b\to x}\frac{f(b)-f(x)}{b-x}$$
$$=\lim_{b\to x}\frac{\frac{3}{b+1}-\frac{3}{x+1}}{b-x}$$
$$=\lim_{b\to x}\frac{\frac{3(x+1)-3(b+1)}{(b+1)(x+1)}}{b-x}$$

$$= \lim_{b \to x} \frac{-3(b-x)}{(b+1)(x+1)(b-x)}$$
$$= \lim_{b \to x} \frac{-3}{(b+1)(x+1)} = \frac{-3}{(x+1)^2}$$

9. $\lim_{b \to x} \frac{f(b) - f(x)}{b - x}$
$$= \lim_{b \to x} \frac{\sqrt{3b+1} - \sqrt{3x+1}}{b-x}$$
Multiplying by
$$\frac{\sqrt{3b+1} + \sqrt{3x+1}}{\sqrt{3b+1} + \sqrt{3x+1}}$$
gives
$$\lim_{b \to x} \frac{(3b+1) - (3x+1)}{(b-x)(\sqrt{3b+1} + \sqrt{3x+1})}$$
$$= \lim_{b \to x} \frac{3(b-x)}{(b-x)(\sqrt{3b+1} + \sqrt{3x+1})}$$
$$= \lim_{b \to x} \frac{3}{(\sqrt{3b+1} + \sqrt{3x+1})}$$
$$= \frac{3}{2\sqrt{3x+1}}$$

11. $\lim_{b \to x} \frac{f(b) - f(x)}{b - x}$
$$= \lim_{b \to x} \frac{b^3 + 2b - 1 - (x^3 + 2x - 1)}{b-x}$$
$$= \lim_{b \to x} \frac{b^3 - x^3 + 2b - 2x}{b-x}$$
$$= \lim_{b \to x} \frac{(b-x)(b^2 + bx + x^2 + 2)}{b-x}$$
$$= \lim_{b \to x} b^2 + bx + x^2 + 2$$
$$= 3x^2 + 2$$

13. The function has negative slope for $x < 0$, positive slope for $x > 0$, and zero slope at $x = 0$. Its slope function (derivative) can only be (c).

15. Here, moving from left to right, the slope goes from negative to positive to negative to positive. Its slope function (derivative) can only be (a) .

17. The graph is increasing to the left of the jump and decreasing to the right. The derivative of this function must be (b) which is postive to the left of the jump and negative to the right.

19. The derivative should look like:

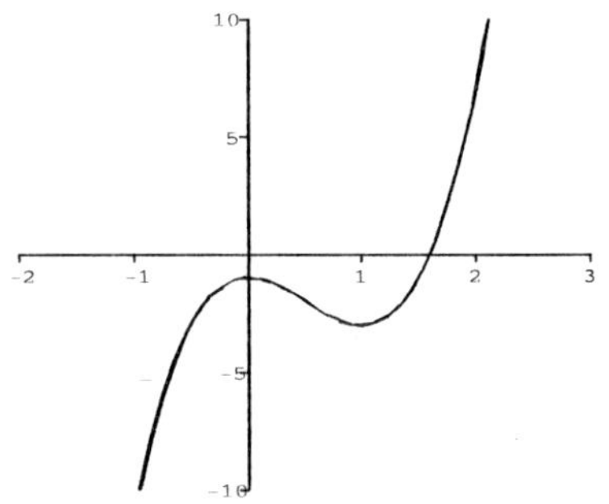

21. The derivative should look like:

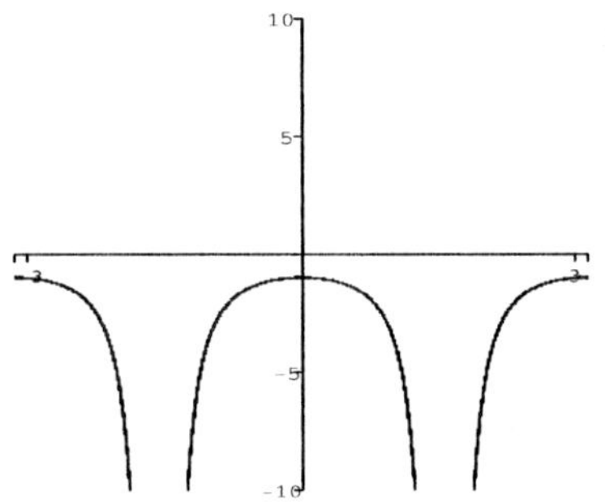

23. One possible graph of $f(x)$:

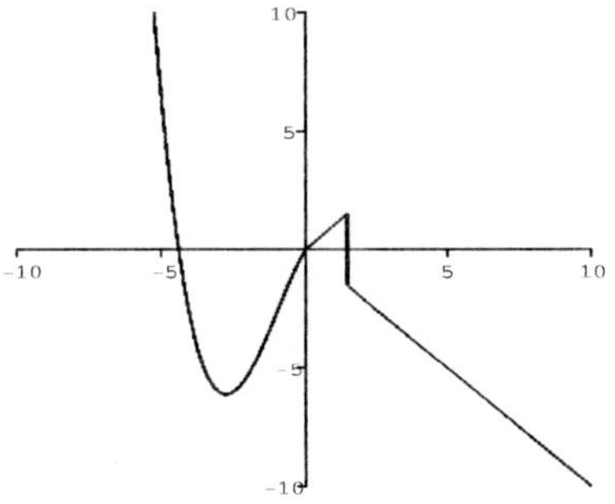

25. $f(x)$ is not differentiable at $x = 0$ or $x = 2$. The graph looks like:

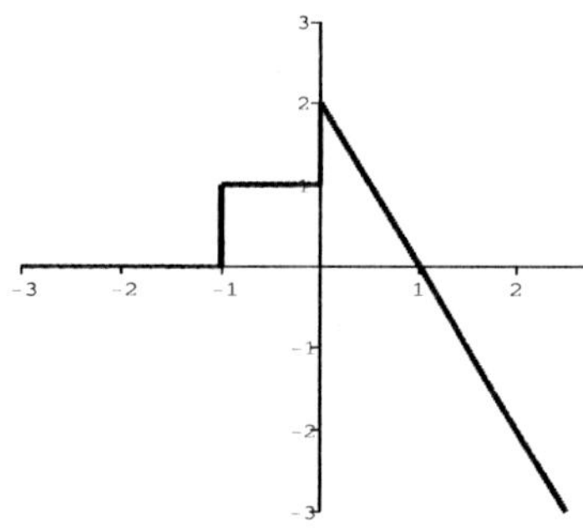

27. $f(x) = x^p \Longrightarrow f'(x) = px^{p-1}$.
If $p \geq 1$, then $p - 1 \geq 0$, so $f'(0) = 0$. Also, if $p = 0$, then $f(x) = 1$, so $f'(0) = 0$. However, if $p < 1$ but $p \neq 0$, then

$$f'(x) = \frac{p}{x^{1-p}}$$

where $1-p \geq 0$, and so $f'(0)$ does not exist.

29. $$\lim_{x\to a} \frac{[f(x)]^2 - [f(a)]^2}{x^2 - a^2}$$
$$= \lim_{x\to a} \frac{[f(x) - f(a)][f(x) + f(a)]}{(x-a)(x+a)}$$
$$= \left[\lim_{x\to a} \frac{f(x) - f(a)}{(x-a)}\right]\left[\lim_{x\to a} \frac{f(x) + f(a)}{(x+a)}\right]$$
$$= f'(a) \cdot \frac{2f(a)}{2a}$$
$$= \frac{f(a)f'(a)}{a}$$

31. We estimate the derivative at $x = 60$ as follows:

$$\frac{3.9 - 2.4}{80 - 40} = \frac{1.5}{40} = 0.0375$$

For every increase of 1 revolution per second of topspin, there is an increase of 0.0375° in margin of error.

33. Compute average velocities:

Time Interval	Average Velocity
(1.7, 2.0)	9.0
(1.8, 2.0)	9.5
(1.9, 2.0)	10.0
(2.0, 2.1)	10.0
(2.0, 2.2)	9.5
(2.0, 2.3)	9.0

Our best estimate of the velocity at $t = 2$ is 10.

35. We compile the rate of change in Ton-MPG over each of the four two-year intervals for which data is given:

intervals	rate of change
(1992,1994)	$\frac{45.7-44.9}{2} = .4$
(1994,1996)	.4
(1996,1998)	.4
(1998,2000)	.2

These rates of change are measured in Ton-MPG per year. Either the first or second (they happen to agree) could be used as an estimate for the one-year interval "1994" while only the last is a promising estimate for the one-year interval "2000". The mere fact that all these numbers are positive suggests that efficiency is improving, but the last number being smaller seems to suggest that the rate of improvement is slipping.

37. We prepare a table of values for the function $f(x) = x^x$ (when x is near 1). Difference quotients based at $x = 1$ are then compiled in the last column.

x	$y = x^x$	$\frac{y-1}{x-1}$
1.1000000	1.1105342	1.1053424
1.0100000	1.0101005	1.0100503
1.0010000	1.0010010	1.0010005
1.0001000	1.0001000	1.0001000
1.0000100	1.0000100	1.0000100
1.0000010	1.0000010	1.0000010
1.0000001	1.0000001	1.0000001

The evidence of this table strongly suggests that the difference quotients (essentially indistinguishable from the values themselves) are heading toward 1. If true, this would mean that $f'(1) = 1$.

39. The left-hand derivative is

$$D_-f(0) = \lim_{h\to 0^-} \frac{f(h)-f(0)}{h} = \lim_{h\to 0^-} \frac{2h+1-1}{h} = 2$$

The right-hand derivative is

$$D_+f(0) = \lim_{h\to 0^+} \frac{f(h)-f(0)}{h} = \lim_{h\to 0^+} \frac{3h+1-1}{h} = 3$$

41. $$D_+f(0) = \lim_{h\to 0^+} \frac{f(h)-f(0)}{h} = \lim_{h\to 0} \frac{k(h)-k(0)}{h} = k'(0).$$

$$D_-f(0) = \lim_{h\to 0^-} \frac{f(h)-f(0)}{h} = \lim_{h\to 0} \frac{g(h)-g(0)}{h} = g'(0)$$

If $f(x)$ has a jump discontinuity at $x = 0$, it would be because its *left limit* at $x = 0$, namely $g(0)$, is not the same as the *value* which is $k(0)$. In that case there could be no left derivative (by Theorem 2.1) and one would have to reject the statement $D_-f(0) = g'(0)$.

43. If $f'(x) > 0$ for all x, then the tangent lines all have positive slope, so the function is always sloping up.

45.

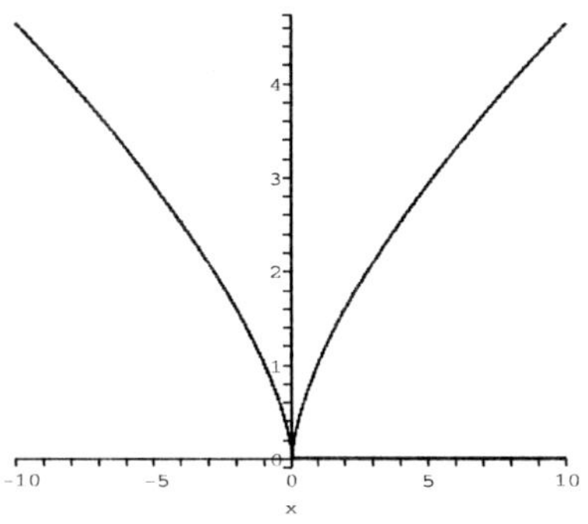

From the graph, we see that $f(x)$ appears continuous at $x = 0$, where it has both *limit* and *value* zero. However, when we try to compute its derivative at $x = 0$, we come to the difference quotient

$$\frac{f(0+h)-f(0)}{h} = \frac{f(h)}{h} = \frac{h^{2/3}}{h} = \frac{1}{h^{1/3}}$$

Clearly this expression has no finite limit as h approaches zero. The numbers get large without bound. We do sometimes say that the vertical line $x = 0$ is the tangent line, but as a line it has no *slope* (just as the function has no derivative).

47. Let $f(x) = -1-x^2$; then for all x, we have $f(x) \le x$. But at $x = -1$, we find $f(-1) = -2$ and

$$\begin{aligned} f'(-1) &= \lim_{h\to 0} \frac{f(-1+h)-f(-1)}{h} \\ &= \lim_{h\to 0} \frac{-1-(-1+h)^2-(-2)}{h} \\ &= \lim_{h\to 0} \frac{1-(1-2h+h^2)}{h} \\ &= \lim_{h\to 0} \frac{2h-h^2}{h} = \lim_{h\to 0} 2-h = 2 \end{aligned}$$

So, $f'(x)$ is not always less than 1.

49. (a) meters per second

(b) items per dollar

51. If $f'(t) < 0$, the function $f(t)$ is negatively sloped and decreasing, meaning the stock is losing value with the passing of time. This may be the basis for selling the stock if the current trend is expected to be a long term one.

53. The following sketches are consistent with the hypotheses of infection rate rising, peaking, and returning to zero. We started with the derivative $I'(t)$ (infection rate) and had to think backwards to construct the function $I(t)$. One can see in $I(t)$ the slope increasing up to the time of peak infection rate, thereafter the *slope* decreasing but not the *values.* They merely level off.

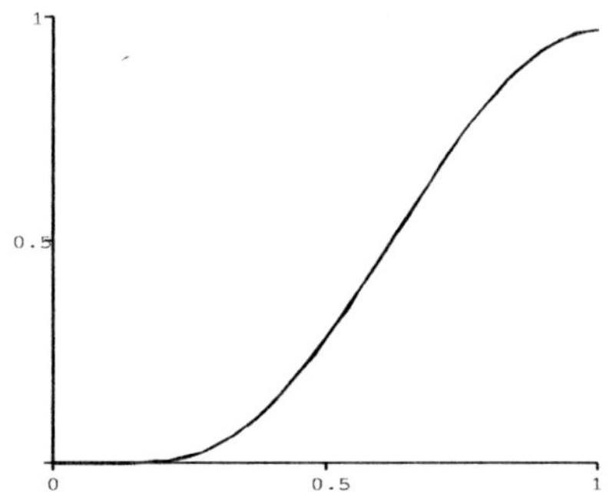

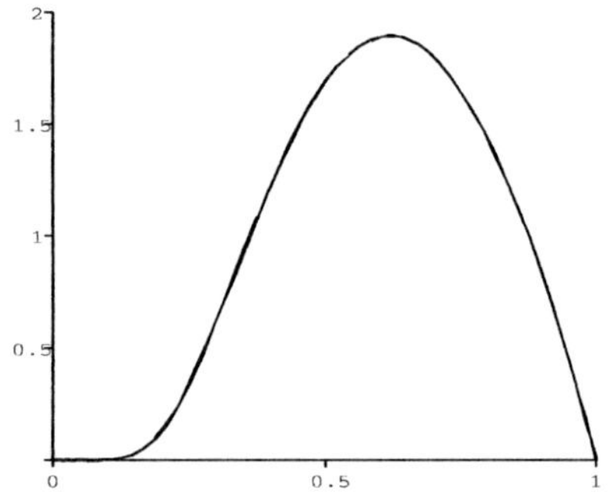

55. Because the curve appears to be bending upward, the slopes of the secant lines (based at $x = 1$ and with upper endpoint beyond 1) will increase with the upper endpoint. This has also the effect that any one of these slopes is greater than the actual derivative. Therefore

$$f'(1) < \frac{f(1.5) - f(1)}{.5} < \frac{f(2) - f(1)}{1}$$

As to where $f(1)$ fits in this list, it seems necessary to read the graph and come up with estimates of $f(1)$ about 4, and $f(2)$ about 7. That would put the third number in the above list at about 3, comfortably less than $f(1)$.

57. This is a tricky one. It happens that for the function $f(x) = x^2 - x$, the value at $x = 1$ is *zero* ($f(1) = 0$)! Because of this fact,

$$\frac{(1 + h)^2 - (1 + h)}{h} = \frac{f(1 + h) - f(1)}{h}$$

and the answer should be:
$f(x) = x^2 - x$ and $a = 1$.

59. $\lim_{h \to 0} \frac{\left(\frac{1}{2+h}\right) - \left(\frac{1}{2}\right)}{h}$ would be $f'(a)$ for $f(x) = \frac{1}{x}$ and $a = 2$.

61. One possible such graph:

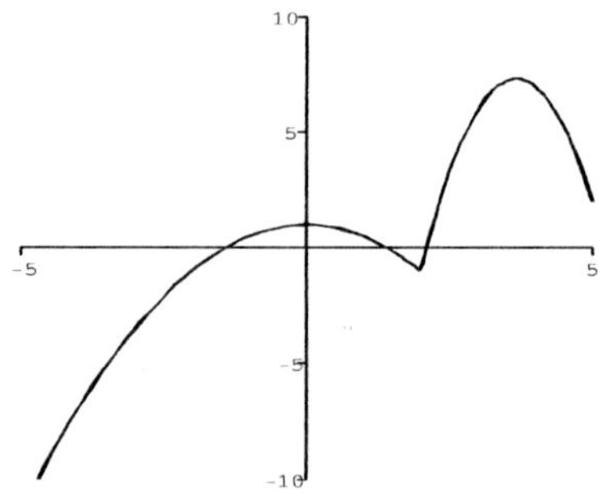

63. We have:

$$f(t) = \begin{cases} 100 & 0 < t \leq 20 \\ 100 + 10(t - 20) & 20 < t \leq 80 \\ 700 + 8(t - 80) & 80 < t < \infty \end{cases}$$

This is another example of a *piecewise linear* function (this one is continuous), and although not differentiable at the transition times $t = 20$

or $t = 80$, elsewhere we have

$$f'(t) = \begin{cases} 0 & 0 < t < 20 \\ 10 & 20 < t < 80 \\ 8 & 80 < t < \infty \end{cases}$$

2.3 Computation of Derivatives: Power Rule

1. $f'(x) = \frac{d}{dx}(x^3) - \frac{d}{dx}(2x) + \frac{d}{dx}(1)$
$= 3x^2 - 2\frac{d}{dx}(x) + 0$
$= 3x^2 - 2(1)$
$= 3x^2 - 2$

3. $f'(t) = \frac{d}{dt}(3t^3) - \frac{d}{dt}\left(2\sqrt{t}\right)$
$= 3\frac{d}{dt}(t^3) - 2\frac{d}{dt}\left(t^{1/2}\right)$
$= 3(3t^2) - 2\left(\frac{1}{2}t^{-1/2}\right)$
$= 9t^2 - \frac{1}{\sqrt{t}}$

5. $f'(x) = \frac{d}{dx}\left(\frac{3}{x}\right) - \frac{d}{dx}(8x) + \frac{d}{dx}(1)$
$= 3\frac{d}{dx}(x^{-1}) - 8\frac{d}{dx}(x) + 0$
$= 3(-x^{-2}) - 8(1)$
$= -\frac{3}{x^2} - 8$

7. $h'(x) = \frac{d}{dx}\left(\frac{10}{\sqrt{x}}\right) - \frac{d}{dx}(2x)$
$= 10\frac{d}{dx}\left(x^{-1/2}\right) - 2\frac{d}{dx}(x)$
$= 10\left(-\frac{1}{2}x^{-3/2}\right) - 2(1)$
$= -5x^{-3/2} - 2$
$= \frac{-5}{x\sqrt{x}} - 2$

9. $f'(s) = \frac{d}{ds}\left(2s^{3/2}\right) - \frac{d}{ds}\left(3s^{-1/3}\right)$
$= 2\frac{d}{ds}\left(s^{3/2}\right) - 3\frac{d}{ds}\left(s^{-1/3}\right)$
$= 2\left(\frac{3}{2}s^{1/2}\right) - 3\left(-\frac{1}{3}s^{-4/3}\right)$
$= 3s^{1/2} + s^{-4/3}$
$= 3\sqrt{s} + \frac{1}{\sqrt[3]{s^4}}$

11. $f'(x) = \frac{d}{dx}\left(2\sqrt[3]{x}\right) + \frac{d}{dx}(3)$
$= 2\frac{d}{dx}\left(x^{1/3}\right) + 0$
$= 2\left(\frac{1}{3}x^{-2/3}\right) = \frac{2}{3}x^{-2/3}$
$= \frac{2}{3\sqrt[3]{x^2}}$

13. $f(x) = x(3x^2 - \sqrt{x}) = 3x^3 - x^{3/2}$ so
$f'(x) = 3\frac{d}{dx}(x^3) - \frac{d}{dx}\left(x^{3/2}\right)$
$= 3(3x^2) - \left(\frac{3}{2}x^{1/2}\right)$
$= 9x^2 - \frac{3}{2}\sqrt{x}$

15. $f(x) = \frac{3x^2 - 3x + 1}{2x}$
$= \frac{3x^2}{2x} - \frac{3x}{2x} + \frac{1}{2x}$
$= \frac{3}{2}x - \frac{3}{2} + \frac{1}{2}x^{-1}$ so
$f'(x) =$
$\frac{d}{dx}\left(\frac{3}{2}x\right) - \frac{d}{dx}\left(\frac{3}{2}\right) + \frac{d}{dx}\left(\frac{1}{2}x^{-1}\right)$
$= \frac{3}{2}\frac{d}{dx}(x) - 0 + \frac{1}{2}\frac{d}{dx}(x^{-1})$
$= \frac{3}{2}(1) + \frac{1}{2}(-1x^{-2})$
$= \frac{3}{2} - \frac{1}{2x^2}$

17. $f'(x) = \frac{d}{dx}(x^4 + 3x^2 - 2) = 4x^3 + 6x$
$f''(x) = \frac{d}{dx}(4x^3 + 6x) = 12x^2 + 6$

19. $f(x) = 2x^4 - 3x^{-1/2}$ so
$\frac{df}{dx} = 8x^3 + \frac{3}{2}x^{-3/2}$
$\frac{d^2f}{dx^2} = 24x^2 - \frac{9}{4}x^{-5/2}$

21. $f'(x) = 4x^3 + 6x$
$f''(x) = 12x^2 + 6$
$f'''(x) = 24x$
$f^{(4)}(x) = 24$

23. $f(x) = \dfrac{x^2 - x + 1}{\sqrt{x}}$
$= x^{3/2} - x^{1/2} + x^{-1/2}$ so
$f'(x) = \dfrac{d}{dx}\left(x^{3/2} - x^{1/2} + x^{-1/2}\right)$
$= \dfrac{3}{2}x^{1/2} - \dfrac{1}{2}x^{-1/2} - \dfrac{1}{2}x^{-3/2}$
$f''(x) =$
$\dfrac{d}{dx}\left(\dfrac{3}{2}x^{1/2} - \dfrac{1}{2}x^{-1/2} - \dfrac{1}{2}x^{-3/2}\right)$
$= \dfrac{3}{4}x^{-1/2} + \dfrac{1}{4}x^{-3/2} + \dfrac{3}{4}x^{-5/2}$
$f'''(x) =$
$\dfrac{d}{dx}\left(\dfrac{3}{4}x^{-1/2} + \dfrac{1}{4}x^{-3/2} + \dfrac{3}{4}x^{-5/2}\right)$
$= -\dfrac{3}{8}x^{-3/2} - \dfrac{3}{8}x^{-5/2} - \dfrac{15}{8}x^{-7/2}$
$= -\dfrac{3(x^2 + x + 5)}{8x^3\sqrt{x}}$

25. $s(t) = -16t^2 + 40t + 10$
$v(t) = s'(t) = -32t + 40$
$a(t) = v'(t) = s''(t) = -32$

27. $s(t) = \sqrt{t} + 2t^2 = t^{1/2} + 2t^2$
$v(t) = s'(t) = \dfrac{1}{2}t^{-1/2} + 4t$
$a(t) = v'(t) = s''(t) = -\dfrac{1}{4}t^{-3/2} + 4$

29. $v(t) = -32t + 40$, $v(1) = 8$, going up. $a(t) = -32$, $a(1) = -32$, speed decreasing.

31. $v(t) = 20t - 24$, $v(2) = 16$, going up. $a(t) = 20$, $a(1) = 20$, speed increasing.

33. $f(x) = 4\sqrt{x} - 2x$, $a = 4$
$f(4) = 4\sqrt{4} - 2(4) = 0$
$f'(x) = \dfrac{d}{dx}\left(4x^{1/2} - 2x\right)$
$= 2x^{-1/2} - 2 = \dfrac{2}{\sqrt{x}} - 2$
$f'(4) = 1 - 2 = -1$
The equation of the tangent line is $y = -1(x - 4) + 0$ or $y = -x + 4$.

35. $f(x) = x^2 - 2$, $a = 2$, $f(2) = 2$
$f'(x) = 2x$
$f'(2) = 4$
The equation of the tangent line is $y = 4(x - 2) + 2$ or $y = 4x - 6$.

37. $f(x) = x^3 - 3x + 1$
$f'(x) = 3x^2 - 3$
The tangent line to $y = f(x)$ is horizontal when $f'(x) = 0$:
$3x^2 - 3 = 0$
$\iff 3(x^2 - 1) = 0$
$\iff 3(x + 1)(x - 1) = 0$
$\iff x = -1$ or $x = 1$.

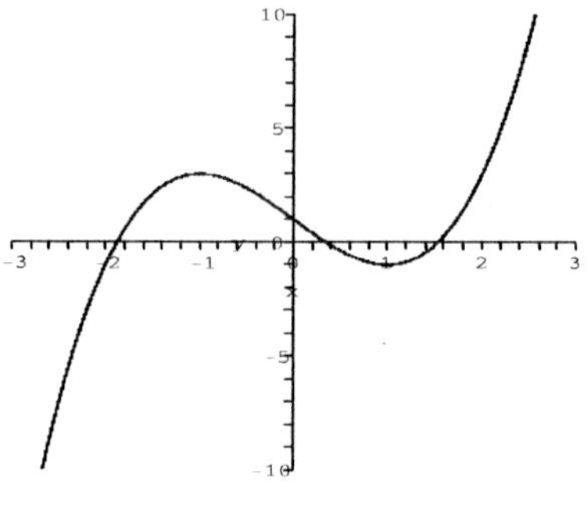

The graph shows that the first is a relative maximum, the second is a relative minimum.

39. $f(x) = x^{2/3}$
$f'(x) = \dfrac{2}{3}x^{-1/3} = \dfrac{2}{3\sqrt[3]{x}}$
The slope of the tangent line to $y = f(x)$ does not exist where the derivative is undefined, which is only when $x = 0$.

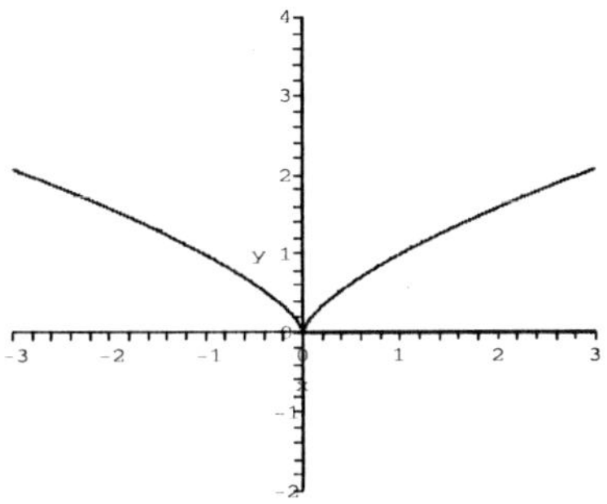

In this case, because the function is continuous, we might say that the tangent line is the vertical line $x = 0$. The feature at $x = 0$ is sometimes known as a *cusp*.

41. As regards the (a) function, its derivative would be negative for all negative x and positive for all positive x. Since no such function appears among the pictures, this (a) function has to be the one whose derivative is absent from the list. There being no f''' in the list, (a) has to be f''.

This same (a) function is negative for a certain interval of the form $(-a, a)$, and the (c) function is decreasing on a similar type of interval. Thus the (a) function (f'') is apparently the derivative of the (c) function. It follows that (c) must be f'.

This leaves (b) for f itself, and our identifications are consistent in every respect.

43. $f(x) = \sqrt{x} = x^{1/2}$

$$f'(x) = \frac{1}{2}x^{-1/2}$$

$$f''(x) = \frac{1}{2}\left(-\frac{1}{2}\right)x^{-3/2}$$

$$f'''(x) = \left(\frac{1}{2}\right)\left(\frac{-1}{2}\right)\left(\frac{-3}{2}\right)x^{-5/2}$$

$$f^{(n)}(x) = (-1)^{n-1}\frac{\Pi_n}{2^n}x^{-(2n-1)/2}$$

in which Π_n is the product of the first $n-1$ *odd* integers (starting from 1 and ending at $2n-3$). Recall that the product of *all* the whole numbers from 1 to n is denoted by $n!$. If one were to multiply Π_n by product of the $n-1$ *even* numbers (from 2 to $2n-2$), one would get $(2n-2)!$ (in the numerator). Of course, one would have to do the same to the denominator, but this product of the new numbers could be written in the form $2^{n-1}(n-1)!$ A final form for an answer could be

$f^{(n)}(x) = (-1)^{n-1}\frac{(2n-2)!}{2^{2n-1}(n-1)!}x^{(2n-1)/2}$.

45. $f(x) = ax^2 + bx + c \Rightarrow f(0) = c$
$f'(x) = 2ax + b \Rightarrow f'(0) = b$
$f''(x) = 2a \Rightarrow f''(0) = 2a$
Given $f''(0) = 3$, we learn $2a = 3$, or $a = 3/2$. Given $f'(0) = 2$ we learn $2 = b$, and given $f(0) = -2$, we learn $c = -2$. In the end
$f(x) = ax^2 + bx + c = \frac{3}{2}x^2 + 2x - 2$.

47. For $y = \frac{1}{x}$, we have $\frac{d}{dx} = -\frac{1}{x^2}$. Thus, the slope of the tangent line at $x = a$ is $-\frac{1}{a^2}$.

When $a = 1$, the slope of the tangent line at $(1, 1)$ is -1, and the equation of the tangent line is $y = -x + 2$. The tangent line intersects the axes at $(0, 2)$ and $(2, 0)$. Thus, the area of the triangle is $\frac{1}{2}(2)(2) = 2$.

When $a = 2$, the slope of the tangent line at $(2, \frac{1}{2})$ is $-\frac{1}{4}$, and the equation of the tangent line is $y = -\frac{1}{4}x + 1$. The tangent line intersects the axes at $(0, 1)$ and $(4, 0)$. Thus, the area of the triangle is $\frac{1}{2}(4)(1) = 2$.

In general, the equation of the tangent line is $y = -\left(\frac{1}{a^2}\right)x + \frac{2}{a}$. The tangent line intersects the axes at $(0, \frac{2}{a})$ and $(2a, 0)$. Thus, the area of the triangle is

$$\frac{1}{2}(2a)\left(\frac{2}{a}\right) = 2$$

49. (a) $$\begin{aligned} g'(x) &= \lim_{h\to 0} \frac{g(x+h) - g(x)}{h} \\ &= \lim_{h\to 0} \frac{1}{h}\left[\max_{a\le t\le x+h} f(t) - \max_{a\le t\le x} f(t)\right] \\ &= \lim_{h\to 0} \frac{1}{h}\left[f(x+h) - f(x)\right] \\ &= f'(x) \end{aligned}$$

(b) $$\begin{aligned} g'(x) &= \lim_{h\to 0} \frac{g(x+h) - g(x)}{h} \\ &= \lim_{h\to 0} \frac{1}{h}\left[\max_{a\le t\le x+h} f(t) - \max_{a\le t\le x} f(t)\right] \\ &= \lim_{h\to 0} \frac{1}{h}\left[f(a) - f(a)\right] \\ &= 0 \end{aligned}$$

51. If $d(t)$ represents the national debt, then $d'(t)$ represents the rate of change of the national debt. The debt itself, by implication, is increasing and therefore $d'(t) > 0$.

Since the rate of increase has been reduced, this implies $d''(t)$ is being reduced. We cannot conclude anything about the size of $d(t)$.

53. $w(b) = cb^{3/2}$

$w'(b) = \dfrac{3c}{2}b^{1/2} = \dfrac{3c\sqrt{b}}{2}$

$w'(b) > 1$ when

$\dfrac{3c\sqrt{b}}{2} > 1,\ \sqrt{b} > \dfrac{2}{3c}$

$b > \dfrac{4}{9c^2}.$

Since c is constant, when b is large enough, b will be greater than $\frac{4}{9c^2}$. After this point, when b increases by 1 unit, the leg width w is increasing by more than 1 unit, so that leg width is increasing faster than body length.

This puts a limitation on the size of land animals since, eventually, the body will not be long enough to accomodate the width of the legs.

55. We can approximate $f'(2000) \approx \frac{9039.5-8690.7}{2001-1999} = 174.4$. This is the rate of change of the GDP in billions of dollars per year.

To approximate $f''(2000)$, we first estimate $f'(1999) \approx \frac{9016.8-8347.3}{2000-1998} = 334.75$ and $f'(1998) \approx \frac{8690.7-8004.5}{1999-1997} = 343.1$.

Since these values are decreasing, $f''(2000)$ is negative. We estimate $f''(2000) \approx \frac{174.4-334.75}{2000-1999} = -160.35$. This represents the rate of change of the rate of change of the GDP over time. In 2000, the GDP is increasing by a rate of 343.1 billion dollars per year, but this increase is decreasing by a rate of 160.35 billion dollars-per-year per year.

57. Newton's Law states that force equals mass times acceleration. That is, if $F(t)$ is the driving force at time t, then $m \cdot f''(t) = m \cdot a(t) = F(t)$ in which m is the mass, appropriately unitized. The third derivative of the distance function is then $f'''(t) = a'(t) = \frac{1}{m}F'(t)$. It is both the derivative of the acceleration and directly proportional to the rate of change in force. Thus an abrupt change in acceleration or "jerk" is the direct consequence of an abrupt change in force.

59–61 Commentary: At this stage, finding a function whose derivative is given, is a matter of thinking backward, or of anticipation. When the derivative is a power, one anticipates that it could have arisen from differentiating a function which was also a power, but whose exponent was one higher. That is, to get to x^p, try cx^{p+1}

where c is some constant. After that, it is a matter of testing and adjusting the constant c. The answer is never unique (why?), but anything offered can always be checked by differentiation.

59. Try $f(x) = cx^4$ for some constant c. Then $f'(x) = 4cx^3$ so c must be 1. One possible answer is x^4.

61. $f'(x) = \sqrt{x} = x^{1/2}$

$f(x) = \dfrac{2}{3}x^{3/2}$ is one possible function.

63.
$$\lim_{h\to 0} \frac{f(a+h) - 2f(a) + f(a-h)}{h^2}$$
$$= \lim_{h\to 0}\left[\frac{f(a+h)-f(a)}{h^2} - \frac{[f(a)-f(a-h)]}{h^2}\right]$$
$$= \lim_{h\to 0}\frac{1}{h}\left[\frac{f(a+h)-f(a)}{h} - \frac{f(a)-f(a-h)}{h}\right]$$
$$= \lim_{h\to 0}\frac{1}{h}\left[\lim_{h\to 0}\frac{f(a+h)-f(a)}{h} - \lim_{h\to 0}\frac{f(a)-f(a-h)}{h}\right]$$
$$= \lim_{h\to 0}\frac{1}{h}\left[f'(a) - f'(a-h)\right]$$

Now let $k = -h$ in the previous equation, to get

$$\lim_{h\to 0} \frac{f(a+h) - 2f(a) + f(a-h)}{h^2}$$
$$= \lim_{k\to 0}\frac{1}{-k}\left[f'(a) - f'(a+k)\right]$$
$$= \lim_{k\to 0}\frac{1}{k}\left[f'(a+k) - f'(a)\right]$$
$$= f''(a)$$

2.4 The Product and Quotient Rules

1. $f(x) = (x^2+3)(x^3-3x+1)$
$$f'(x) = \tfrac{d}{dx}(x^2+3)\cdot(x^3-3x+1) + (x^2+3)\cdot\tfrac{d}{dx}(x^3-3x+1)$$
$$= (2x)(x^3-3x+1) + (x^2+3)(3x^2-3)$$

3. $f(x) = \left(\sqrt{x}+3x\right)\left(5x^2 - \tfrac{3}{x}\right)$
$$= (x^{1/2}+3x)(5x^2-3x^{-1})$$
$$f'(x) = \left(\tfrac{1}{2}x^{-1/2}+3\right)\left(5x^2-3x^{-1}\right) + (x^{1/2}+3x)(10x+3x^{-2})$$

5. $f(x) = \frac{3x-2}{5x+1}$
$$f'(x) = \frac{\left((5x+1)\frac{d}{dx}(3x-2)-(3x-2)\frac{d}{dx}(5x+1)\right)}{(5x+1)^2}$$
$$= \frac{3(5x+1)-(3x-2)5}{(5x+1)^2}$$
$$= \frac{15x+3-15x+10}{(5x+1)^2} = \frac{13}{(5x+1)^2}$$

7. $f(x) = \frac{3x-6\sqrt{x}}{5x^2-2} = \frac{3(x-2x^{1/2})}{5x^2-2}$
$$f'(x) = 3\frac{\left((5x^2-2)\frac{d}{dx}(x-2x^{1/2})-(x-2x^{1/2})\frac{d}{dx}(5x^2-2)\right)}{(5x^2-2)^2}$$
$$= 3\frac{\left((5x^2-2)(1-x^{-1/2})-(x-2x^{1/2})(10x)\right)}{(5x^2-2)^2}$$
$$= 3\frac{\left((5x^2-2-5x^{3/2}+2x^{-1/2})-(10x^2-20x^{3/2})\right)}{(5x^2-2)^2}$$
$$= \frac{3(-5x^2+15x^{3/2}+2x^{-1/2}-2)}{(5x^2-2)^2}$$

9. $f(x) = \frac{(x+1)(x-2)}{x^2-5x+1} = \frac{x^2-x-2}{x^2-5x+1}$
$$f'(x) = \frac{\left((x^2-5x+1)\frac{d}{dx}(x^2-x-2)-(x^2-x-2)\frac{d}{dx}(x^2-5x+1)\right)}{(x^2-5x+1)^2}$$
$$= \frac{\left((x^2-5x+1)(2x-1)-(x^2-x-2)(2x-5)\right)}{(x^2-5x+1)^2}$$
$$= \frac{-4x^2+6x-11}{(x^2-5x+1)^2}$$

11. We do not recommend treating this one as a quotient, but advise preliminary simplification.
$$f(x) = \frac{x^2+3x-2}{\sqrt{x}}$$
$$= \frac{x^2}{\sqrt{x}} + \frac{3x}{\sqrt{x}} - \frac{2}{\sqrt{x}}$$
$$= x^{3/2} + 3x^{1/2} - 2x^{-1/2}$$
$$f'(x) = \tfrac{3}{2}x^{1/2} + \tfrac{3}{2}x^{-1/2} + x^{-3/2}$$

13. We simplify instead of using the product rule.
$f(x) = x\left(\sqrt[3]{x} + 3\right) = x^{4/3} + 3x$
$f'(x) = \frac{4}{3}x^{1/3} + 3$

15. $f(x) = (x^2 - 1)\frac{x^3+3x^2}{x^2+2}$
$f'(x) =$
$\frac{d}{dx}(x^2 - 1) \cdot \left(\frac{x^3+3x^2}{x^2+2}\right)$
$+ (x^2 - 1) \cdot \frac{d}{dx}\left(\frac{x^3+3x^2}{x^2+2}\right)$

We have

$\frac{d}{dx}\left(\frac{x^3+3x^2}{x^2+2}\right) =$
$\frac{(x^2+2)\frac{d}{dx}(x^3+3x^2)-(x^3+3x^2)\frac{d}{dx}(x^2+2)}{(x^2+2)^2}$
$= \frac{(x^2+2)\cdot(3x^2+6x)-(x^3+3x^2)\cdot(2x)}{(x^2+2)^2}$
$= \frac{3x^4+6x^2+6x^3+12x-(2x^4+6x^3)}{(x^2+2)^2}$
$= \frac{x^4+6x^2+12x}{(x^2+2)^2}$

so $f'(x) =$
$(2x) \cdot \left(\frac{x^3+3x^2}{x^2+2}\right) + (x^2 - 1) \cdot \frac{x^4+6x^2+12x}{(x^2+2)^2}$

17. $\frac{d}{dx}\left[f(x)g(x)h(x)\right]$
$= \frac{d}{dx}\left[\left(f(x)g(x)\right)h(x)\right]$
$= \left(f(x)g(x)\right)h'(x)+h(x)\frac{d}{dx}\left(f(x)g(x)\right)$
$= \left(f(x)g(x)\right)h'(x)$
$+ h(x)\left(f(x)g'(x) + g(x)f'(x)\right)$
$= f'(x)g(x)h(x)$
$+ f(x)g'(x)h(x) + f(x)g(x)h'(x)$

In the general case of a product of n functions, the derivative will have n terms to be added, each term a product of all but one of the functions multiplied by the derivative of the missing function.

19. $f'(x) = \left[\frac{d}{dx}(x^{2/3})\right](x^2-2)(x^3-x+1)$
$+x^{2/3}\left[\frac{d}{dx}(x^2-2)\right](x^3-x+1)$
$+ x^{2/3}(x^2-2)\frac{d}{dx}(x^3-x+1)$
$= \frac{2}{3}x^{-1/3}(x^2-2)(x^3-x+1)$
$+ x^{2/3}(2x)(x^3-x+1)$
$+ x^{2/3}(x^2-2)(3x^2-1)$

21. $h(x) = f(x)g(x)$
$h'(x) = f'(x)g(x) + f(x)g'(x)$

(a) $h(1) = f(1)g(1)$
$= (-2)(1) = -2$
$h'(1) = f'(1)g(1) + f(1)g'(1)$
$= (3)(1) + (-2)(-2) = 7$
So the equation of the tangent line is
$y = 7(x - 1) - 2.$

(b) $h(0) = f(0)g(0)$
$= (-1)(3) = -3$
$h'(0) = f'(0)g(0) + f(0)g'(0)$
$= (-1)(3) + (-1)(-1)$
$= -2$
So the equation of the tangent line is
$y = -2x - 3.$

23. $h(x) = x^2f(x)$
$h'(x) = 2xf(x) + x^2f'(x)$

(a) $h(1) = 1^2f(1) = -2$
$h'(1) = 2(1)f(1) + 1^2f'(1)$
$= (2)(-2) + 3 = -1$
So the equation of the tangent line is
$y = -(x - 1) - 2.$

(b) $h(0) = 0^2f(0) = 0$
$h'(0) = 2(0)f(0) + 0^2f'(0) = 0$
So the equation of the tangent line is
$y = 0.$

25. The rate at which the quantity Q changes is Q'. Since the amount is said to be "decreasing at a rate of 4%" we have to ask "4% of *what*?" The answer in this type of context is usually 4% of *itself*. In other words, $Q' = -.04Q$. As for P, the 3% rate of increase would translate as $P' = .03P$. By the product rule, with $R = PQ$, we have:
$R' = (PQ)' = P'Q + PQ'$
$= (.03P)Q + P(-.04Q)$
$= -(.01)PQ = (-.01)R.$

In other words, revenue is decreasing at a rate of 1%.

27. $R' = Q'P + QP'$
At a certain moment of time (call it t_0) we are given $P(t_0) = 20$ (\$/item)
$Q(t_0) = 20{,}000$ (items)
$P'(t_0) = 1.25$ (\$/item/year)
$Q'(t_0) = 2{,}000$ (items/year)
$\Rightarrow R'(t_0) = 2{,}000(20) + (20{,}000)1.25$
$= 65{,}000$ \$/year
So revenue is increasing by \$65,000/year at the time t_0.

29. If $u(m) = \dfrac{82.5m - 6.75}{m + .15}$ then using the quotient rule,

$$\frac{du}{dm} = \frac{(m+.15)(82.5) - (82.5m - 6.75)1}{(m+.15)^2} = \frac{19.125}{(m+.15)^2}$$

which is clearly positive. It seems to be saying that initial ball speed is an increasing function of the mass of the bat. Meanwhile,

$$u'(1) = \frac{19.125}{1.15^2} \approx 14.46$$
$$u'(1.2) = \frac{19.125}{1.35^2} \approx 10.49,$$

which suggests that the rate at which this speed is increasing is decreasing.

31. If $u(m) = \dfrac{14.11}{m + .05} = \dfrac{282.2}{20m+1}$, then

$$\frac{du}{dm} = \frac{(20m+1)\cdot 0 - 282.2(20)}{(20m+1)^2} = \frac{-5644}{(20m+1)^2}$$

This is clearly negative, which means that impact speed of the ball is a decreasing function of the weight of the club. It appears that the explanation may have to do with the stated fact that the speed of the club is inversely proportional to its mass. Although the lesson of Example 4.6 was that a heavier club makes for greater ball velocity, that was assuming a fixed club speed, quite a different assumption from this problem.

33. $f'(x) = \lim_{h\to 0} \dfrac{f(x+h) - f(x)}{h}$

$$f'(0) = \lim_{h\to 0} \frac{f(h) - f(0)}{h} = \lim_{h\to 0} \frac{hg(h) - 0}{h} = \lim_{h\to 0} \frac{hg(h)}{h} = \lim_{h\to 0} g(h) = g(0)$$

since g is continuous at $x = 0$.
When $g(x) = |x|$, $g(x)$ is continuous but not differentiable at $x = 0$. We have

$f(x) = x|x| = \begin{cases} -x^2 & x < 0 \\ x^2 & x \geq 0. \end{cases}$ This is differentiable at $x = 0$.

35. Answers depend on CAS.

37. For any constant k, the derivative of $\sin kx$ is $k \cos kx$.

Graph of $\frac{d}{dx} \sin x$:

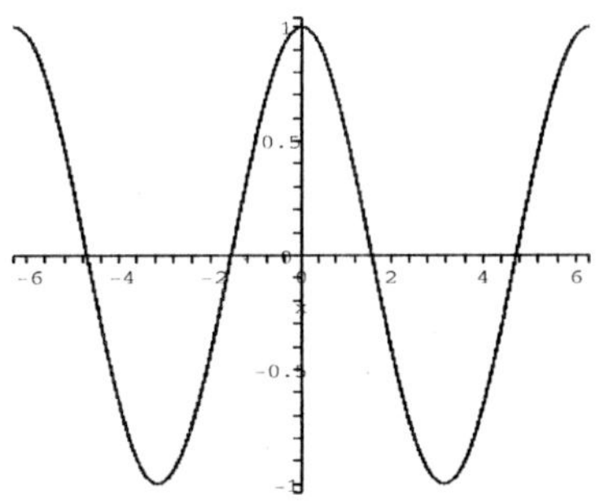

Graph of $\frac{d}{dx} \sin 2x$:

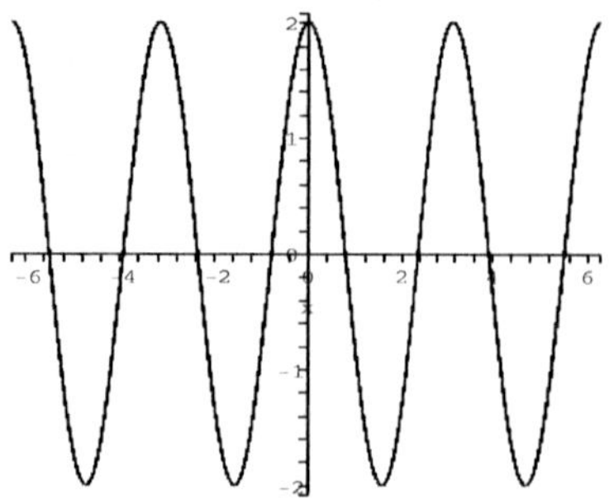

Graph of $\frac{d}{dx}\sin 3x$:

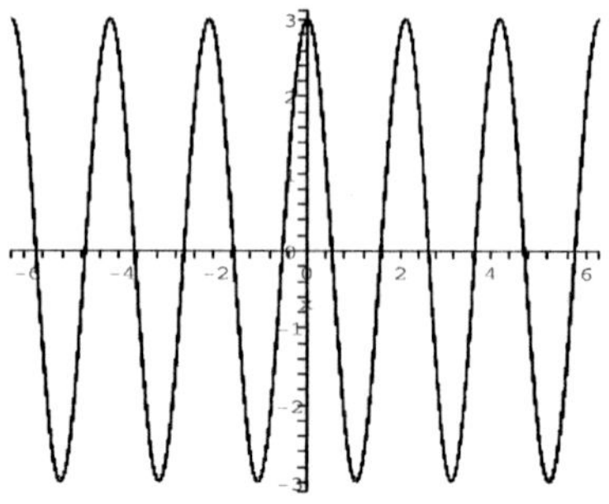

39. Using the quotient rule, we got a derivative in the form $\dfrac{3x}{2\sqrt{3x^3+x^2}}$ which could be written $\dfrac{3x}{2\sqrt{x^2(3x+1)}}$. One *could* then factor $\sqrt{x^2}$ out of the denominator as $|x|$ and use $\dfrac{x}{|x|} = \begin{cases} 1 & x>0 \\ -1 & x<0 \end{cases}$ to rewrite the function as in the problem. CAS answers may vary.

41. If $F(x) = f(x)g(x)$ then

$$F'(x) = f'(x)g(x) + f(x)g'(x) \text{ and}$$

$$\begin{aligned} F''(x) &= f''(x)g(x) + f'(x)g'(x) \\ &\quad + f'(x)g'(x) + f(x)g''(x) \\ &= f''(x)g(x) + 2f'(x)g'(x) \\ &\quad + f(x)g''(x) \end{aligned}$$

$$\begin{aligned} F'''(x) &= f'''(x)g(x) + f''(x)g'(x) \\ &\quad + 2f''(x)g'(x) + 2f'(x)g''(x) \\ &\quad + f'(x)g''(x) + f(x)g'''(x) \\ &= f'''(x)g(x) + 3f''(x)g'(x) \\ &\quad + 3f'(x)g''(x) + f(x)g'''(x) \end{aligned}$$

One can see obvious parallels to the binomial coefficients as they come from Pascal's Triangle:

$(a+b)^2 = a^2 + 2ab + b^2$

$(a+b)^3 = a^3 + 3a^2b + 3ab^2 + b^3$.

On this basis, one could correctly predict the pattern of the fourth or any higher derivative.

43. If $g(x) = [f(x)]^2 = f(x)f(x)$, then

$$\begin{aligned} g'(x) &= f'(x)f(x) + f(x)f'(x) \\ &= 2f(x)f'(x). \end{aligned}$$

45. $\left(P + \dfrac{n^2a}{V^2}\right)(V - nb) = nRT$

$$P + \frac{n^2a}{V^2} = \frac{nRT}{V-nb}$$

$$P = \frac{nRT}{V-nb} - \frac{n^2a}{V^2}$$

From this, we find with some difficulty

$$P'(V) = \frac{-nRT}{(V-nb)^2} + \frac{2n^2a}{V^3}$$

$$P''(V) = \frac{2nRT}{(V-nb)^3} + \frac{6n^2a}{V^4}.$$

Obviously, if $P'(V) = 0$, then

$$\frac{2na}{V^3} = \frac{RT}{(V-nb)^2} (= X)$$

in which X is a temporary name. If $P''(V)$ is *also* zero, then

$$0 = P''(V) = \frac{2nX}{(V-nb)} - \frac{3nX}{V}$$

$$= nX\left[\frac{2}{V-nb} - \frac{3}{V}\right] = \frac{nX(3nb-V)}{V(V-nb)},$$

$\Rightarrow V = 3nb$, so $V - nb = 2nb$, and

$$X = \frac{2na}{V^3} = \frac{2a}{27n^2b^3}.$$

$$RT = (V-nb)^2X = 4n^2b^2X = \frac{8a}{27b},$$

so $T = \dfrac{8a}{27bR}$, and since

$P = \dfrac{nRT}{V-nb} - \dfrac{n^2a}{V^2}$, we have

$$P = \frac{8an}{27b(2nb)} - \frac{n^2a}{9n^2b^2} = \frac{a}{27b^2}.$$

In summary,

$$(T_c, P_c, V_c) = \left(\frac{8a}{27bR}, \frac{a}{27b^2}, 3nb\right)$$

Substitute in the given numbers; in particular $T_c = 647°$ (Kelvin).

47. $f(x) = \dfrac{x^{2.7}}{1+x^{2.7}}$

$$f'(x) = \frac{(1+x^{2.7})\cdot 2.7x^{1.7} - 2.7x^{1.7}\cdot(x^{2.7})}{(1+x^{2.7})^2} = \frac{2.7x^{1.7}}{(1+x^{2.7})^2}$$

The fact that $0 < f(x) < 1$ when $x > 0$ suggest to us that f may be some kind of concentration ratio or percentage-of-presence of the allosteric enzyme in some system. If so, the derivative would be interpreted as the rate of change in the concentration per unit of activator.

49. $\frac{d}{dx}[x^3 f(x)] = 3x^2 \cdot f(x) + x^3 f'(x)$

51. Utilizing $\frac{d}{dx}(\sqrt{x}) = \frac{1}{2\sqrt{x}}$ (which is a special case of the power rule), we find

$$\frac{d}{dx}\left(\frac{\sqrt{x}}{f(x)}\right) = \frac{f(x)\frac{1}{2\sqrt{x}} - \sqrt{x}f'(x)}{[f(x)]^2} = \frac{f(x) - 2xf'(x)}{2\sqrt{x}[f(x)]^2}.$$

2.5 The Chain Rule

1. $f(x) = (x^3-1)^2$

Using the chain rule:

$f'(x) = 2(x^3-1)(3x^2) = 6x^2(x^3-1)$

Using the product rule:

$f(x) = (x^3-1)(x^3-1)$

$$\begin{aligned} f'(x) &= (3x^2)(x^3-1) + (x^3-1)(3x^2) \\ &= 2(3x^2)(x^3-1) \\ &= 6x^2(x^3-1) \end{aligned}$$

Using preliminary multiplication:

$f(x) = x^6 + 2x^3 + 1$

$$\begin{aligned} f'(x) &= 6x^5 + 6x^2 \\ &= 6x^2(x^3-1) \end{aligned}$$

3. $f(x) = (x^2+1)^3$

Chain rule:

$f'(x) = 3(x^2+1)^2 \cdot 2x$

Using preliminary multiplication:

$f(x) = x^6 + 3x^4 + 3x^2 + 1$

$f'(x) = 6x^5 + 12x^3 + 6x$

5. $f(x) = \sqrt{x^2+4}$

$$f'(x) = \frac{1}{2\sqrt{x^2+4}} \cdot 2x = \frac{x}{\sqrt{x^2+4}}$$

7. $f(x) = x^5\sqrt{x^3+2}$

$$f'(x) = x^5 \frac{1}{2\sqrt{x^3+2}} 3x^2 + 5x^4\sqrt{x^3+2} = \frac{3x^7 + 10x^4(x^3+2)}{2\sqrt{x^3+2}} = \frac{13x^7 + 20x^4}{2\sqrt{x^3+2}}$$

9. $f(x) = \dfrac{x^3}{(x^2+4)^2}$

$$f'(x) = \frac{3x^2(x^2+4)^2 - 2(x^2+4)(2x)x^3}{(x^2+4)^4} = \frac{3x^4 + 12x^2 - 4x^4}{(x^2+4)^3} = \frac{x^2(12-x^2)}{(x^2+4)^3}$$

11. $f(x) = \dfrac{6}{\sqrt{x^2+4}} = 6(x^2+4)^{-1/2}$

$$f'(x) = -3(x^2+4)^{-3/2} \cdot 2x = \frac{-6x}{(x^2+4)^{3/2}}$$

13. $f(x) = (\sqrt{x}+3)^{4/3}$

$$f'(x) = \frac{4(\sqrt{x}+3)^{1/3}}{3} \cdot \frac{1}{2\sqrt{x}} = \frac{2(\sqrt{x}+3)^{1/3}}{3\sqrt{x}}$$

15. $f(x) = \left(\sqrt{x^3+2} + 2x\right)^{-2}$

$f'(x) =$

$$-2\left(\sqrt{x^3+2}+2x\right)^{-3}\left[\frac{3x^2}{2\sqrt{x^3+2}}+2\right]$$
$$= -\frac{3x^2+4\sqrt{x^3+2}}{(\sqrt{x^3+2}+2x)^3\cdot\sqrt{x^3+2}}$$

17. $f(x) = \dfrac{x}{\sqrt{x^2+1}}$

$$f'(x) = \frac{\sqrt{x^2+1} - x\left(\frac{1}{2\sqrt{x^2+1}}\right)2x}{x^2+1}$$
$$= \frac{1}{(x^2+1)\sqrt{x^2+1}}$$

19. $f(x) = \sqrt{\dfrac{x}{x^2+1}}$

$$f'(x) = \frac{1}{2\sqrt{\frac{x}{x^2+1}}}\cdot\frac{(x^2+1)-2x^2}{(x^2+1)^2}$$
$$= \frac{1-x^2}{2\sqrt{x}(x^2+1)^{3/2}}$$

21. $f(x) = \sqrt[3]{x\sqrt{x^4+2x\sqrt[4]{\dfrac{8}{x+2}}}}$

$$f(x) = \left(x\left[x^4+2x\left(\tfrac{8}{x+2}\right)^{1/4}\right]^{1/2}\right)^{1/3}$$
$$f'(x) = \tfrac{1}{3}\left(x\left[x^4+2x\left(\tfrac{8}{x+2}\right)^{1/4}\right]^{1/2}\right)^{-2/3}\cdot$$
$$\left(\left[x^4+2x\left(\tfrac{8}{x+2}\right)^{1/4}\right]^{1/2}+\right.$$
$$+x\left(\tfrac{1}{2}\right)\left[x^4+2x\left(\tfrac{8}{x+2}\right)^{1/4}\right]^{-1/2}\cdot$$
$$\left[4x^3+2\left(\tfrac{8}{x+2}\right)^{1/4}\right.$$
$$\left.\left.+2x\left(\tfrac{1}{4}\right)\left(\tfrac{8}{x+2}\right)^{-3/4}\left(\tfrac{-8}{(x+2)^2}\right)\right]\right)$$

23. $f(x) = \sqrt{x^2+16},\ a = 3,\ f(3) = 5$

$$f'(x) = \frac{1}{2\sqrt{x^2+16}}(2x) = \frac{x}{\sqrt{x^2+16}}$$
$$f'(3) = \frac{3}{\sqrt{3^2+16}} = \frac{3}{5}$$

So the tangent line is $y = \dfrac{3}{5}(x-3)+5$ or $y = \dfrac{3}{5}x + \dfrac{16}{5}$.

25. $s(t) = \sqrt{t^2+8}$

$v(t) = s'(t) = \dfrac{2t}{2\sqrt{t^2+8}} = \dfrac{t}{\sqrt{t^2+8}}$ m/s

$v(2) = \dfrac{2}{\sqrt{12}} = \dfrac{1}{\sqrt{3}} = \dfrac{\sqrt{3}}{3}$ m/s

27. For higher derivatives, fractional exponents will be required.

$$f(x) = \sqrt{2x+1} = (2x+1)^{1/2}$$
$$f'(x) = \frac{1}{2}(2x+1)^{-1/2}\cdot 2 = (2x+1)^{-1/2}$$
$$f''(x) = -\frac{1}{2}(2x+1)^{-3/2}(2)$$
$$= -(2x+1)^{-3/2}$$
$$f'''(x) = -\left(-\frac{3}{2}\right)(2x+1)^{-5/2}\cdot 2$$
$$= 3(2x+1)^{-5/2}$$
$$f^{(4)}(x) = 3\left(-\frac{5}{2}\right)(2x+1)^{-7/2}\cdot 2$$
$$= -15(2x+1)^{-7/2}$$
$$f^{(n)}(x) = (-1)^{n+1}1\cdot 3\ldots(2n-3)(2x+1)^{-(2n-1)/2}$$

29. $h'(1) = f'(g(1))g'(1)$
$g(1) = 4$, so $h'(1) = f'(4)g'(1)$.

From the table, we have:

$f'(4) \approx \dfrac{2-(-2)}{5-3} = 2$, and

$g'(1) \approx \dfrac{6-2}{2-0)} = 2$ so

$h'(1) \approx 4$.

31. $k'(3) = g'(f(3))f'(3)$
$f(3) = -2$, so $k'(3) = g'(-2)f'(3)$.

From the table, we have:

$f'(3) \approx \dfrac{0-(-3)}{4-2} = \dfrac{3}{2}$, and

$g'(-2) \approx \dfrac{2-6}{-1-(-3)} = -2$ so

$k'(1) \approx -3$.

33. $h'(x) = f'(g(x))g'(x)$
$$h'(1) = f'(g(1))g'(1)$$
$$= f'(2)\cdot(-2) = -6$$

35. $f(x) = x^3+4x-1$ is a one-to-one function with $f(0) = -1$ and $f'(0) = 4$. Therefore $g(-1) = 0$ and

$$g'(-1) = \frac{1}{f'(g(-1))} = \frac{1}{f'(0)} = \frac{1}{4}.$$

37. $f(x) = x^5 + 3x^3 + x$ is a one-to-one function with $f(1) = 5$ and $f'(1) = 5 + 9 + 1 = 15$. Therefore $g(5) = 1$ and

$$g'(5) = \frac{1}{f'(g(5))} = \frac{1}{f'(1)} = \frac{1}{15}.$$

39. $f(x) = \sqrt{x^3 + 2x + 4}$ is a one-to-one function and $f(0) = 2$ so $g(2) = 0$. Meanwhile,

$$f'(x) = \frac{1}{2\sqrt{x^3 + 2x + 4}}(3x^2 + 2)$$
$$f'(0) = 1/2$$
$$g'(2) = \frac{1}{f'(g(2))} = \frac{1}{f'(0)} = 2.$$

41. $f(x) = (x^2 + 3)^2 \cdot 2x$
Recognizing the "$2x$" as the derivative of x^2+3, we guess $g(x) = c(x^2 + 3)^3$ where c is some constant.
$g'(x) = 3c(x^2 + 3)^2 \cdot 2x$
which will be $f(x)$ only if $3c = 1$, so $c = 1/3$, and

$$g(x) = \frac{(x^2+3)^3}{3}.$$

43. $f(x) = \dfrac{x}{\sqrt{x^2+1}}$.
Recognizing the "x" as half the derivative of $x^2 + 1$, and knowing that differentiation throws the square root into the denominator, we guess $g(x) = c\sqrt{x^2+1}$ where c is some constant and find that

$$g'(x) = \frac{c}{2\sqrt{x^2+1}}(2x)$$

will match $f(x)$ if $c = 1$, so

$$g(x) = \sqrt{x^2+1}.$$

45. As a temporary device given *any* f, set $g(x) = f(-x)$. Then by the chain rule,

$$g'(x) = f'(-x)(-1) = -f'(-x).$$

In the even case ($g = f$) this reads $f'(-x) = -f'(x)$ and shows f' is odd. In the odd case ($g = -f$ and therefore $g' = -f'$), this reads $-f'(x) = -f'(-x)$ or $f'(x) = f'(-x)$ and shows f' is even.

47.
$$\frac{d}{dx}f(\sqrt{x}) = f'(\sqrt{x}) \cdot \frac{d}{dx}\sqrt{x} = f'(\sqrt{x}) \cdot \frac{1}{2\sqrt{x}}$$

49.
$$\frac{d}{dx}\left(\frac{1}{1+[f(x)]^2}\right) = -\left(\frac{1}{1+[f(x)]^2}\right)^2 \cdot \frac{d}{dx}\left(1+[f(x)]^2\right) = -\frac{1}{(1+[f(x)]^2)^2} \cdot 2f(x) \cdot f'(x)$$

51. $f'(x) = b'(a(x))a'(x)$.
$a(2) = 0$, $b'(0) = -3$, $a'(2) = 2$, so $f'(2) = -3 \cdot 2 = -6$.

53. $f'(x) = c'(a(x))a'(x)$.
$a(-1) = 0$, $c'(0) = -3$, $a'(-1) = -2$, so $f'(-1) = -3 \cdot -2 = 6$.

55. $f(x) = (x^3 - 3x^2 + 2x)^{1/3}$
$f'(x) =$
$\frac{1}{3}(x^3 - 3x^2 + 2x)^{-2/3} \cdot (3x^2 - 6x + 2)$
The derivative of f does not exist at values of x for which

$$0 = x^3 - 3x^2 + 2x = x(x^2 - 3x + 2) = x(x-1)(x-2)$$

Thus, the derivative of f does not exist for $x = 0, 1, 2$. The derivative fails to exist at these points because the tangent lines at these points are vertical.

2.6 Derivatives of Trigonometric Functions

1. The peaks and valleys of $\cos(x)$ (e.g., 0, π, 2π, etc.) are matched with the zeros of $\sin(x)$, and the decreasing intervals for $\cos(x)$ (e.g., $[0, \pi]$) correspond to the intervals where $\sin(x)$ is positive, hence where $-\sin(x)$ is negative. These features lend credibility to the notion that $-\sin(x)$ might be the derivative of $\cos(x)$.

3. $f(x) = 4\sin x - x$
$f'(x) = 4\cos x - 1$

5. $f(x) = \tan^3 x - \csc^4 x$
$f'(x) = 3\tan^2 x \sec^2 x$
$\quad + 4\csc^3 x \csc x \cot x$
$= 3\tan^2 x \sec^2 x + 4\csc^4 x \cot x$

7. $f(x) = x\cos 5x^2$
$f'(x) = (1)\cos 5x^2 + x(-\sin 5x^2)\cdot 10x$
$= \cos 5x^2 - 10x^2 \sin 5x^2$

9. $f(x) = \sin(\tan(x^2))$
$f'(x) = \cos(\tan(x^2))\cdot\sec^2(x^2)\cdot 2x$

11. $f(x) = \dfrac{\sin(x^2)}{x^2}$
$f'(x) = \dfrac{x^2\cos(x^2)\cdot 2x - \sin(x^2)\cdot 2x}{x^4}$
$= \dfrac{2x[x^2\cos(x^2)\cdot 2x - \sin(x^2)]}{x^4}$
$= \dfrac{2[x^2\cos(x^2) - \sin(x^2)]}{x^3}$

13. $f(t) = \sin t \sec t = \tan t$
$f'(t) = \sec^2 t$

15. $f(x) = \dfrac{1}{\sin(4x)} = \csc(4x)$
$f'(x) = -\csc(4x)\cot(4x)\cdot(4)$
$= -4\csc(4x)\cot(4x)$
$= \dfrac{-4\cos(4x)}{\sin^2(4x)}$

17. $f(x) = 2\sin x \cos x$
$f'(x) = 2\cos x\cdot\cos x + 2\sin x(-\sin x)$
$= 2\cos^2 x - 2\sin^2 x$

19. $f(x) = \tan\sqrt{x^2+1}$
$f'(x) = (\sec^2\sqrt{x^2+1})\cdot$
$\left(\dfrac{1}{2}\right)(x^2+1)^{-1/2}(2x)$

21. Answers depend on CAS.

23. Answers depend on CAS.

25. $f(x) = \sin 4x,\ a = \dfrac{\pi}{8},$
$f\left(\dfrac{\pi}{8}\right) = \sin\dfrac{\pi}{2} = 1$
$f'(x) = 4\cos 4x$
$f'\left(\dfrac{\pi}{8}\right) = 4\cos\dfrac{\pi}{2} = 0$
So the equation of the tangent line is
$y - 1 = 0\left(x - \dfrac{\pi}{8}\right)$ or $y = 1$.

27. $f(x) = \cos x,\ a = \dfrac{\pi}{2},$
$f\left(\dfrac{\pi}{2}\right) = \cos\dfrac{\pi}{2} = 0$
$f'(x) = -\sin x$
$f'\left(\dfrac{\pi}{2}\right) = -\sin\dfrac{\pi}{2} = -1$
So the equation of the tangent line is
$y - 0 = -1\left(x - \dfrac{\pi}{2}\right)$ or $y = -x + \pi/2$.

29. $s(t) = t^2 - \sin(2t),\ t_0 = 0$
$v(t) = s'(t) = 2t - 2\cos(2t)$
$v(0) = 0 - 2\cos(0) = 0 - 2 = -2$ ft/s

31. $s(t) = \dfrac{\cos t}{t},\ t_0 = \pi$
$v(t) = s'(t)$
$= \dfrac{-1}{t^2}\cos t + \dfrac{1}{t}(-\sin t)$
$v(\pi) = -\dfrac{\cos\pi}{\pi^2} - \dfrac{\sin\pi}{\pi}$
$= \dfrac{1}{\pi^2} - \dfrac{1}{\pi}(0) = \dfrac{1}{\pi^2}$ ft/s

33. $f(t) = 4\sin 3t$
$f'(t) = 12\cos 3t$

The maximum speed of 12 occurs when the vertical position is zero.

35. $Q(t) = 3\sin 2t + t + 4$
$I(t) = \frac{dQ}{dt} = 6\cos 2t + 1$
At time $t = 0$, $I(0) = 7$ amps. At time $t = 1$, $I(1) = 6\cos 2 + 1 \approx -1.497$ amps.

37. $f(x) = \sin x$
$f'(x) = \cos x$
$f''(x) = -\sin x$
$f'''(x) = -\cos x$
$f^{(4)}(x) = \sin x = f(x)$

$$\begin{aligned}\Rightarrow f^{(75)}(x) &= (f^{(72)})^{(3)}(x)\\ &= (f^{(18\cdot 4)})^{(3)}(x)\\ &= f''' = -\cos x\end{aligned}$$

$$\begin{aligned}f^{(150)}(x) &= (f^{(148)})^{(2)}(x)\\ &= (f^{(37\cdot 4)})^{(2)}(x)\\ &= f'' = -\sin x\end{aligned}$$

39. Since $0 \le \sin\theta \le \theta$, we have
$-\theta \le -\sin(\theta) \le 0$ which implies
$-\theta \le \sin(-\theta) \le 0$
so for $-\frac{\pi}{2} \le \theta \le 0$ we have
$\theta \le \sin\theta \le 0$.
We also know that
$\lim_{\theta\to 0^-} \theta = 0 = \lim_{\theta\to 0^-} 0$,
so the Squeeze Theorem implies that
$\lim_{\theta\to 0^-} \sin\theta = 0$.

41. If $f(x) = \cos(x)$, then

$$\begin{aligned}&\frac{f(x+h) - f(x)}{h}\\ &= \frac{\cos(x+h) - \cos(x)}{h}\\ &= \frac{\cos x\cos h - \sin x \sin h - \cos x}{h}\\ &= (\cos x)\frac{(\cos h - 1)}{h} - (\sin x)\left(\frac{\sin h}{h}\right).\end{aligned}$$

Taking the limit according to Lemma 6.1

$$\begin{aligned}f'(x) &= \lim_{h\to 0}\frac{f(x+h) - f(x)}{h}\\ &= (\cos x)\cdot\lim_{h\to 0}\frac{\cos(h) - 1}{h}\\ &\quad - (\sin x)\cdot\lim_{h\to 0}\frac{\sin(h)}{h}\\ &= \cos x\cdot 0 - \sin x\cdot 1\\ &= -\sin x.\end{aligned}$$

43. (a) $$\begin{aligned}\lim_{x\to 0}\frac{\sin 3x}{x} &= \lim_{x\to 0}\frac{3\sin 3x}{3x}\\ &= 3\cdot\lim_{x\to 0}\frac{\sin(3x)}{(3x)}\\ &= 3\cdot 1 = 3\end{aligned}$$

(b) $$\begin{aligned}\lim_{t\to 0}\frac{\sin t}{4t} &= \frac{1}{4}\lim_{t\to 0}\frac{\sin t}{t}\\ &= \frac{1}{4}\cdot 1 = \frac{1}{4}\end{aligned}$$

(c) $$\begin{aligned}&\lim_{x\to 0}\frac{\cos x - 1}{5x}\\ &= \frac{1}{5}\lim_{x\to 0}\frac{\cos x - 1}{x} = 0\end{aligned}$$

(d) Let $u = x^2$: then $u \to 0$ as $x \to 0$, and

$$\lim_{x\to 0}\frac{\sin x^2}{x^2} = \lim_{u\to 0}\frac{\sin u}{u} = 1$$

45. If $x \ne 0$, then f is continuous by Theorem 4.2 in Section 1.4, and f is differentiable by the Quotient rule (Theorem 4.2 in Section 2.4). Thus, we only need to check $x = 0$. To see that f is continuous at $x = 0$:

$$\lim_{x\to 0} f(x) = \lim_{x\to 0}\frac{\sin x}{x} = 1$$

(by Lemma 6.3)
Since $\lim_{x\to 0} f(x) = f(0)$, f is continuous at $x = 0$.

To see that f is differentiable at $x = 0$:

$$f'(a) = \lim_{x\to a}\frac{f(x) - f(a)}{x - a}$$

$$\begin{aligned}f'(0) &= \lim_{x\to 0}\frac{f(x) - f(0)}{x - 0}\\ &= \lim_{x\to 0}\frac{(\sin x)/x - 1}{x}\end{aligned}$$

In the proof of Lemma 6.3, equation

6.8 was derived:

$$1 > \frac{\sin x}{x} > \cos x$$

Thus,

$$0 > \frac{\sin x}{x} - 1 > \cos x - 1$$

and therefore, if $x > 0$,

$$0 > \frac{\frac{\sin x}{x} - 1}{x} > \frac{\cos x - 1}{x}$$

and if $x < 0$,

$$0 < \frac{\frac{\sin x}{x} - 1}{x} < \frac{\cos x - 1}{x}$$

By Lemma 6.4,

$$\lim_{x\to 0} \frac{\cos x - 1}{x} = 0$$

Applying the squeeze theorem to the previous two inequalities, we obtain

$$\lim_{x\to 0} \frac{\frac{\sin x}{x} - 1}{x} = 0$$

and so $f'(0) = 0$.

47. For $x \neq 0$,

$$f'(x) = \frac{x\cos x - \sin x}{x^2}$$

$$\begin{aligned} f''(x) &= \frac{x^2(\cos x - x\sin x - \cos x)}{x^4} - \frac{2x(x\cos x - \sin x)}{x^4} \\ &= \frac{-x^3\sin x - 2x^2\cos x + 2x\sin x}{x^4} \\ &= \frac{(2-x^2)\sin x - 2x\cos x}{x^3} \end{aligned}$$

Thus, $f''(x)$ exists and is continuous for all $x \neq 0$. For $x = 0$,

$$\begin{aligned} f''(0) &= \lim_{x\to 0} \frac{f'(x) - f'(0)}{x - 0} \\ &= \lim_{x\to 0} \frac{\frac{x\cos x - \sin x}{x^2} - 0}{x - 0} \\ &= \lim_{x\to 0} \frac{x\cos x - \sin x}{x^3} \end{aligned}$$

Applying L'Hospital's rule, one obtains

$$\begin{aligned} f''(0) &= \lim_{x\to 0} \frac{\cos x - x\sin x - \cos x}{3x^2} \\ &= -\frac{1}{3}\lim_{x\to 0} \frac{\sin x}{x} = -\frac{1}{3} \end{aligned}$$

Finally, applying L'Hospital's rule to $f''(x)$, one obtains

$$\begin{aligned} &\lim_{x\to 0} f''(x) \\ &= \lim_{x\to 0} \frac{(2-x^2)\sin x - 2x\cos x}{x^3} \\ &= \lim_{x\to 0} \left[\frac{(2-x^2)\cos x - 2x\sin x}{3x^2} + \frac{2x\sin x - 2\cos x}{3x^2}\right] \\ &= \lim_{x\to 0} \frac{-x^2\cos x}{3x^2} \\ &= -\frac{1}{3}\lim_{x\to 0} \cos x = -\frac{1}{3} \end{aligned}$$

Thus, $\lim_{x\to 0} f''(x) = f''(0)$, and so f'' is continuous at $x = 0$.

49. The sketch: $y = x$ and $y = \sin(x)$

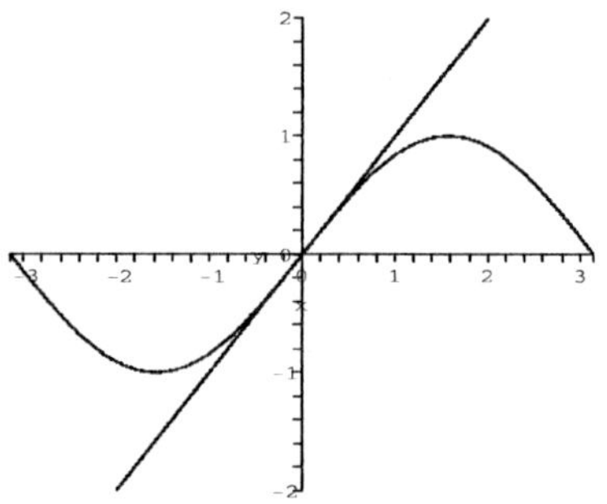

It is not possible visually to either detect or rule out intersections near $x = 0$ (other than zero itself).

We have that $f'(x) = \cos x$, which is less than 1 for $0 < x < 1$. If $\sin x \geq x$ for some x in the interval $(0, 1)$, then there would be a point on the graph of $y = \sin x$ which lies above the line $y = x$, but then (since $\sin x$ is continuous) the slope of the tangent line of $\sin x$ would have to be greater than or equal to 1 at some point in that interval, contradicting $f'(x) < 1$. Since

$\sin x < x$ for $0 < x < 1$, we have $-\sin x > -x$ for $0 < x < 1$. Then $-\sin x = \sin(-x)$ so $\sin(-x) > -x$ for $0 < x < 1$ which is the same as saying $\sin x > x$ for $-1 < x < 0$.

Since $-1 \le \sin x \le 1$, the only interval on which $y = \sin x$ might intersect $y = x$ is $[-1, 1]$. We know they intersect at $x = 0$ and we just showed that they do not intersect on the intervals $(-1, 0)$ and $(0, 1)$. So the only other points they might intersect are $x = \pm 1$, but we know that $\sin(\pm 1) \ne \pm 1$, so these graphs intersect only at $x = 0$.

51. As seen from the graphs, changing the scale on the x-axis increases the number of oscillations or periods on the display. As the number of periods on the display increase, the graph looks more and more like a bunch of line segments. Its inflection points and concavity are no longer detectable.

2.7 Derivatives of Exponential and Logarithmic Functions

1. $f'(x) = 3x^2 \cdot e^x + x^3 \cdot e^x = e^x x^2 (x+3)$

3. $f'(x) = 1 + 2^x \ln 2$

5. $f'(x) = 2e^{4x+1} \cdot 4 = 8e^{4x+1}$

7. $f'(x) = (1/3)^{x^2} \cdot \ln(1/3) \cdot 2x$
$= -2x \ln(3)(1/3)^{x^2}$

9. $f'(x) = 4^{-3x+1} \cdot \ln 4 \cdot (-3)$
$= -6 \ln(2) 4^{-3x+1}$

11. $f'(x) = \dfrac{x \cdot 4e^{4x} - e^{4x} \cdot 1}{x^2}$
$= \dfrac{e^{4x}(4x-1)}{x^2}$

13. $f'(x) = \dfrac{1}{2x} \cdot (2) = \dfrac{1}{x}$

15. $f'(x) = \dfrac{3x^2+3}{x^3+3x} = \dfrac{3(x^2+1)}{x(x^2+3)}$

17. $f'(x) = \dfrac{1}{\cos x} \cdot -\sin x = -\tan x$

19. $f'(x) = \cos\left[\ln(\cos x^3)\right] \cdot \dfrac{1}{\cos x^3} \cdot (-\sin x^3) \cdot 3x^2$
$= -3x^2 \cdot \cos\left[\ln(\cos x^3)\right] \cdot \tan x^3$

21. $f(x) = \dfrac{\sqrt{\ln x^2}}{x} = \dfrac{\sqrt{2 \ln x}}{x}$, so
$f'(x) = \dfrac{x \cdot \frac{1}{2\sqrt{2\ln x}} \cdot \frac{2}{x} - \sqrt{2\ln x} \cdot 1}{x^2}$
$= \dfrac{1 - 2\ln x}{x^2\sqrt{2\ln x}} = \dfrac{1 - \ln x^2}{x^2\sqrt{\ln x^2}}$

23. $f'(x) = \dfrac{\sec x \tan x + \sec^2 x}{\sec x + \tan x} = \sec x$

25. $f(1) = 3e^1 = 3e$
$f'(x) = 3e^x$
$f'(1) = 3e^1 = 3e$
So the equation of the tangent line is $y - 3e = 3e(x-1)$ or $y = 3ex$.

27. $f(1) = 3$
$f'(x) = 3^x \ln 3$
$f'(1) = 3 \cdot \ln 3$
So the equation of the tangent line is $y = (3 \cdot \ln 3)(x-1) + 3$.

29. $f(1) = 0$
$f'(x) = 2x \ln x + x^2 \cdot \dfrac{1}{x} = 2x \ln x + x$
$f'(1) = 2 \cdot 1 \ln 1 + 1 = 2 \cdot 0 + 1 = 1$
So the equation of the tangent line is $y = 1(x-1) + 0$ or $y = x - 1$.

31. $v'(t) = 100 \cdot 3^t \ln 3$
$\dfrac{v'(t)}{v(t)} = \dfrac{100 \cdot 3^t \ln 3}{100 \cdot 3^t} = \ln 3 \approx 1.10$
So the percentage change is about 110%.

33. $v(t) = 100e^t$
$v'(t) = 100e^t$
$\dfrac{v'(t)}{v(t)} = \dfrac{100e^t}{100e^t} = 1$
So the percentage change is 100%.

35. $p(t) = 200 \cdot 3^t$
$\ln(p(t)) = \ln(200) + t\ln(3)$
$\dfrac{p'(t)}{p(t)} = \dfrac{d}{dt}[\ln(p(t)] = \ln 3 \approx 1.099$,
so the rate of change of population is about 110% per unit of time.

37. $f(t) = Ae^{rt}$
$APY = \dfrac{f(1) - A}{A} = \dfrac{Ae^r - A}{A} = e^r - 1$

(a) $APY = e^{0.05} - 1 \approx .05127$ (5.1%)

(b) $APY = e^{0.1} - 1 \approx .10517$ (10.5%)

(c) $APY = e^{0.2} - 1 \approx .22140$ (22.1%)

(d) $APY = e^{\ln 2} - 1 = 1$ (100%)

(e) $APY = e^1 - 1 \approx 1.71828$ (172%)

39. $f(x) = x^{\sin x}$
$\ln f(x) = \sin x \cdot \ln x$
$$\frac{f'(x)}{f(x)} = \frac{d}{dx}(\sin x \cdot \ln x) = \cos x \cdot \ln x + \frac{\sin x}{x}$$
$$f'(x) = x^{\sin x}\left(\frac{x\cos x \cdot \ln x + \sin x}{x}\right)$$

41. $f(x) = (\sin x)^x$
$\ln f(x) = x \cdot \ln(\sin x)$
$$\frac{f'(x)}{f(x)} = \frac{d}{dx}(x \cdot \ln(\sin x)) = \frac{x\cos x}{\sin x} + \ln(\sin x) = x\cot x + \ln(\sin x)$$
$$f'(x) = (\sin x)^x \cdot (x\cot x + \ln(\sin x))$$

43. $f(x) = x^{\ln x}$
$\ln f(x) = \ln x \cdot \ln x = \ln^2 x$
$\dfrac{f'(x)}{f(x)} = \dfrac{d}{dx}\left(\ln^2 x\right) = \dfrac{2\ln x}{x}$
$f'(x) = x^{\ln x}\left[\dfrac{2\ln x}{x}\right] = 2x^{[(\ln x)-1]}\ln x$

45. $f(t) = e^{-t}\cos t$
$v(t) = f'(t) = -e^{-t}\cos t + e^{-t}(-\sin t)$
$= -e^{-t}(\cos t + \sin t)$
If the velocity is zero, it is because $\cos t = -\sin t$, so
$t = \dfrac{3\pi}{4}, \dfrac{7\pi}{4}, \ldots, \dfrac{(3+4n)\pi}{4}, \ldots$
Position when velocity is zero:
$f(3\pi/4) = e^{-3\pi/4}\cos(3\pi/4)$
$= e^{-3\pi/4}(-1/\sqrt{2}) \approx -.067020$
$f(7\pi/4) = e^{-7\pi/4}\cos(7\pi/4)$
$= e^{-7\pi/4}(1/\sqrt{2}) \approx .002896$
Graph of the velocity function:

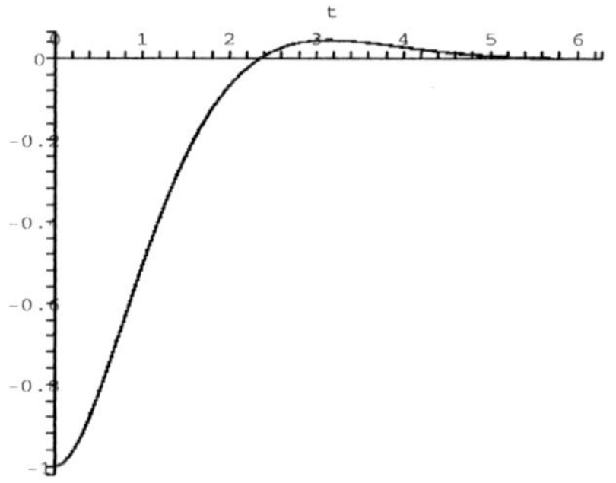

47. Graphically, the maximum velocity seems to occur at $t = \pi$.

49. $f(x) = \sinh x = \dfrac{e^x - e^{-x}}{2}$
$f'(x) = \dfrac{e^x + e^{-x}}{2} = \cosh x$

$g(x) = \cosh x = \dfrac{e^x + e^{-x}}{2}$

$g'(x) = \dfrac{e^x - e^{-x}}{2} = \sinh x$

51. If $f(x) = \sinh x$, then $f'(x) = \cosh x$ and $f''(x) = \sinh x = f(x)$.

If $f(x) = \cosh x$, then $f'(x) = \sinh x$ and $f''(x) = \cosh x = f(x)$.

53. Let $(a, \ln a)$ be the point of intersection of the tangent line and the graph of $y = f(x)$.

$f(x) = \ln x$

$f'(x) = \dfrac{1}{x}$

$m = f'(a) = \dfrac{1}{a}$

Since the tangent line passes through the origin, the equation of the tangent line is

$$y = mx = \frac{1}{a}x$$

Since $(a, \ln a)$ is a point on the tangent line,

$$\ln a = \frac{1}{a}a = 1$$

so $a = e$.

55. $f(x) = e^{\ln x^2}$

$f'(x) = e^{\ln x^2} \cdot \dfrac{d}{dx} \ln x^2$

$= e^{\ln x^2} \cdot \dfrac{2}{x} = 2x$

Much easier if one noticed at the outset that $f(x) = x^2$.

57. $f(x) = \ln \sqrt{4e^{3x}} = \dfrac{1}{2}\left[\ln\left(4 \cdot e^{3x}\right)\right]$

$= \dfrac{1}{2}\left[\ln 4 + \ln e^{3x}\right] = \dfrac{\ln 4 + 3x}{2}$

$f'(x) = \dfrac{3}{2}$

59. We approximate $\lim\limits_{h \to 0} \frac{a^h - 1}{h}$ for $a = 3$.

h	$\frac{a^h-1}{h}$
0.01	1.10466919
0.001	1.09921598
0.0001	1.09867264
0.00001	1.09861832
−0.01	1.09259958
−0.001	1.09800903
−0.0001	1.09855194

The limit seems to be approaching approximately 1.0986, which is very close to $\ln 3 \approx 1.09861$.

61. $x(t) = \dfrac{6}{2e^{-8t} + 1} = 6(2e^{-8t} + 1)^{-1}$

$x'(t) = -6(2e^{-8t} + 1)^{-2} \cdot (-16e^{-8t})$

$= \dfrac{96e^{-8t}}{(2e^{-8t} + 1)^2}$

Since $e^{-8t} > 0$ for any t, both numerator and denominator are positive, so that $x'(t) > 0$. Then, since $x(t)$ is an increasing function with a limiting value of 6 (as t goes to infinity), the concentration never exceeds (indeed, never reaches) the value of 6.

63. If $g(x) = e^x$, then $g'(x) = e^x$ and $g''(x) = e^x$ so $g(0) = g'(0) = g''(0) = e^0 = 1$.

If $f(x) = \dfrac{a + bx}{1 + cx}$, then $f(0) = a$,

$f'(x) = \dfrac{b(1 + cx) - (a + bx)(c)}{(1 + cx)^2}$

$= \dfrac{b - ac}{(1 + cx)^2} = (b-ac)(1+cx)^{-2}$

$f'(0) = b - ac$

$f''(x) = (b - ac)(-2)(1 + cx)^{-3}c$

$= \dfrac{-2c(b - ac)}{(1 + cx)^3}$

$f''(0) = -2c(b - ac)$

$1 = g(0) = f(0) = a$ so $a = 1$.

$1 = g'(0) = f'(0) = b - ac = b - c$

$1 = g''(0) = f''(0) = -2c(b - ac) = -2c$

so $c = -1/2$ and $b = 1 + c = 1 - 1/2 = 1/2$

so $b = 1/2$.
In summary, $a = 1$, $b = 1/2$, $c = -1/2$ and
$$g(x) = \frac{1 + (x/2)}{1 - (x/2)} = \frac{2 + x}{2 - x}.$$

65. $f(x) = e^{-x^2/2}$
$f'(x) = e^{-x^2/2} \cdot (-2x/2)$
$= -xe^{-x^2/2}$
$f''(x) = -\left[x(-xe^{-x^2/2}) + 1 \cdot e^{-x^2/2}\right]$
$= e^{-x^2/2}(x^2 - 1)$
This will be zero only when $x = \pm 1$.

67. It helps immensely to leave the name f as it was in #65, and give a new name g to the new function here, so that
$$g(x) = e^{-(x-m)^2/2c^2} = f(u)$$
in which $u = \frac{x-m}{c}$. Then
$$g'(x) = f'(u)\frac{du}{dx} = \frac{f'(u)}{c} = \frac{-uf(u)}{c}$$
$$= \frac{-(x-m)e^{-(x-m)^2/2c^2}}{c^2},$$
$$g''(x) = \frac{d}{dx}\left(\frac{f'(u)}{c}\right) = \frac{f''(u)\frac{du}{dx}}{c}$$
$$= \frac{f''(u)}{c^2} = \frac{(u^2 - 1)f(u)}{c^2}$$
$$= \frac{((x-m)^2 - c^2)e^{-(x-m)^2/2c^2}}{c^4}$$
This will be zero only when $x = m \pm c$.

2.8 Implicit Differentiation and Inverse Trigonometric Functions

1. Explicitly:
$4y^2 = 8 - x^2$
$y^2 = \frac{8-x^2}{4}$
$y = \pm\frac{\sqrt{8-x^2}}{2}$ (choose plus to fit (2,1))

For $y = \dfrac{\sqrt{8 - x^2}}{2}$,
$$y' = \frac{1}{2}\frac{(-2x)}{2\sqrt{8 - x^2}} = \frac{-x}{2\sqrt{8 - x^2}},$$
$y'(2) = -1/2$.

Implicitly:
$$\frac{d}{dx}(x^2 + 4y^2) = \frac{d}{dx}(8)$$
$2x + 8y \cdot y' = 0$
$$y' = \frac{-2x}{8y} = \frac{-x}{4y}$$
$$\text{at } (2,1) : y' = \frac{-2}{4 \cdot 1} = -\frac{1}{2}$$

3. Explicitly:
$y(1 - 3x^2) = \cos x$
$$y = \frac{\cos x}{1 - 3x^2}$$
$$y'(x) = \frac{(1 - 3x^2)(-\sin x) - \cos x(-6x)}{(1 - 3x^2)^2}$$
$$= \frac{-\sin x + 3x^2 \sin x + 6x \cos x}{(1 - 3x^2)^2}$$
$y'(0) = 0$

Implicitly:
$$\frac{d}{dx}(y - 3x^2y) = \frac{d}{dx}(\cos x)$$
$y' - (6xy + 3x^2y') = -\sin x$
$y'(1 - 3x^2) = 6xy - \sin x$
$$y' = \frac{6xy - \sin x}{1 - 3x^2}$$
at $(0,1)$: $y' = 0$ (again).

5. $$\frac{d}{dx}(x^2y^2 + 3y) = \frac{d}{dx}(4x)$$
$2xy^2 + x^2 2y \cdot y' + 3y' = 4$
$y'(2x^2y + 3) = 4 - 2xy^2$
$$y' = \frac{4 - 2xy^2}{2x^2y + 3}$$

7. $$\frac{d}{dx}(\sqrt{xy} - 4y^2) = \frac{d}{dx}(12)$$
$$\frac{1}{2\sqrt{xy}} \cdot \frac{d}{dx}(xy) - 8y \cdot y' = 0$$

$$\frac{1}{2\sqrt{xy}} \cdot (xy' + y) - 8y \cdot y' = 0$$
$$(xy' + y) - 16y \cdot y'\sqrt{xy} = 0$$
$$y'\,(x - 16y\sqrt{xy}) = -y$$
$$y' = \frac{-y}{(x - 16y\sqrt{xy})} = \frac{y}{16y\sqrt{xy} - x}$$

9. $x + 3 = 4xy + y^3$
$$1 = \frac{d}{dx}\left(4xy + y^3\right) = 4(xy'+y)+3y^2y'$$
$$1 - 4y = y'(3y^2 + 4x)$$
$$y' = \frac{1-4y}{3y^2+4x}$$

11. $\frac{d}{dx}(e^{x^2y} - e^y) = \frac{d}{dx}(x)$
$$e^{x^2y}\frac{d}{dx}(x^2y) - e^y y' = 1$$
$$e^{x^2y}(2xy + x^2y') - e^y y' = 1$$
$$y'(x^2e^{x^2y} - e^y) = 1 - 2xye^{x^2y}$$
$$y' = \frac{1 - 2xye^{x^2y}}{x^2e^{x^2y} - e^y}$$

13. $\frac{d}{dx}\left(\sqrt{x+y} - 4x^2\right) = \frac{d}{dx}(y)$
$$\frac{1}{2\sqrt{x+y}} \cdot (1 + y') - 8x = y'$$
$$y'\left(\frac{1}{2\sqrt{x+y}} - 1\right) = \frac{-1}{2\sqrt{x+y}} + 8x$$
$$y'\left(\frac{1 - 2\sqrt{x+y}}{2\sqrt{x+y}}\right) = \frac{16x\sqrt{x+y} - 1}{2\sqrt{x+y}}$$
$$y' = \frac{16x\sqrt{x+y} - 1}{1 - 2\sqrt{x+y}}$$

15. $\frac{d}{dx}\left(e^{4y} - \ln y\right) = \frac{d}{dx}(2x)$
$$e^{4y} \cdot 4y' - \frac{1}{y} \cdot y' = 2$$
$$y'\left(4e^{4y} - \frac{1}{y}\right) = 2$$
$$y'\left(\frac{4ye^{4y} - 1}{y}\right) = 2$$
$$y' = \frac{2y}{4ye^{4y} - 1}$$

17. Rewrite: $x^2 = 4y^3$

Differentiate by x: $2x = 12y^2 \cdot y'$
$$y' = \frac{2x}{12y^2} = \frac{x}{6y^2}$$
at $(2,1)$: $y' = \frac{2}{6 \cdot 1^2} = \frac{1}{3}$

The equation of the tangent line is
$y - 1 = \frac{1}{3}(x - 2)$ or $y = \frac{1}{3}(x + 1)$.

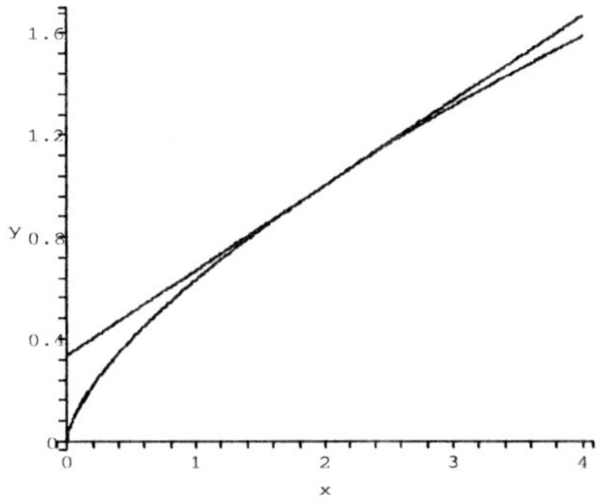

19. This one has $y = 0$ as part of the curve(s), but our point of reference is not on that part, so we can assume y is not zero, cancel it, and come to $x^2y = 4$
$$\frac{d}{dx}(x^2y) = \frac{d}{dx}(4)$$
$$2xy + x^2 \cdot y' = 0$$
$$y' = \frac{-2y}{x}$$
at $(2,1)$: $y' = -2/2 = -1$.

The equation of the tangent line is
$y - 1 = (-1)(x - 2)$ or $y = -x + 3$.

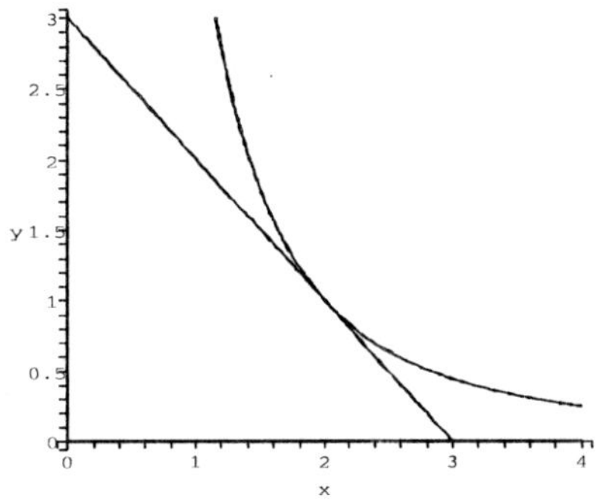

21. $4y^2 = 4x^2 - x^4$
$$8yy' = 8x - 4x^3$$
$$y' = \frac{x(2 - x^2)}{2y}$$
The slope of the tangent line at

$(1, \sqrt{3}/2)$ is

$$m = \frac{(1)(2-1^2)}{2\left(\frac{\sqrt{3}}{2}\right)} = \frac{1}{\sqrt{3}} = \frac{\sqrt{3}}{3}.$$

The equation of the tangent line is

$$y - \frac{\sqrt{3}}{2} = \frac{\sqrt{3}}{3}(x-1)$$
$$y = \frac{\sqrt{3}}{3}x + \frac{\sqrt{3}}{2} - \frac{\sqrt{3}}{3}$$
$$y = \frac{\sqrt{3}}{3}x + \frac{\sqrt{3}}{6}.$$

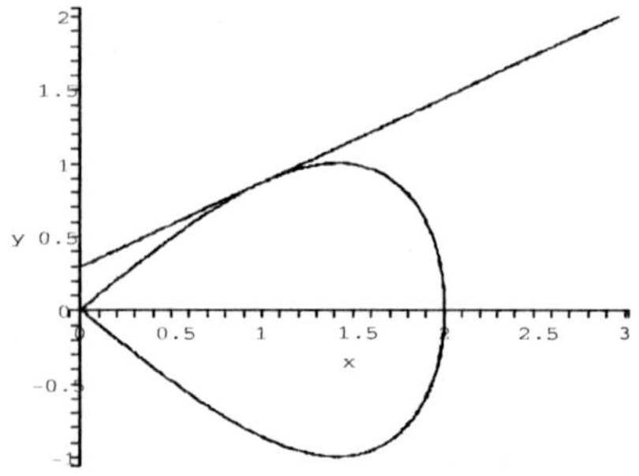

23. $\frac{d}{dx}(x^2 + y^3 - 3y) = \frac{d}{dx}(4)$
$$2x + 3y^2y' - 3y' = 0$$
$$y'(3y^2 - 3) = -2x$$
$$y' = \frac{2x}{3 - 3y^2}$$

Horizontal tangents:
From the formula, $y' = 0$ only when $x = 0$. When $x = 0$, we have $0^2 + y^3 - 3y = 4$. Using a CAS to solve this, we find that

$$y = \left(2 - \sqrt{3}\right)^{1/3} + \left(2 + \sqrt{3}\right)^{1/3} \approx 2.2$$

is a horizontal tangent line, tangent to the curve at the (approximate) point $(0, 2.2)$.

Vertical tangents: the denominator in y' must be zero.
$$3 - 3y^2 = 0$$
$$y^2 = 1 \quad \text{or} \quad y = \pm 1.$$
When $y = 1$ we have
$$x^2 + (1)^3 - 3(1) = 4$$
$$x^2 = 6 \quad \text{or} \quad x = \pm\sqrt{6} \approx \pm 2.4.$$
Also, when $y = -1$, we have
$$x^2 + (-1)^3 - 3(-1) = 4$$
$$x^2 = 2$$
$$x = \pm\sqrt{2} \approx \pm 1.4.$$
Thus, we find 4 vertical tangent lines: $x = -\sqrt{6}$, $x = -\sqrt{2}$, $x = \sqrt{2}$, $x = \sqrt{6}$, tangent to the curve (respectively) at the points $(-\sqrt{6}, 1)$, $(-\sqrt{2}, -1)$, $(\sqrt{2}, -1)$, and $(\sqrt{6}, 1)$.

25. $\frac{d}{dx}(x^2y^2 + 3x - 4y) = \frac{d}{dx}(5)$
$$x^2 2yy' + 2xy^2 + 3 - 4y' = 0$$
Differentiate both sides of this with respect to x:
$$\frac{d}{dx}(x^2 2yy' + 2xy^2 + 3 - 4y') = \frac{d}{dx}(0)$$
$$2(2xyy' + x^2(y')^2 + x^2yy'') + 2(2xyy' + y^2) - 4y'' = 0$$
$$2xyy' + x^2(y')^2 + x^2yy'' + 2xyy' + y^2 - 2y'' = 0$$
$$4xyy' + x^2(y')^2 + y^2 = y''(2 - x^2y)$$
$$y'' = \frac{4xyy' + x^2(y')^2 + y^2}{2 - x^2y}$$

27. $\frac{d}{dx}(y^2) = \frac{d}{dx}(x^3 - 6x + 4\cos y)$
$$2yy' = 3x^2 - 6 - 4\sin y \cdot y'$$
Differentiating again with respect to x: $2[yy'' + (y')^2]$
$$= 6x - 4[\sin y \cdot y'' + \cos y \cdot (y')^2],$$
$$yy'' + (y')^2 = 3x - 2\sin y \cdot y'' - 2\cos y \cdot (y')^2,$$
$$y''(y + 2\sin y) = 3x - [2\cos y + 1](y')^2$$
$$y'' = \frac{3x - [2\cos y + 1](y')^2}{y + 2\sin y}$$

29. $f'(x) = \frac{1}{1 + (\sqrt{x})^2} \cdot \frac{d}{dx}\sqrt{x} = \frac{1}{2(1+x)\sqrt{x}}$

31. $f'(x) = \frac{1}{1 + (\cos x)^2} \cdot \frac{d}{dx}\cos x$

$$= \frac{-\sin x}{1 + (\cos x)^2}$$

33. $f'(x) = 4\sec(x^4)\tan(x^4) \cdot 4x^3$

35. $f'(x) = e^{\tan^{-1} x} \dfrac{d}{dx} \tan^{-1} x$

$$= \frac{e^{\tan^{-1} x}}{1 + x^2}$$

37. $f'(x) = \dfrac{(x^2+1)\frac{1}{x^2+1} - \tan^{-1} x(2x)}{(x^2+1)^2}$

$$= \frac{1 - 2x\tan^{-1} x}{(x^2+1)^2}$$

39. $x^2 + y^3 - 2y = 3$

$y' = \dfrac{-2x}{3y^2 - 2}$

If $x = 1.9$, solving for y requires solving the equation $y^3 - 2y + 0.61 = 0$. Using the equation of the tangent line found in Example 8.1, $y = -4x + 9$, $y(1.9) \approx 1.4$.

If $x = 2.1$, solving for y requires solving the equation $y^3 - 2y + 1.41 = 0$. Using the equation of the tangent line found in Example 8.1, $y = -4x + 9$, $y(2.1) \approx 0.6$.

41. Both of the points $(-3, 0)$ and $(0, 3)$ are on the curve:

$0^2 = (-3)^3 - 6(-3) + 9 = -27 + 18 + 9$

$3^2 = (0)^3 - 6(0) + 9 = 9$

The equation of the line through these points has slope

$\dfrac{0-3}{-3-0} = \dfrac{-3}{-3} = 1$

and y-intercept 3, so $y = x + 3$. This line intersects the curve at:

$y^2 = x^3 - 6x + 9$

$(x+3)^2 = x^3 - 6x + 9$

$x^2 + 6x + 9 = x^3 - 6x + 9$

$x^3 - 12x - x^2 = 0$

$x(x^2 - x - 12) = 0$

Therefore, $x = 0, -3$ or 4 and so the third point is $(4, 7)$.

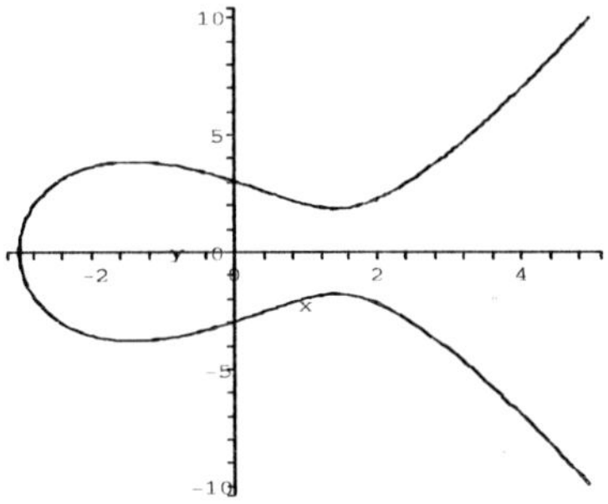

43. For the inverse hyperbolic tangent function,

$y = \tanh^{-1} x \iff x = \tanh y$

Differentiating both sides of $x = \tanh y$ implicitly, we obtain

$$1 = \frac{(e^y + e^{-y})^2 - (e^y - e^{-y})^2}{(e^y + e^{-y})^2} y'$$

$$= \left(1 - \frac{(e^y - e^{-y})^2}{(e^y + e^{-y})^2}\right) y'$$

$$= \left(1 - \left[\frac{e^y - e^{-y}}{e^y + e^{-y}}\right]^2\right) y'$$

$$= \left(1 - [\tanh y]^2\right) y'$$

$$= \left(1 - x^2\right) y'$$

$$y' = \frac{1}{1 - x^2}$$

For the inverse hyperbolic cotangent function,

$y = \coth^{-1} x \iff x = \coth y$

Differentiating both sides of $x = \coth y$ implicitly, we obtain

$$1 = \frac{(e^y - e^{-y})^2 - (e^y + e^{-y})^2}{(e^y - e^{-y})^2} y'$$

$$= \left(1 - \frac{(e^y + e^{-y})^2}{(e^y - e^{-y})^2}\right) y'$$

$$= \left(1 - \left[\frac{e^y + e^{-y}}{e^y - e^{-y}}\right]^2\right) y'$$

$$= \left(1 - [\coth y]^2\right) y'$$

$$= \left(1 - x^2\right) y'$$

$$y' = \frac{1}{1 - x^2}$$

The derivative formulas are not identical because their domains are different. The domain of the inverse hyperbolic tangent function and its derivative is $|x| < 1$, and the domain of the inverse hyperbolic cotangent function and its derivative is $|x| > 1$.

45. $y = \sin^{-1} x + \cos^{-1} x$

$\dfrac{dy}{dx} = \dfrac{1}{\sqrt{1-x^2}} + \dfrac{-1}{\sqrt{1-x^2}} = 0$

Therefore, $y = c$, where c is a constant. To determine c, substitute any convenient value of x, such as $x = 0$.

$\sin^{-1} x + \cos^{-1} x = c$

$\sin^{-1} 0 + \cos^{-1} 0 = c$

$0 + \dfrac{\pi}{2} = c$

Thus,

$\sin^{-1} x + \cos^{-1} x = \dfrac{\pi}{2}$.

47. $\dfrac{d}{dx}(x^2 y - 2y) = \dfrac{d}{dx}(4)$

$2xy + x^2 y' - 2y' = 0$

$y'(x^2 - 2) = -2xy$

$y' = \dfrac{-2xy}{x^2 - 2}$

The derivative is undefined at $x = \pm\sqrt{2}$, suggesting that there might be vertical tangent lines at these points. Similarly, $y' = 0$ at $y = 0$, suggesting that there might be a horizontal tangent line at this point.

However, plugging $x = \pm\sqrt{2}$ into the original equation gives $0 = 4$, a contradiction which shows that there are no points on this curve with x value $\pm\sqrt{2}$. Likewise, plugging $y = 0$ into the original equation gives $0 = 4$. Again, this is a contradiction which shows that there are no points on the graph with y value of 4.

Sketching the graph, we see that there is a horizontal asymptote at $y = 0$ and vertical asymptotes at $x = \pm\sqrt{2}$.

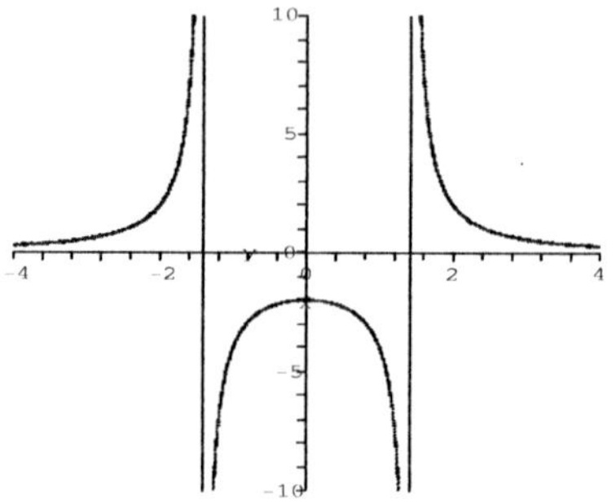

49. If $y_1 = c/x$, then $y_1' = -c/x^2 = -y_1/x$. If $y_2^2 = x^2 + k$, then $2y_2(y_2') = 2x$ and $y_2' = x/y_2$. If we are at a particular point (x_0, y_0) on both graphs, this means $y_1(x_0) = y_0 = y_2(x_0)$ and

$y_1' \cdot y_2' = \left(\dfrac{-y_0}{x_0}\right) \cdot \left(\dfrac{x_0}{y_0}\right) = -1$

This means that the slopes are negative reciprocals and the curves are orthogonal.

51. For the first type of curve, $y' = 3cx^2$.

For the second type of curve, $2x + 6yy' = 0$, and

$y' = \dfrac{-2x}{6y} = \dfrac{-x}{3y}$

$= \dfrac{-x}{3cx^3} = \dfrac{-1}{3cx^2}$.

These are negative reciprocals of each other, so the families of curves are orthogonal.

53. Conjecture: The family of functions $\{y_1 = cx^n\}$ is orthogonal to the family of functions $\{x^2 + ny_2^2 = k\}$ whenever $n \neq 0$.

If $y_1 = cx^n$, then $y_1' = cnx^{n-1} = ny_1/x$. If $ny_2^2 = -x^2 + k$, then $2ny_2(y_2') = -2x$ and $y_2' = -x/ny_2$. If we are at a particular point (x_0, y_0) on both graphs, this means $y_1(x_0) = y_0 = y_2(x_0)$ and

$y_1' \cdot y_2' = \left(\dfrac{ny_0}{x_0}\right) \cdot \left(-\dfrac{x_0}{ny_0}\right) = -1.$

This means that the slopes are negative reciprocals and the curves are orthogonal.

55. In example 8.6, we are given
$\theta'(d) = \dfrac{2(-130)}{4+d^2}$.
Setting this equal to -3 and solving for d gives $d^2 = 82 \Rightarrow d \approx 9$ft. The batter can track the ball after they would have to start swinging (when the ball is 30 feet away), but not all the way to home plate.

57. The viewing angle is given by the formula
$\theta(x) = \tan^{-1}(3/x) - \tan^{-1}(1/x)$.
This will be maximum where the derivative is zero.
$\theta'(x) = \dfrac{1}{1+(3/x)^2} \cdot \dfrac{-3}{x^2} - \dfrac{1}{1+1/x^2} \cdot \dfrac{-1}{x^2}$
$= \dfrac{1}{1+x^2} - \dfrac{3}{9+x^2}$.
This is zero when

$$\frac{1}{1+x^2} = \frac{3}{9+x^2} \Rightarrow x^2 = 3 \Rightarrow x = \sqrt{3}.$$

2.9 The Mean Value Theorem

1. $f(x) = x^2 + 1, [-2, 2]$
$f(-2) = 5 = f(2)$
As a polynomial, $f(x)$ is continuous on $[-2, 2]$, differentiable on $(-2, 2)$, and the conditions of Rolle's Theorem hold. There exists $c \in (-2, 2)$ such that $f'(c) = 0$. But $f'(c) = 2c$, $\Rightarrow c = 0$.

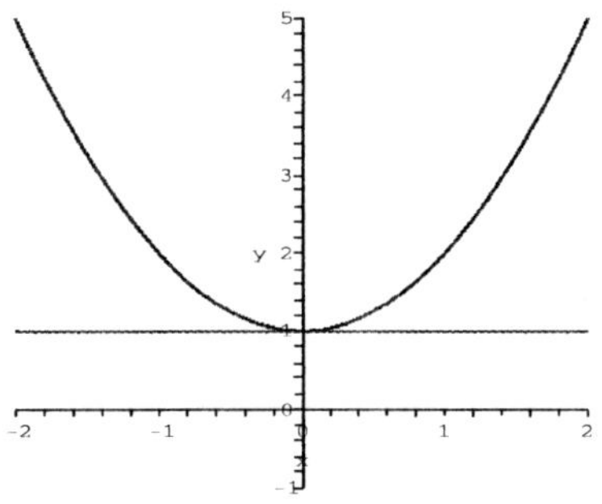

3. $f(x) = x^3 + x^2$, on $[0, 1]$, with $f(0) = 0$, $f(1) = 2$. As a polynomial $f(x)$ is continuous on $[0, 1]$ and differentiable on $(0, 1)$. Since the conditions of the Mean Value Theorem hold there exists a number $c \in (0, 1)$ such that
$f'(c) = \dfrac{f(1) - f(0)}{1 - 0} = \dfrac{2-0}{1-0} = 2$.
But $f'(c) = 3c^2 + 2c$.
$\Rightarrow 3c^2 + 2c = 2$,
$3c^2 + 2c - 2 = 0$.
By the quadratic formula
$c = \dfrac{-2 \pm \sqrt{2^2 - 4(3)(-2)}}{2(3)}$
$= \dfrac{-2 \pm \sqrt{28}}{6}$
$= \dfrac{-2 \pm 2\sqrt{7}}{6} = \dfrac{-1 \pm \sqrt{7}}{3}$
$\Rightarrow c \approx -1.22$ or $c \approx 0.55$
But since $-1.22 \notin (0, 1)$ we accept only the other alternative:
$c = \dfrac{-1+\sqrt{7}}{3} \approx 0.55$

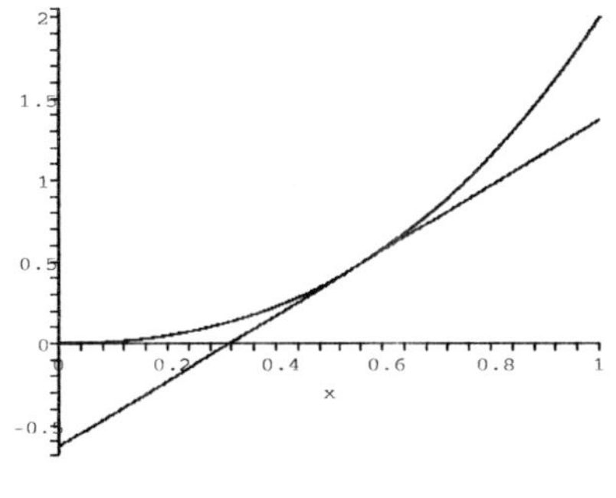

5. $f(x) = \sin x$, $[0, \pi/2]$,

$f(0) = 0$, $f(\pi/2) = 1$.
As a trig function, $f(x)$ is continuous on $[0, \pi/2]$ and differentiable on $(0, \pi/2)$. The conditions of the Mean Value Theorem hold, and there exists $c \in (0, \pi/2)$ such that

$$f'(c) = \frac{f\left(\frac{\pi}{2}\right) - f(0)}{\frac{\pi}{2} - 0} = \frac{1-0}{\frac{\pi}{2}-0} = \frac{2}{\pi}.$$

But $f'(c) = \cos(c)$ and c is to be in the first quadrant, therefore

$$c = \cos^{-1}\left(\frac{2}{\pi}\right) \approx .88$$

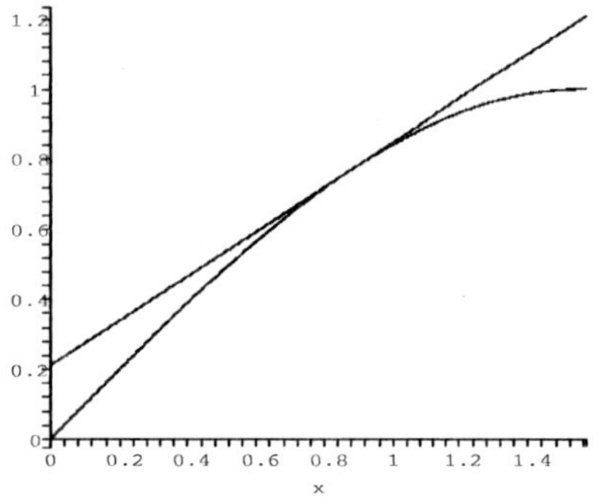

7. If $f'(x) > 0$ for all x then for each (a, b) with $a < b$ we know there exists a $c \in (a, b)$ such that

$$\frac{f(b) - f(a)}{b - a} = f'(c) > 0.$$

$a < b$ makes the denominator positive, and so we must have the numerator also positive, which implies $f(a) < f(b)$.

9. $f'(x) = 3x^2 + 5$. This is positive for all x, so $f(x)$ is increasing.

11. $f'(x) = -3x^2 - 3$. This is negative for all x, so $f(x)$ is decreasing.

13. $f'(x) = e^x$. This is positive for all x, so $f(x)$ is increasing.

15. $f'(x) = \frac{1}{x}$

$f'(x) > 0$ for $x > 0$, that is, for all x in the domain of f. So $f(x)$ is increasing.

17. Let $f(x) = x^3 + 5x + 1$. As a polynomial, $f(x)$ is continuous and differentiable for all x, with $f'(x) = 3x^2 + 5$, which is positive for all x so $f(x)$ is strictly increasing for all x. Therefore the equation can have at most one solution.

Since $f(x)$ is negative at $x = -1$ and positive at $x = 1$, and $f(x)$ is continuous, there must be a solution to $f(x) = 0$.

19. Let $f(x) = x^4 + 3x^2 - 2$. The derivative is $f'(x) = 4x^3 + 6x$. This is negative for negative x, and positive for positive x so $f(x)$ is strictly decreasing on $(-\infty, 0)$ and strictly increasing on $(0, \infty)$. Since $f(0) = -2 \neq 0$, $f(x)$ can have at most one zero for $x < 0$ and one zero for $x > 0$. The function is continuous everywhere and $f(-1) = 2 = f(1)$, therefore $f(x) = 0$ has exactly one solution between $x = -1$ and $x = 0$, exactly one solution between $x = 0$ and $x = 1$, and no other solutions.

21. $f(x) = x^3 + ax + b$, $a > 0$. Any cubic (actually any *odd degree*) polynomial heads in opposite directions $(\pm\infty)$ as x goes to the oppositely signed infinities, and therefore by the Intermediate Value Theorem has at least one root. For the uniqueness, we look at the derivative, in this case $3x^2 + a$. Because $a > 0$ by assumption, this expression is strictly positive. The function is strictly increasing and can have at most one root.

23. $f(x) = x^5 + ax^3 + bx + c$, $a > 0$, $b > 0$

Here is another odd degree polynomial (see #21) with at least one root.

$f'(x) = 5x^4 + 3ax^2 + b$ is evidently strictly positive because of our assumptions about a, b. Exactly as in #21, there can be at most one root.

25. The average velocity on $[a, b]$ is $\frac{s(b) - s(a)}{b - a}$

By the Mean Value theorem, there exists a $c \in (a, b)$ such that

$s'(c) = \frac{s(b) - s(a)}{b - a}$

Thus, the instantaneous velocity at $t = c$ is equal to the average velocity between times $t = a$ and $t = b$.

27. Define $h(x) = f(x) - g(x)$. Then h is differentiable because f and g are, and $h(a) = h(b) = 0$. Apply Rolle's theorem to h on $[a, b]$ to conclude that there exists $c \in (a, b)$ such that $h'(c) = 0$. Thus, $f'(c) = g'(c)$, and so f and g have parallel tangent lines at $x = c$.

29. $f(x) = x^2$

One candidate: $g_0(x) = kx^3$

Because we require $x^2 = g_0'(x) = 3kx^2$, we must have $3k = 1$, $k = 1/3$.

Most general solution:

$g(x) = g_0(x) + c = x^3/3 + c$

where c is an arbitrary constant.

31. Although the obvious first candidate is $g_0(x) = -1/x$,due to the disconnection of the domain by the discontinuity at $x = 0$, we could add *different* constants, one for negative x, another for positive x. Thus the most general solution is:

$$g(x) = \begin{cases} -1/x + a & \text{for } x > 0 \\ -1/x + b & \text{for } x < 0. \end{cases}$$

33. If $g'(x) = \sin x$, then $g(x) = -\cos x + c$ for any constant c.

35. $f(x) = 1/x$ on $[-1, 1]$. We easily see that $f(1) = 1$, $f(-1) = -1$, and $f'(x) = -1/x^2$. If we try to find the c in the interval $(-1, 1)$ for which

$$f'(c) = \frac{f(1) - f(1)}{1 - (-1)} = \frac{1 - (-1)}{1 - (-1)} = 1,$$

the equation would be $-1/c^2 = 1$ or $c^2 = -1$. There is of course no such c, and the explanation is that the function is not defined for $x = 0 \in (-1, 1)$ and so the function is not continuous.

The hypotheses for the Mean Value Theorem are not fulfilled.

37. $f(x) = \tan x$ on $[0, \pi]$, $f'(x) = \sec^2(x)$. We know the tangent has a massive discontinuity at $x = \pi/2$, so as in #35, we should not be surprised if the Mean Value Theorem does not apply. As applied to the interval $[0, \pi]$ it would say

$$\sec^2(c) = f'(c) = \frac{f(\pi) - f(0)}{\pi - 0} = \frac{\tan \pi - \tan 0}{\pi - 0} = 0.$$

But secant $= 1/$cosine is never 0 in the interval $(-1, 1)$, so no such c exists.

39. If a derivative g' is positive at a single point $x = b$, then $g(x)$ is an increasing function for x sufficiently near b, i.e., $g(x) > g(b)$ for $x > b$ but sufficiently near b. In this problem, we will apply that remark to f' at $x = 0$, and conclude from $f''(0) > 0$ that $f'(x) > f'(0) = 0$ for $x > 0$ but sufficiently small. This being true about the derivative f', it tells us that f itself is increasing on some interval $(0, a)$ and in particular that $f(x) > f(0) = 0$ for $0 < x < a$. On the other side (the negative side) f' is negative, f is decreasing (to zero) and therefore likewise positive. In summary, $x = 0$ is a genuine relative minimum.

41. Consider the function $g(x) = x - \sin(x)$, obviously with $g(0) = 0$ and $g'(x) = 1 - \cos(x)$. If there was ever a point $a > 0$ with $\sin(a) \geq a$, ($g(a) \leq 0$), then by the MVT applied to g on the interval $[0, a]$, there would be a point c ($0 < c < a$) with
$g'(c) = \dfrac{g(a) - g(0)}{a - 0} = \dfrac{g(a)}{a} \leq 0.$
This would read $1 - \cos(c) = g'(c) \leq 0$ or $\cos(c) \geq 1$. The latter condition is possible only if $\cos(c) = 1$ and $\sin(c) = 0$, in which case c (being positive) would be *at minimum* π. But even in this unlikely case we still would have $\sin(a) \leq 1 < \pi \leq c < a$.

Since $\sin a < a$ for all $a > 0$, we have $-\sin a > -a$ for all $a > 0$, but $-\sin a = \sin(-a)$ so we have $\sin(-a) > -a$ for all $a > 0$. This is the same as saying $\sin a > a$ for all $a < 0$ so in absolute value we have $|\sin a| < |a|$ for all $a \neq 0$.

Thus the only possible solution to the equation $\sin x = x$ is $x = 0$, which we know to be true.

43. Since the inverse sine function is increasing on the interval $[0, 1)$ (it has a positive derivative) we start from the previously proven inequality $\sin(x) < x$ for $0 < x$. If indeed $0 < x < 1$, we can apply the inverse sine and conclude
$x = \sin^{-1}(\sin(x)) < \sin^{-1}(x)$.

45. $f(x) = \begin{cases} 2x & x \leq 0 \\ 2x - 4 & x > 0 \end{cases}$

$f(x) = 2x - 4$ is continuous and differentiable on $(0, 2)$. Also, $f(0) = 0 = f(2)$. But $f'(x) \equiv 2$ on $(0, 2)$, so there is no c such that $f'(c) = 0$. Rolle's Theorem requires that $f(x)$ be continuous on the closed interval, but we have a jump discontinuity at $x = 0$, which is enough to preclude the applicability of Rolle's.

Ch. 2 Review Exercises

1. $\dfrac{3.4 - 2.6}{1.5 - 0.5} = \dfrac{0.8}{1} = 0.8$

3.
$$\begin{aligned} f'(2) &= \frac{f(2+h) - f(2)}{h} \\ &= \lim_{h \to 0} \frac{(2+h)^2 - 2(2+h) - (0)}{h} \\ &= \lim_{h \to 0} \frac{4 + 4h + h^2 - 4 - 2h}{h} \\ &= \lim_{h \to 0} \frac{2h + h^2}{h} \\ &= \lim_{h \to 0} 2 + h = 2 \end{aligned}$$

5.
$$\begin{aligned} f'(1) &= \lim_{h \to 0} \frac{f(1+h) - f(1)}{h} \\ &= \lim_{h \to 0} \frac{\sqrt{1+h} - 1}{h} \\ &= \lim_{h \to 0} \frac{\sqrt{1+h} - 1}{h} \cdot \frac{\sqrt{1+h} + 1}{\sqrt{1+h} + 1} \\ &= \lim_{h \to 0} \frac{1 + h - 1}{h(\sqrt{1+h} + 1)} \\ &= \lim_{h \to 0} \frac{1}{\sqrt{1+h} + 1} = \frac{1}{2} \end{aligned}$$

7.
$$\begin{aligned} f'(x) &= \lim_{h \to 0} \frac{f(x+h) - f(x)}{h} \\ &= \lim_{h \to 0} \frac{(x+h)^3 + (x+h) - (x^3 + x)}{h} \\ &= \lim_{h \to 0} \frac{3x^2h + 3xh^2 + h^3 + h}{h} \\ &= \lim_{h \to 0} 3x^2 + 3xh + h^2 + 1 \\ &= 3x^2 + 1 \end{aligned}$$

9. The point is $(1, 0)$. $y' = 4x^3 - 2$ so the slope at $x = 1$ is 2, and the equation of the tangent line is $y - 0 = 2(x - 1)$ or $y = 2x - 2$.

11. The point is $(0, 3)$. $y' = 6e^{2x}$, so the slope at $x = 0$ is 6, and the equation of the tangent line is $y - 3 = 6(x - 0)$ or $y = 6x + 3$.

13. Find the slope to $y - x^2y^2 = x - 1$ at $(1, 1)$.

$\frac{d}{dx}(y - x^2y^2) = \frac{d}{dx}(x - 1)$

$y' - 2xy^2 - x^2 2y \cdot y' = 1$

$y'(1 - x^2 2y) = 1 + 2xy^2$

$y' = \frac{1 + 2xy^2}{1 - 2x^2y}$

At $(1, 1)$:

$y' = \frac{1 + 2(1)(1)^2}{1 - 2(1)^2(1)} = \frac{3}{-1} = -3$

The equation of the tangent line is $y - 1 = -3(x - 1)$ or $y = -3x + 4$.

15. $s(t) = -16t^2 + 40t + 10$

$v(t) = s'(t) = -32t + 40$

$a(t) = v'(t) = -32$

17. $s(t) = 10e^{-2t}\sin 4t$

$v(t) = s'(t)$

$= 10\left(-2e^{-2t}\sin 4t + 4e^{-2t}\cos 4t\right)$

$a(t) = v'(t)$

$= 10\cdot(-2)\left[-2e^{-2t}\sin 4t + e^{-2t}4\cos 4t\right]$

$+10(4)\cdot\left[-2e^{-2t}\cos 4t - e^{-2t}4\sin 4t\right]$

$= 160e^{-2t}\cos 4t - 120e^{-2t}\sin 4t$

19. $v(t) = s'(t) = -32t + 40$

$v(1) = -32(1) + 40 = 8$

The ball is rising.

$v(2) = -32(2) + 40 = -24$

The ball is falling.

21. (a) $m_{\text{sec}} = \frac{f(2) - f(1)}{2 - 1}$

$= \frac{\sqrt{3} - \sqrt{2}}{1} \approx .318$

(b) $m_{\text{sec}} = \frac{f(1.5) - f(1)}{1.5 - 1}$

$= \frac{\sqrt{2.5} - \sqrt{2}}{.5} \approx .334$

(c) $m_{\text{sec}} = \frac{f(1.1) - f(1)}{1.1 - 1}$

$= \frac{\sqrt{2.1} - \sqrt{2}}{.1} \approx .349$

Best estimate for the slope of the tangent line: (c) (approximately .349).

23. $f'(x) = 4x^3 - 9x^2 + 2$

25. $f'(x) = -\frac{3}{2}x^{-3/2} - 10x^{-3}$

$= \frac{-3}{2x\sqrt{x}} - \frac{10}{x^3}$

27. $f'(t) = 2t(t + 2)^3 + t^2 \cdot 3(t + 2)^2 \cdot 1$

$= 2t(t + 2)^3 + 3t^2(t + 2)^2$

$= t(t + 2)^2(5t + 4)$

29. $g'(x) = \frac{(3x^2 - 1) \cdot 1 - x(6x)}{(3x^2 - 1)^2}$

$= \frac{3x^2 - 1 - 6x^2}{(3x^2 - 1)^2}$

$= -\frac{3x^2 + 1}{(3x^2 - 1)^2}$

31. $f'(x) = 2x\sin x + x^2\cos x$

33. $f'(x) = \sec^2\sqrt{x} \cdot \frac{1}{2\sqrt{x}}$

35. $f'(t) = \csc t \cdot 1 + t \cdot (-\csc t \cdot \cot t)$

$= \csc t - t\csc t\cot t$

37. $u'(x) = 2e^{-x^2}(-2x) = -4xe^{-x^2}$

39. $f'(x) = 1 \cdot \ln x^2 + x \cdot \frac{1}{x^2} \cdot 2x$

$= \ln x^2 + 2$

41. $f'(x) = \frac{1}{2\sqrt{\sin 4x}} \cdot \cos 4x \cdot 4$

43. $f'(x) = 2\left(\frac{x + 1}{x - 1}\right)\frac{d}{dx}\left(\frac{x + 1}{x - 1}\right)$

$= 2\left(\frac{x + 1}{x - 1}\right)\frac{(x - 1) - (x + 1)}{(x - 1)^2}$

$= 2\left(\frac{x + 1}{x - 1}\right)\frac{-2}{(x - 1)^2}$

$= \frac{-4(x + 1)}{(x - 1)^3}$

45. $f'(t) = e^{4t} \cdot 1 + te^{4t} \cdot 4 = (1+4t)e^{4t}$

47. $\dfrac{1}{\sqrt{1-(2x)^2}} \cdot 2$

49. $\dfrac{1}{1+(\cos 2x)^2} \cdot (-2\sin 2x)$

51. The derivative should look roughly like:

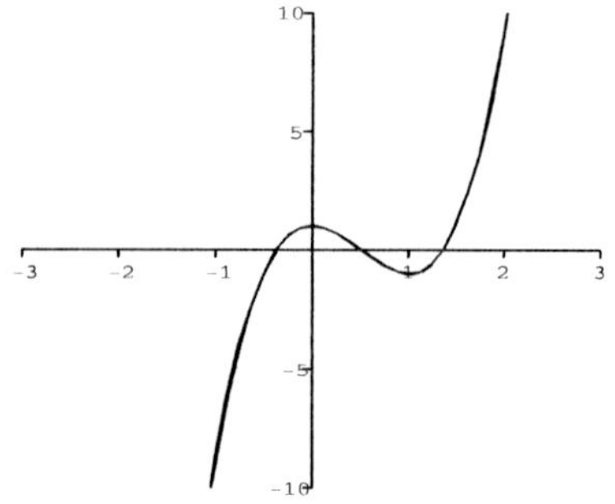

53. $f(x) = x^4 - 3x^3 + 2x^2 - x - 1$
$f'(x) = 4x^3 - 9x^2 + 4x - 1$
$f''(x) = 12x^2 - 18x + 4$

55. $f(x) = xe^{2x}$
$f'(x) = 1 \cdot e^{2x} + xe^{2x} \cdot 2 = e^{2x} + 2xe^{2x}$
$f''(x) = e^{2x} \cdot 2 + 2 \cdot (e^{2x} + 2xe^{2x})$
$= 4e^{2x} + 4xe^{2x}$
$f'''(x) = 4e^{2x} \cdot 2 + 4\,(e^{2x} + 2xe^{2x})$
$= 12e^{2x} + 8xe^{2x}$

57. $f(x) = \tan x$
$f'(x) = \sec^2 x$
$f''(x) = 2\sec x \cdot \sec x \tan x$
$= 2\sec^2 x \tan x$

59. $f(x) = \sin 3x$
$f'(x) = \cos 3x \cdot 3 = 3\cos 3x$
$f''(x) = 3(-\sin 3x \cdot 3) = -9\sin 3x$
$f'''(x) = -9\cos 3x \cdot 3 = -27\cos 3x$
$f^{(26)}(x) = -3^{26}\sin 3x$

61. $R(t) = P(t)Q(t)$
$R'(t) = Q'(t) \cdot P(t) + Q(t) \cdot P'(t)$
$P(0) = 2.4(\$)$
$Q(0) = 12$ (thousands)
$Q'(t) = -1.5$ (thousands per year)
$P'(t) = 0.1$ ($ per year)
$R'(0) = (-1.5) \cdot (2.4) + 12 \cdot (0.1)$
$= -2.4$ (thousand $ per year)
Revenue is decreasing at a rate of $2400 per year.

63. $f(t) = 4\cos 2t$
$v(t) = f'(t) = 4(-\sin 2t) \cdot 2$
$= -8\sin 2t$

(a) The velocity is zero when $v(t) = -8\sin 2t = 0$, i.e., when $2t = 0, \pi, 2\pi, \ldots$ so when $t = 0, \pi/2, \pi, 3\pi/2, \ldots$
$f(t) = 4$ for $t = 0, \pi, 2\pi, \ldots$
$f(t) = 4\cos 2t = -4$ for $t = \pi/2, 3\pi/2, \ldots$
The position of the spring when the velocity is zero is 4 or -4.

(b) The velocity is a maximum when $v(t) = -8\sin 2t = 8$, i.e., when $2t = 3\pi/2, 7\pi/2, \ldots$ so $t = 3\pi/4, 7\pi/4, \ldots$
$f(t) = 4\cos 2t = 0$ for $t = 3\pi/4, 7\pi/4, \ldots$
The position of the spring when the velocity is at a maximum is zero.

(c) Velocity is at a minimum when $v(t) = -8\sin 2t = -8$, i.e., when $2t = \pi/2, 5\pi/2, \ldots$ so $t = \pi/4, 5\pi/4, \ldots$
$f(t) = 4\cos 2t = 0$ for $t = \pi/4, 5\pi/4, \ldots$
The position of the spring when the velocity is at a minimum is also zero.

65. $\dfrac{d}{dx}(x^2y - 3y^3) = \dfrac{d}{dx}(x^2+1)$
$2xy + x^2y' - 3 \cdot 3y^2 \cdot y' = 2x$
$y'(x^2 - 9y^2) = 2x - 2xy$
$y' = \dfrac{2x(1-y)}{x^2 - 9y^2}$

67. $\dfrac{d}{dx}\left(\dfrac{y}{x+1} - 3y\right) = \dfrac{d}{dx}\tan x$

$\dfrac{(x+1)y' - y\cdot(1)}{(x+1)^2} - 3y' = \sec^2 x$

$y'(x+1) - y = (x+1)^2(3y' + \sec^2 x)$

$y' = \dfrac{\sec^2 x(x+1)^2 + y}{(x+1)[1 - 3(x+1)]}$

69. When $x = 0$, $-3y^3 = 1$, $y = \frac{-1}{\sqrt[3]{3}}$ (call this a).

From our formula (#65), we find $y' = 0$ at this point. To find y'', implicitly differentiate the first derivative (second line in #65):

$$2(xy' + y) + (2xy' + x^2y'') - 9\left[2y(y')^2 + y^2y''\right] = 2$$

At $(0, a)$ with $y' = 0$, we find

$2a - 9a^2y'' = 2$,

$y'' = \dfrac{-2\sqrt[3]{3}}{9}\left(\sqrt[3]{3} + 1\right)$

Below is a sketch of the graph of $x^2y - 3y^3 = x^2 + 1$.

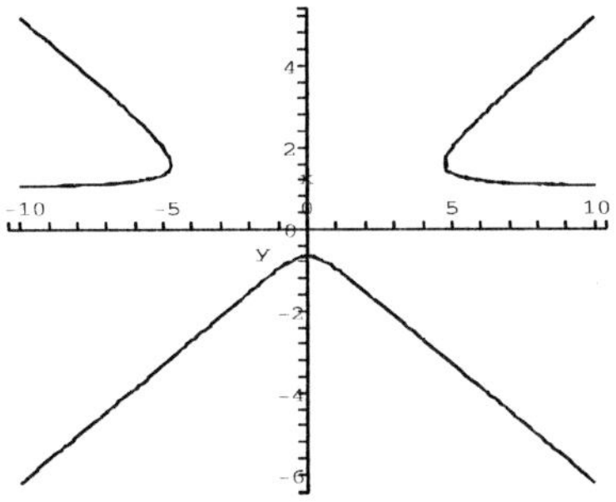

71. $y' = 3x^2 - 12x = 3x(x-4)$

(a) $y' = 0$ for $x = 0$ $(y = 1)$, and $x = 4$ $(y = -31)$ so there are horizontal tangent lines at $(0,1)$ and $(4,-31)$.

(b) y' is defined for all x, so there are no vertical tangent lines.

73. $\dfrac{d}{dx}(x^2y - 4y) = \dfrac{d}{dx}x^2$

$2xy + x^2y' - 4y' = 2x$

$y'(x^2 - 4) = 2x - 2xy$

$y' = \dfrac{2x - 2xy}{x^2 - 4} = \dfrac{2x(1-y)}{x^2-4}$

(a) $y' = 0$ when $x = 0$ or $y = 1$.
At $y = 1$, $x^2 \cdot 1 - 4 \cdot 1 = x^2$
$x^2 - 4 = x^2$
This is impossible, so there is no x for which $y = 1$.
At $x = 0$, $0^2 \cdot y - 4y = 0^2$, so $y = 0$.
Therefore, there is a horizontal tangent line at $(0,0)$.

(b) y' is not defined when $x^2 - 4 = 0$, or $x = \pm 2$. At $x = \pm 2$, $4y - 4y = 4$ so the function is not defined at $x = \pm 2$. There are no vertical tangent lines.

75. $f(x)$ is continuous and differentiable for all x, and $f'(x) = 3x^2 + 7$, which is positive for all x. By Theorem 9.2, if the equation $f(x) = 0$ has two solutions, then $f'(x) = 0$ would have at least one solution, but it has none. We discussed at length (Section 2.9) why every odd degree polynomial has at least one root, so in this case there is exactly one root.

77. $f(x) = x^5 + 2x^3 - 1$ is a one-to-one function with $f(1) = 2$, $f'(1) = 11$. If g is the name of the inverse, then $g(2) = 1$ and

$g'(2) = \dfrac{1}{f'(g(2))} = \dfrac{1}{f'(1)} = \dfrac{1}{11}$.

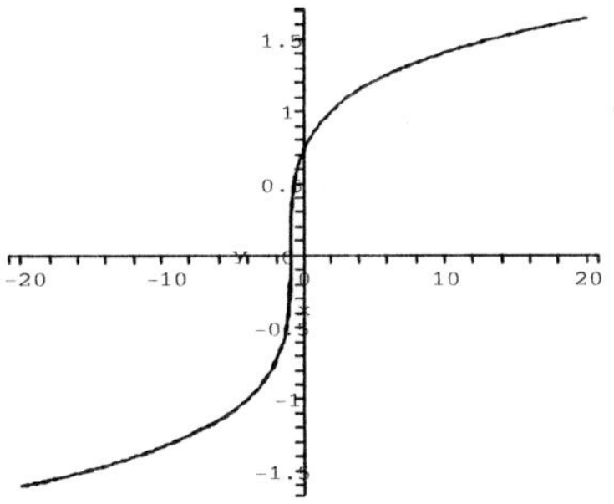

79. Let $a > 0$. We know that $f(x) = \cos x - 1$ is continuous and differentiable on the interval $(0, a)$. Also $f'(x) = \sin x \leq 1$ for all x. The Mean Value Theorem implies that there exists some c in the interval $(0, a)$ such that $f'(c) = \sin c$. But

$$f'(c) = \frac{\cos a - 1 - (\cos 0 - 1)}{a - 0} = \frac{\cos a - 1}{a}.$$

Since this is equal to $\sin c$ and $\sin c \leq 1$ for any c, we get that

$$\cos a - 1 \leq a$$

as desired. This works for all positive a, but since $\cos x - 1$ is symmetric about the y axis, we get

$$|\cos x - 1| \leq |x|.$$

They are actually equal at $x = 0$.

81. To show that $g(x)$ is continuous at $x = a$, we need to show that the limit as x approaches a of $g(x)$ exists and is equal to $g(a)$. But

$$\lim_{x \to a} g(x) = \lim_{x \to a} \frac{f(x) - f(a)}{x - a},$$

which is the definition of the derivative of $f(x)$ at $x = a$. Since $f(x)$ is differentiable at $x = a$, we know this limit exists and is equal to $f'(a)$, which, in turn, is equal to $g(a)$. Thus $g(x)$ is continuous at $x = a$.

83. $f(x) = x^2 - 2x$ on $[0, 2]$
$f(2) = 0 = f(0)$
If $f'(c) = \dfrac{f(2) - f(0)}{2 - 0} = \dfrac{0 - 0}{2} = 0$
then $2c - 2 = f'(c) = 0$ so $c = 1$.

85. $f(x) = 3x^2 - \cos x$
One trial: $g_o(x) = kx^3 - \sin x$
$g_o'(x) = 3kx^2 - \cos x$
Need $3k = 3$, $k = 1$, and the general solution is
$g(x) = g_o(x) + c = x^3 - \sin x + c$
for c an arbitrary constant.

87. $x = 1$ is to be double root of

$$\begin{aligned} f(x) &= (x^3 + 1) - [m(x - 1) + 2] \\ &= (x^3 + 1 - 2) - m(x - 1) \\ &= (x^3 - 1) - m(x - 1) \\ &= (x - 1)\left[x^2 + x + 1 - m\right] \end{aligned}$$

Let $g(x) = x^2 + x + 1 - m$. Then $x = 1$ is a *double* root of f only if $(x - 1)$ is a *factor* of g, in which case $g(1) = 0$. Therefore we require $0 = g(1) = 3 - m$ or $m = 3$. Now $g(x) = x^2 + x - 2 = (x - 1)(x + 2)$, $f(x) = (x - 1)g(x) = (x - 1)^2(x + 2)$ and $x = 1$ is a double root.

The line tangent to the curve $y = x^3 + 1$ at the point $(1, 2)$ has slope $y' = 3x^2 = 3(1) = 3(= m)$. The equation of the tangent line is $y - 2 = 3(x - 1)$ or $y = 3x - 1(= m(x - 1) + 2)$.

Chapter 3

Applications of Differentiation

3.1 Linear Approximations and Newton's Method

1. $f(x_0) = f(1) = \sqrt{1} = 1$
$f'(x) = \frac{1}{2}x^{-1/2}$
$f'(x_0) = f'(1) = \frac{1}{2}$
So
$$\begin{aligned} L(x) &= f(x_0) + f'(x_0)(x - x_0) \\ &= 1 + \tfrac{1}{2}(x - 1) \\ &= \tfrac{1}{2} + \tfrac{1}{2}x \end{aligned}$$

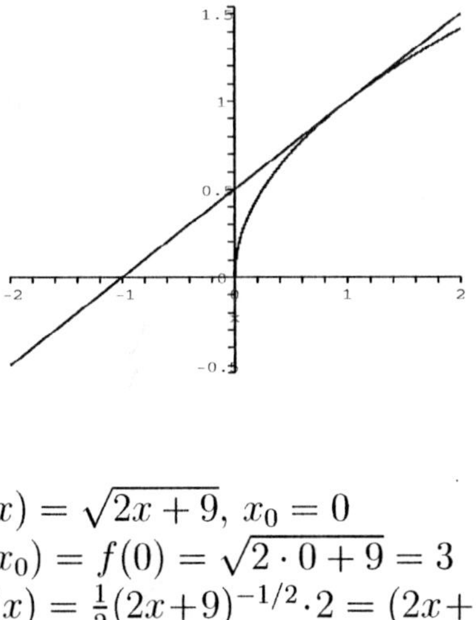

3. $f(x) = \sqrt{2x+9}$, $x_0 = 0$
$f(x_0) = f(0) = \sqrt{2 \cdot 0 + 9} = 3$
$f'(x) = \frac{1}{2}(2x+9)^{-1/2} \cdot 2 = (2x+9)^{-1/2}$
$f'(x_0) = f'(0) = (2 \cdot 0 + 9)^{-1/2} = \frac{1}{3}$
So
$$\begin{aligned} L(x) &= f(x_0) + f'(x_0)(x - x_0) \\ &= 3 + \tfrac{1}{3}(x - 0) \\ &= 3 + \tfrac{1}{3}x \end{aligned}$$

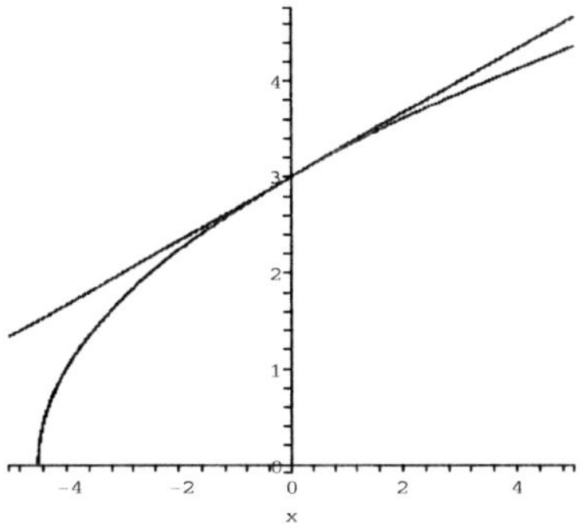

5. $f(x) = \sin 3x$, $x_0 = 0$
$f(x_0) = f(0) = \sin(3 \cdot 0) = \sin 0 = 0$
$f'(x) = 3\cos 3x$
$f'(x_0) = f'(0) = 3\cos 3 \cdot 0 = 3$
$$\begin{aligned} L(x) &= f(x_0) + f'(x_0)(x - x_0) \\ &= 0 + 3(x - 0) \\ &= 3x \end{aligned}$$

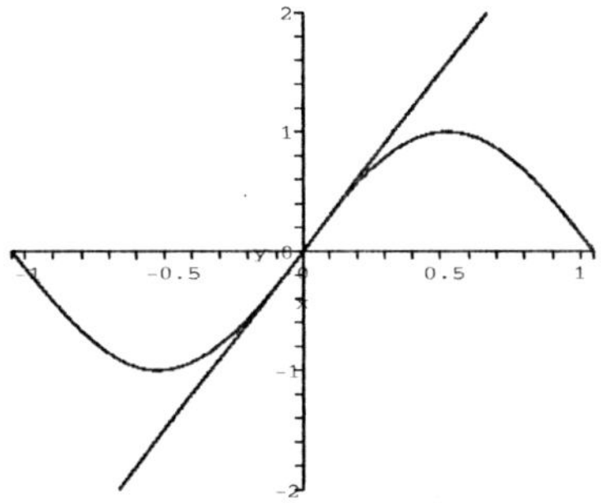

7. (a) $f(0) = g(0) = h(0) = 1$, so all pass through the point $(0, 1)$.
$f'(0) = 2(0 + 1) = 2$,
$g'(0) = 2\cos(2 \cdot 0) = 2$, and
$h'(0) = 2e^{2 \cdot 0} = 2$,
so all have slope 2 at $x = 0$.
The linear approximation at $x = 0$ for all three functions is
$L(x) = 1 + 2x$.

(b) Graph of $f(x) = (x + 1)^2$:

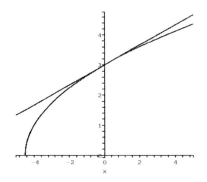

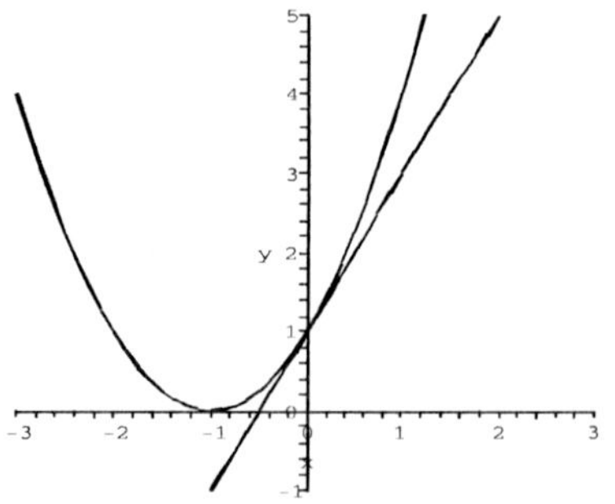

Graph of $f(x) = 1 + \sin(2x)$:

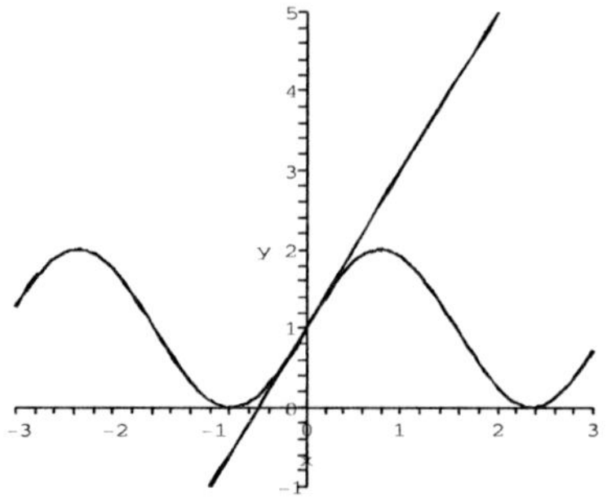

Graph of $f(x) = e^{2x}$:

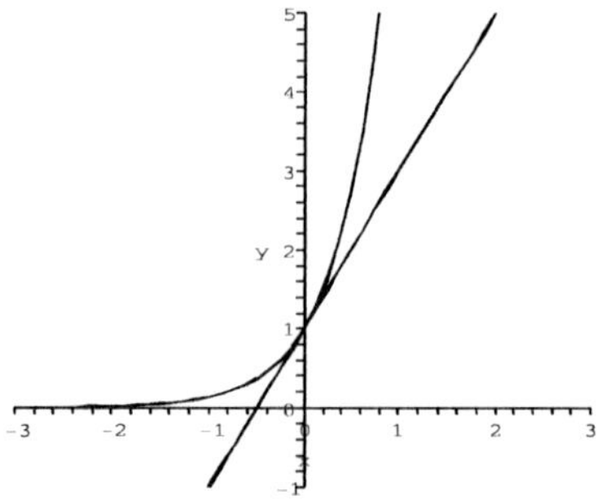

9. (a) $f(x) = \sqrt[4]{16 + x}, x_0 = 0$
$f(0) = \sqrt[4]{16 + 0} = 2$
$f'(x) = \frac{1}{4}(16 + x)^{-3/4}$
$f'(0) = \frac{1}{4}(16 + 0)^{-3/4} = \frac{1}{32}$
$L(x) = f(0) + f'(0)(x - 0)$
$= 2 + \frac{1}{32}x$
$L(0.04) = 2 + \frac{1}{32}(0.04) = 2.00125$

(b) $L(0.08) = 2 + \frac{1}{32}(0.08) = 2.0025$

(c) $L(0.16) = 2 + \frac{1}{32}(0.16) = 2.005$

11. (a) $\sqrt[4]{16.04} = 2.0012488$
$L(0.04) = 2.00125$
$|2.0012488 - 2.00125|$
$= .00000117$

(b) $\sqrt[4]{16.08} = 2.0024953$
$L(.08) = 2.0025$
$|2.0024953 - 2.0025|$
$= .00000467$

(c) $\sqrt[4]{16.16} = 2.0049814$
$L(.16) = 2.005$
$|2.0049814 - 2.005| = .0000186$

13. (a) $L(x) = f(20) + \frac{18-14}{20-30}(x - 20)$
$L(24) \approx 18 - \frac{4}{10}(24 - 20)$
$= 18 - 0.4(4)$
$= 16.4$ games

(b) $L(x) = f(40) + \frac{14-12}{30-40}(x - 40)$
$f(36) \approx 12 - \frac{2}{10}(36 - 40)$
$= 12 - 0.2(-4)$
$= 12.8$ games

15. (a) $L(x) = f(200) + \frac{142-128}{220-200}(x - 200)$
$L(208) = 128 + \frac{14}{20}(208 - 200)$
$= 128 + 0.7(8) = 133.6$

(b) $L(x) = f(240) + \frac{142-136}{220-240}(x - 240)$
$L(232) = 136 - \frac{6}{20}(232 - 240)$
$= 136 - 0.3(-8) = 138.4$

17. The first tangent line intersects the x-axis at a point a little to the right of 1. So x_1 is about 1.25 (very roughly). The second tangent line intersects the x-axis at a point between 1 and x_1, so x_2 is about 1.1 (very roughly). Newton's Method will converge to the zero at $x = 1$.

19. It wouldn't work because $f'(0) = 0$.

21. $f(x) = x^3 + 3x^2 - 1 = 0$, $x_0 = 1$
$f'(x) = 3x^2 + 6x$

(a) $x_1 = x_0 - \dfrac{f(x_0)}{f'(x_0)}$
$= 1 - \dfrac{1^3 + 3 \cdot 1^2 - 1}{3 \cdot 1^2 + 6 \cdot 1}$
$= 1 - \dfrac{3}{9} = \dfrac{2}{3}$

$$\begin{aligned} x_2 &= x_1 - \frac{f(x_1)}{f'(x_1)} \\ &= \frac{2}{3} - \frac{\left(\frac{2}{3}\right)^3 + 3\left(\frac{2}{3}\right)^2 - 1}{3\left(\frac{2}{3}\right)^2 + 6\left(\frac{2}{3}\right)} \\ &= \frac{79}{144} \approx 0.5486 \end{aligned}$$

(b) 0.53209

23. $f(x) = x^4 - 3x^2 + 1 = 0$, $x_0 = 1$
$f'(x) = 4x^3 - 6x$

$$\begin{aligned} \text{(a) } x_1 &= x_0 - \frac{f(x_0)}{f'(x_0)} \\ &= 1 - \left(\frac{1^4 - 3\cdot 1^2 + 1}{4\cdot 1^3 - 6\cdot 1}\right) = \frac{1}{2} \\ x_2 &= x_1 - \frac{f(x_1)}{f'(x_1)} \\ &= \frac{1}{2} - \left(\frac{\left(\frac{1}{2}\right)^4 - 3\left(\frac{1}{2}\right)^2 + 1}{4\left(\frac{1}{2}\right)^3 - 6\left(\frac{1}{2}\right)}\right) \\ &= \frac{5}{8} \end{aligned}$$

(b) 0.61803

25. Use $x_{i+1} = x_i - \frac{f(x_i)}{f'(x_i)}$ with
$f(x) = x^3 + 4x^2 - 3x + 1$, and
$f'(x) = 3x^2 + 8x - 3$.

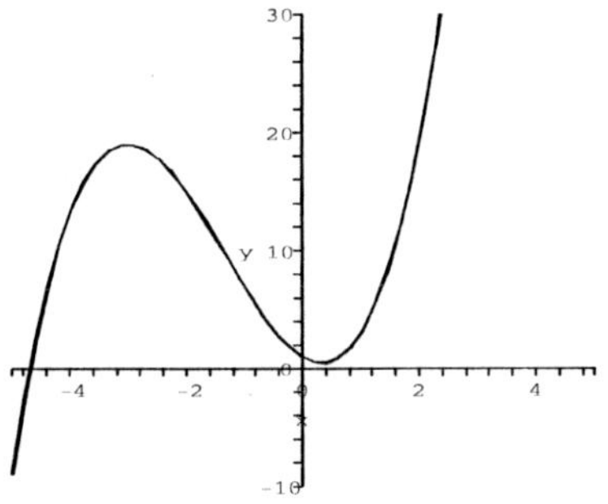

Start with $x_0 = -5$ to find the root near -5:
$x_1 = -4.718750$, $x_2 = -4.686202$,
$x_3 = -4.6857796$, $x_4 = -4.6857795$

27. Use $x_{i+1} = x_i - \frac{f(x_i)}{f'(x_i)}$ with
$f(x) = x^5 + 3x^3 + x - 1$, and
$f'(x) = 5x^4 + 9x^2 + 1$.

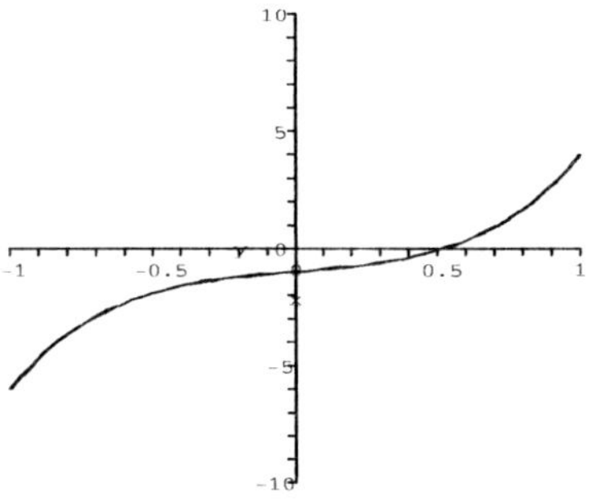

Start with $x_0 = 0.5$ to find the root near 0.5:
$x_1 = 0.526316$, $x_2 = 0.525262$,
$x_3 = 0.525261$, $x_4 = 0.525261$

29. Use $x_{i+1} = x_i - \frac{f(x_i)}{f'(x_i)}$ with
$f(x) = \sin x - x^2 + 1$, and
$f'(x) = \cos x - 2x$

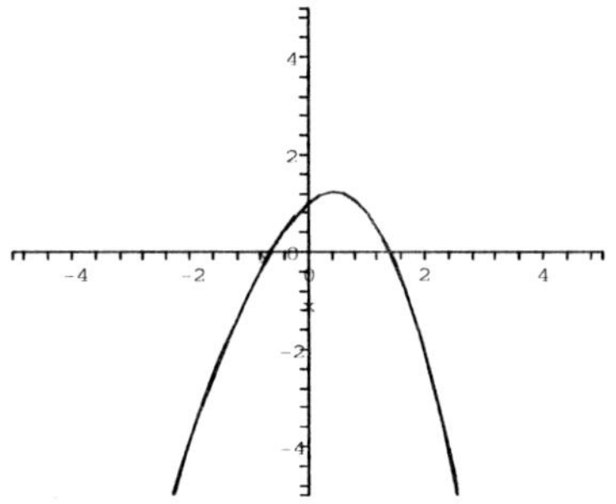

Start with $x_0 = -0.5$ to find the root near -0.5:
$x_1 = -0.644108$, $x_2 = -0.636751$
$x_3 = -0.636733$, $x_4 = -0.636733$

Start with $x_0 = 1.5$ to find the root near 1.5:
$x_1 = 1.413799$, $x_2 = 1.409634$
$x_3 = 1.409624$, $x_4 = 1.409624$

31. Use $x_{i+1} = x_i - \frac{f(x_i)}{f'(x_i)}$ with
$f(x) = e^x + x$, and
$f'(x) = e^x + 1$

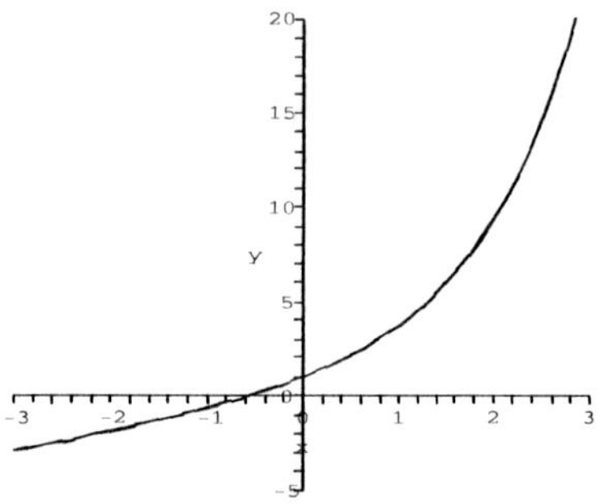

Start with $x_0 = -0.5$ to find the root between 0 and -1:
$x_1 = -0.566311$, $x_2 = -0.567143$
$x_3 = -0.567143$, $x_4 = -0.567143$

33. $$\begin{aligned} x_{n+1} &= x_n - \frac{f(x_n)}{f'(x_n)} \\ &= x_n - \left(\frac{x_n^2 - c}{2x_n}\right) \\ &= x_n - \frac{x_n^2}{2x_n} + \frac{c}{2x_n} \\ &= \frac{x_n}{2} + \frac{c}{2x_n} \\ &= \frac{1}{2}\left(x_n + \frac{c}{x_n}\right) \end{aligned}$$
If $x_0 < \sqrt{a}$, then $a/x_0 > \sqrt{a}$, so $x_0 < \sqrt{a} < a/x_0$.

35. $f(x) = x^2 - 11$; $x_0 = 3$; $\sqrt{11} \approx 3.316625$

37. $f(x) = x^3 - 11$; $x_0 = 2$; $\sqrt[3]{11} \approx 2.22398$

39. $f(x) = x^{4.4} - 24$; $x_0 = 2$; $\sqrt[4.4]{24} \approx 2.059133$

41. $f(x) = 4x^3 - 7x^2 + 1 = 0$, $x_0 = 0$
$f'(x) = 12x^2 - 14x$
$x_1 = x_0 - \frac{f(x_0)}{f'(x_0)} = 0 - \frac{1}{0}$
The method fails because $f'(x_0) = 0$
Roots are 0.3454, 0.4362, 1.659.

43. $f(x) = x^2 + 1$, $x_0 = 0$
$f'(x) = 2x$
$x_1 = x_0 - \frac{f(x_0)}{f'(x_0)} = 0 - \frac{1}{0}$
The method fails because $f'(x_0) = 0$. There are no roots.

45. $f(x) = \dfrac{4x^2 - 8x + 1}{4x^2 - 3x - 7} = 0$, $x_0 = -1$

Note: $f(x_0) = f(-1)$ is undefined, so Newton's Method fails because x_0 is not in the domain of f. Notice that $f(x) = 0$ only when $4x^2 - 8x + 1 = 0$. So using Newton's Method on $g(x) = 4x^2 - 8x + 1$ with $x_0 = -1$ leads to $x \approx .1339$. The other root is $x \approx 1.8660$.

47. (a) With $x_0 = 1.2$,
$x_1 = 0.800000000$,
$x_2 = 0.950000000$,
$x_3 = 0.995652174$,
$x_4 = 0.999962680$,
$x_5 = 0.999999997$,
$x_6 = 1.000000000$,
$x_7 = 1.000000000$

(b) With $x_0 = 2.2$,
$x_0 = 2.200000, x_1 = 2.107692$,
$x_2 = 2.056342, x_3 = 2.028903$,
$x_4 = 2.014652, x_5 = 2.007378$,
$x_6 = 2.003703, x_7 = 2.001855$,
$x_8 = 2.000928, x_9 = 2.000464$,
$x_{10} = 2.000232, x_{11} = 2.000116$,
$x_{12} = 2.000058, x_{13} = 2.000029$,
$x_{14} = 2.000015, x_{15} = 2.000007$,
$x_{16} = 2.000004, x_{17} = 2.000002$,
$x_{18} = 2.000001, x_{19} = 2.000000$,
$x_{20} = 2.000000$

The convergence is much faster with $x_0 = 1.2$.

49. (a) With $x_0 = -1.1$
$x_1 = -1.0507937$,
$x_2 = -1.0256065$,
$x_3 = -1.0128572$,
$x_4 = -1.0064423$,
$x_5 = -1.0032246$,
$x_6 = -1.0016132$,

$x_7 = -1.0008068,$
$x_8 = -1.0004035,$
$x_9 = -1.0002017,$
$x_{10} = -1.0001009,$
$x_{11} = -1.0000504,$
$x_{12} = -1.0000252,$
$x_{13} = -1.0000126,$
$x_{14} = -1.0000063,$
$x_{15} = -1.0000032,$
$x_{16} = -1.0000016,$
$x_{17} = -1.0000008,$
$x_{18} = -1.0000004,$
$x_{19} = -1.0000002,$
$x_{20} = -1.0000001,$
$x_{21} = -1.0000000,$
$x_{22} = -1.0000000$

(b) With $x_0 = 2.1$
$x_0 = 2.100000000,$
$x_1 = 2.006060606,$
$x_2 = 2.000024340,$
$x_3 = 2.000000000,$
$x_4 = 2.000000000$

The rate of convergence in (a) is slower than the rate of convergence in (b).

51. $f(x) = \tan x,\ f(0) = \tan 0 = 0$
$f'(x) = \sec^2 x,\ f'(0) = \sec^2 0 = 1$
$L(x) = f(0) + f'(0)(x-0)$
$= 0 + 1(x-0) = x$
$L(0.01) = 0.01$
$f(0.01) = \tan 0.01 \approx 0.0100003$
$L(0.1) = 0.1$
$f(0.1) = \tan(0.1) \approx 0.1003$
$L(1) = 1$
$f(1) = \tan 1 \approx 1.557$

53. $f(x) = \sqrt{4+x}$
$f(0) = \sqrt{4+0} = 2$
$f'(x) = \frac{1}{2}(4+x)^{-1/2}$
$f'(0) = \frac{1}{2}(4+0)^{-1/2} = \frac{1}{4}$
$L(x) = f(0) + f'(0)(x-0) = 2 + \frac{1}{4}x$
$L(0.01) = 2 + \frac{1}{4}(0.01) = 2.0025$
$f(0.01) = \sqrt{4+0.01} \approx 2.002498$
$L(0.1) = 2 + \frac{1}{4}(0.1) = 2.025$
$f(0.1) = \sqrt{4+0.1} \approx 2.0248$
$L(1) = 2 + \frac{1}{4}(1) = 2.25$
$f(1) = \sqrt{4+1} \approx 2.2361$

55. If you graph $|\tan x - x|$, you see that the difference is less than .01 on the interval $-.306 < x < .306$ (In fact, a slightly larger interval would work as well.)

57. For small x we approximate e^x by $x+1$ (see exercise 54).

$$\frac{Le^{2\pi d/L} - e^{-2\pi d/L}}{e^{2\pi d/L} + e^{-2\pi d/L}} \approx \frac{L\left[\left(1+\frac{2\pi d}{L}\right) - \left(1 - \frac{2\pi d}{L}\right)\right]}{\left(1+\frac{2\pi d}{L}\right) + \left(1-\frac{2\pi d}{L}\right)} \approx \frac{L\left(\frac{4\pi d}{L}\right)}{2} = 2\pi d$$

$$f(d) \approx \frac{4.9}{\pi}\cdot 2\pi d = 9.8d$$

59. The smallest positive solution of the first equation is 0.132782, and for the second equation the smallest positive solution is 1, so the species modeled by the second equation is certain to go extinct. This is consistent with the models, since the expected number of offspring for the population modeled by the first equation is 2.2, while for the second equation it is only 1.3.

61. The only positive solution is 0.6407.

63. $W(x) = \dfrac{PR^2}{(R+x)^2}, x_0 = 0$

$$W'(x) = \frac{-2PR^2}{(R+x)^3}$$

$$L(x) = W(x_0) + W'(x_0)(x - x_0) = \frac{PR^2}{(R+0)^2} + \left(\frac{-2PR^2}{(R+0)^3}\right)(x-0) = P - \frac{2Px}{R}$$

$$L(x) = 120 - .01(120) = P - \frac{2Px}{R} = 120 - \frac{2\cdot 120x}{R}$$

$$.01 = \frac{2x}{R}$$
$$x = .005R = .005(20{,}900{,}000)$$
$$= 104{,}500 \text{ ft}$$

65. To find the smallest positive solution of $\tan(\sqrt{x}) = \sqrt{x}$, plot $f(x) = \tan(\sqrt{x}) - \sqrt{x}$ to see that it crosses the x-axis at approximately $x = 20$. Newton's method (3 iterations) leads to $L \approx 20.19$.
$y = \sqrt{L} - \sqrt{L}x - \sqrt{L}\cos\sqrt{L}x + \sin\sqrt{L}x$
$= 4.493 - 4.493x - 4.493\cos 4.493x + \sin 4.493x$

67. The linear approximation for the inverse tangent function at $x = 0$ is
$f(x) \approx f(0) + f'(0)(x - 0)$
$\tan^{-1}(x) \approx \tan^{-1}(0) + \frac{1}{1+0^2}(x - 0)$
$\tan^{-1}(x) \approx x$
Using this approximation,
$$\phi = \tan^{-1}\left(\frac{3[1 - d/D] - w/2}{D - d}\right)$$
$$\phi \approx \frac{3[1 - d/D] - w/2}{D - d}$$
If $d = 0$, then $\phi \approx \frac{3-w/2}{D}$. Thus, if w or D increase, then ϕ decreases.

69. (a) As we should expect, when we start with $x_0 = 0.1$, Newton's method converges to 0.

(b) When we start with $x_0 = 1.1$, Newton's method converges to 1.

(c) When we start with $x_0 = 2.1$, Newton's method converges to 2.

3.2 Indeterminate Forms and L'Hôpital's Rule

1. $$\lim_{x\to-2} \frac{x+2}{x^2-4} = \lim_{x\to-2} \frac{x+2}{(x+2)(x-2)} = \lim_{x\to-2} \frac{1}{x-2} = -\frac{1}{4}$$

3. $$\lim_{x\to\infty} \frac{3x^2+2}{x^2-4} = \lim_{x\to\infty} \frac{3+\frac{2}{x^2}}{1-\frac{4}{x^2}} = \frac{3}{1} = 3$$

5. $\lim_{x\to0} \frac{e^{2x}-1}{x}$ is type $\frac{0}{0}$;
we apply L'Hôpital's Rule to get
$$\lim_{x\to0} \frac{2e^{2x}}{1} = \frac{2}{1} = 2.$$

7. $\lim_{x\to0} \frac{\tan^{-1}x}{\sin x}$ is type $\frac{0}{0}$;
we apply L'Hôpital's Rule to get
$$\lim_{x\to0} \frac{1/(1+x^2)}{\cos x} = \lim_{x\to0} \frac{1}{1} = 1.$$

9. $\lim_{x\to\pi} \frac{\sin 2x}{\sin x}$ is type $\frac{0}{0}$;
we apply L'Hôpital's Rule to get
$$\lim_{x\to\pi} \frac{2\cos 2x}{\cos x} = \frac{2(1)}{-1} = -2.$$

11. $\lim_{x\to0} \frac{\sin x - x}{x^3}$ is type $\frac{0}{0}$;
we apply L'Hôpital's Rule thrice to get
$$= \lim_{x\to0} \frac{\cos x - 1}{3x^2} = \lim_{x\to0} \frac{-\sin x}{6x}$$
$$= \lim_{x\to0} \frac{-\cos x}{6} = \frac{-1}{6}.$$

13. $$\lim_{x\to1} \frac{\sqrt{x}-1}{x-1} = \lim_{x\to1} \frac{\sqrt{x}-1}{x-1}\frac{\sqrt{x}+1}{\sqrt{x}+1} = \lim_{x\to1} \frac{x-1}{(x-1)(\sqrt{x}+1)} = \lim_{x\to1} \frac{1}{\sqrt{x}+1} = \frac{1}{2}$$

15. $\lim_{x\to\infty} \frac{x^3}{e^x}$ is type $\frac{\infty}{\infty}$;
we apply L'Hôpital's Rule thrice to get

$$\lim_{x\to\infty} \tfrac{3x^2}{e^x} = \lim_{x\to\infty} \tfrac{6x}{e^x} = \lim_{x\to\infty} \tfrac{6}{e^x} = 0.$$

17. $\lim_{x\to 0} \frac{x\cos x - \sin x}{x\sin^2 x}$ is type $\frac{\infty}{\infty}$;
we apply L'Hôpital's Rule twice to get

$$\lim_{x\to 0} \frac{\cos x - x\sin x - \cos x}{\sin^2 x + 2x\sin x\cos x} = \lim_{x\to 0} \frac{-x\sin x}{\sin x\,(\sin x + 2x\cos x)} = \lim_{x\to 0} \frac{-x}{\sin x + 2x\cos x} = \lim_{x\to 0} \frac{-1}{\cos x + 2\cos x - 2x\sin x} = -\frac{1}{3}.$$

19. $\lim_{x\to 1} \frac{\sin \pi x}{x-1}$ is type $\frac{\infty}{\infty}$;
we apply L'Hôpital's Rule to get

$$\lim_{x\to 1} \frac{\pi\cos\pi x}{1} = \frac{\pi(-1)}{1} = -\pi.$$

21. $\lim_{x\to\infty} \frac{\ln x}{x^2}$ is type $\frac{\infty}{\infty}$;
we apply L'Hôpital's Rule to get

$$\lim_{x\to\infty} \frac{1/x}{2x} = \lim_{x\to\infty} \frac{1}{2x^2} = 0.$$

23. $\lim_{x\to\infty} xe^{-x} = \lim_{x\to\infty} \frac{x}{e^x}$ is type $\frac{0}{0}$;
we apply L'Hôpital's Rule to get

$$\lim_{x\to\infty} \frac{1}{e^x} = 0.$$

25. As x approaches 1 from below, $\ln x$ is a small negative number. Hence $\ln(\ln x)$ is undefined, so the limit is undefined.

27. $\lim_{x\to 0^+} x\ln x = \lim_{x\to 0^+} \frac{\ln x}{1/x}$ is type $\frac{\infty}{\infty}$;
we apply L'Hôpital's Rule to get

$$\lim_{x\to 0^+} \frac{1/x}{-1/x^2} = \lim_{x\to 0^+} -x = 0.$$

29. $\lim_{x\to 0^+} \frac{\ln x}{\cot x}$ is type $\frac{\infty}{\infty}$;
we apply L'Hôpital's Rule to get

$$\lim_{x\to 0^+} \frac{1/x}{-\csc^2 x} = \lim_{x\to 0^+} \frac{-\sin^2 x}{x} = \lim_{x\to 0^+} \left(-\sin x \frac{\sin x}{x}\right) = (0)(1) = 0.$$

31. $\lim_{x\to\infty} \left(\sqrt{x^2+1} - x\right)$

$$= \lim_{x\to\infty} \left(\left(\sqrt{x^2+1} - x\right) \frac{\sqrt{x^2+1}+x}{\sqrt{x^2+1}+x}\right) = \lim_{x\to\infty} \left(\frac{x^2+1-x^2}{\sqrt{x^2+1}+x}\right) = \lim_{x\to\infty} \frac{1}{\sqrt{x^2+1}+x} = 0.$$

33. Let $y = \left(1+\frac{1}{x}\right)^x$.
Then $\ln y = x\ln\left(1+\frac{1}{x}\right)$. Then

$$\lim_{x\to\infty} \ln y = \lim_{x\to\infty} x\ln\left(1+\frac{1}{x}\right) = \lim_{x\to\infty} \frac{\ln\left(1+\frac{1}{x}\right)}{1/x} = \lim_{x\to\infty} \frac{\frac{1}{1+\frac{1}{x}}\left(\frac{-1}{x^2}\right)}{-1/x^2} = \lim_{x\to\infty} \frac{1}{1+\frac{1}{x}} = 1.$$

Hence $\lim_{x\to\infty} y = \lim_{x\to\infty} e^{\ln y} = e$.

35. $\lim_{x\to 0^+} \left(\frac{1}{\sqrt{x}} - \frac{\sqrt{x}}{\sqrt{x+1}}\right)$

$$= \lim_{x\to 0^+} \left(\frac{\sqrt{x+1} - (\sqrt{x})^2}{\sqrt{x}\sqrt{x+1}}\right) = \lim_{x\to 0^+} \left(\frac{\sqrt{x+1} - x}{\sqrt{x}\sqrt{x+1}}\right) = \infty.$$

37. Let $y = (1/x)^x$. Then $\ln y = x\ln(1/x)$. Then

$$\lim_{x\to 0^+} \ln y = \lim_{x\to 0^+} x\ln(1/x) = 0,$$

by Exercise 27. Thus

$$\lim_{x\to 0^+} y = \lim_{x\to 0^+} e^{\ln y} = 1.$$

39. L'Hôpital's rule does not apply. As $x \to 0$, the numerator gets close to 1 and the denominator is small and positive. Hence the limit is ∞.

41. L'Hôpital's rule does not apply. As $x \to 0$, the numerator is small and positive while the denominator goes to $-\infty$. Hence the limit is 0. Also $\lim_{x\to 0} \frac{2x}{2/x}$, which equals $\lim_{x\to 0} x^2$, is not of the form $\frac{0}{0}$ so L'Hôpital's rule doesn't apply here either.

43. Starting with $\lim_{x\to 0} \frac{\sin 3x}{\sin 2x}$, we cannot "cancel sin" to get $\lim_{x\to 0} \frac{3x}{2x}$. We can cancel the x's in the last limit to get the final answser of 3/2. The first step is likely to give a correct answer because the linear approximation of $\sin 3x$ is $3x$, and the linear approximation of $\sin 2x$ is $2x$. The linear approximations are better the closer x is to zero, so the limits are likely to be the same.

45. (a) $$\lim_{x\to 0} \frac{\sin x^2}{x^2} = \lim_{x\to 0} \frac{2x\cos x^2}{2x} = \lim_{x\to 0} \cos x^2 = 1,$$

which is the same as $\lim_{x\to 0} \frac{\sin x}{x}$.

(b) $$\lim_{x\to 0} \frac{1-\cos x^2}{x^4} = \lim_{x\to 0} \frac{2x\sin x^2}{4x^3} = \lim_{x\to 0} \frac{\sin x^2}{2x^2}$$

$$= \frac{1}{2}\lim_{x\to 0} \frac{\sin x^2}{x^2} = (1/2)\,(1) = 1/2 \text{ (by part (a))},$$

while

$$\lim_{x\to 0} \frac{1-\cos x}{x^2} = \lim_{x\to 0} \frac{\sin x}{2x} = \frac{1}{2}\lim_{x\to 0} \frac{\sin x}{x} = \frac{1}{2}(1) = \frac{1}{2}$$

so both of these limits are the same.

47. $$\lim_{x\to 0} \frac{\sin kx^2}{x^2} = \lim_{x\to 0} \frac{2kx\cos kx^2}{2x} = \lim_{x\to 0} k\cos kx^2 = k(1) = k$$

49. $\lim_{x\to\infty} e^x = \lim_{x\to\infty} x^n = \infty$

$\lim_{x\to\infty} \frac{e^x}{x^n} = \infty$ since n applications of L'Hôpital's rule yields

$$\lim_{x\to\infty} \frac{e^x}{n!} = \infty.$$

Hence e^x dominates x^n.

51. $$\lim_{x\to 0} \frac{e^{cx}-1}{x} = \lim_{x\to 0} \frac{ce^{cx}}{1} = c$$

53. If $x \to 0$, then $x^2 \to 0$, so if $\lim_{x\to 0} \frac{f(x)}{g(x)} = L$, then $\lim_{x\to 0} \frac{f(x^2)}{g(x^2)} = L$ (but not conversely). If $a \neq 0$ or 1, then $\lim_{x\to a} \frac{f(x)}{g(x)}$ involves the behavior of the quotient near a, while $\lim_{x\to a} \frac{f(x^2)}{g(x^2)}$ involves the behavior of the quotient near the different point a^2.

55. $$\lim_{\omega\to 0} \frac{2.5(4\omega t - \sin 4\omega t)}{4\omega^2} = \lim_{\omega\to 0} \frac{2.5(4t - 4t\cos 4\omega t)}{8\omega} = \lim_{\omega\to 0} \frac{2.5(16t^2\sin 4\omega t)}{8} = 0$$

57. (a) $\dfrac{(x+1)(2+\sin x)}{x(2+\cos x)}$

(b) $\dfrac{x}{e^x}$

(c) $\dfrac{3x+1}{x-7}$

(d) $\dfrac{3-8x}{1+2x}$

59. The area of triangular region 1 is (1/2)(base)(height) $= (1/2)(1-\cos\theta)(\sin\theta)$.
Let P be the center of the circle. The area of region 2 equals the area of sector APC minus the area of triangle APB. The area of the sector is $\theta/2$, while the area of triangle APB is (1/2)(base)(height) $= (1/2)(\cos\theta)(\sin\theta)$.
Hence the area of region 1 divided by the area of region 2 is

$$\frac{(1/2)(1-\cos\theta)(\sin\theta)}{\theta/2-(1/2)(\cos\theta)(\sin\theta)} = \frac{(1-\cos\theta)(\sin\theta)}{\theta-\cos\theta\sin\theta} = \frac{\sin\theta-\cos\theta\sin\theta}{\theta-\cos\theta\sin\theta} = \frac{\sin\theta-(1/2)\sin 2\theta}{\theta-(1/2)\sin 2\theta}$$

Then

$$\lim_{\theta\to 0}\frac{\sin\theta-(1/2)\sin 2\theta}{\theta-(1/2)\sin 2\theta} = \lim_{\theta\to 0}\frac{\cos\theta-\cos 2\theta}{1-\cos 2\theta} = \lim_{\theta\to 0}\frac{-\sin\theta+2\sin 2\theta}{2\sin 2\theta} = \lim_{\theta\to 0}\frac{-\cos\theta+4\cos 2\theta}{4\cos 2\theta} = \frac{-1+4(1)}{4(1)} = \frac{3}{4}$$

3.3 Maximum and Minimum Values

1. (a) No absolute extrema.

(b) $f(0) = -1$ is absolute max. There is no absolute minimum (vertical asymptotes at $x = \pm 1$).

(c) No absolute extrema. (They would be at the endpoints which are not included in the interval.)

3. (a) $f\left(\frac{\pi}{2}+2n\pi\right) = 1$ for any integer n is abs max;
$f\left(\frac{3\pi}{2}+2n\pi\right) = -1$ for any integer n is abs min

(b) $f(0) = 0$ is abs min; $f(\pi/4) = \dfrac{\sqrt{2}}{2}$ is abs max

(c) $f(\pi/2) = 1$ is abs max; there is no abs min, which would occur at both endpoints (not included in the interval).

5. $f(x) = x^2+5x-1$
$f'(x) = 2x+5$
$2x+5 = 0$
$x = -5/2$ is a critical number. This is a parabola opening upward, so we have a minimum.

7. $f(x) = x^3-3x+1$
$f'(x) = 3x^2-3$
$3x^2-3 = 3(x^2-1)$
$= 3(x+1)(x-1) = 0$
$x = \pm 1$ are critical numbers.
This is a cubic with a positive leading coefficient so $x = -1$ is a local max, $x = 1$ is a local min.

9. $f(x) = x^3-3x^2+6x$
$f'(x) = 3x^2-6x+6$
$3x^2-6x+6 = 3(x^2-2x+2) = 0$
We can use the quadratic formula to find the roots, which are $x = 1\pm\sqrt{-1}$.
These are imaginary so there are no real critical numbers.

11. $f(x) = x^4-3x^3+2$
$f'(x) = 4x^3-9x^2$

$4x^3 - 9x^2 = x^2(4x - 9) = 0$
$x = 0, 9/4$ are critical numbers
$x = 9/4$ is a local min; $x = 0$ is neither a local max nor min.

13. $f(x) = x^{3/4} - 4x^{1/4}$
$f'(x) = \frac{3}{4x^{1/4}} - \frac{1}{x^{3/4}}$
If $x \neq 0$, $f'(x) = 0$ when $3x^{3/4} = 4x^{1/4}$ $x = 0, 16/9$ are critical numbers.
$x = 16/9$ is a local min, $x = 0$ is neither a local max nor min.

15. $f(x) = \sin x \cos x$ on $[0, 2\pi]$
$f'(x) = \cos x \cos x + \sin x(-\sin x)$
$= \cos^2 x - \sin^2 x$

$\cos^2 x - \sin^2 x = 0$
$\cos^2 x = \sin^2 x$
$\cos x = \pm \sin x$

$x = \pi/4, 3\pi/4, 5\pi/4, 7\pi/4$ are critical numbers.
$x = \pi/4, 5\pi/4$ are local max, $x = 3\pi/4, 7\pi/4$ are local min.

17. $f(x) = \frac{x^2 - 2}{x + 2}$
Note that $x = -2$ is not in the domain of f.
$f'(x) = \frac{(2x)(x+2) - (x^2 - 2)(1)}{(x+2)^2}$
$= \frac{2x^2 + 4x - x^2 + 2}{(x+2)^2}$
$= \frac{x^2 + 4x + 2}{(x+2)}$

$f'(x) = 0$ when $x^2 + 4x + 2 = 0$, so the critical numbers are $x = -2 \pm \sqrt{2}$.
$x = -2 + \sqrt{2}$ is a local min; $x = -2 + \sqrt{2}$ is a local max.

19. $f(x) = \frac{x}{x^2 + 1}$
$f'(x) = \frac{1(x^2 + 1) - x(2x)}{(x^2 + 1)^2}$
$= \frac{x^2 + 1 - 2x^2}{(x^2 + 1)^2}$
$= \frac{1 - x^2}{(x^2 + 1)^2}$

$f'(x) = 0$ for $1 - x^2 = 0, x = 1, -1$;
$f'(x)$ is defined for all x, so $x = 1, -1$ are the critical numbers.
$x = -1$ is local min, $x = 1$ is local max.

21. $f(x) = \frac{e^x + e^{-x}}{2}$
$f'(x) = \frac{e^x - e^{-x}}{2}$
$f'(x) = 0$ when $e^x = e^{-x}$, that is, $x = 0$.
$f'(x)$ is defined for all x, so $x = 0$ is a critical number.
$x = 0$ is a local min.

23. $f(x) = x^{4/3} + 4x^{1/3} + 4x^{-2/3}$
f is not defined at $x = 0$.
$f'(x) = \frac{4}{3}x^{1/3} + \frac{4}{3}x^{-2/3} - \frac{8}{3}x^{-5/3}$
$= \frac{4}{3}x^{-5/3}(x^2 + x - 2)$
$= \frac{4}{3}x^{-5/3}(x - 1)(x + 2)$

$x = -2, 1$ are critical numbers.
$x = -2$ and $x = 1$ are local minima.

25. $f(x) = 2x\sqrt{x+1} = 2x(x+1)^{1/2}$
Domain of f is all $x \geq -1$.
$f'(x) = 2(x+1)^{1/2} + 2x\left(\frac{1}{2}(x+1)^{-1/2}\right)$
$= \frac{2(x+1) + x}{\sqrt{x+1}}$
$= \frac{3x + 2}{\sqrt{x+1}}$
$f'(x) = 0$ for $3x + 2 = 0$, $x = -2/3$.
$f'(x)$ is undefined for $\sqrt{x+1} = 0$, $x = -1$ so $x = -2/3, -1$ are critical numbers.
$x = -2/3$ is a local min. $x = -1$ is an endpoint so is neither a local min nor a local max, though it is a maximum on the interval $[-1, 0)$.

27. Because of the absolute value sign, there may be critical numbers where the function $x^2 - 1$ changes sign; that is, at $x = \pm 1$. For $x > 1$ and for $x < -1$, $f(x) = x^2 - 1$ and $f'(x) = 2x$, so there are no critical numbers on these intervals. For $-1 < x < 1$, $f(x) = 1 - x^2$ and $f'(x) = -2x$, so 0 is a critical number. A graph confirms this analysis and shows there is a local max at $x = 0$ and local min at $x = \pm 1$.

29. First, let's find the critical numbers for $x < 0$. In this case,
$f(x) = x^2 + 2x - 1$
$f'(x) = 2x + 2 = 2(x + 1)$
so the only critical number in this interval is $x = -1$ and it is a local minimum.
Now for $x > 0$,
$f(x) = x^2 - 4x + 3$
$f'(x) = 2x - 4 = 2(x - 2)$
so the only critical number is $x = 2$ and it is a local minimum.
Finally, $x = 0$ is also a critical number, since f is not continuous and hence not differentiable at $x = 0$. Indeed, $x = 0$ is a local maximum.

31. $f(x) = x^3 - 3x + 1$
$f'(x) = 3x^2 - 3 = 3(x^2 - 1)$
$f'(x) = 0$ for $x = \pm 1$.

(a) On $[0, 2]$, 1 is the only critical number. We calculate:
$f(0) = 1$
$f(1) = -1$ is the abs min.
$f(2) = 3$ is the abs max.

(b) On the interval $[-3, 2]$, we have both 1 and -1 as critical numbers. We calculate:
$f(-3) = -17$ is the abs min.
$f(-1) = 3$ is the abs max.
$f(1) = -1$
$f(2) = 3$ is also the abs max.

33. $f(x) = x^{2/3}$
$f'(x) = \frac{2}{3}x^{-1/3} = \frac{2}{3\sqrt[3]{x}}$
$f'(x) \neq 0$ for any x, but $f'(x)$ undefined for $x = 0$, so $x = 0$ is critical number.

(a) On $[-4, -2]$:
$0 \notin [-4, -2]$ so we only look at endpoints.
$f(-4) = \sqrt[3]{16} \approx 2.52$
$f(-2) = \sqrt[3]{4} \approx 1.59$
So $f(-4) = \sqrt[3]{16}$ is the abs max and $f(-2) = \sqrt[3]{4}$ is the abs min.

(b) On $[-1, 3]$, we have 0 as a critical number.
$f(-1) = 1$
$f(0) = 0$ is the abs min.
$f(3) = 3^{2/3}$ is the abs max.

35. $f(x) = e^{-x^2}$
$f'(x) = -2xe^{-x^2}$
Hence $x = 0$ is the only critical number.

(a) On $[0, 2]$:
$f(0) = 1$ is the abs max.
$f(2) = e^{-4}$ is the abs min.

(b) On $[-3, 2]$:
$f(-3) = e^{-9}$ is the abs min.
$f(0) = 1$ is the abs max.
$f(2) = e^{-4}$

37. $f(x) = \dfrac{3x^2}{x - 3}$
Note that $x = 3$ is not in the domain of f.

$$f'(x) = \frac{6x(x-3) - 3x^2(1)}{(x-3)^2} = \frac{6x^2 - 18x - 3x^2}{(x-3)^2} = \frac{3x^2 - 18x}{(x-3)^2}$$

$$= \frac{3x(x-6)}{(x-3)^2}$$

The critical points are $x = 0$, $x = 6$.

(a) On $[-2, 2]$:
$f(-2) = -12/5$
$f(2) = -12$
$f(0) = 0$
Hence abs max is $f(0) = 0$ and abs min is $f(2) = -12$.

(b) On $[2, 8]$, the function is not continuous and in fact has no absolute max or min.

39. $f'(x) = 4x^3 - 6x + 2 = 0$ at about $x = 0.3660$, -1.3660 and at $x = 1$.

(a) $f(-1) = 3$, $f(1) = 1$.
The absolute min is $(-1, 3)$ and the absolute max is approximately $(0.3660, 1.3481)$.

(b) The absolute min is approximately $(-1.3660, -3.8481)$ and the absolute max is $(-3, 49)$.

41. $f'(x) = 2x - 3\cos x + 3x \sin x = 0$ at about $x = 0.6371$, -1.2269 and -2.8051.

(a) The absolute min is approximately $(0.6371, -1.1305)$ and the absolute max is approximately $(-1.2269, 2.7463)$.

(b) The absolute min is approximately $(-2.8051, -0.0748)$ and the absolute max is approximately $(-5, 29.2549)$.

43. $f'(x) = \sin x + x \cos x = 0$ at $x = 0$ and about 2.0288 and 4.9132.

(a) The absolute min is $(0, 3)$ and the absolute max is $(\pm\pi/2, 3 + \pi/2)$.

(b) The absolute min is approximately $(4.9132, -1.814)$ and the absolute max is approximately $(2.0288, 4.820)$.

45. If an absolute max or min occurs only at the endpoint of a closed interval, then there will be no absolute max or min on the open interval.

31) on $(0, 2)$, $f(1) = -1$ is min, no max.
on $(-3, 2)$, $f(-1) = 3$ is max, no min.

32) on $(-3, 1)$, $f(-2) = -14$ is min, no max.
on $(-1, 3)$, $f(2) = -14$ is min, no max.

33) on $(-4, -2)$, no max or min.
on $(-1, 3)$, $f(0) = 0$ is min, no max.

34) on $(0, 2\pi)$, $f(5\pi/4) = -\sqrt{2}$ is min, $f(\pi/4) = \sqrt{2}$ is max.
on $(\pi/2, \pi)$, no max or min.

35) on $(0, 2)$, no max or min
on $(-3, 2)$, $f(0) = 1$ is max; no min

36) on $(-2, 0)$, no min or max.
on $(0, 4)$, $f(1/2) = e^{-2}/4$ is max, no min.

37) on $(-2, 2)$, $f(0) = 0$ is max; no min
on $(2, 8)$, no max or min

38) on $(0, 1)$, no min or max.
on $(-3, 4)$, no max, $f(0) = 0$ is min.

47. On $[-2, 2]$, the absolute maximum is 3 and the absolute minimum doesn't exist.

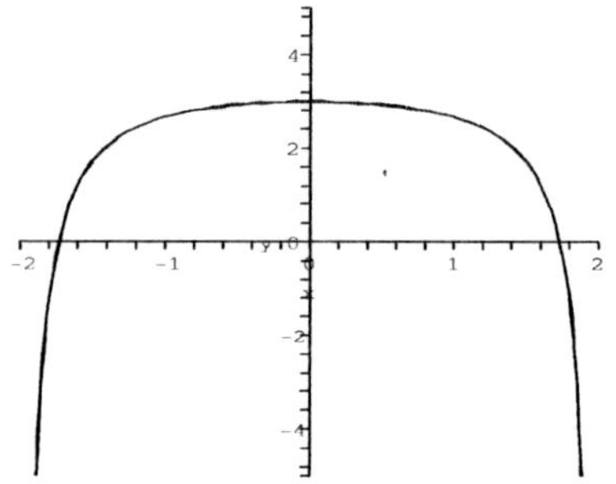

49. On $(-2, 2)$ the absolute maximum is 4 and the absolute minimum is 2.

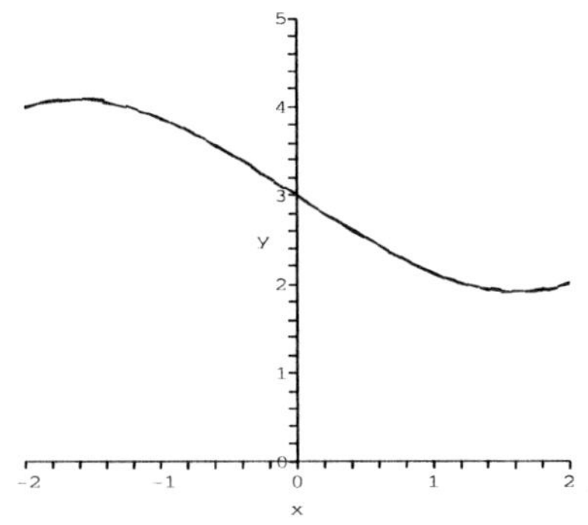

51. You will not be able to construct an example with a continuous function, but there are many examples using a function with a discontinuity, for example $f(x) = \sec^2 x$.

53. $f(x) = x^3 + cx + 1$
$f'(x) = 3x^2 + c$
We know (perhaps from a pre-calculus course) that for any cubic polynomial with positive leading coefficient, when x is large and positive the value of the polynomial is very large and positive, and when x is large and negative, the value of the polynomial is very large and negative.

Type 1: $c > 0$. There are no critical numbers. As you move from left to right, the graph of f is always rising.

Type 2: $c < 0$ There are two critical numbers $x = \pm\sqrt{-c/3}$. As you move from left to right, the graph rises until we get to the first critical number, then the graph must fall until we get to the second critical number, and then the graph rises again. So the critical number on the left is a local maximum and the critical number on the right is a local minimum.

Type 3: $c = 0$. There is only one critical number, which is neither a local max nor a local min.

55. $f(x) = x^3 + bx^2 + cx + d$
$f'(x) = 3x^2 + 2bx + c$
The quadratic formula says that the critical numbers are

$$x = \frac{-2b \pm \sqrt{4b^2 - 12c}}{6} = \frac{-b \pm \sqrt{b^2 - 3c}}{3}.$$

So if $c < 0$, the quantity under the square root is positive and there are two critical numbers. This is like the Type 2 cubics in Exercise 53. We know that as x goes to infinity, the polynomial $x^3 + bx^2 + cx + d$ gets very large and positive, and when x goes to minus infinity, the polynomial is very large but negative. Therefore, the critical number on the left must be a local max, and the critical number on the right must be a local min.

57. $f(x) = x^4 + cx^2 + 1$
$f'(x) = 4x^3 + 2cx = 2x(2x^2 + c)$
So $x = 0$ is always a critical number.

Case 1: $c \geq 0$. The only solution to $2x(2x^2 + c) = 0$ is $x = 0$, so $x = 0$ is the only critical number. This must be a minimum, since we know that the function $x^4 + cx^2 + 1$ is large and positive when $|x|$ is large (so the graph is roughly U-shaped). We could also note that $f(0) = 1$, and 1 is clearly the absolute minimum of this function if $c \geq 0$.

Case 2: $c < 0$. Then there are two other critical numbers $x = \pm\sqrt{-c/2}$. Now $f(0)$ is still equal to 1, but the value of f at both new critical numbers is less than 1. Hence $f(0)$ is a local max, and both new critical numbers are local minimums.

59. With $t = 90$ and $r = 1/30$, we have

$$P(n) = \frac{3^n}{n!}e^{-3}.$$

We compute P for the first few values of n:

n	P
0	e^{-3}
1	$3e^{-3}$
2	$4.5e^{-3}$
3	$4.5e^{-3}$
4	$3.375e^{-3}$

Once $n > 3$, the values of P will decrease as n increases. This is due to the fact that to get $P(n+1)$ from $P(n)$, we multiply $P(n)$ by $3/(n+1)$. Since $n > 3$, $3/(n+1) < 1$ and so $P(n+1) < P(n)$. Thus we see from the table that P is maximized at $n = 3$ (it is also maximized at $n = 2$). It makes sense that P would be maximized at $n = 3$ because

$$(90 \text{ mins})\left(\frac{1}{30}\text{ goals/min}\right) = 3 \text{ goals}.$$

61. Since f is differentiable on (a, b), it is continuous on the same interval. Since f is decreasing at a and increasing at b, f must have a local minimum for some value c, where $a < c < b$. By Fermat's theorem, c is a critical number for f. Since f is differentiable at c, $f'(c)$ exists, and therefore $f'(c) = 0$.

63. Graph of $f(x) = \dfrac{x^2}{x^2+1}$:

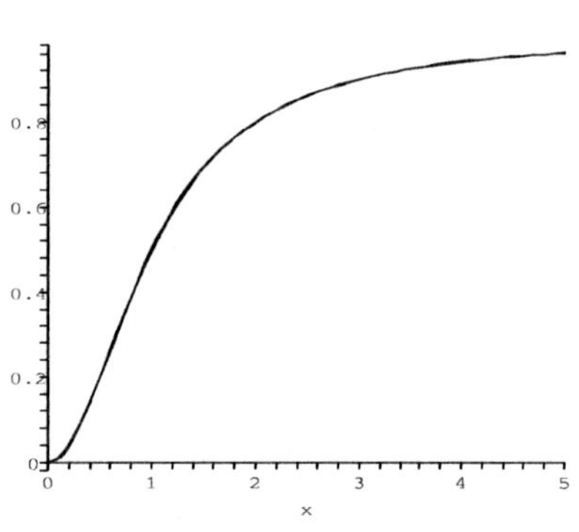

$$f'(x) = \frac{2x(x^2+1) - x^2(2x)}{(x^2+1)^2} = \frac{2x}{(x^2+1)^2}$$

$$f''(x) = \frac{2(x^2+1)^2 - 2x \cdot 2(x^2+1) \cdot 2x}{(x^2+1)^4} = \frac{2(x^2+1)\left[(x^2+1) - 4x^2\right]}{(x^2+1)^4} = \frac{2\left[1 - 3x^2\right]}{(x^2+1)^3}$$

$f''(x) = 0$ for $x = \pm\dfrac{1}{\sqrt{3}}$,

$x = -\dfrac{1}{\sqrt{3}} \notin (0, \infty)$

$x = \dfrac{1}{\sqrt{3}}$ is steepest point.

65. $y = x^5 - 4x^3 - x + 10$, $x \in [-2, 2]$
$y' = 5x^4 - 12x^2 - 1$
$x = -1.575, 1.575$ are critical numbers of y. There is a local max at $x = -1.575$, local min at $x = 1.575$. $x = -1.575$ represents the top and $x = 1.575$ represents the bottom of the roller coaster.
$y''(x) = 20x^3 - 24x = 4x(5x^2 - 6) = 0$
$x = 0, \pm\sqrt{6/5}$ are critical numbers of y'. We calculate y' at the critical numbers and at the endpoints $x = \pm 2$:
$y'(0) = -1$
$y'\left(\pm\sqrt{6/5}\right) = -41/5$
$y'(\pm 2) = 31$
So the points where the roller coaster is making the steepest descent are $x = \pm\sqrt{6/5}$, but the steepest part of the roller coast is during the ascents at ± 2.

67. $W(t) = a \cdot e^{-be^{-t}}$
as $t \to \infty$, $-be^{-t} \to 0$, so $W(t) \to a$.
$W'(t) = a \cdot e^{-be^{-t}} \cdot be^{-t}$
as $t \to \infty$, $be^{-t} \to 0$, so $W'(t) \to 0$.
$W''(t) = (a \cdot e^{-be^{-t}} \cdot be^{-t}) \cdot be^{-t} + (a \cdot e^{-be^{-t}}) \cdot (-be^{-t})$

$= a \cdot e^{-be^{-t}} \cdot be^{-t}\left[be^{-t} - 1\right]$
$W''(t) = 0$ when $be^{-t} = 1$
$e^{-t} = b^{-1}$
$-t = \ln b^{-1}$
$t = \ln b$
$W'(\ln b) = a \cdot e^{-be^{-\ln b}} \cdot be^{-\ln b}$
$= a \cdot e^{-b\left(\frac{1}{b}\right)} \cdot b \cdot \frac{1}{b} = ae^{-1}$
Maximum growth rate is ae^{-1} when $t = \ln b$.

69. Label the triangles as illustrated.

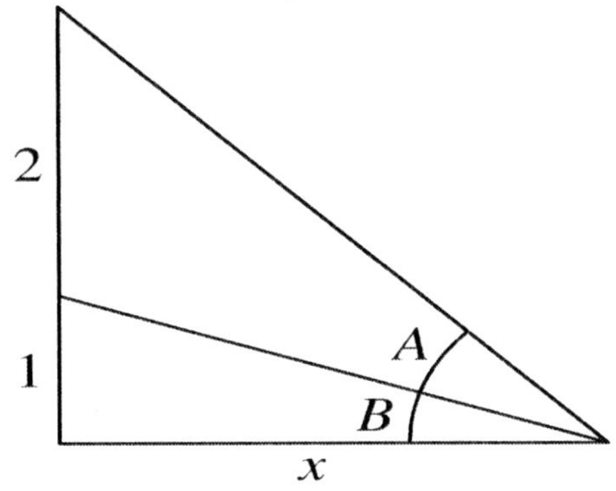

$\tan(A + B) = 3/x$
$A + B = \tan^{-1}(3/x)$
$\tan B = 1/x$
$B = \tan^{-1}(1/x)$
Therefore,
$A = (A + B) - B$
$A = \tan^{-1}(3/x) - \tan^{-1}(1/x)$

$$\frac{dA}{dx} = \frac{-3/x^2}{1 + (3/x)^2} - \frac{-1/x^2}{1 + (1/x)^2}$$
$$= \frac{1}{x^2 + 1} - \frac{3}{x^2 + 9}$$

The maximum viewing angle will occur at a critical value.

$$\frac{dA}{dx} = 0$$
$$\frac{1}{x^2 + 1} = \frac{3}{x^2 + 9}$$
$x^2 + 9 = 3x^2 + 3$
$2x^2 = 6$
$x^2 = 3$
$x = \sqrt{3}$ ft ≈ 1.73 ft
This is a maximum because when x is large and when x is a little bigger than 0, the angle is small.

3.4 Increasing and Decreasing Functions

1. $y = x^3 - 3x + 2$
$y' = 3x^2 - 3 = 3(x^2 - 1)$
$= 3(x + 1)(x - 1)$

$x = \pm 1$ are critical numbers.
$(x + 1) > 0$ on $(-1, \infty)$, $(x + 1) < 0$ on $(-\infty, -1)$
$(x - 1) > 0$ on $(1, \infty)$, $(x - 1) < 0$ on $(-\infty, -1)$
$3(x + 1)(x - 1) > 0$ on $(1, \infty) \cup (-\infty, -1)$ so y is increasing on $(1, \infty)$ and on $(-\infty, -1)$
$3(x + 1)(x - 1) < 0$ on $(-1, 1)$, so y is decreasing on $(-1, 1)$.

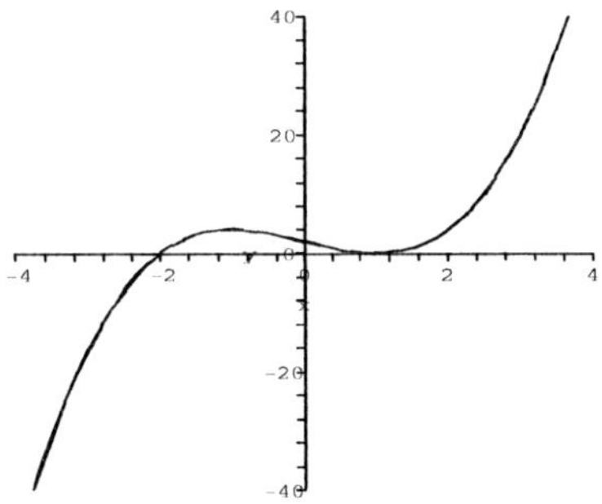

3. $y = x^4 - 8x^2 + 1$
$y' = 4x^3 - 16x = 4x(x^2 - 4)$
$= 4x(x - 2)(x + 2)$
$x = 0, 2, -2$

$4x > 0$ on $(0, \infty)$, $4x < 0$ on $(-\infty, 0)$
$(x - 2) > 0$ on $(2, \infty)$, $(x - 2) < 0$ on $(-\infty, 2)$
$(x + 2) > 0$ on $(-2, \infty)$, $(x + 2) < 0$ on $(-\infty, -2)$
$4(x - 2)(x + 2) > 0$ on $(-2, 0) \cup (2, \infty)$, so the function is increasing on $(-2, 0)$ and on $(2, \infty)$.
$4(x - 2)(x + 2) < 0$ on $(-\infty, -2) \cup (0, 2)$, so y is decreasing on $(-\infty, -2)$

and on $(0, 2)$.

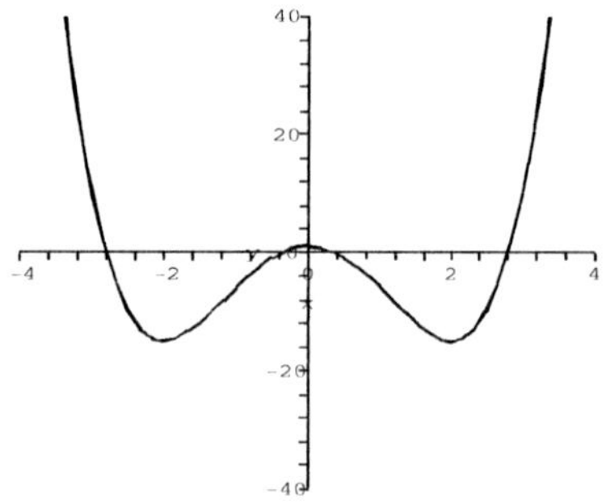

5. $y = (x+1)^{2/3}$

$y' = \frac{2}{3}(x+1)^{-1/3} = \frac{2}{3\sqrt[3]{x+1}}$

y' is not defined for $x = -1$

$\frac{2}{3\sqrt[3]{x+1}} > 0$ on $(-1, \infty)$, y is increasing

$\frac{2}{3\sqrt[3]{x+1}} < 0$ on $(-\infty, -1)$, y is decreasing

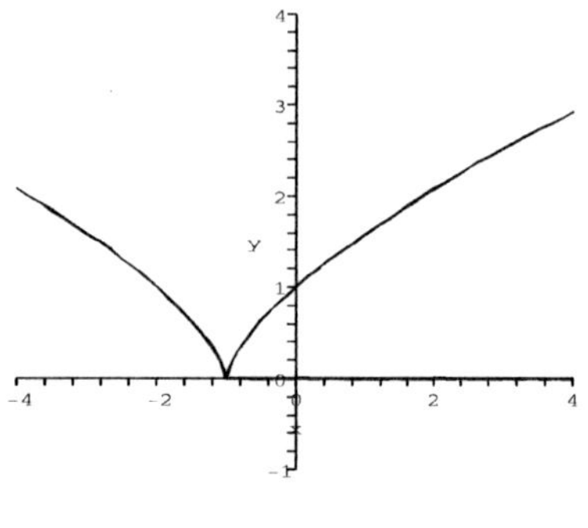

7. $y = \sin x + \cos x$

$y' = \cos x - \sin x = 0$

$\cos x = \sin x$

$x = \pi/4,\ 5\pi/4,\ 9\pi/4$, etc. $\cos x - \sin x > 0$ on $(-3\pi/4, \pi/4) \cup (5\pi/4, 9\pi/4) \cup \ldots$

$\cos x - \sin x < 0$ on $(\pi/4, 5\pi/4) \cup (9\pi/4, 13\pi/4) \cup \ldots$

So $y = \sin x + \cos x$ is decreasing on $(\pi/4, 5\pi/4), (9\pi/4, 13\pi/4)$, etc., and is increasing on $(-3\pi/4, \pi/4), (5\pi/4, 9\pi/4)$, etc.

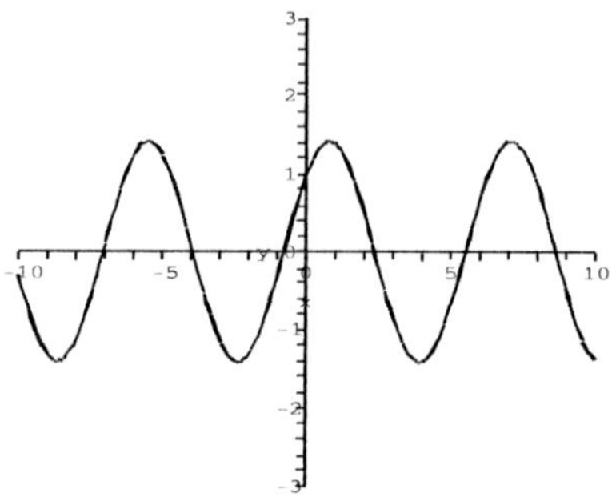

9. $y = e^{x^2-1}$

$y' = e^{x^2-1} \cdot 2x = 2xe^{x^2-1}$

$x = 0$

$2xe^{x^2-1} > 0$ on $(0, \infty)$, y is increasing

$2xe^{x^2-1} < 0$ on $(-\infty, 0)$, y is decreasing

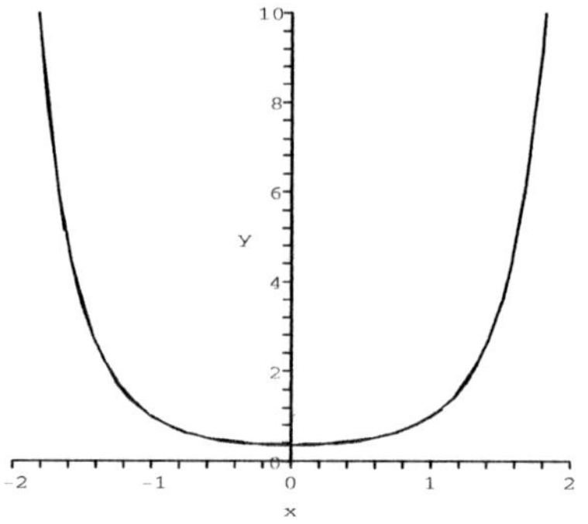

11. $y = x^4 + 4x^3 - 2$

$y' = 4x^3 + 12x^2 = 4x^2(x+3)$

Critical numbers are $x = 0$, $x = -3$.

$4x^2(x+3) > 0$ on $(-3, 0) \cup (0, \infty)$

$4x^2(x+3) < 0$ on $(-\infty, -3)$

Hence $x = -3$ is a local minimum and $x = 0$ is not an extremum.

13. $y = xe^{-2x}$

$y' = 1 \cdot e^{-2x} + x \cdot e^{-2x}(-2)$

$= e^{-2x} - 2xe^{-2x}$

$= e^{-2x}(1 - 2x)$

$x = \frac{1}{2}$

$e^{-2x}(1-2x) > 0$ on $(-\infty, 1/2)$

$e^{-2x}(1-2x) < 0$ on $(1/2, \infty)$

So $y = xe^{-2x}$ has a local maximum at $x = 1/2$.

15. $y = \tan^{-1}(x^2)$

$y' = \dfrac{2x}{1+x^4}$

Critical number is $x = 0$.

$\dfrac{2x}{1+x^4} > 0$ for $x > 0$

$\dfrac{2x}{1+x^4} < 0$ for $x < 0$.

Hence $x = 0$ is a local minimum.

17. $y = \dfrac{x}{1+x^3}$

Note that the function is not defined for $x = -1$.

$$y' = \frac{1(1+x^3) - x(3x^2)}{(1+x^3)} = \frac{1+x^3-3x^3}{(1+x^3)^2} = \frac{1-2x^3}{(1+x^3)^2}$$

Critical number is $x = \sqrt[3]{1/2}$

$y' > 0$ on $(-\infty, -1) \cup (-1, -\sqrt[3]{1/2})$

$y' < 0$ on $(\sqrt[3]{1/2}, \infty)$

Hence $x = \sqrt[3]{1/2}$ is a local max.

19. $y = \sqrt{x^3+3x^2} = (x^3+3x^2)^{1/2}$

Domain is all $x \geq -3$.

$$y' = \frac{1}{2}(x^3+3x^2)^{-1/2}(3x^2+6x) = \frac{3x^2+6x}{2\sqrt{x^3+3x^2}} = \frac{3x(x+2)}{2\sqrt{x^3+3x^2}}$$

$x = 0, -2, -3$ are critical numbers.

y' undefined at $x = 0, -3$

$y' > 0$ on $(-3, -2) \cup (0, \infty)$

$y' < 0$ on $(-2, 0)$

So $y = \sqrt{x^3+3x^2}$ has local max at $x = -2$, local min at $x = 0$. $x = -3$ is an endpoint, and so is not a local extremum.

21. $y = \dfrac{x}{x^2-1}$

$$y' = \frac{x^2-1-x(2x)}{(x^2-1)^2} = -\frac{x^2+1}{(x^2-1)^2}$$

There are no values of x for which $y' = 0$. There are no critical points, because the values for which y' does not exist (that is, $x = \pm 1$) are not in the domain.

There are vertical asymptotes at $x = \pm 1$, and a horizontal asymptote at $y = 0$. This can be verified by calculating the following limits:

$$\lim_{x\to\pm\infty} \frac{x}{x^2-1} = 0$$

$$\lim_{x\to -1} \frac{x}{x^2-1} = \infty$$

$$\lim_{x\to 1} \frac{x}{x^2-1} = -\infty$$

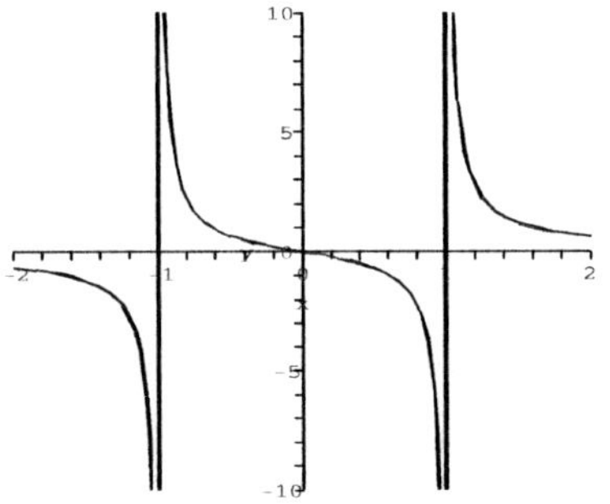

23. $y = \dfrac{x^2}{x^2-4x+3} = \dfrac{x^2}{(x-1)(x-3)}$

Vertical asymptotes $x = 1$, $x = 3$. When $|x|$ is large, the function approaches the value 1, so $y = 1$ is a horizontal asymptote.

$$y' = \frac{2x(x^2-4x+3) - x^2(2x-4)}{(x^2-4x+3)^2} = \frac{2x^3-8x^2+6x-2x^3+4x^2}{(x^2-4x+3)^2} = \frac{-4x^2+6x}{(x^2-4x+3)^2} = \frac{2x(-2x+3)}{(x^2-4x+3)^2} = \frac{2x(-2x+3)}{[(x-3)(x-1)]^2}$$

Critical numbers are $x = 0$ (local min) and $x = 3/2$ (local max).

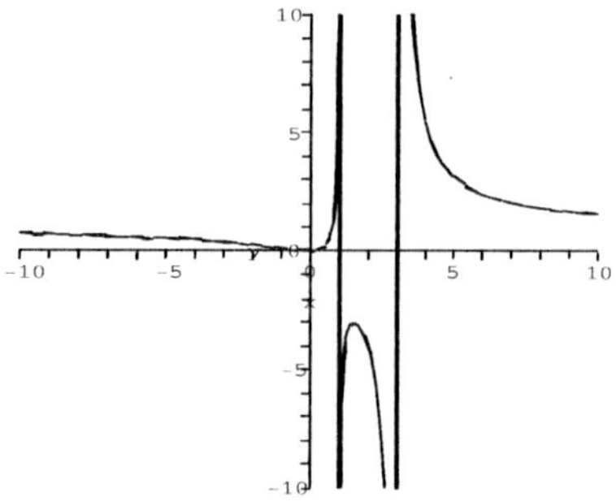

25. $y = \dfrac{x}{\sqrt{x^2+1}}$

$$y' = \frac{\sqrt{x^2+1} - x^2/\sqrt{x^2+1}}{x^2+1} = \frac{1}{(x^2+1)^{3/2}}$$

The derivative is never zero, so there are no critical points. To verify that there are horizontal asymptotes at $y = \pm 1$:

$$y = \frac{x}{\sqrt{x^2+1}} = \frac{x}{\sqrt{x^2}\sqrt{1+\frac{1}{x^2}}} = \frac{x}{|x|\sqrt{1+\frac{1}{x^2}}}$$

Thus,

$$\lim_{x\to\infty} \frac{x}{|x|\sqrt{1+\frac{1}{x^2}}} = 1$$

$$\lim_{x\to-\infty} \frac{x}{|x|\sqrt{1+\frac{1}{x^2}}} = -1$$

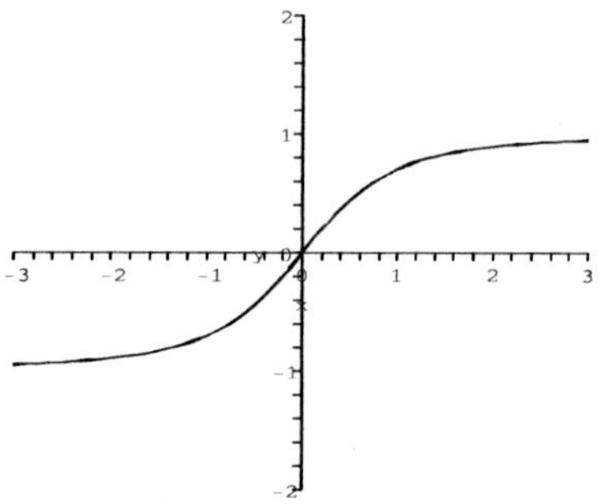

27. $y' = 3x^2 - 26x - 10 = 0$ when $x = \dfrac{26 \pm \sqrt{796}}{6}$. Local max at $x = -0.3689$; local min at $x = 9.0356$.

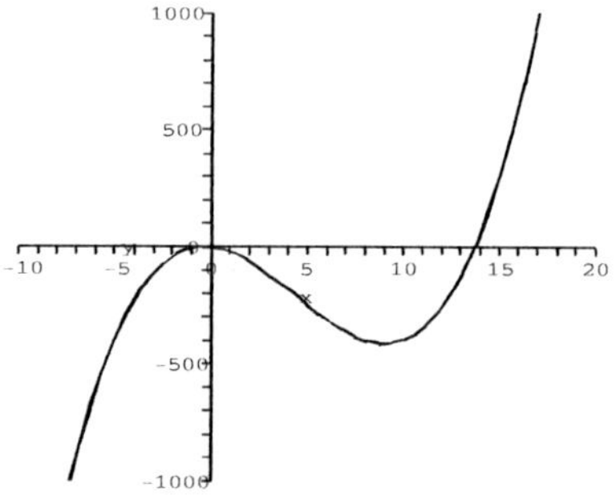

29. $y' = 4x^3 - 45x^2 - 4x + 40$
Local minima at $x = -0.9474, 11.2599$; local max at 0.9374.
Local behavior near $x = 0$ looks like

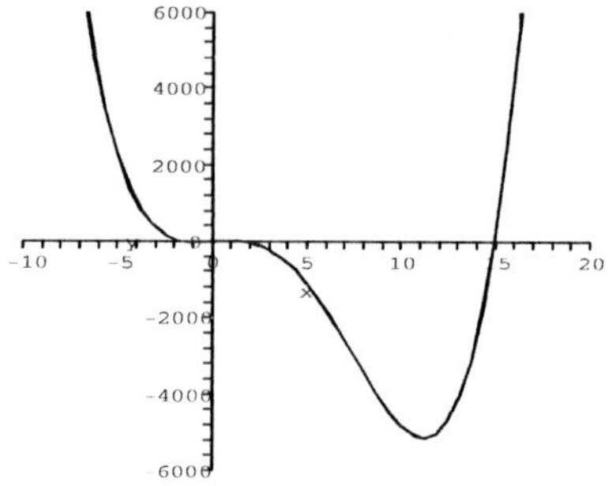

Global behavior of the function looks like

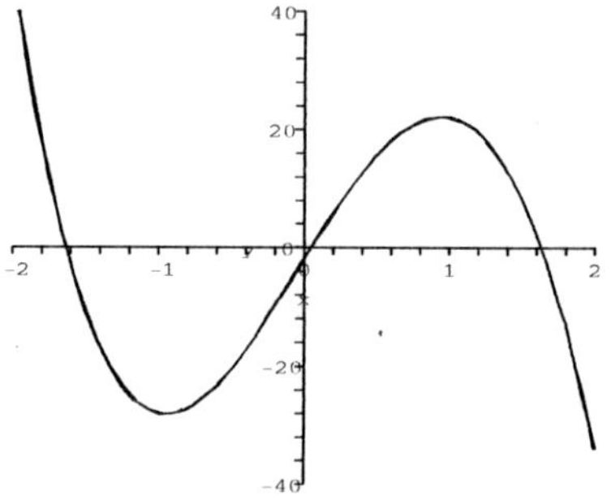

31. $y' = 5x^4 - 600x + 605$
Local minima at $x = -1.0084, 10.9079$; local maxima at $x = -10.9079, 1.0084$.
Local behavior near $x = 0$ looks like

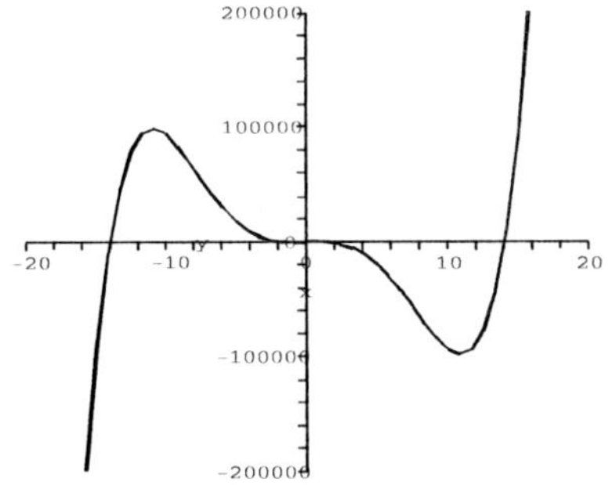

Global behavior of the function looks like

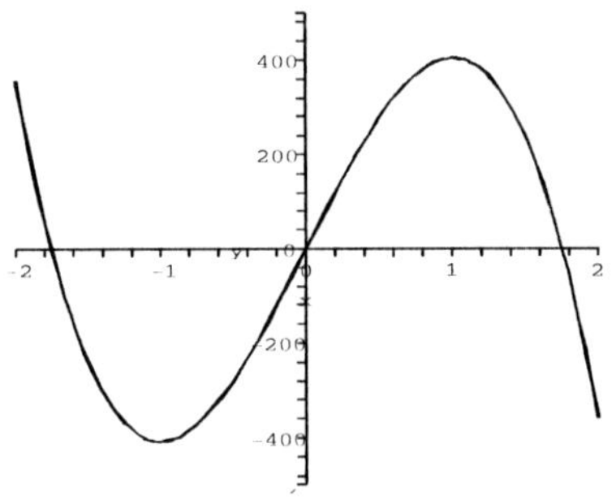

33. $y' = (2x+1)e^{-2x}$
$\quad + (x^2 + x + 0.45)(-2)e^{-2x}$
Local min at $x = -0.2236$; local max at $x = 0.2236$. Local behavior near $x = 0$ looks like

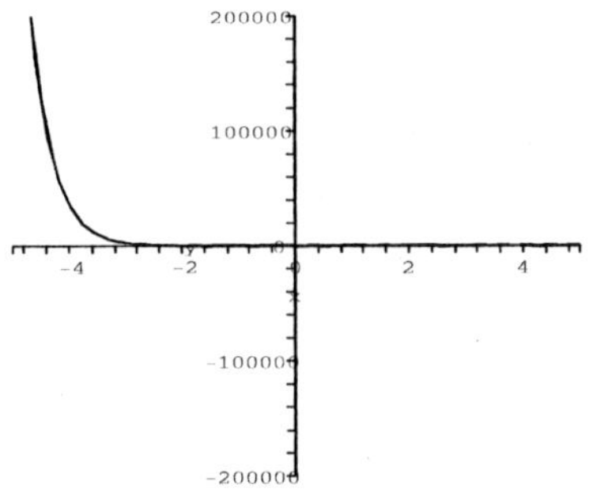

Global behavior of the function looks like

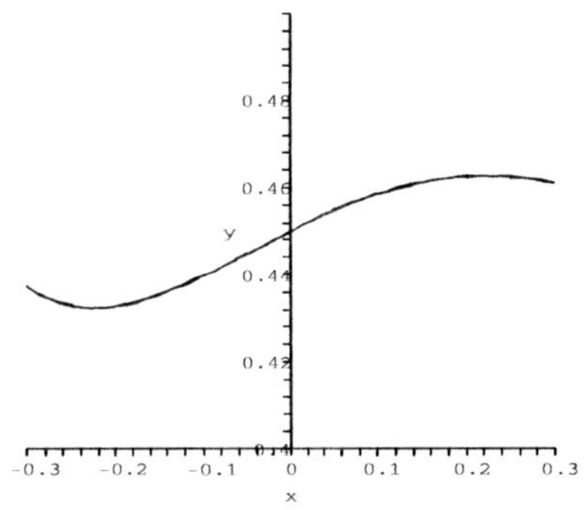

35. One possible graph:

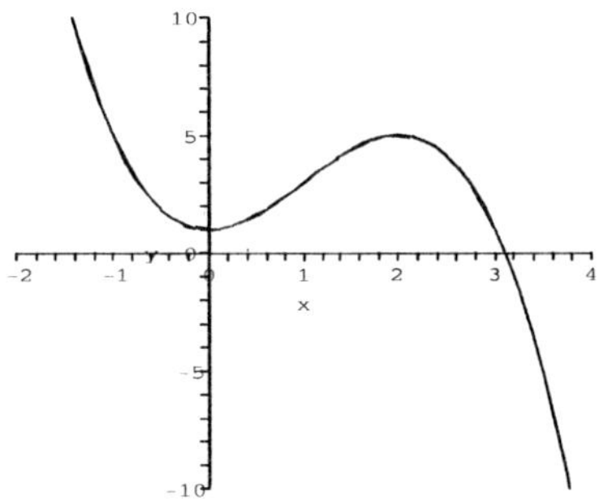

37. One possible graph:

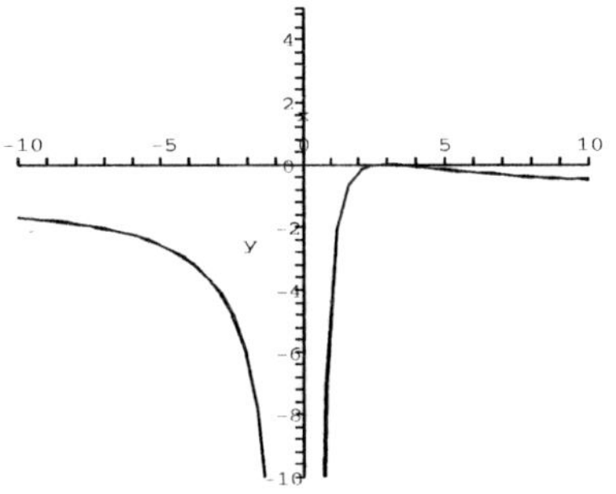

39. The derivative is

$$y' = \frac{-3x^4 + 120x^3 - 1}{(x^4 - 1)^2}.$$

We estimate the critical numbers to be approximately 0.2031 and 39.999. The following graph shows global behavior:

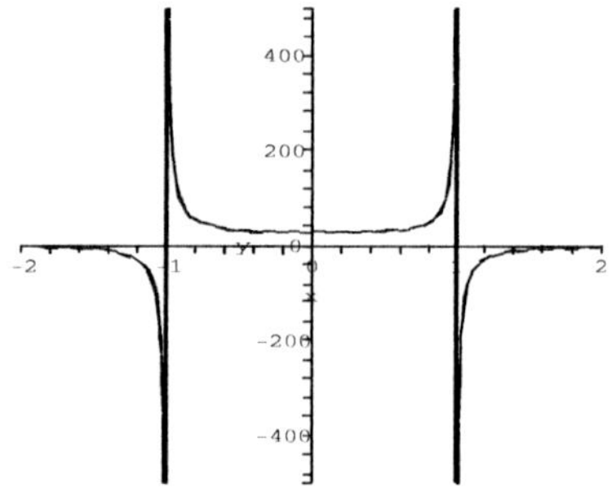

The following graphs show local behavior:

The following graphs show local behavior:

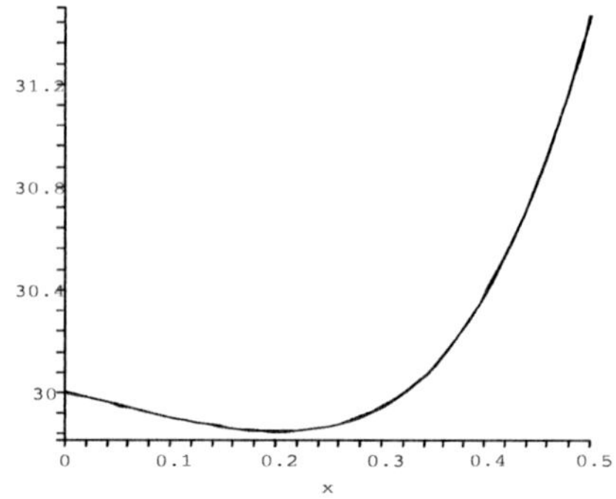

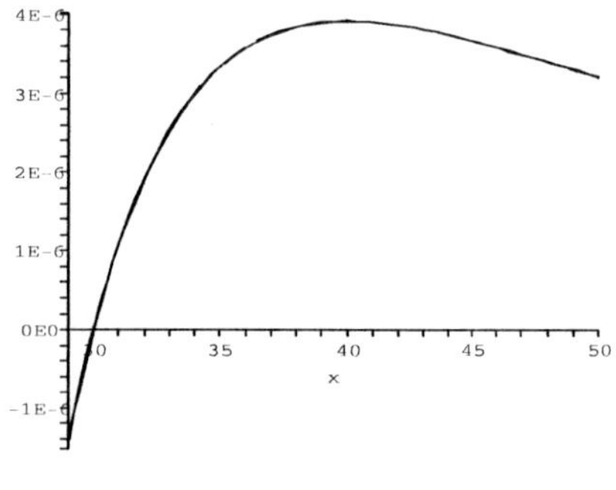

41. The derivative is

$$y' = \frac{-x^2 - 120x + 1}{(x^2 + 1)^2}.$$

We estimate the critical numbers to be approximately 0.008 and -120.008.
The following graph shows global behavior:

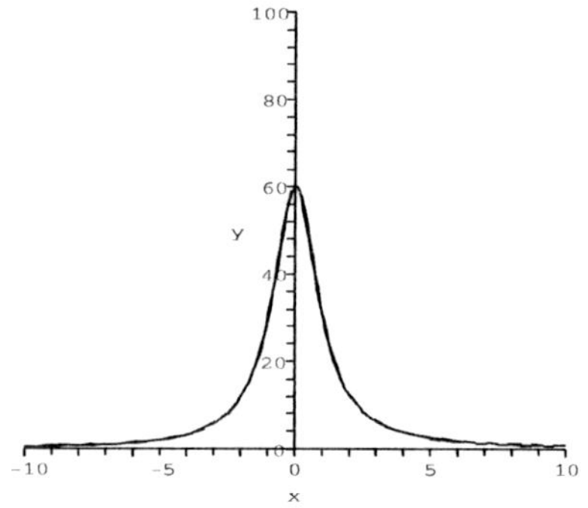

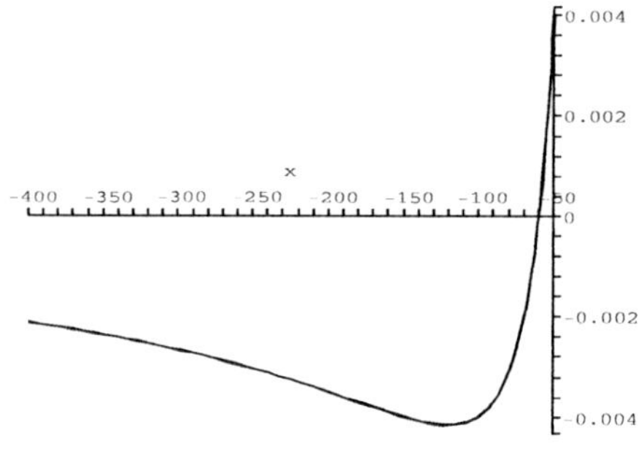

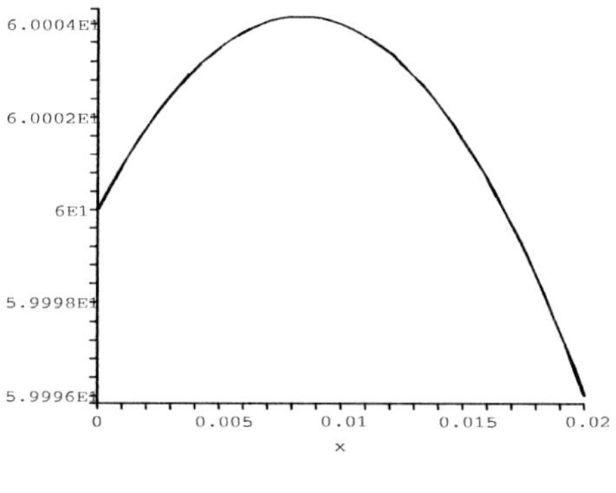

43. $f'(0) = \lim_{x \to 0} \dfrac{f(x) - f(0)}{x - 0}$

$= \lim_{x \to 0} \dfrac{f(x)}{x}$

$= \lim_{x \to 0} \left[1 + 2x \sin\left(\dfrac{1}{x}\right) \right] = 1$

For $x \neq 0$,
$f'(x) =$
$1 + 2\left[2x \sin\left(\dfrac{1}{x}\right) + x^2 \left(\dfrac{-1}{x^2}\right) \cos\left(\dfrac{1}{x}\right) \right]$

$= 1 + 4x \sin\left(\dfrac{1}{x}\right) - 2 \cos\left(\dfrac{1}{x}\right)$

For values of x close to the origin, the

middle term of the derivative is small, and since the last term $-2\cos(1/x)$ reaches its minimum value of -2 in every neighborhood of the origin, f' has negative values on every neighborhood of the origin. Thus, f is not increasing on any neighborhood of the origin.

This conclusion does not contradict Theorem 4.1 because the theorem states that if a function's derivative is positive for all values in an interval, then it is increasing in that interval. In this example, the derivative is not positive throughout any interval containing the origin.

45. f is continuous on $[a, b]$, and $c \in (a, b)$ is a critical number.

(i) If $f'(x) > 0$ for all $x \in (a, c)$ and $f'(x) < 0$ for all $x \in (c, b)$, by Theorem 3.1, f is increasing on (a, c) and decreasing on (c, b), so $f(c) > f(x)$ for all $x \in (a, c)$ and $x \in (c, b)$. Thus $f(c)$ is a local max.

(ii) If $f'(x) < 0$ for all $x \in (a, c)$ and $f'(x) > 0$ for all $x \in (c, b)$, by Theorem 3.1, f is decreasing on (a, c) and increasing on (c, b). So $f(c) < f(x)$ for all $x \in (a, c)$ and $x \in (c, b)$. Thus $f(c)$ is a local min.

(iii) If $f'(x) > 0$ on (a, c) and (c, b), then $f(c) > f(x)$ for all $x \in (a, c)$ and $f(c) < f(x)$ for all $x \in (c, b)$, so c is not a local extremum.

If $f'(x) < 0$ on (a, c) and (c, b), then $f(c) < f(x)$ for all $x \in (a, c)$ and $f(c) > f(x)$ for all $x \in (c, b)$, so c is not a local extremum.

47. Let $f(x) = 2\sqrt{x}$, $g(x) = 3 - 1/x$. Then $f(1) = 2\sqrt{1} = 2$, and $g(1) = 3 - 1 = 2$, so $f(1) = g(1)$.
$f'(x) = \dfrac{1}{\sqrt{x}}$
$g'(x) = \dfrac{1}{x^2}$
So $f'(x) > g'(x)$ for all $x > 1$, and
$f(x) = 2\sqrt{x} > 3 - \dfrac{1}{x} = g(x)$
for all $x > 1$.

49. Let $f(x) = e^x$, $g(x) = x + 1$. Then $f(0) = e^0 = 1$, $g(0) = 0+1 = 1$, so $f(0) = g(0)$.
$f'(x) = e^x$, $g'(x) = 1$
So $f'(x) > g'(x)$ for $x > 0$.
Thus $f(x) = e^x > x + 1 = g(x)$ for $x > 0$.

51. Let $f(x) = 3 + e^{-x}$; then $f(0) = 4$, $f'(x) = -e^{-x} < 0$, so f is decreasing. But $f(x) = 3 + e^{-x} = 0$ has no solution.

53. The domain of $\sin^{-1} x$ is the interval $[-1, 1]$. The function is increasing on the entire domain.

55. TRUE. If $x_1 < x_2$, then $g(x_1) < g(x_2)$ since g is increasing, and then $f(g(x_1)) < f(g(x_2))$ since f is increasing.

57. $s(t) = \sqrt{t+4} = (t+4)^{1/2}$
$s'(t) = \dfrac{1}{2}(t+4)^{-1/2} = \dfrac{1}{2\sqrt{t+4}} > 0$
So total sales are always increasing at the rate of $\dfrac{1}{2\sqrt{t+4}}$ thousand dollars per month.

59. If the roots of the derivative are very close together, then the extrema will be very close together and difficult to see on a graph showing global behavior of the function. One function with the given derivative is

$f(x) = \frac{1}{3}x^3 - 0.01x + 2$

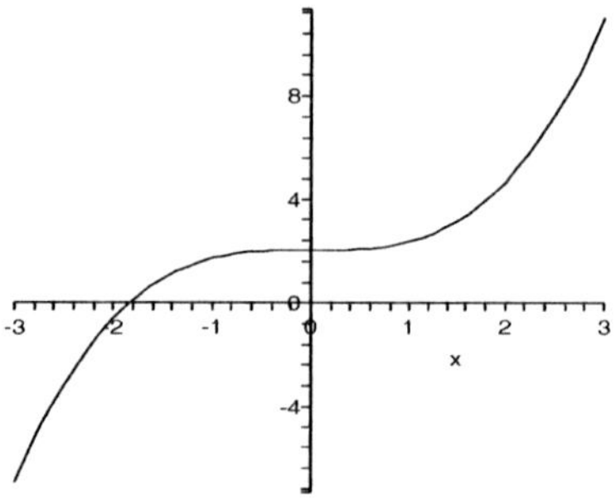

The two extreme points near $x = 0$ are impossible to detect from the graph using a usual scale.
To construct a degree 4 polynomial with two hidden extrema near $x = 1$ and another extrema (not hidden) near $x = 0$, we start with a derivative,

$g'(x) = x(x - 0.9)(x - 1.1)$
$= x^3 - 2x^2 + 0.99x.$

A function with this derivative is
$g(x) = \frac{1}{4}x^4 - \frac{2}{3}x^3 + \frac{0.99}{2}x^2 - 3$

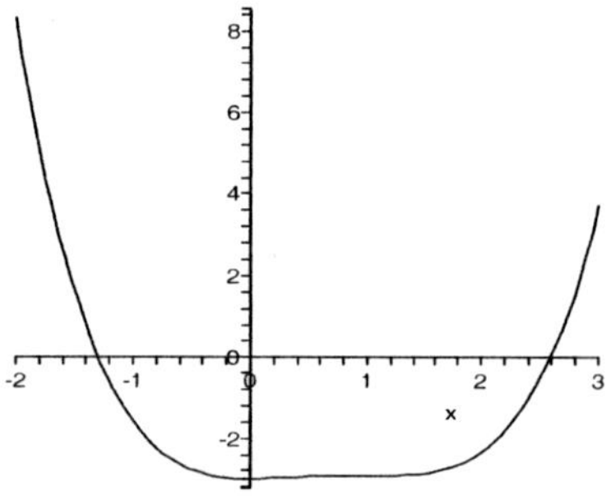

61. $f(x) = x^3 + bx^2 + cx + d$
$f'(x) = 3x^2 + 2bx + c$
$f'(x) \geq 0$ for all x if and only if
$(2b)^2 - 4(3)(c) \leq 0$
if and only if $4b^2 \leq 12c$
if and only if $b^2 \leq 3c$.

63. (a)

$$\mu'(-10) \approx \frac{0.0048 - 0.0043}{-12 - (-8)} = \frac{0.0005}{-4} = -0.000125$$

(b)

$$\mu'(-6) \approx \frac{0.0048 - 0.0043}{-4 - (-8)} = \frac{0.0005}{4} = 0.000125$$

Whether the warming of the ice due to skating makes it easier or harder depends on the current temperature of the ice. As seen from these examples, the coefficient of friction μ is decreasing when the temperature is $-10°$ and increasing when the temperature is $-6°$.

3.5 Concavity and the Second Derivative Test

1. $f'(x) = 3x^2 - 6x + 4$
$f''(x) = 6x - 6 = 6(x - 1)$
$f''(x) > 0$ on $(1, \infty)$
$f''(x) < 0$ on $(-\infty, 1)$
So f is concave down on $(-\infty, 1)$ and concave up on $(1, \infty)$.

3. $f(x) = x + \frac{1}{x} = x + x^{-1}$
$f'(x) = 1 - x^{-2}$
$f''(x) = 2x^{-3}$
$f''(x) > 0$ on $(0, \infty)$
$f''(x) < 0$ on $(-\infty, 0)$
So f is concave up on $(0, \infty)$ and concave down on $(-\infty, 0)$.

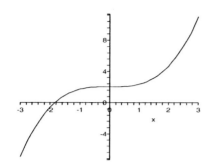

5. $f'(x) = \cos x + \sin x$
$f''(x) = -\sin x + \cos x$
$f''(x) < 0$ on $\ldots \left(\frac{\pi}{4}, \frac{5\pi}{4}\right) \cup \left(\frac{9\pi}{4}, \frac{13\pi}{4}\right) \ldots$
$f''(x) > 0$ on $\ldots \left(\frac{3\pi}{4}, \frac{\pi}{4}\right) \cup \left(\frac{5\pi}{4}, \frac{9\pi}{4}\right) \ldots$
f is concave down on $\ldots \left(\frac{\pi}{4}, \frac{5\pi}{4}\right) \cup \left(\frac{9\pi}{4}, \frac{13\pi}{4}\right) \ldots,$
concave up on $\ldots \left(\frac{3\pi}{4}, \frac{\pi}{4}\right) \cup \left(\frac{5\pi}{4}, \frac{9\pi}{4}\right) \ldots$

7. $f'(x) = \frac{4}{3}x^{1/3} + \frac{4}{3}x^{-2/3}$
$f''(x) = \frac{4}{9}x^{-2/3} + \frac{8}{9}x^{-5/3}$
$= \frac{4}{9x^{2/3}}\left(1 - \frac{2}{x}\right)$
The quantity $\dfrac{4}{9x^{2/3}}$ is never negative, so the sign of the second derivative is the same as the sign of $1 - \dfrac{2}{x}$. Hence the function is concave up for $x > 2$ and $x < 0$, and is concave down for $0 < x < 2$.

9. $f(x) = x^4 + 4x^3 - 1$
$f'(x) = 4x^3 + 12x^2 = x^2(4x + 12)$
So the critical numbers are $x = 0$ and $x = -3$.
$f''(x) = 12x^2 + 24x$
$f''(0) = 0$ so the second derivative test for $x = 0$ is inconclusive.
$f''(-3) = 36 > 0$ so $x = -3$ is a local minimum.

11. $f(x) = xe^{-x}$
$f'(x) = e^{-x} - xe^{-x} = e^{-x}(1 - x)$
So the only critical number is $x = 1$.
$f''(x) = -e^{-x} - e^{-x} + xe^{-x} = e^{-x}(-2 + x)$
$f''(1) = e^{-1}(-1) < 0$ so $x = 1$ is a local maximum.

13. $f(x) = \dfrac{x^2 - 5x + 4}{x}$

$$f'(x) = \frac{(2x - 5)x - (x^2 - 5x + 4)(1)}{x^2}$$
$$= \frac{x^2 - 4}{x^2}$$

So the critical numbers are $x = \pm 2$.

$$f''(x) = \frac{(2x)(x^2) - (x^2 - 4)(2x)}{x^4} = \frac{8x}{x^4}$$

$f''(2) = 1 > 0$ so $x = 2$ is a local minimum.
$f''(-2) = -1 < 0$ so $x = -2$ is a local maximum.

15. $y = (x^2 + 1)^{2/3}$

$$y' = \frac{2}{3}(x^2 + 1)^{-1/3}(2x)$$
$$= \frac{4x(x^2 + 1)^{-1/3}}{3}$$

So the only critical number is $x = 0$.

$$y'' = \frac{4}{3}\left[(x^2 + 1)^{-1/3} + \left(\frac{-2x^2}{3}\right)(x^2 + 1)^{-4/3}\right]$$
$$= \frac{4}{3}\frac{\left(x^2 + 1 - \frac{2x^2}{3}\right)}{(x^2 + 1)^{4/3}} = \frac{4}{9}\frac{(3x^2 + 3 - 2x^2)}{(x^2 + 1)^{4/3}}$$
$$= \frac{4}{9}\frac{(x^2 + 3)}{(x^2 + 1)^{4/3}}$$

So the function is concave up everywhere, decreasing for $x < 0$, and increasing for $x > 0$. Also $x = 0$ is a local min.

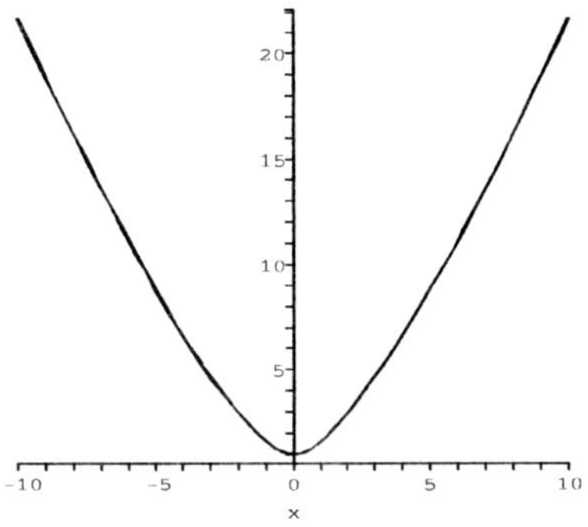

17. $f(x) = \dfrac{x^2}{x^2 - 9}$

$$f'(x) = \frac{2x(x^2 - 9) - x^2(2x)}{(x^2 - 9)^2}$$
$$= \frac{-18x}{(x^2 - 9)^2}$$
$$= \frac{-18x}{\{(x + 3)(x - 3)\}^2}$$
$$f''(x) = \frac{-18(x^2 - 9)^2 + 18x \cdot 2(x^2 - 9) \cdot 2x}{(x^2 - 9)^4}$$
$$= \frac{54x^2 + 162}{(x^2 - 9)^3}$$

$$= \frac{54(x^2+3)}{(x^2-9)^3}$$

$f'(x) > 0$ on $(-\infty, -3) \cup (-3, 0)$
$f'(x) < 0$ on $(0, 3) \cup (3, \infty)$
$f''(x) > 0$ on $(-\infty, -3) \cup (3, \infty)$
$f''(x) < 0$ on $(-3, 3)$
$f''(0) = \dfrac{162}{(-9)^3}$
f is increasing on $(-\infty, -3) \cup (-3, 0)$, decreasing on $(0, 3) \cup (3, \infty)$, concave up on $(-\infty, -3) \cup (3, \infty)$, concave down on $(-3, 3)$, $x = 0$ is a local max. f has a horizontal asymptote of $y = 1$ and vertical asymptotes at $x = \pm 3$.

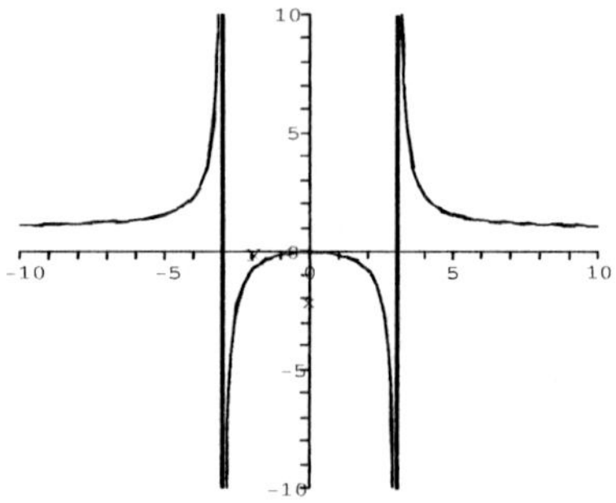

19. $f(x) = \sin x + \cos x$
$f'(x) = \cos x - \sin x$
$f''(x) = -\sin x - \cos x$
$f'(x) = 0$ when $x = \pi/4 + k\pi$ for all integers k. When k is even, $f''(\pi/4 + k\pi) = -\sqrt{2} < 0$ so $f(x)$ has a local maximum. When k is odd, $f''(\pi/4 + k\pi) = \sqrt{2} > 0$ so $f(x)$ has a local minimum.
$f'(x) < 0$ on the intervals of the form $(\pi/4 + 2k\pi, \pi/4 + (2k+1)\pi)$, so $f(x)$ is decreasing on these intervals.
$f'(x) > 0$ on the intervals of the form $(\pi/4 + (2k+1)\pi, \pi/4 + (2k+2)\pi)$, so $f(x)$ is increasing on these intervals.
$f''(x) > 0$ on the intervals of the form $(3\pi/4 + 2k\pi, 3\pi/4 + (2k+1)\pi)$ so $f(x)$ is concave up on these intervals.
$f''(x) < 0$ on the intervals of the form $(3\pi/4 + (2k+1)\pi, 3\pi/4 + (2k+2)\pi)$ so $f(x)$ is concave down on these intervals.

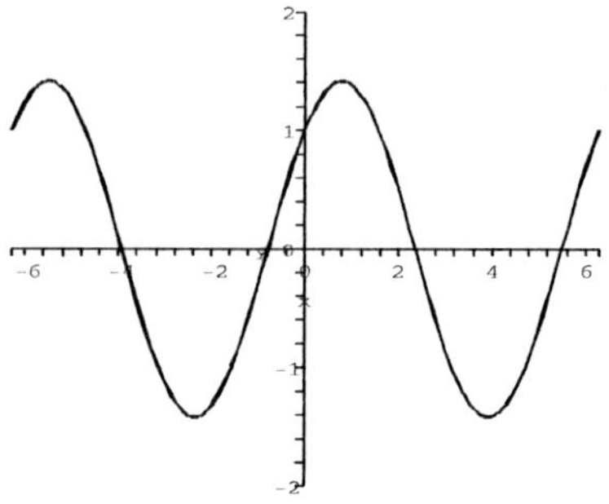

21. $f(x) = x^{3/4} - 4x^{1/4}$
Domain of $f(x)$ is $\{x | x \geq 0\}$.

$$f'(x) = \frac{3}{4}x^{-1/4} - x^{-3/4} = \frac{\frac{3}{4}\sqrt{x} - 1}{x^{3/4}}$$

So $x = 0$ and $x = 16/9$ are critical points, but because of the domain we only need to really consider the latter.
$f'(1) = -1/4$ so $f(x)$ is decreasing on $(0, 16/9)$.
$f'(4) = \dfrac{0.5}{4^{3/4}} > 0$ so $f(x)$ is increasing on $(16/9, \infty)$.
Thus $x = 16/9$ is the location of a local minimum for $f(x)$.
$f''(x) = \frac{-3}{16}x^{-5/4} + \frac{3}{4}x^{-7/4}$
$= \dfrac{\frac{-3}{16}\sqrt{x} + \frac{3}{4}}{x^{7/4}}$
The critical number here is $x = 16$. We find that $f''(x) > 0$ on the interval $(0, 16)$ (so $f(x)$ is concave up on this interval) and $f''(x) < 0$ on the interval $(16, \infty)$ (so $f(x)$ is concave down on this interval).

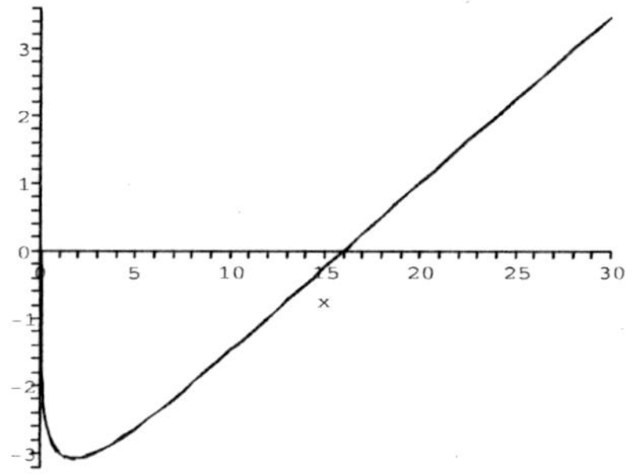

23. The easiest way to sketch this graph is to notice that

$$f(x) = x|x| = \begin{cases} x^2 & x \geq 0 \\ -x^2 & x < 0 \end{cases}$$

Since

$$f'(x) = \begin{cases} 2x & x \geq 0 \\ -2x & x < 0 \end{cases}$$

there is a critical point at $x = 0$. However, it is neither a local maximum nor a local minimum. Since

$$f''(x) = \begin{cases} 2 & x > 0 \\ -2 & x < 0 \end{cases}$$

there is an inflection point at the origin. Note that the second derivative does not exist at $x = 0$.

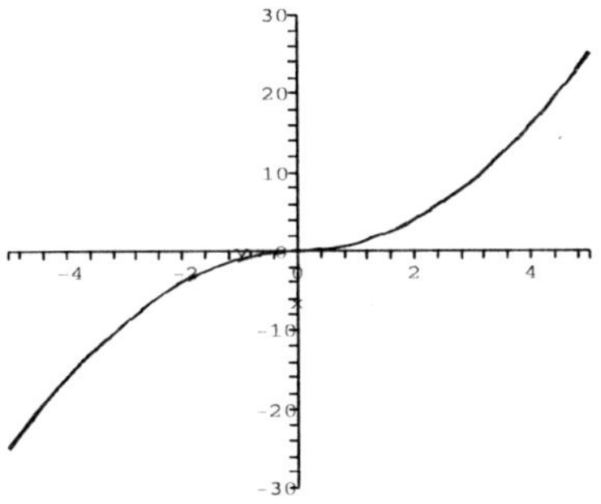

25. $f(x) = x^{1/5}(x+1) = x^{6/5} + x^{1/5}$

$$f'(x) = \tfrac{6}{5}x^{1/5} + \tfrac{1}{5}x^{-4/5}$$
$$= \tfrac{1}{5}x^{-4/5}(6x+1)$$
$$f''(x) = \tfrac{6}{25}x^{-4/5} - \tfrac{4}{25}x^{-9/5}$$
$$= \tfrac{2}{25}x^{-9/5}(3x-2)$$

Note that $f(0) = 0$, and yet the derivatives do not exist at $x = 0$. This means that there is a vertical tangent line at $x = 0$. The first derivative is negative for $x < -1/6$ and positive for $-1/6 < x < 0$ and $x > 0$. The second derivative is positive for $x < 0$ and $x > 2/3$, and negative for $0 < x < 2/3$. Thus, there is a local minimum at $x = -1/6$ and inflection points at $x = 0$ and $x = 2/3$.

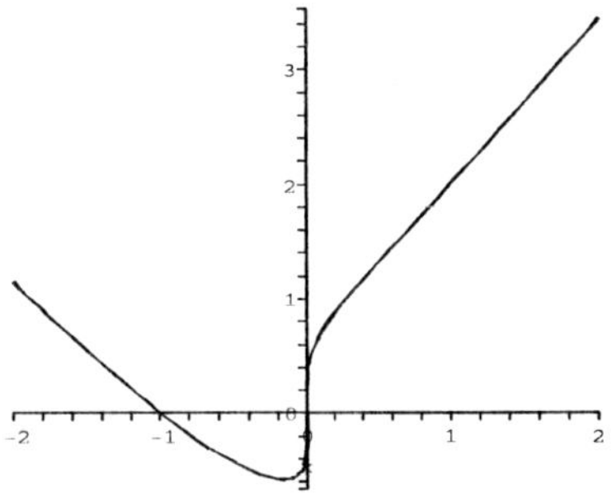

27. $f(x) = x^4 - 26x^3 + x$

$f'(x) = 4x^3 - 78x^2 + 1$

The critical numbers are approximately -0.1129, 0.1136 and 19.4993.

$f'(-1) < 0$ implies $f(x)$ is decreasing on $(-\infty, -0.1129)$.

$f'(0) > 0$ implies $f(x)$ is increasing on $(-0.1129, 0.1136)$.

$f'(1) < 0$ implies $f(x)$ is decreasing on $(0.1136, 19.4993)$.

$f'(20) > 0$ implies $f(x)$ is increasing on $(19.4993, \infty)$.

Thus $f(x)$ has local minimums at $x = -0.1129$ and $x = 19.4993$ and a local maximum at $x = 0.1136$.

$f''(x) = 12x^2 - 156x = x(12x - 156)$

The critical numbers are $x = 0$ and $x = 13$.

$f''(-1) > 0$ implies $f(x)$ is concave up on $(-\infty, 0)$.

$f''(1) < 0$ implies $f(x)$ is concave down on $(0, 13)$.

$f''(20) > 0$ implies $f(x)$ is concave up on $(13, \infty)$.

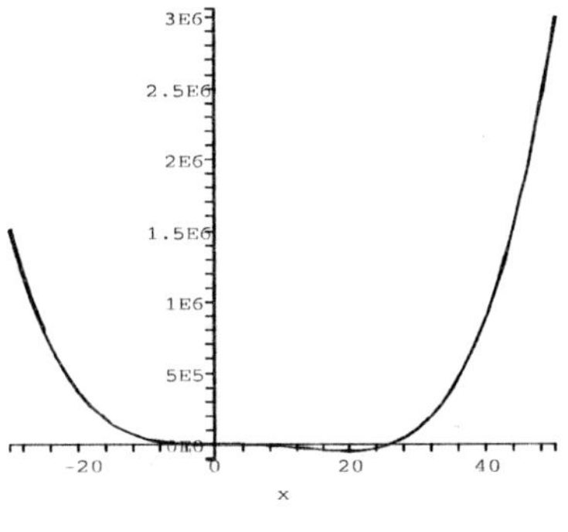

29. $y = \sqrt[3]{x^2 - 1}$
$y' = \dfrac{4x}{3(2x^2-1)^{2/3}} = 0$ at $x = 0$ and is undefined at $x = \pm\sqrt{1/2}$.
$y'' = \dfrac{-4(2x^2+3)}{9(2x^2-1)^{5/3}}$ is never 0, and is undefined where y' is.
The function changes concavity at $x = \pm\sqrt{1/2}$, so these are inflection points. The slope does not change at these values, so they are not extrema. The Second Derivative Test shows that $x = 0$ is a minimum.

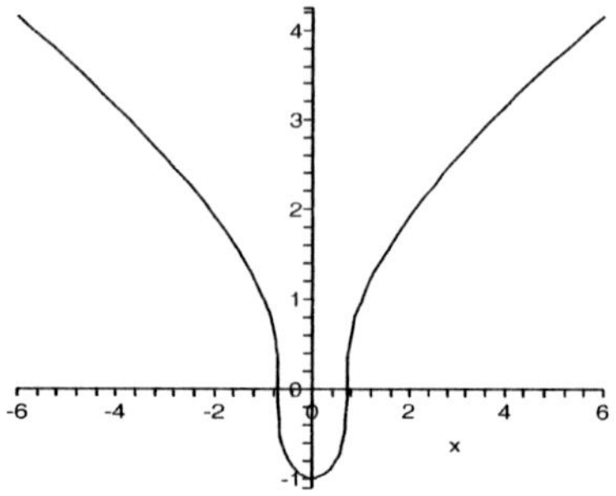

31. $f(x) = x^4 - 16x^3 + 42x^2 - 39.6x + 14$
$f'(x) = 4x^3 - 48x^2 + 84x - 39.6$
$f''(x) = 12x^2 - 96x + 84$
$= 12(x^2 - 8x + 7)$
$= 12(x-7)(x-1)$
$f'(x) > 0$ on $(.8952, 1.106) \cup (9.9987, \infty)$
$f'(x) < 0$ on $(-\infty, .8952) \cup (1.106, 9.9987)$
$f''(x) > 0$ on $(-\infty, 1) \cup (7, \infty)$
$f''(x) < 0$ on $(1, 7)$
f is increasing on $(.8952, 1.106)$ and on $(9.9987, \infty)$, decreasing on $(-\infty, .8952)$ and on $(1.106, 9.9987)$, concave up on $(-\infty, 1) \cup (7, \infty)$, concave down on $(1, 7)$, $x = .8952, 9.9987$ are local min, $x = 1.106$ is local max, $x = 1, 7$ are inflection points.

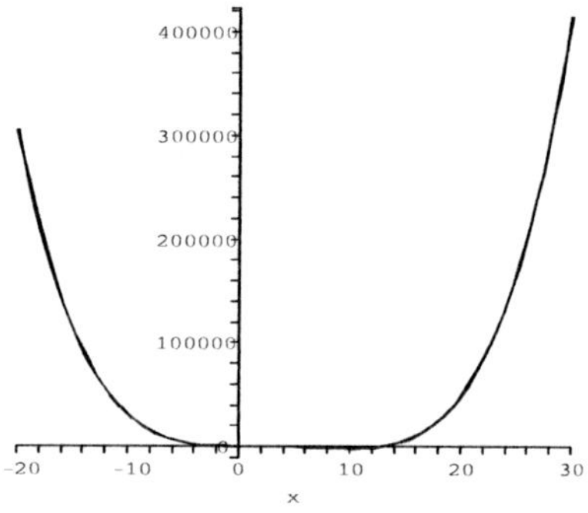

33. $f(x) = x\sqrt{x^2 - 4}$; f undefined on $(-2, 2)$
$f'(x) = \sqrt{x^2-4} + x\left(\frac{1}{2}\right)(x^2-4)^{-1/2}(2x)$
$= \sqrt{x^2-4} + \dfrac{x^2}{\sqrt{x^2-4}}$
$= \dfrac{2x^2-4}{\sqrt{x^2-4}}$
$f''(x) = \dfrac{4x\sqrt{x^2-4} - (2x^2-4)\frac{1}{2}(x^2-4)^{-1/2}(2x)}{x^2-4}$
$= \dfrac{4x(x^2-4)x - (2x^2-4)}{(x^2-4)^{3/2}}$
$= \dfrac{2x^3-12x}{(x^2-4)^{3/2}} = \dfrac{2x(x^2-6)}{(x^2-4)^{3/2}}$
$f'(x) > 0$ on $(-\infty, -2) \cup (2, \infty)$
$f''(x) > 0$ on $\left(-\sqrt{6}, 2\right) \cup \left(\sqrt{6}, \infty\right)$
$f''(x) < 0$ on $\left(-\infty, -\sqrt{6}\right) \cup \left(2, \sqrt{6}\right)$
f is increasing on $(-\infty, -2)$ and on $(2, \infty)$, concave up on $\left(-\sqrt{6}, -2\right) \cup \left(\sqrt{6}, \infty\right)$, concave down on $\left(-\infty, -\sqrt{6}\right) \cup \left(2, \sqrt{6}\right)$, $x = \pm\sqrt{6}$ are inflection points.

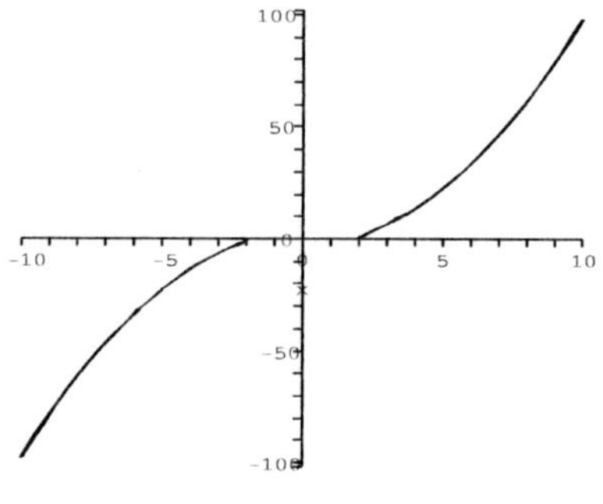

35. The function has horizontal asymp-

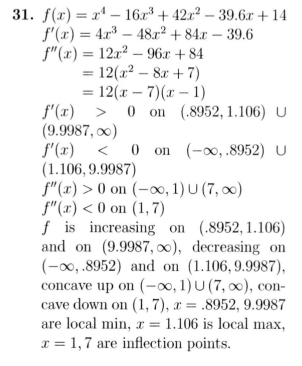

tote $y = 0$, and is undefined at $x = \pm 1$.
$y' = \dfrac{-2x}{x^4 - 2x^2 + 2} = 0$
only when $x = 0$.
$y'' = \dfrac{2(3x^4 - 2x^2 - 2)}{(x^4 - 2x^2 + 2)^2} = 0$
at approximately $x = \pm 1.1024$ and changes sign there, so these are inflection points (very easy to miss by looking at the graph). The Second Derivative Test shows that $x = 0$ is a local maximum.

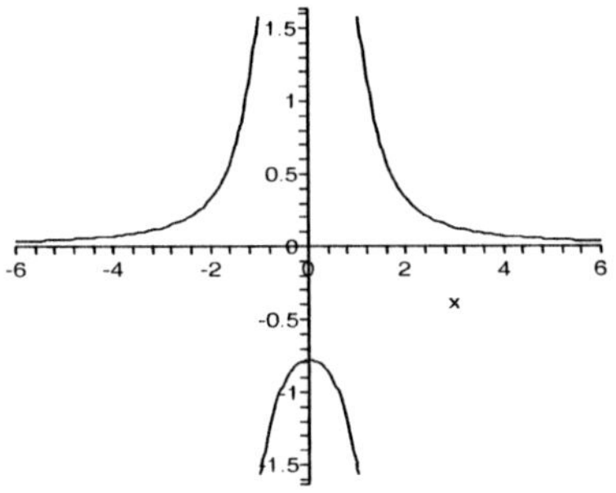

37. $f(x)$ is concave up on $(-\infty, -0.5)$ and $(0.5, \infty)$;
$f(x)$ is concave down on $(-0.5, 0.5)$.

39. $f(x)$ is concave up on $(1, \infty)$;
$f(x)$ is concave down on $(-\infty, 1)$.

41. One possible graph:

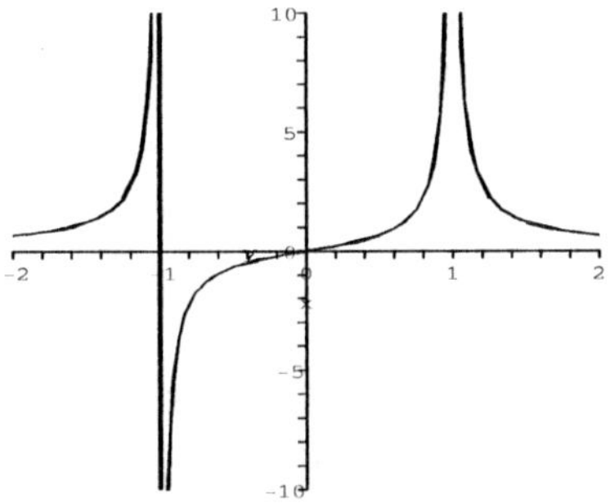

43. One possible graph:

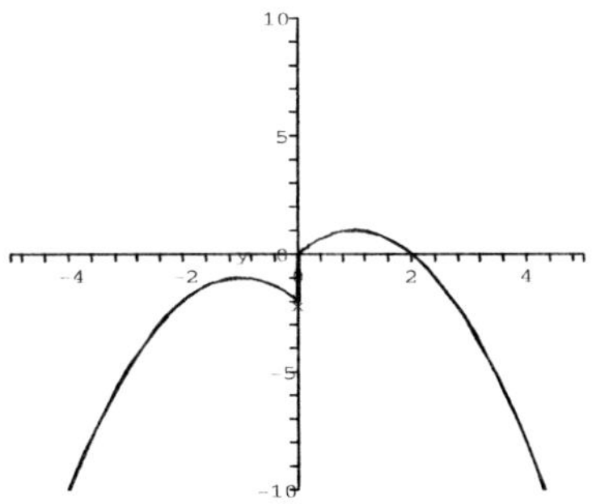

45. $f(x) = ax^3 + bx^2 + cx + d$
$f'(x) = 3ax^2 + 2bx + c$
$f''(x) = 6ax + 2b$
Thus, $f''(x) = 0$ for $x = -b/3a$. Since f'' changes sign at this point, f has an inflection point at $x = -b/3a$. Note that $a \neq 0$.

For the quartic function (where again $a \neq 0$),
$f(x) = ax^4 + bx^3 + cx^2 + dx + e$
$f'(x) = 4ax^3 + 3bx^2 + 2cx + d$
$f''(x) = 12ax^2 + 6bx + 2c$
$\quad = 2(6ax^2 + 3bx + c)$
The second derivative is zero when
$x = \dfrac{-3b \pm \sqrt{9b^2 - 24ac}}{12a}$
$= \dfrac{-3b \pm \sqrt{3(3b^2 - 8ac)}}{12a}$
There are two distinct solutions to the previous equation (and therefore two inflection points) if and only if $3b^2 - 8ac > 0$.

47. The function has the following properties:
increasing on $(0, \infty)$;
decreasing on $(-\infty, 0)$;
local minimum at $x = 0$;
concave up on $(-\infty, \infty)$;
no inflection points.

49. For #47:
increasing on $(-\infty, -1)$ and $(1, \infty)$;
decreasing on $(-1, 1)$;
local maximum at $x = -1$;

local minimum at $x = 1$;
concave up on $(0, \infty)$;
concave down on $(-\infty, 0)$;
inflection point at $x = 0$.

For # 48:
increasing on $(0, 2)$ and $(2, \infty)$;
decreasing on $(-\infty, 0)$;
local minimum at $x = 0$;
concave up on $(-\infty, 1)$ and $(2, \infty)$;
concave down on $(1, 2)$;
inflection points at $x = 1$ and $x = 2$.

51. We need to know $w'(0)$ to know if the depth is increasing.

53. $s(x) = -3x^3 + 270x^2 - 3600x + 18000$
$s'(x) = -9x^2 + 540x - 3600$
$s''(x) = -18x + 540 = 0$
$x = 30$. This is a max because the graph of $s'(x)$ is a parabola opening down. So spend \$30,000 on advertising to maximize the rate of change of sales. This is also the inflection point of $s(x)$.

55. $C(x) = .01x^2 + 40x + 3600$
$\overline{C}(x) = \dfrac{C(x)}{x} = .01x + 40 + 3600x^{-1}$
$\overline{C}'(x) = .01 - 3600x^{-2} = 0$
$x = 600$. This is a min because $\bar{C}''(x) = 7200x^{-3} > 0$ for $x > 0$, so the graph is concave up. So manufacture 600 units to minimize average cost.

57. Both functions are increasing for $x > 0$ and have the same asymptote, $y = 1$, so that is no help. However, if
$$f(x) = \frac{x}{27 + x},$$
then
$$f'(x) = \frac{27}{(27 + x)^2}.$$
Hence $f'(x)$ is decreasing for $x > 0$ and so f has no inflection points on this interval. On the other hand, if
$$f(x) = \frac{x^3}{c^3 + x^3},$$
then
$$f'(x) = \frac{3cx^2}{(c^3 + x^3)^2}$$
and
$$f''(x) = \frac{6c^3x(c^3 - 2x^3)}{(c^3 + x^3)^3}$$
and so there is an inflection point at $x = c/\sqrt[3]{2}$. When $c = 27$, $27/\sqrt[3]{2} \approx 21.4$, in excellent agreement with the given graph.

59. Let $f(x) = -1 - x^2$. Then
$f'(x) = -2x$
$f''(x) = -2$
so f is concave down for all x, but $-1 - x^2 = 0$ has no solution.

61. Since the tangent line points above the sun, the sun appears higher in the sky than it really is.

3.6 Overview of Curve Sketching

1. $f(x) = x^3 - 3x^2 + 3x$
$= x(x^2 - 3x + 3)$
The only x-intercept is $x = 0$; the y-intercept is $(0, 0)$.
$f'(x) = 3x^2 - 6x + 3$
$= 3(x^2 - 2x + 1) = 3(x - 1)^2$
$f'(x) > 0$ for all x, so $f(x)$ is increasing for all x and has no local extrema.
$f''(x) = 6x - 6 = 6(x - 1)$
There is an inflection point at $x = 1$: $f(x)$ is concave down on $(-\infty, 1)$ and concave up on $(1, \infty)$.
Finally, $f(x) \to \infty$ as $x \to \infty$ and $f(x) \to -\infty$ as $x \to -\infty$.

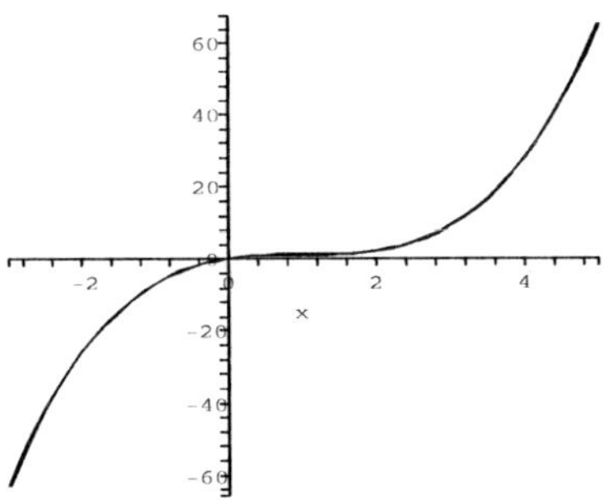

3. $f(x) = x^5 - 2x^3 + 1$
The x-intercepts are $x = 1$ and $x \approx -1.5129$; the y-intercept is $(0, 1)$.
$f'(x) = 5x^4 - 6x^2 = x^2(5x^2 - 6)$
The critical numbers are $x = 0$ and $x = \pm\sqrt{6/5}$. Plugging values from each of the intervals into $f'(x)$, we find that $f'(x) > 0$ on $(-\infty, -\sqrt{6/5})$ and $(\sqrt{6/5}, \infty)$ so $f(x)$ is increasing on these intervals. $f'(x) < 0$ on $(-\sqrt{6/5}, 0)$ and $(0, \sqrt{6/5})$ so $f(x)$ is decreasing on these intervals. Thus $f(x)$ has a local maximum at $-\sqrt{6/5}$ and a local minimum at $\sqrt{6/5}$.
$f''(x) = 20x^3 - 12x = 4x(5x^2 - 3)$
The critical numbers are $x = 0$ and $x = \pm\sqrt{3/5}$. Plugging values from each of the intervals into $f''(x)$, we find that $f''(x) > 0$ on $(-\sqrt{3/5}, 0)$ and $(\sqrt{3/5}, \infty)$ so $f(x)$ is concave up on these intervals. $f''(x) < 0$ on $(-\infty, -\sqrt{3/5})$ and $(0, \sqrt{3/5})$ so $f(x)$ is concave down on these intervals. Thus $f(x)$ has inflection points at all three of these critical numbers.
Finally, $f(x) \to \infty$ as $x \to \infty$ and $f(x) \to -\infty$ as $x \to -\infty$.

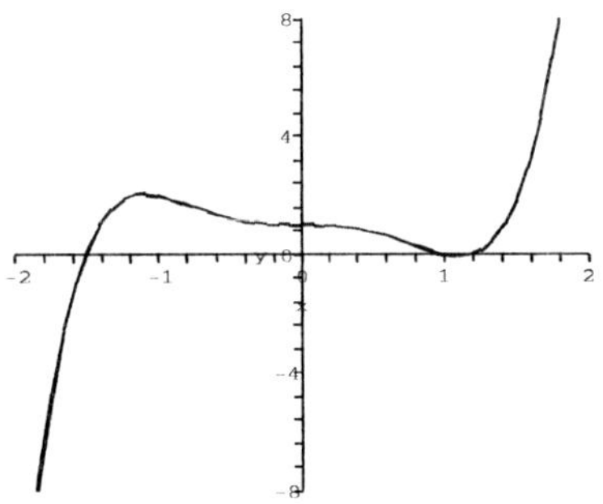

5. $f(x) = x + \dfrac{4}{x} = \dfrac{x^2 + 4}{x}$
This function has no x- or y-intercepts. The domain is $\{x | x \neq 0\}$. $f(x)$ has a vertical asymptote at $x = 0$ such that $f(x) \to -\infty$ as $x \to 0^-$ and $f(x) \to \infty$ as $x \to 0^+$.

$$f'(x) = 1 - 4x^{-2} = \frac{x^2 - 4}{x^2}$$

The critical numbers are $x = \pm 2$. We find that $f'(x) > 0$ on $(-\infty, -2)$ and $(2, \infty)$ so $f(x)$ is increasing on these intervals. $f'(x) < 0$ on $(-2, 0)$ and $(0, 2)$, so $f(x)$ is decreasing on these intervals. Thus $f(x)$ has a local maximum at $x = -2$ and a local minimum at $x = 2$.
$f''(x) = 8x^{-3}$
$f''(x) < 0$ on $(-\infty, 0)$ so $f(x)$ is concave down on this interval and $f''(x) > 0$ on $(0, \infty)$ so $f(x)$ is concave up on this interval, but $f(x)$ has an asymptote (not an inflection point) at $x = 0$.
Finally, $f(x) \to -\infty$ as $x \to -\infty$ and $f(x) \to \infty$ as $x \to \infty$.

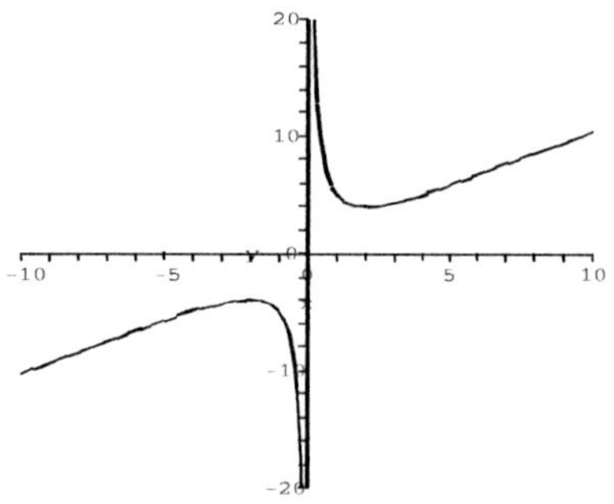

7. $f(x) = x \ln x$
The domain is $\{x|x > 0\}$. There is an x-intercept at $x = 1$ and no y-intercept.
$f'(x) = \ln x + 1$
The only critical number is $x = e^{-1}$. $f'(x) < 0$ on $(0, e^{-1})$ and $f'(x) > 0$ on (e^{-1}, ∞) so $f(x)$ is decreasing on $(0, e^{-1})$ and increasing on (e^{-1}, ∞). Thus $f(x)$ has a local minimum at $x = e^{-1}$.
$f''(x) = 1/x$, which is positive for all x in the domain of f, so $f(x)$ is always concave up.
$f(x) \to \infty$ as $x \to \infty$.

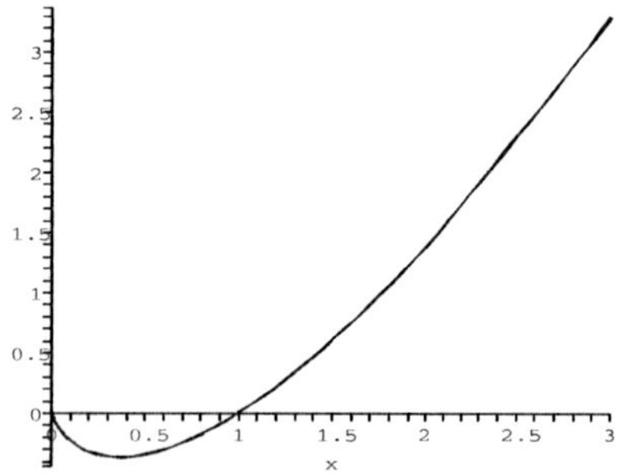

9. $f(x) = \sqrt{x^2+1}$
The y-intercept is $(0, 1)$. There are no x-intercepts.

$$f'(x) = \frac{1}{2}(x^2+1)^{-1/2}2x = \frac{x}{\sqrt{x^2+1}}$$

The only critical number is $x = 0$. $f'(x) < 0$ when $x < 0$ and $f'(x) > 0$ when $x > 0$ so $f(x)$ is increasing on $(0, \infty)$ and decreasing on $(-\infty, 0)$. Thus $f(x)$ has a local minimum at $x = 0$.

$$f''(x) = \frac{\sqrt{x^2+1} - x\frac{1}{2}(x^2+1)^{-1/2}2x}{x^2+1}$$
$$= \frac{1}{(x^2+1)^{3/2}}$$

Since $f''(x) > 0$ for all x, we see that $f(x)$ is concave up for all x.
$f(x) \to \infty$ as $x \to \pm\infty$.

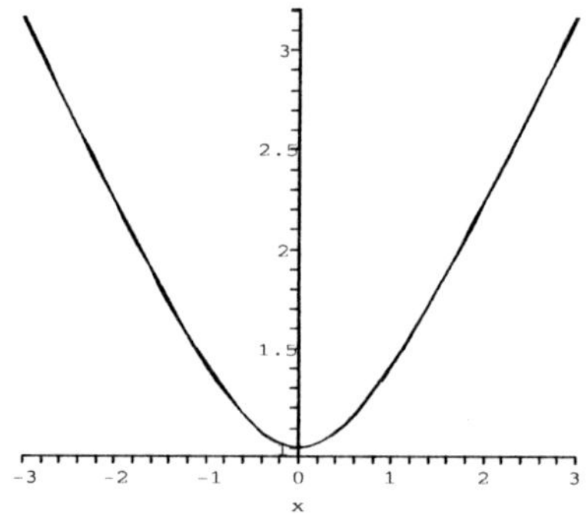

11. $f(x) = \dfrac{4x}{x^2 - x + 1}$
The function has horizontal asymptote at $y = 0$.
$$f'(x) = \frac{4(1-x^2)}{(x^2-x+1)^2}$$
There are critical numbers at $x = \pm 1$.
$$f''(x) = \frac{8(x^3 - 3x + 1)}{(x^2-x+1)^3}$$
with critical numbers at approximately $x = -1.8793$, 0.3473, and 1.5321. $f''(x)$ changes sign at these values, so these are inflection points. The Second Derivative test shows that $x = -1$ is a minimum, and $x = 1$ is a maximum.

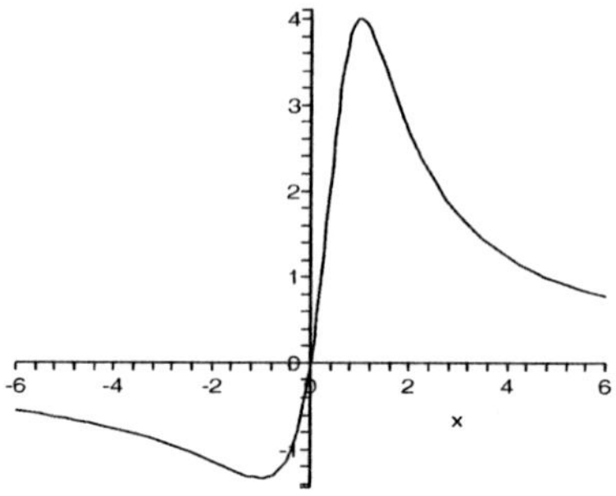

13. $f(x) = (x^3 - 3x^2 + 2x)^{1/3}$

$$f'(x) = \frac{3x^2 - 6x + 2}{3(x^3 - 3x^2 + 2x)^{2/3}}$$

There are critical numbers at $x = \dfrac{3 \pm \sqrt{3}}{3}$, 0, 1 and 2.

$$f''(x) = \frac{-6x^2 + 12x - 8}{9(x^3 - 3x^2 + 2x)^{5/3}}$$

with critical numbers $x = 0$, 1 and 2. $f''(x)$ changes sign at these values, so these are inflection points. The Second Derivative test shows that $x = \dfrac{3+\sqrt{3}}{3}$ is a minimum, and $x = \dfrac{3-\sqrt{3}}{3}$ is a maximum.

$f(x) \to -\infty$ as $x \to -\infty$ and $f(x) \to \infty$ as $x \to \infty$.

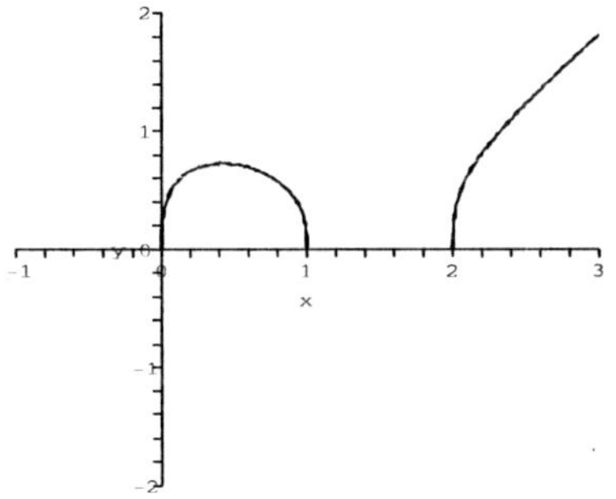

15. $f(x) = x^5 - 5x = x(x^4 - 5)$

x-intercepts are $x = 0$ and $x = \pm\sqrt[4]{5}$. The y-intercept is $(0, 0)$.

$f'(x) = 5x^4 - 5 = 5(x^4 - 1)$

The critical numbers are $x = \pm 1$.

$f''(x) = 20x^3$ so $x = -1$ is a local maximum and $x = 1$ is a local minimum. $f(x)$ is increasing on $(-\infty, -1)$ and $(1, \infty)$ and decreasing on $(-1, 1)$. It is concave up on $(0, \infty)$ and concave down on $(-\infty, 0)$, with an inflection point at $x = 0$.

$f(x) \to -\infty$ as $x \to -\infty$ and $f(x) \to \infty$ as $x \to \infty$.

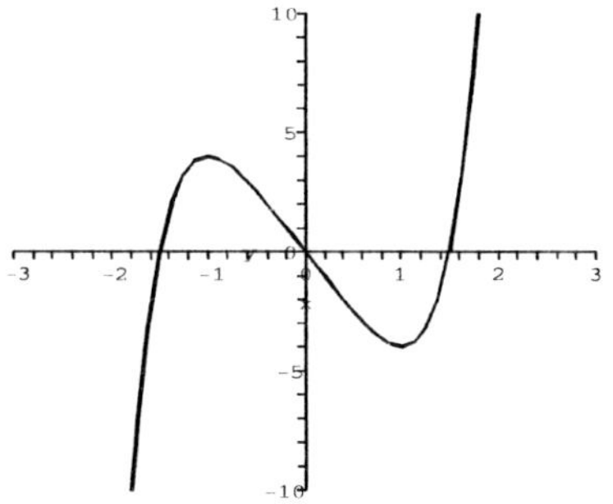

17.

$$f(x) = e^{-2/x}$$

$$f'(x) = e^{-2/x}\left(\frac{2}{x^2}\right) = \frac{2}{x^2}e^{-2/x}$$

$$f''(x) = \frac{-4}{x^3}e^{-2/x} + \frac{2}{x^2}e^{-2/x}\left(\frac{2}{x^2}\right)$$

$$= \frac{4}{x^4}e^{-2/x} - \frac{4}{x^3}e^{-2/x}$$

$f'(x) > 0$ on $(-\infty, 0) \cup (0, \infty)$
$f''(x) > 0$ on $(-\infty, 0) \cup (0, 1)$
$f''(x) < 0$ on $(1, \infty)$
f increasing on $(-\infty, 0)$ and on $(0, \infty)$, concave up on $(-\infty, 0) \cup (0, 1)$, concave down on $(1, \infty)$, inflection point at $x = 1$. f is undefined at $x = 0$.

$\lim_{x \to 0^+} e^{-2/x} = \lim_{x \to 0^+} \dfrac{1}{e^{2/x}} = 0$ and
$\lim_{x \to 0^-} e^{-2/x} = \infty$

So f has a vertical asymptote at $x = 0$. $\lim_{x \to \infty} e^{-2/x} = \lim_{x \to -\infty} e^{-2/x} = 1$

So f has a horizontal asymptote at $y = 1$.

Global graph of $f(x)$:

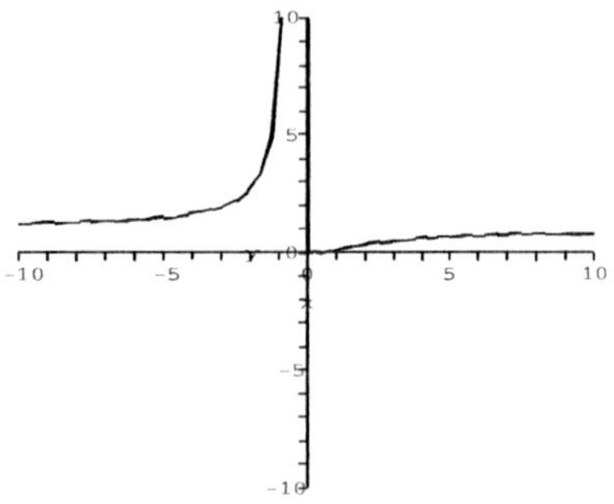

Local graph of $f(x)$:

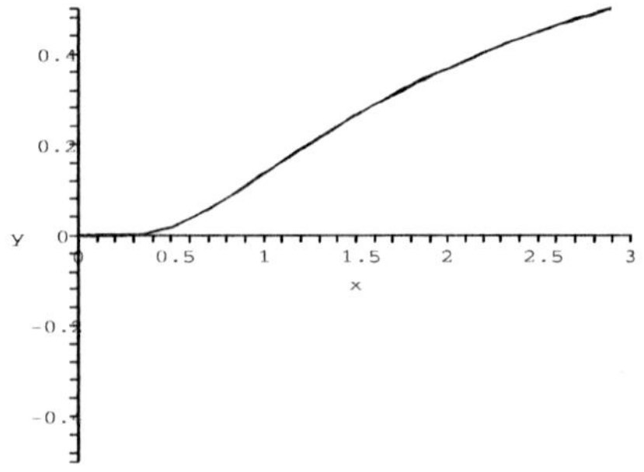

19. $f(x) = (x^3 - 3x^2 + 2x)^{2/3}$
$f'(x) = \dfrac{2(3x^2 - 6x + 2)}{3(x^3 - 3x^2 + 2x)^{1/3}}$
There are critical numbers at $x = \dfrac{3 \pm \sqrt{3}}{3}$, 0, 1 and 2.
$f''(x) = \dfrac{18x^4 - 72x^3 + 84x^2 - 24x - 8}{9(x^3 - 3x^2 + 2x)^{4/3}}$
with critical numbers $x = 0$, 1 and 2 and $x \approx -0.1883$ and 2.1883. $f''(x)$ changes sign at these last two values, so these are inflection points. The Second Derivative test shows that $x = \dfrac{3 \pm \sqrt{3}}{3}$ are both maxima. Local minima occur at $x = 0$, 1 and 2.
$f(x) \to \infty$ as $x \to \pm\infty$.

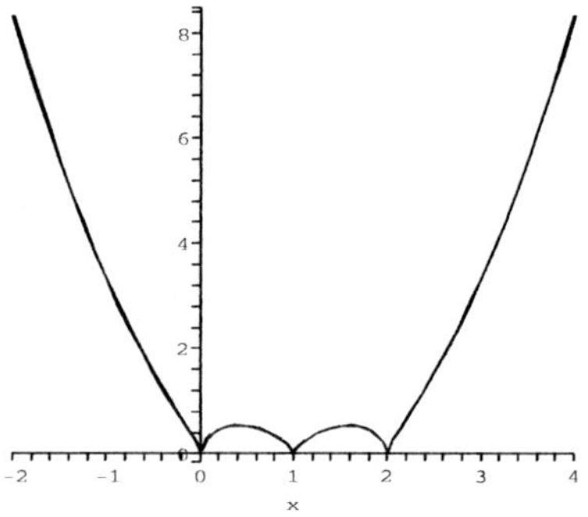

21. $f(x) = \dfrac{x^2 + 1}{3x^2 - 1}$
Note that $x = \pm\sqrt{1/3}$ are not in the domain of the function, but yield vertical asymptotes.

$$f'(x) = \frac{2x(3x^2 - 1) - (x^2 + 1)(6x)}{(3x^2 - 1)^2} = \frac{(6x^3 - 2x) - (6x^3 + 6x)}{(3x^2 - 1)^2} = \frac{-8x}{(3x^2 - 1)^2}$$

So the only critical point is $x = 0$.

$f'(x) > 0$ for $x < 0$
$f'(x) < 0$ for $x > 0$
so f is increasing on $(-\infty, -\sqrt{1/3})$ and on $(-\sqrt{1/3}, 0)$; decreasing on $(0, \sqrt{1/3})$ and on $(\sqrt{1/3}, \infty)$. Thus there is a local max at $x = 0$.
$f''(x) = 8 \cdot \dfrac{9x^2 + 1}{(3x^2 - 1)^3}$
$f''(x) > 0$ on $(-\infty, -\sqrt{1/3}) \cup (\sqrt{1/3}, \infty)$
$f''(x) < 0$ on $(-\sqrt{1/3}, \sqrt{1/3})$
Hence f is concave up on $(-\infty, -\sqrt{1/3})$ and on $(\sqrt{1/3}, \infty)$; concave down on $(-\sqrt{1/3}, \sqrt{1/3})$.

Finally, when $|x|$ is large, the function approached 1/3, so $y = 1/3$ is a horizontal asymptote.

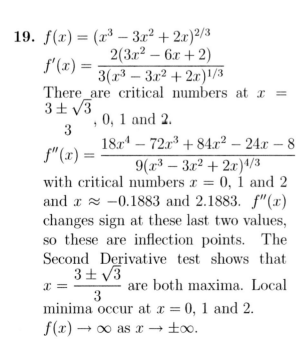

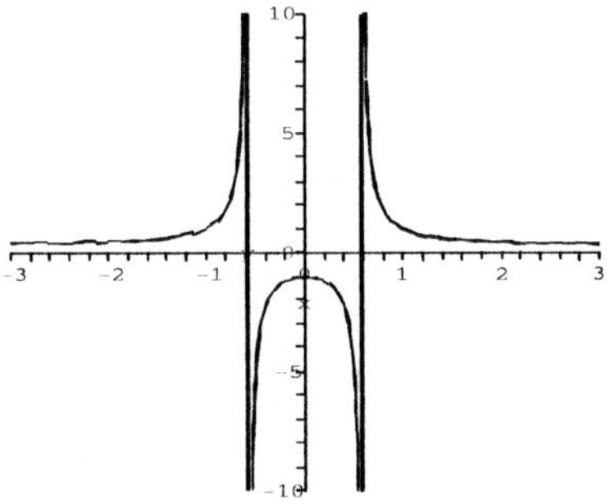

23. $f(x) = \dfrac{5x}{x^3 - x + 1}$
Looking at the graph of $x^3 - x + 1$, we see that there is one real root, at approximately -1.325; so the domain of the function is all x except for this one point, and $x = -1.325$ will be a vertical asymptote. There is a horizontal asymptote of $y = 0$.

$$f'(x) = 5\frac{1 - 2x^3}{(x^3 - x - 1)^2}$$

The only critical point is $x = \sqrt[3]{1/2}$. By the first derivative test, this is a local max.

$$f''(x) = 10\frac{3x^5 + x^3 - 6x^2 + 1}{(x^3 - x + 1)^3}$$

The numerator of f'' has three real roots, which are approximately $x = -.39018$, $x = .43347$, and $x = 1.1077$.
$f''(x) > 0$ on $(-\infty, -1.325) \cup (-.390, .433) \cup (1.108, \infty)$
$f''(x) < 0$ on $(-1.325, -.390) \cup (.433, 1.108)$
So f is concave up on $(-\infty, -1.325) \cup (-.390, .433) \cup (1.108, \infty)$ and concave down on $(-1.325, -.390) \cup (.433, 1.108)$. Hence $x = -.39018$, $x = .43347$, and $x = 1.1077$ are inflection points.

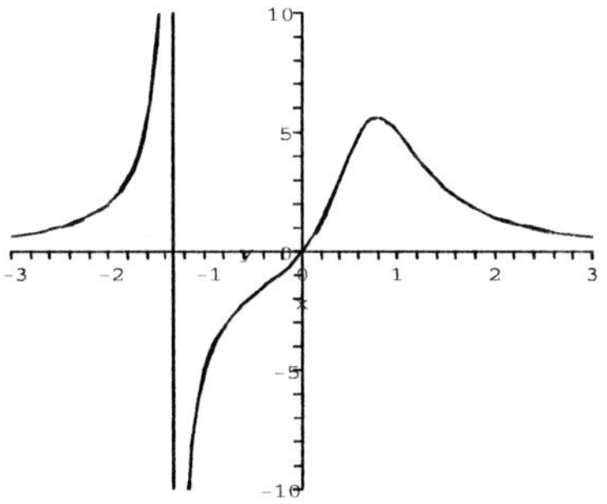

25. $f(x) = x^2\sqrt{x^2 - 9}$
f is undefined on $(-3, 3)$.
$f'(x) =$

$$2x\sqrt{x^2 - 9} + x^2\left(\frac{1}{2}(x^2 - 9)^{-1/2} \cdot 2x\right)$$
$$= 2x\sqrt{x^2 - 9} + \frac{x^3}{\sqrt{x^2 - 9}}$$
$$= \frac{2x(x^2 - 9) + x^3}{\sqrt{x^2 - 9}}$$
$$= \frac{3x^3 - 18x}{\sqrt{x^2 - 9}} = \frac{3x(x^2 - 6)}{\sqrt{x^2 - 9}}$$
$$= \frac{3x(x + \sqrt{6})(x - \sqrt{6})}{\sqrt{x^2 - 9}}$$

Critical points ± 3. (Note that f is undefined at $x = 0$, $\pm\sqrt{6}$.)

$$f''(x) = \frac{(9x^2 - 18)\sqrt{x^2 - 9}}{x^2 - 9} - \frac{(3x^3 - 18x) \cdot \frac{1}{2}(x^2 - 9)^{-1/2} \cdot 2x}{x^2 - 9}$$
$$= \frac{(9x^2 - 18)(x^2 - 9) - x(3x^3 - 18x)}{(x^2 - 9)^{3/2}}$$
$$= \frac{(6x^4 - 81x^2 + 162)}{(x^2 - 9)^{3/2}}$$

$f''(x) = 0$ when

$$x^2 = \frac{81 \pm \sqrt{81^2 - 4(6)(162)}}{2(6)}$$
$$= \frac{81 \pm \sqrt{2673}}{12} = \frac{1}{4}(27 \pm \sqrt{297})$$

So $x \approx \pm 3.325$ or $x \approx \pm 1.562$, but these latter values are not in the same domain. So only ± 3.325 are potential

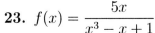

inflection points.
$f'(x) > 0$ on $(3, \infty)$
$f'(x) < 0$ on $(-\infty, -3)$
$f''(x) > 0$ on $(-\infty, -3.3) \cup (3.3, \infty)$
$f''(x) < 0$ on $(-3.3, 3) \cup (3, 3.3)$
f is increasing on $(3, \infty)$, decreasing on $(-\infty, -3)$, concave up on $(-\infty, -3.3) \cup (3.3, \infty)$, concave down on $(-3.3, -3) \cup (3, 3.3)$. $x = \pm 3.3$ are inflection points.
Global graph of $f(x)$:

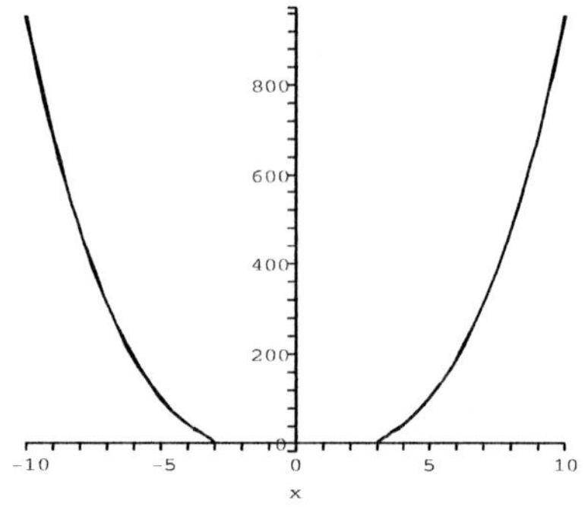

Local graph of $f(x)$:

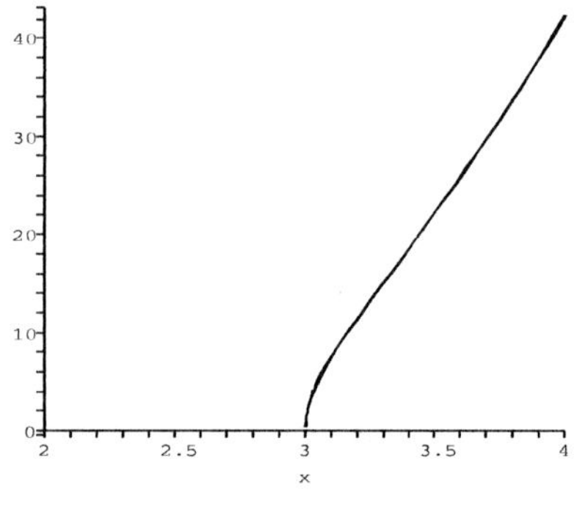

27. $f(x) = e^{-2x} \sin x$
$f'(x) = e^{-2x}(\cos x - 2\sin x)$
$f''(x) = e^{-2x}(3\sin x - 4\cos x)$
$f'(x) = 0$ when $\cos x = 2\sin x$; that is, when $\tan x = 1/2$; that is, when $x = k\pi + \tan^{-1}(1/2)$, where k is any integer.
$f'(x) < 0$, and f is decreasing, on intervals of the form $(2k\pi + \tan^{-1}(\frac{1}{2}), (2k+1)\pi + \tan^{-1}(\frac{1}{2}))$
$f'(x) > 0$ and f is increasing, on intervals of the form $((2k-1)\pi + \tan^{-1}(\frac{1}{2}), 2k\pi + \tan^{-1}(\frac{1}{2}))$
Hence f has a local max at $x = 2k\pi + \tan^{-1}(1/2)$ and a local min at $x = (2k+1)\pi + \tan^{-1}(1/2)$.
$f''(x) = 0$ when $3\sin x = 4\cos x$; that is, when $\tan x = 4/3$; that is, when $x = k\pi + \tan^{-1}(4/3)$. The sign of f'' changes at each of these points, so all of them are inflection points.

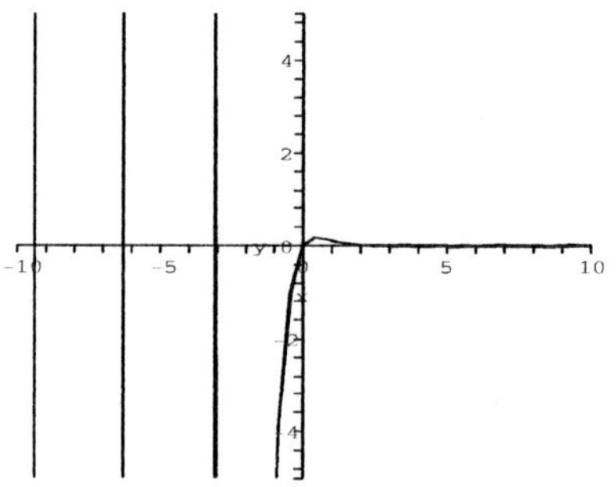

29. $f(x) = x^4 - 16x^3 + 42x^2 - 39.6x + 14$
$f'(x) = 4x^3 - 48x^2 + 84x - 39.6$
$f''(x) = 12x^2 - 96x + 84$
$= 12(x^2 - 8x + 7)$
$= 12(x-7)(x-1)$
$f'(x) > 0$ on $(.8952, 1.106) \cup (9.9987, \infty)$
$f'(x) < 0$ on $(-\infty, .8952) \cup (1.106, 9.9987)$
$f''(x) > 0$ on $(-\infty, 1) \cup (7, \infty)$
$f''(x) < 0$ on $(1, 7)$
f is increasing on $(.8952, 1.106)$ and on $(9.9987, \infty)$, decreasing on $(-\infty, .8952)$ and on $(1.106, 9.9987)$, concave up on $(-\infty, 1) \cup (7, \infty)$, concave down on $(1, 7)$, $x = .8952, 9.9987$ are local min, $x = 1.106$ is local max, $x = 1, 7$ are inflection points.
$f(x) \to \infty$ as $x \to \pm\infty$.
Global graph of $f(x)$:

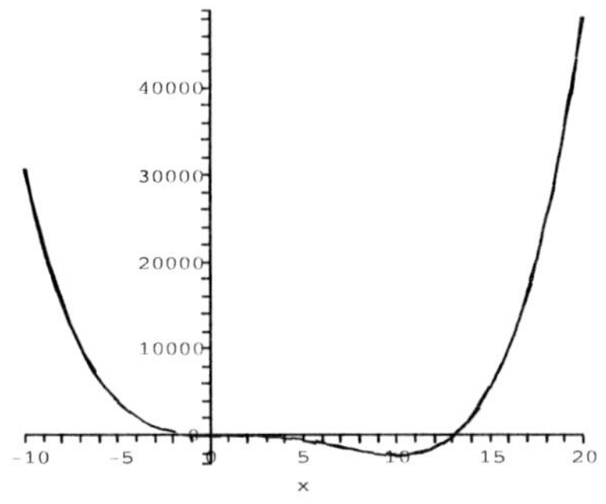

Local graph of $f(x)$:

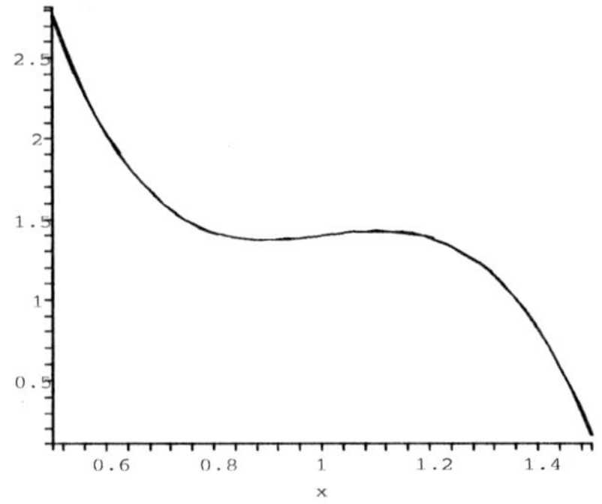

31. $f(x) = \dfrac{25 - 50\sqrt{x^2 + 0.25}}{x}$

$= 25\left(\dfrac{1 - 2\sqrt{x^2+0.25}}{x}\right)$

$= 25\left(\dfrac{1 - \sqrt{4x^2+1}}{x}\right)$

Note that $x = 0$ is not in the domain of the function.

$$f'(x) = 25\left(\frac{1 - \sqrt{4x^2+1}}{x^2\sqrt{4x^2+1}}\right)$$

We see that there are no critical points. Indeed, $f' < 0$ wherever f is defined. One can verify that

$f''(x) > 0$ on $(0, \infty)$
$f''(x) < 0$ on $(-\infty, 0)$
Hence the function is concave up on $(0, \infty)$ and concave down on $(-\infty, 0)$.

$\displaystyle\lim_{x\to\infty} \frac{25 - 50\sqrt{x^2+0.25}}{x}$

$\displaystyle= \lim_{x\to\infty} \frac{25}{x} - \frac{50\sqrt{x^2+0.25}}{x}$

$\displaystyle= \lim_{x\to\infty} 0 - 50\frac{x\sqrt{1+\frac{0.25}{x^2}}}{x}$

$\displaystyle= \lim_{x\to\infty} -50\sqrt{1+\frac{0.25}{x^2}} = -50$

$\displaystyle\lim_{x\to-\infty} \frac{25 - 50\sqrt{x^2+0.25}}{x}$

$\displaystyle= \lim_{x\to\infty} \frac{25}{x} - \frac{50\sqrt{x^2+0.25}}{x}$

$\displaystyle= \lim_{x\to-\infty} 0 - 50\frac{(-x)\sqrt{1+\frac{0.25}{x^2}}}{x}$

$\displaystyle= \lim_{x\to\infty} 50\sqrt{1+\frac{0.25}{x^2}} = 50$

So f has horizontal asymptotes at $y = 50$ and $y = -50$.

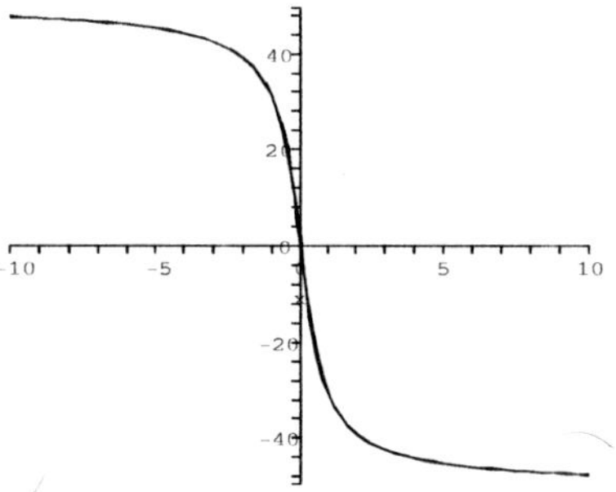

33. $f(x) = x^4 + cx^2$
$f'(x) = 4x^3 + 2cx$
$f''(x) = 12x^2 + 2c$

$c = 0$: 1 extremum, 0 inflection points
$c < 0$: 3 extrema, 2 inflection points
$c > 0$: 1 extremum, 0 inflection points
$c \to -\infty$: the graph widens and lowers
$c \to +\infty$: the graph narrows

35. $f(x) = \dfrac{x^2}{x^2 + c^2}$

$f'(x) = \dfrac{2c^2x}{(x^2+c^2)^2}$

$f''(x) = \dfrac{2c^4 - 6c^2x^2}{(x^2+c^2)^3}$

If $c = 0$: $f(x) = 1$, except that f is undefined at $x = 0$.

$c < 0$, $c > 0$: horizontal asymptote at $y = 1$, local min at $x = 0$, since the derivative changes sign from negative to positive at $x = 0$; also there are inflection points at $x = \pm c/\sqrt{3}$.

As $c \to -\infty$, $c \to +\infty$: the graph widens.

37. When $c = 0$, $f(x) = \sin(0) = 0$.

Since $\sin x$ is an odd function, $\sin(-cx) = -\sin(cx)$. Thus negative values of c give the reflection through the x-axis of their positive counterparts. For large values of c, the graph looks just like $\sin x$, but with a very small period.

39. $f(x) = xe^{-bx}$
$f(0) = 0$
$f(x) > 0$ for $x > 0$

$\lim_{x\to\infty} xe^{-bx} = \lim_{x\to\infty} \frac{x}{e^{bx}} = \lim_{x\to\infty} \frac{1}{be^{bx}} = 0$
(by L'Hôspital's rule)
$f'(x) = e^{-bx}(1 - bx)$, so there is a unique critical point at $x = 1/b$, which must be the maximum. The bigger b is, the closer the max is to the origin. For time since conception, $1/b$ represents the most common gestation time. For survival time, $1/b$ represents the most common life span.

41. No: Let $f(x) = \dfrac{x+1}{x^2+1}$. The roots of the denominator are complex, so there are no vertical asymptotes.

No: Let $f(x) = \dfrac{x^4 - 2x + 3}{x^2+1}$. This function goes to ∞ as $x \to \pm\infty$.

43. $f(x) = \dfrac{3x^2 - 1}{x} = 3x - \dfrac{1}{x}$

$y = 3x$ is a slant asymptote.

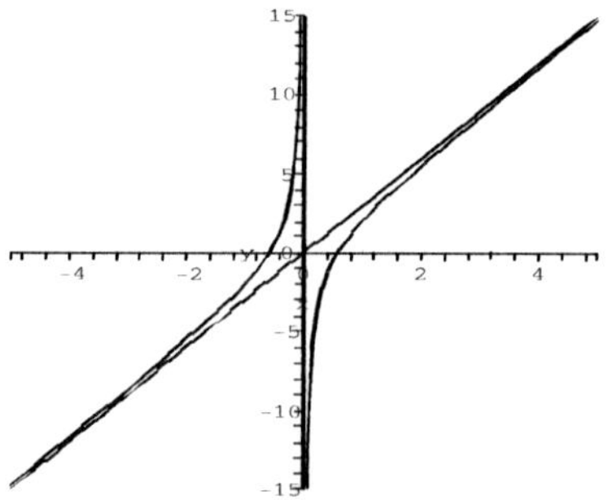

45. $f(x) = \dfrac{x^3 - 2x^2 + 1}{x^2} = x - 2 + \dfrac{1}{x^2}$

$y = x - 2$ is a slant asymptote.

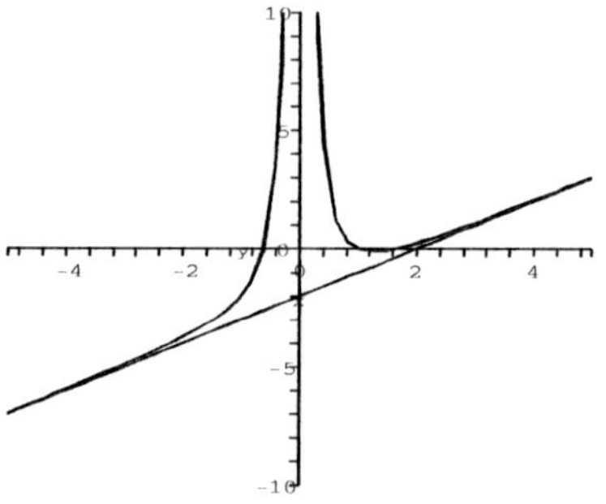

47. $f(x) = \dfrac{x^4}{x^3+1} = x - \dfrac{x}{x^3+1}$

$y = x$ is a slant asymptote.

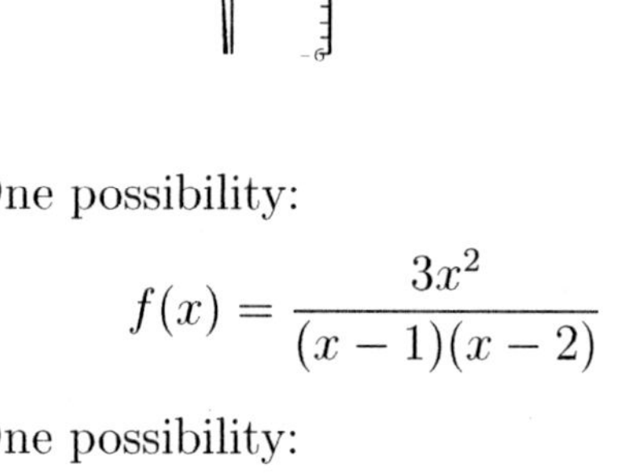

49. One possibility:

$$f(x) = \frac{3x^2}{(x-1)(x-2)}$$

51. One possibility:

$$f(x) = \frac{2x}{\sqrt{(x-1)(x+1)}}$$

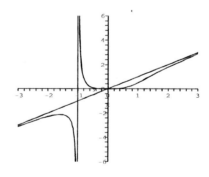

53. $f(x) = \sinh x = \dfrac{e^x - e^{-x}}{2}$

$f'(x) = \dfrac{e^x + e^{-x}}{2}$

$f'(x) > 0$ for all x so $f(x)$ is always increasing and has no extrema.

$f''(x) = \dfrac{e^x - e^{-x}}{2}$

$f''(x) = 0$ only when $x = 0$ and changes sign here, so $f(x)$ has an inflection point at $x = 0$.

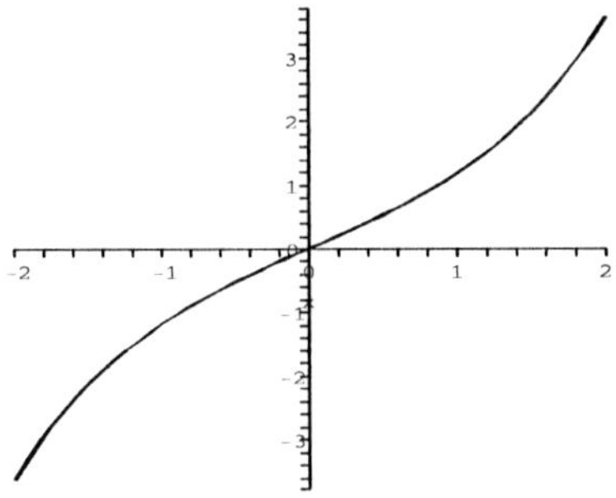

$f(x) = \cosh x = \dfrac{e^x + e^{-x}}{2}$

$f'(x) = \dfrac{e^x - e^{-x}}{2}$

$f'(x) = 0$ only when $x = 0$.

$f''(x) = \dfrac{e^x + e^{-x}}{2}$

$f''(x) > 0$ for all x, so $f(x)$ has no inflection points, but $x = 0$ is a minimum.

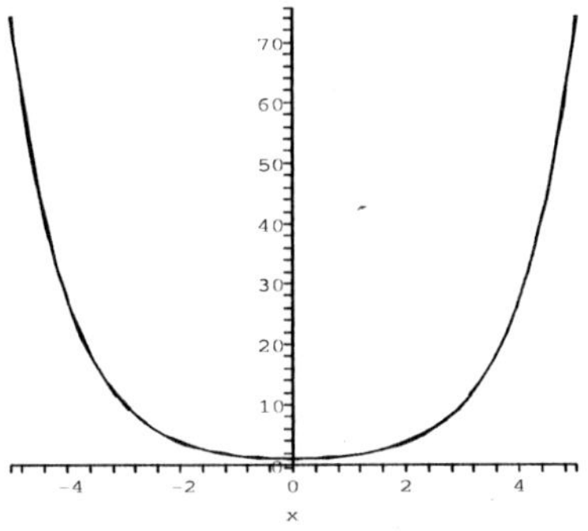

3.7 Optimization

1. $f(x) = x^2 + 1$ has a minimum at $x = 0$, while $\sin(x^2 + 1)$ has minima where $x^2 + 1 = 3\pi/2 + 2np\pi$.

3.

$$A = xy = 1800$$
$$y = \frac{1800}{x}$$
$$P = 2x + y = 2x + \frac{1800}{x}$$
$$P' = 2 - \frac{1800}{x^2} = 0$$
$$2x^2 = 1800$$
$$x = 30$$

$P'(x) > 0$ for $x > 30$
$P'(x) < 0$ for $0 < x < 30$
So $x = 30$ is min.

$$y = \frac{1800}{x} = \frac{1800}{30} = 60$$

So the dimensions are $30' \times 60'$ and the minimum perimeter is 120 ft.

5.

$$P = 2x + 3y = 120$$
$$3y = 120 - 2x$$
$$y = 40 - \frac{2}{3}x$$
$$A = xy$$
$$A(x) = x\left(40 - \frac{2}{3}x\right)$$
$$A'(x) = 1\left(40 - \frac{2}{3}x\right) + x\left(-\frac{2}{3}\right)$$
$$= 40 - \frac{4}{3}x = 0$$
$$40 = \frac{4}{3}x$$
$$x = 30$$

$A'(x) > 0$ for $0 < x < 30$
$A'(x) < 0$ for $x > 30$

So $x = 30$ is max, $y = 40 - \frac{2}{3} \cdot 30 = 20$

So the dimensions are $20' \times 30'$.

7.

$$\begin{aligned}
A &= xy \\
P &= 2x + 2y \\
2y &= P - 2x \\
y &= \frac{P}{2} - x \\
A(x) &= x\left(\frac{P}{2} - x\right) \\
A'(x) &= 1 \cdot \left(\frac{P}{2} - x\right) + x(-1) \\
&= \frac{P}{2} - 2x = 0 \\
P &= 4x \\
x &= \frac{P}{4}
\end{aligned}$$

$A'(x) > 0$ for $0 < x < P/4$
$A'(x) < 0$ for $x > P/4$

So $x = P/4$ is max,

$$y = \frac{P}{2} - x = \frac{P}{2} - \frac{P}{4} = \frac{P}{4}$$

So the dimensions are $\frac{P}{4} \times \frac{P}{4}$. Thus we have a square.

9.

$$\begin{aligned}
d &= \sqrt{(x-0)^2 + (y-1)^2} \\
y &= x^2 \\
d &= \sqrt{x^2 + (x^2 - 1)^2} \\
&= (x^4 - x^2 + 1)^{1/2} \\
d'(x) &= \frac{1}{2}(x^4 - x^2 + 1)^{-1/2}(4x^3 - 2x) \\
&= \frac{2x(2x^2 - 1)}{2\sqrt{x^4 - x^2 + 1}} = 0
\end{aligned}$$

$x = 0, \pm\sqrt{1/2}$;
$f(0) = 1$, $f(\sqrt{1/2}) = 3/4$, $f(-\sqrt{1/2}) = \frac{3}{4}$;
Thus $x = \pm\sqrt{1/2}$ are min, and the points on $y = x^2$ closest to $(0, 1)$ are $(\sqrt{1/2}, 1/2)$ and $(-\sqrt{1/2}, 1/2)$.

11.

$$\begin{aligned}
d &= \sqrt{(x-0)^2 + (y-0)^2} \\
y &= \cos x \\
d &= \sqrt{x^2 + \cos^2 x} \\
d'(x) &= \frac{2x - 2\cos x \sin x}{2\sqrt{x^2 + \cos^2 x}} = 0 \\
x &= \cos x \sin x \\
x &= 0
\end{aligned}$$

So $x = 0$ is min and the point on $y = \cos x$ closest to $(0, 0)$ is $(0, 1)$.

13. For $(0, 1)$, $(\sqrt{1/2}, 1/2)$ on $y = x^2$, we have
$y' = 2x$, $y'(\sqrt{1/2}) = 2 \cdot \sqrt{1/2} = \sqrt{2}$ and

$$m = \frac{\frac{1}{2} - 1}{-\sqrt{\frac{1}{2}} - 0} = \frac{1}{\sqrt{2}}.$$

For $(0, 1)$, $(-\sqrt{1/2}, 1/2)$ on $y = x^2$, we have
$y'(-\sqrt{1/2}) = 2(-\sqrt{1/2}) = -\sqrt{2}$ and

$$m = \frac{\frac{1}{2} - 1}{-\sqrt{\frac{1}{2}} - 0} = \frac{1}{\sqrt{2}}.$$

For $(3, 4)$, $(2.06, 4.2436)$ on $y = x^2$, we have $y'(2.06) = 2(2.06) = 4.12$ and

$$m = \frac{4.2436 - 4}{2.06 - 3} = -0.2591 \approx -\frac{1}{4.12}.$$

15.

$$\begin{aligned} V &= l \cdot w \cdot h \\ V(x) &= (10-2x)(6-2x)\cdot x,\ 0 \le x \le 3 \\ V'(x) &= -2(6-2x)\cdot x + (10-2x)(-2)\cdot x \\ &\quad + (10-2x)(6-2x) \\ &= 60 - 64x + 12x^2 \\ &= 4(3x^2 - 16x + 15) \\ &= 0 \\ x &= \frac{16 \pm \sqrt{(-16)^2 - 4\cdot 3\cdot 15}}{6} \\ &= \frac{8}{3} \pm \frac{\sqrt{19}}{3} \\ x &= \frac{8}{3} + \frac{\sqrt{19}}{3} > 3. \end{aligned}$$

$V'(x) > 0$ for $x < 8/3 - \sqrt{19}/3$
$V'(x) < 0$ for $x > 8/3 - \sqrt{19}/3$

So $x = \dfrac{8}{3} - \dfrac{\sqrt{19}}{3}$ is a max.

17. Let x be the distance from the connection point to the easternmost development. Then $0 \le x \le 5$.

$$\begin{aligned} f(x) &= \sqrt{3^2 + (5-x)^2} + \sqrt{4^2 + x^2}, \\ &\quad 0 \le x \le 5 \\ f'(x) &= -(9 + (5-x)^2)^{-1/2}(5-x) \\ &\quad + \frac{1}{2}(16 + x^2)^{-1/2}(2x) \\ &= \frac{x-5}{\sqrt{9 + (5-x)^2}} + \frac{x}{\sqrt{16+x^2}} \\ &= 0 \end{aligned}$$

$$x = \frac{20}{7} \approx 2.857$$

$$f(0) = 4 + \sqrt{34} \approx 9.831$$

$$f\left(\frac{20}{7}\right) = \sqrt{74} \approx 8.602$$

$$f(5) = 3 + \sqrt{41} \approx 9.403$$

So $x = 20/7$ is minimum. The length of new line at this point is approximately 8.6 miles. Since $f(0) \approx 9.8$ and $f(5) \approx 9.4$, the water line should be 20/7 miles west of the second development.

19.

$$\begin{aligned} C(x) &= 5\sqrt{16 + x^2} + 2\sqrt{36 + (8-x)^2} \\ &\quad 0 \le x \le 8 \\ C(x) &= 5\sqrt{16+x^2} + 2\sqrt{100 - 16x + x^2} \\ C'(x) &= 5\left(\frac{1}{2}\right)(16 + x^2)^{-1/2}\cdot 2x \\ &\quad + 2\left(\frac{1}{2}\right)(100 - 16x + x^2)^{-1/2}(2x - 16) \\ &= \frac{5x}{\sqrt{16+x^2}} + \frac{2x-16}{\sqrt{100 - 16x + x^2}} \\ &= 0 \end{aligned}$$

$x \approx 1.2529$
$C(0) = 40$
$C(1.2529) \approx 39.0162$
$C(8) \approx 56.7214$

The highway should emerge from the marsh 1.2529 miles east of the bridge. If we build a straight line to the interchange, we have $x = (3.2)$.

Since $C(3.2) - C(1.2529) \approx 1.963$, we save \$1.963 million.

21.

$$\begin{aligned} C(x) &= 5\sqrt{16 + x^2} + 3\sqrt{36 + (8-x)^2} \\ &\quad 0 \le x \le 8 \\ C'(x) &= \frac{5x}{\sqrt{16+x^2}} + \frac{3x - 24}{\sqrt{100 - 16x + x^2}} \end{aligned}$$

Setting $C'(x) = 0$ yields
$x \approx 1.8941$
$C(0) = 50$
$C(1.8941) \approx 47.8104$
$C(8) \approx 62.7214$

The highway should emerge from the marsh 1.8941 miles east of the bridge. So if we must use the path from exercise 21, the extra cost is

$C(1.2529) - C(1.8941)$
$= 48.0452 - 47.8104 = 0.2348$
or about \$234.8 thousand.

23.

$$T(x) = \frac{\sqrt{1+x^2}}{v_1} + \frac{\sqrt{1+(2-x)^2}}{v_2}$$

$$T'(x) = \frac{1}{v_1} \cdot \frac{1}{2}(1+x^2)^{-1/2} \cdot 2x$$
$$+ \frac{1}{v_2}(1+(2-x)^2)^{-1/2} \cdot (2-x)(-1)$$
$$= \frac{x}{v_1\sqrt{1+x^2}} + \frac{x-2}{v_2\sqrt{1+(2-x)^2}}$$

Note that

$$T'(x) = \frac{1}{v_1} \cdot \frac{x}{\sqrt{1+x^2}}$$
$$- \frac{1}{v_2} \cdot \frac{(2-x)}{\sqrt{1+(2-x)^2}}$$
$$= \frac{1}{v_1}\sin\theta_1 - \frac{1}{v_2}\sin\theta_2$$

When $T'(x) = 0$, we have

$$\frac{1}{v_1}\sin\theta_1 = \frac{1}{v_2}\sin\theta_2$$
$$\frac{\sin\theta_1}{\sin\theta_2} = \frac{v_1}{v_2}$$

25. Cost: $C = 2(2\pi r^2) + 2\pi rh$
Convert from fluid ounces to cubic inches:
12 fl oz $=$ 12 fl oz $\cdot$ 1.80469 in^3/fl oz
$= 21.65628$ in^3
Volume: $V = \pi r^2 h$ so

$$h = \frac{V}{\pi r^2} = \frac{21.65628}{\pi r^2}$$

$$C = 4\pi r^2 + 2\pi r\left(\frac{21.65628}{\pi r^2}\right)$$
$$C(r) = 4\pi r^2 + 43.31256 r^{-1}$$
$$C'(r) = 8\pi r - 43.31256 r^{-2}$$
$$= \frac{8\pi r^3 - 43.31256}{r^2}$$

$$r = \sqrt[3]{\frac{43.31256}{8\pi}} = 1.1989''$$

when $C'(r) = 0$.
$C'(r) < 0$ on $(0, 1.1989)$
$C'(r) > 0$ on $(1.1989, \infty)$
Thus $r = 1.1989$ minimizes the cost.

$$h = \frac{21.65628}{\pi(1.1989)^2} = 4.7957''$$

27.

$$V(r) = cr^2(r_0 - r)$$
$$V'(r) = 2cr(r_0 - r) + cr^2(-1)$$
$$= 2crr_0 - 3cr^2$$
$$= cr(2r_0 - 3r)$$

$V'(r) = 0$ when $r = 2r_0/3$
$V'(r) > 0$ on $(0, 2r_0/3)$
$V'(r) < 0$ on $(2r_0/3, \infty)$
Thus $r = 2r_0/3$ maximizes the velocity.
$r = 2r_0/3 < r_0$, so the windpipe contracts.

29. $p(x) = \dfrac{V^2 x}{(R+x)^2}$

$$p'(x) = \frac{V^2(R+x)^2 - V^2 x \cdot 2(R+x)}{(R+x)^4}$$
$$= \frac{V^2R^2 - V^2x^2}{(R+x)^4}$$

$p'(x) = 0$ when $x = R$
$p'(x) > 0$ on $(0, R)$
$p'(x) < 0$ on (R, ∞)

Thus $x = R$ maximizes the power absorbed.

31. $\pi r + 4r + 2w = 8 + \pi$

$$w = \frac{8 + \pi - r(\pi + 4)}{2}$$

$$A(r) = \frac{\pi r^2}{2} + 2rw$$
$$= \frac{\pi r^2}{2} + r(8 + \pi - r(\pi + 4))$$
$$= r^2\left(-4 - \frac{\pi}{2}\right) + r(8 + \pi)$$
$$A'(r) = -2r\left(4 + \frac{\pi}{2}\right) + (8 + \pi) = 0$$

$A'(r) = 0$ when $r = 1$

$A'(r) > 0$ on $(0, 1)$
$A'(r) < 0$ on $(1, \infty)$

Thus $r = 1$ maximizes the area so $w = \dfrac{8 + \pi - (\pi + 4)}{2} = 2$. The dimensions of the rectangle are 2×2.

33. $l \times w = 92$, $w = 92/l$

$$\begin{aligned} A(l) &= (l+4)(w+2) \\ &= (l+4)(92/l+2) \\ &= 92 + 368/l + 2l + 8 \\ &= 100 + 368l^{-1} + 2l \end{aligned}$$

$$A'(l) = -368l^{-2} + 2 = \frac{2l^2 - 368}{l^2}$$

$A'(l) = 0$ when $l = \sqrt{184} = 2\sqrt{46}$
$A'(l) < 0$ on $(0, 2\sqrt{46})$
$A'(l) > 0$ on $(2\sqrt{46}, \infty)$

So $l = 2\sqrt{46}$ minimizes the total area. When $l = 2\sqrt{46}$, $w = \frac{92}{2\sqrt{46}} = \sqrt{46}$.

For the minimum total area, the printed area has width $\sqrt{46}$ in. and length $2\sqrt{46}$ in., and the advertisement has overall width $\sqrt{46} + 2$ in. and overall length $2\sqrt{46} + 4$ in.

35. Let L represent the length of the ladder. Then from the diagram, it follows that
$L = a\sec\theta + b\csc\theta$.
Therefore,

$$\frac{dL}{d\theta} = a\sec\theta\tan\theta - b\csc\theta\cot\theta$$
$$0 = a\sec\theta\tan\theta - b\csc\theta\cot\theta$$
$$a\sec\theta\tan\theta = b\csc\theta\cot\theta$$
$$\begin{aligned} \frac{b}{a} &= \frac{\sec\theta\tan\theta}{\csc\theta\cot\theta} \\ &= \frac{1}{\cos\theta}\frac{\sin\theta}{\cos\theta}\frac{\sin\theta}{1}\frac{\sin\theta}{\cos\theta} \\ &= \tan^3\theta \end{aligned}$$

Thus,
$\tan\theta = \sqrt[3]{b/a}$
$\theta = \tan^{-1}\left(\sqrt[3]{b/a}\right)$
$= \tan^{-1}\left(\sqrt[3]{4/5}\right)$
≈ 0.748 rad or 42.87 degrees

Thus, the length of the longest ladder that can fit around the corner is approximately

$$\begin{aligned} L &= a\sec\theta + b\csc\theta \\ &= 5\sec(0.748) + 4\csc(0.748) \\ &\approx 12.7 \text{ ft} \end{aligned}$$

37. Using the result of exercise 36 and solving for b:

$$\begin{aligned} L &= (a^{2/3} + b^{2/3})^{3/2} \\ L^{2/3} &= a^{2/3} + b^{2/3} \\ b^{2/3} &= L^{2/3} - a^{2/3} \\ b &= (L^{2/3} - a^{2/3})^{3/2} \\ &= (8^{2/3} - 5^{2/3})^{3/2} \\ &\approx 1.16 \text{ ft} \end{aligned}$$

39.

$$\begin{aligned} R(x) &= \frac{35x - x^2}{x^2 + 35} \\ R'(x) &= -35\frac{x^2 + 2x - 35}{(x^2+35)^2} \\ &= -35\frac{(x-5)(x+7)}{(x^2+35)^2} \end{aligned}$$

Hence the only critical number for $x \geq 0$ is $x = 5$ (that is, 5000 items). This must correspond to the absolute maximum, since $R(0) = 0$ and $R(x)$ is negative for large x. So maximum revenue is $R(5) = 2.5$ (that is, \$2500).

41. $Q'(t)$ is efficiency because it represents the number of additional items produced per unit time.

$$\begin{aligned} Q(t) &= -t^3 + 12t^2 + 60t \\ Q'(t) &= -3t^2 + 24t + 60 \\ &= 3(-t^2 + 8t + 20) \end{aligned}$$

This is the quantity we want to maximize.

$Q''(t) = 3(-2t+8)$ so the only critical number is $t = 4$ hours. This must be the maximum since the function $Q'(t)$ is a parabola opening down.

43. Let $C(t)$ be the total cost of the tickets. Then
$C(t) =$(price per ticket)(# of tickets)
$$C(t) = (40 - (t - 20))(t)$$
$$= (60 - t)(t) = 60t - t^2$$
for $20 < t < 50$. Then $C'(t) = 60 - 2t$, so $t = 30$ is the only critical number. This must correspond to the maximum since $C(t)$ is a parabola opening down.

45.

$$R = \frac{2v^2\cos^2\theta}{g}(\tan\theta - \tan\beta)$$
$$R'(\theta) = \frac{2v^2}{g}\left[2\cos\theta(-\sin\theta)(\tan\theta - \tan\beta) + \cos^2\theta\cdot\sec^2\theta\right]$$
$$= \frac{2v^2}{g}\left[-2\cos\theta\sin\theta\cdot\frac{\sin\theta}{\cos\theta} + 2\cos\theta\sin\theta\tan\beta + \cos^2\theta\cdot\frac{1}{\cos^2\theta}\right]$$
$$= \frac{2v^2}{g}\left[-2\sin^2\theta + \sin(2\theta)\tan\beta + 1\right]$$
$$= \frac{2v^2}{g}\left[-2\sin^2\theta + \sin(2\theta)\tan\beta + (\sin^2\theta + \cos^2\theta)\right]$$
$$= \frac{2v^2}{g}\left[\sin(2\theta)\tan\beta + (\cos^2\theta - \sin^2\theta)\right]$$
$$= \frac{2v^2}{g}\left[\sin(2\theta)\tan\beta + \cos(2\theta)\right]$$

$R'(\theta) = 0$ when
$$\tan\beta = \frac{-\cos(2\theta)}{\sin(2\theta)} = -\cot(2\theta)$$
$$= -\tan\left(\frac{\pi}{2} - 2\theta\right)$$
$$= \tan\left(2\theta - \frac{\pi}{2}\right)$$

Hence $\beta = 2\theta - \pi/2$, so
$$\theta = \frac{1}{2}\left(\beta + \frac{\pi}{2}\right)$$
$$= \frac{\beta}{2} + \frac{\pi}{4} = \frac{\beta°}{2} + 45°$$

(a) $\beta = 10°$, $\theta = 50°$

(b) $\beta = 0°$, $\theta = 45°$

(c) $\beta = -10°$, $\theta = 40°$

47. $T = \dfrac{-1}{c}\ln\left(1 - c\cdot\dfrac{b-a}{v_0}\right)$
$b = 300$, $a = 0$, $v_0 = 125$, $c = 0.1$
$$T = \frac{-1}{0.1}\ln\left(1 - 0.1\cdot\frac{300-0}{125}\right)$$
$$= 2.744 \text{ sec}$$
$$T(x) = -10\ln(1 - 0.0008(300 - x)) - 10\ln(1 - 0.0008x) + 0.1$$
$$T'(x) = -10\left(\frac{0.0008}{0.76 + 0.0008x} - \frac{0.0008}{1 - 0.0008x}\right) = 0$$
$0.0008(1 - 0.0008x) = 0.0008(0.76 + 0.0008x)$
$T'(x) = 0$ when $x = \dfrac{1 - 0.76}{0.0016} = 150$ ft. $T'(x) < 0$ on $(0, 150)$
$T'(x) > 0$ on $(150, 300)$
Hence $x = 150$ minimizes the total time.
$$T(150) = -10\ln(1 - 0.0008(300 - 150)) - 10\ln(1 - 0.0008(150)) + 0.1$$
$$= 2.656 \text{ sec.}$$
So the relay is faster.

If the delay is 0.2 sec, the relay takes longer.

49. $T(x) = -10\ln\left(1 - 0.1\dfrac{300 - x}{125}\right)$
$$- 10\ln\left(1 - 0.1\frac{x}{100}\right) + 0.1$$
$$= -10(\ln(1 - 0.0008(300 - x)) - 10\ln(1 - .001x) + 0.01$$
$$T'(x) = -10\left(\frac{0.0008}{0.76 + 0.0008x} - \frac{0.001}{1 - 0.001x}\right)$$

$= 0$

$0.0008(1 - 0.001x) = 0.001(0.76 + 0.0008x)$

$T'(x) = 0$ when $x = 25$ ft.

$T'(x) < 0$ on $(0, 25)$

$T'(x) > 0$ on $(25, 300)$

Hence $x = 25$ minimizes the total time.

$$T(25) = -10\ln(1 - 0.0008(300 - 25)) - 10\ln(1 - 0.001(25)) + 0.1 = 2.838 \text{ sec.}$$

So the relay takes longer. Without the delay, the relay would take 2.738 sec, so a delay of $2.744 - 2.738 = .006$ sec makes the two times equal.

51. $A = 4xy$

$\dfrac{dA}{dx} = 4\left(xy' + y\right)$

To determine $y' = \dfrac{dy}{dx}$, use the equation for the ellipse:

$1 = \dfrac{x^2}{a^2} + \dfrac{y^2}{b^2}$

$0 = \dfrac{2x}{a^2} + \dfrac{2yy'}{b^2}$

$\dfrac{2yy'}{b^2} = -\dfrac{2x}{a^2}$

$y' = -\dfrac{b^2}{a^2}\dfrac{x}{y}$

Substituting this expression for y' into the expression for $\dfrac{dA}{dx}$, we get

$$\frac{dA}{dx} = xy' + y = x\left(-\frac{b^2}{a^2}\frac{x}{y}\right) + y = -\frac{b^2}{a^2}\frac{x^2}{y} + y$$

The area is maximized when its derivative is zero:

$$0 = -\frac{b^2}{a^2}\frac{x^2}{y} + y$$

$$\frac{b^2}{a^2}\frac{x^2}{y} = y$$

$$\frac{x^2}{a^2} = \frac{y^2}{b^2}$$

Substituting the previous relationship into the equation for the ellipse, we get

$$\frac{x^2}{a^2} = \frac{y^2}{b^2} = \frac{1}{2}$$

and therefore,

$$x = \frac{a}{\sqrt{2}} \quad \text{and} \quad y = \frac{b}{\sqrt{2}}$$

Thus, the maximum area is

$$A = 4\frac{a}{\sqrt{2}}\frac{b}{\sqrt{2}} = 2ab$$

Since the area of the circumscribed rectangle is $4ab$, the required ratio is

$$2ab : \pi ab : 4ab = 1 : \frac{\pi}{2} : 2$$

53. Suppose that $a = b$ in the isosceles triangle, so that

$$A^2 = s(s-a)(s-b)(s-c) = s(s-a)^2(s-c)$$

Since $s = \frac{1}{2}(a+b+c)$, it follows that $s = \frac{1}{2}(2a + c) = a + \frac{c}{2}$, so that $s - a = \frac{c}{2}$. Thus,

$$A^2 = s\left(\frac{c^2}{4}\right)(s - c) = \frac{s}{4}\left(sc^2 - c^3\right)$$

Since s is a constant (it's half of the perimeter), we can now differentiate to get

$$2A\frac{dA}{dc} = \frac{s}{4}\left(2sc - 3c^2\right)$$

$$0 = c(2s - 3c)$$

Thus, the area is maximized when $2s - 3c = 0$, which means $c = \frac{2}{3}s$. Solving for a, we get

$$a = s - \frac{c}{2} = s - \frac{s}{3} = \frac{2}{3}s.$$

Thus, the area is maximized when $a = b = c$; in other words the area is maximized when the triangle is equilateral.

The maximum area is

$$A = \sqrt{s(s-c)^3} = \sqrt{s\left(\frac{s}{3}\right)^3}$$
$$= \frac{s^2}{9}\sqrt{3} = \frac{p^2}{36}\sqrt{3}$$

3.8 Related Rates

1. $V(t) = (\text{depth})(\text{area}) = \frac{\pi}{48}\left[r(t)\right]^2$ (units in cubic feet per min)

$$V'(t) = \frac{\pi}{48}2r(t)r'(t) = \frac{\pi}{24}r(t)r'(t)$$

We are given $V'(t) = \frac{120}{7.5} = 16$.

Hence $16 = \dfrac{\pi}{24}r(t)r'(t)$ so

$$r'(t) = \frac{(16)(24)}{\pi r(t)}.$$

(a) When $r = 100$,
$$r'(t) = \frac{(16)(24)}{100\pi} = \frac{96}{25\pi}$$
$$\approx 1.2223 \text{ ft/min},$$

(b) When $r = 200$,
$$r'(t) = \frac{(16)(24)}{200\pi} = \frac{48}{25\pi}$$
$$\approx 0.61115 \text{ ft/min}$$

3. From #1,

$$V'(t) = \frac{\pi}{48}2r(t)r'(t) = \frac{\pi}{24}r(t)r'(t),$$

so $\dfrac{g}{7.5} = \dfrac{\pi}{24}(100)(.6) = 2.5\pi$,

so $g = (7.5)(2.5)\pi$

$= 18.75\pi \approx 58.905$ gal/min.

5. t = hours elapsed since injury
r = radius of the infected area
A = area of the infection
$A = \pi r^2$
$A'(t) = 2\pi r(t) \cdot r'(t)$
When $r = 3$ mm, $r' = 1$ mm/hr,
$A' = 2\pi(3)(1) = 6\pi$ mm^2/hr

7. $V(t) = \dfrac{4}{3}\pi[r(t)]^3$
$V'(t) = 4\pi[r(t)]^2 r'(t) = Ar'(t)$
If $V'(t) = kA(t)$, then
$$r'(t) = \frac{V'(t)}{A(t)} = \frac{kA(t)}{A(t)} = k.$$

9.

$$10^2 = x^2 + y^2$$
$$0 = 2x\frac{dx}{dt} + 2y\frac{dy}{dt}$$
$$\frac{dy}{dt} = -\frac{x}{y}\frac{dx}{dt}$$
$$= -\frac{6}{8}(3)$$
$$= -2.25 \text{ ft/s}$$

11.

$$\theta = \pi - \tan^{-1}\left(\frac{40}{60-x}\right) - \tan^{-1}\left(\frac{20}{x}\right)$$
$$\frac{d\theta}{dx} = -\frac{40\left(\frac{1}{60-x}\right)^2}{1+\left(\frac{40}{60-x}\right)^2} + \frac{\frac{20}{x^2}}{1+\left(\frac{20}{x}\right)^2}$$

When $x = 30$, this becomes

$$\frac{d\theta}{dx} = -\frac{40\left(\frac{1}{30}\right)^2}{1+\left(\frac{40}{30}\right)^2} + \frac{\frac{20}{900}}{1+\left(\frac{20}{30}\right)^2}$$
$$= -\frac{1}{1625} \text{ rad/ft}$$
$$\frac{d\theta}{dt} = \frac{d\theta}{dx}\frac{dx}{dt}$$
$$= \left(-\frac{1}{1625}\right)(4)$$
$$\approx -0.00246 \text{ rad/s}$$

13. We know $[x(t)]^2 + 4^2 = [s(t)]^2$. Hence $2x(t)x'(t) = 2s(t)s'(t)$, so

$$x'(t) = \frac{s(t)s'(t)}{x(t)} = \frac{-240s(t)}{x(t)}.$$

When $x = 40$, $s = \sqrt{40^2+4^2} = 4\sqrt{101}$, so at that moment

$$x'(t) = \frac{(-240)(4\sqrt{101})}{40} = -24\sqrt{101}.$$

So the speed is $24\sqrt{101} \approx 241.2$mph.

15. If the police car is not moving, then $x'(t) = 0$, but all the other data are unchanged. So

$$\begin{aligned} d'(t) &= \frac{x(t)x'(t) + y(t)y'(t)}{\sqrt{[x(t)]^2 + [y(t)]^2}} \\ &= \frac{-(1/2)(50)}{\sqrt{1/4 + 1/16}} \\ &= \frac{-100}{\sqrt{5}} \approx -44.721. \end{aligned}$$

This is more accurate.

17. $$\begin{aligned} d'(t) &= \frac{x(t)x'(t) + y(t)y'(t)}{\sqrt{[x(t)]^2 + [y(t)]^2}} \\ &= \frac{-(1/2)(\sqrt{2}-1)(50) - (1/2)(50)}{\sqrt{1/4 + 1/4}} \\ &= -50. \end{aligned}$$

19. $\overline{C}(x) = 10 + \dfrac{100}{x}$
$\overline{C}'(x(t)) = \dfrac{-100}{x^2} \cdot x'(t)$
$\overline{C}'(10) = -1(2) = -2$ dollars per item, so average cost is decreasing at the rate of \$2 per year.

21. From the table, we see that the recent trend is for advertising to increase by \$2000 per year. A good estimate is then $x'(2) \approx 2$ (in units of thousands). Starting with the sales equation
$s(t) = 60 - 40e^{-0.05x(t)}$,
we use the chain rule to obtain
$s'(t) = -40e^{-0.05x(t)}[-0.05x'(t)]$
$= 2x'(t)e^{-0.05x(t)}$.
Using our estimate that $x'(2) \approx 2$ and since $x(2) = 20$, we get $s'(2) \approx 2(2)e^{-1} \approx 1.471$. Thus, sales are increasing at the rate of approximately \$1471 per year.

23. We have $\tan\theta = \dfrac{x}{2}$, so

$$\begin{aligned} \frac{d}{dt}(\tan\theta) &= \frac{d}{dt}\left(\frac{x}{2}\right) \\ \sec^2\theta \cdot \theta' &= \frac{1}{2}x' \\ \theta' &= \frac{1}{2\sec^2\theta} \cdot x' = \frac{x'\cos^2\theta}{2} \end{aligned}$$

at $x = 0$, we have $\tan\theta = \dfrac{x}{2} = \dfrac{0}{2}$ so $\theta = 0$ and we have $x' = -130\text{ft/s}$ so

$$\theta' = \frac{(-130)\cdot\cos^2 0}{2} = -65 \text{ rad/s}.$$

25. t = number of seconds since launch
x = height of rocket in miles after t seconds
θ = camera angle in radians after t seconds

$$\begin{aligned} \tan\theta &= \frac{x}{2} \\ \frac{d}{dx}(\tan\theta) &= \frac{d}{dx}\left(\frac{x}{2}\right) \\ \sec^2\theta \cdot \theta' &= \frac{1}{2}x' \\ \theta' &= \frac{\cos^2\theta \cdot x'}{2} \end{aligned}$$

When $x = 3$, $\tan\theta = 3/2$, so $\cos\theta = 2/\sqrt{13}$.

$$\theta' = \frac{\left(\frac{2}{\sqrt{13}}\right)^2 (.2)}{2} \approx .03 \text{ rad/s}$$

27. Let θ be the angle between the end of the shadow and the top of the lamppost. Then $\tan\theta = \dfrac{6}{s}$ and $\tan\theta =$

$\dfrac{18}{s+x}$, so

$$\begin{aligned}\frac{x+s}{18} &= \frac{s}{6}\\ \frac{d}{dx}\left(\frac{x+s}{18}\right) &= \frac{d}{dx}\left(\frac{s}{6}\right)\\ \frac{x'+s'}{18} &= \frac{s'}{6}\\ x'+s' &= 3s'\\ s' &= \frac{x'}{2}\end{aligned}$$

Since $x' = 2$, $s' = 2/2 = 1$ ft/s.

29. $P(t) \cdot V'(t) + P'(t)V(t) = 0$

$$\frac{P'(t)}{V'(t)} = -\frac{P(t)}{V(t)} = -\frac{c}{V(t)^2}$$

31. Let $r(t)$ be the length of the rope at time t and $x(t)$ be the distance (along the water) between the boat and the dock.

$$\begin{aligned}r(t)^2 &= 36 + x(t)^2\\ 2r(t)r'(t) &= 2x(t)x'(t)\\ x'(t) &= \frac{r(t)r'(t)}{x(t)} = \frac{-2r(t)}{x(t)}\\ &= \frac{-2\sqrt{36+x^2}}{x}\end{aligned}$$

When $x = 20$, $x' = -2.088$; when $x = 10$, $x' = -2.332$.

33. $f(t) = \dfrac{1}{2L(t)}\sqrt{\dfrac{T}{\rho}} = \dfrac{110}{L(t)}$.

$f'(t) = \dfrac{-110}{L(t)^2}L'(t)$.

When $L = 1/2$, $f(t) = 220$ cycles per second. If $L' = -4$ at this time, then $f'(t) = 1760$ cycles per second per second. It will only take $1/8$ second at this rate for the frequency to go from 220 to 440, and raise the pitch one octave.

35. Let R represent the radius of the circular surface of the water in the tank.

$$\begin{aligned}V(R) &= \pi\left[60^2(60^2-R^2)^{1/2} - \frac{1}{3}(60^2-R^2)^{3/2} + \frac{2}{3}60^3\right]\\ \frac{dV}{dR} &= \pi\left[60^2\left(\frac{1}{2}\right)(60^2-R^2)^{-1/2}(-2R) - \frac{1}{3}\left(\frac{3}{2}\right)(60^2-R^2)^{1/2}(-2R)\right]\\ &= \pi\left[\frac{-60^2R}{\sqrt{60^2-R^2}} + R\sqrt{60^2-R^2}\right]\\ &= \pi R\left[\frac{-60^2+60^2-R^2}{\sqrt{60^2-R^2}}\right]\\ &= \frac{-\pi R^3}{\sqrt{60^2-R^2}}\\ \frac{dR}{dt} &= \frac{dV/dt}{dV/dR}\\ &= \frac{10}{dV/dR}\\ &= \frac{-10\sqrt{60^2-R^2}}{\pi R^3}\end{aligned}$$

(a) Substituting $R = 60$ into the previous equation, we get $\frac{dR}{dt} = 0$.

(b) We need to determine the value of R when the tank is three-quarters full. The volume of the spherical tank is $\frac{4}{3}\pi 60^3$, so when the tank is three-quarters full, $V(R) = \pi 60^3$. Substituting this value into the formula for $V(R)$ and solving for R (using a CAS, for example) we get $R \approx 56.265$. Substituting this value into the formula for dR/dt,

we get $$\begin{aligned}\frac{dR}{dt} &= \frac{-10\sqrt{60^2-R^2}}{\pi R^3}\\ &\approx \frac{-10\sqrt{60^2-56.265^2}}{\pi 56.265^3}\\ &\approx -0.00037 \text{ ft/s}\end{aligned}$$

37. The volume of the conical pile is $V = \frac{1}{3}\pi r^2 h$. Since $h = 2r$, we can write the volume as

$$V = \frac{1}{3}\pi\left(\frac{h}{2}\right)^2 h = \frac{1}{12}\pi h^3$$

Thus,

$$\frac{dV}{dt} = \frac{\pi h^2}{4}\cdot\frac{dh}{dt}$$
$$20 = \frac{\pi 6^2}{4}\cdot\frac{dh}{dt}$$
$$\frac{dh}{dt} = \frac{20}{9\pi}$$
$$\frac{dr}{dt} = \frac{10}{9\pi}$$

39.

$$\theta = \tan^{-1}\left(\frac{2s}{vT}\right)$$
$$\frac{d\theta}{dt} = \frac{\left(-\frac{2s}{T}\right)v^{-2}v'(t)}{1+\left(\frac{2s}{vT}\right)^2}$$
$$= \frac{-2sv'(t)}{Tv^2\left[1+\frac{4s^2}{v^2T^2}\right]}$$
$$= \frac{-2sTv'(t)}{T^2v^2+4s^2}$$

For $T = 1$, $s = 0.6$ and $v'(t) = 1$,

$$\frac{d\theta}{dT} = \frac{-1.2}{v^2+1.44}$$

(a) $v = 1$ m/s $\Rightarrow$

$$\frac{d\theta}{dT} = \frac{-1.2}{2.44} \approx -0.4918 \text{ rad/s}$$

(b) $v = 2$ m/s $\Rightarrow$

$$\frac{d\theta}{dT} = \frac{-1.2}{5.44} \approx -0.2206 \text{ rad/s}$$

3.9 Rates of Change in Economics and the Sciences

1. The marginal cost function is $C'(x) = 3x^2 + 40x + 90$.
The marginal cost at $x = 50$ is $C'(50) = 9590$. The cost of producing the 50th item is $C(50) - C(49) = 9421$.

3. The marginal cost function is $C'(x) = 3x^2 + 42x + 110$.
The marginal cost at $x = 100$ is $C'(100) = 34310$. The cost of producing the 100th item is $C(100) - C(99) = 33990$.

5. $C'(x) = 3x^2 - 60x + 300$
$C''(x) = 6x - 60 = 0$
$x = 10$ is the inflection point because $C''(x)$ changes from negative to positive at this value. After this point, cost rises more sharply.

7. $\overline{C}(x) = C(x)/x = 0.1x + 3 + \dfrac{2000}{x}$
$\overline{C}'(x) = 0.1 - \dfrac{2000}{x^2}$
Critical number is $x = 100\sqrt{2} \approx 141.4$.

$\overline{C}'(x)$ is negative to the left of the critical number and positive to the right, so this must be the minimum.

9. $\overline{C}(x) = C(x)/x = 10\dfrac{e^{0.02x}}{x}$
$\overline{C}'(x) = 10e^{.02x}\left(\dfrac{.02x-1}{x^2}\right)$
Critical number is $x = 50$. $\overline{C}'(x)$ is negative to the left of the critical number and positive to the right, so this must be the minimum.

11.

$$C(x) = 0.01x^2 + 40x + 3600$$
$$C'(x) = 0.02x + 40$$
$$\overline{C}(x) = \frac{C(x)}{x} = 0.01x + 40 + \frac{3600}{x}$$
$$C'(100) = 42$$
$$\overline{C}(100) = 77$$
so $C'(100) < \overline{C}(100)$
$$\overline{C}(101) = 76.65 < \overline{C}(100)$$

13.

$$\overline{C}'(x) = 0.01 - \frac{3600}{x^2} = 0$$
so $x = 600$ is min and
$$C'(600) = 52$$
$$\overline{C}(600) = 52$$

15. $P(x) = R(x) - C(x)$
$P'(x) = R'(x) - C'(x) = 0$
$R'(x) = C'(x)$

17. $E = \frac{p}{f(p)} f'(p)$
$= \frac{p}{200(30-p)}(-200) = \frac{p}{p-30}$

To solve $\frac{p}{p-30} < -1$, multiply both sides by the negative quantity $p - 30$, to get $p > (-1)(p-30)$ or $p > 30 - p$, so $2p > 30$, so $15 < p < 30$.

19. $f(p) = 100p(20 - p) = 100(20p - p^2)$

$E = \frac{p}{f(p)} f'(p)$
$= \frac{p}{100p(20-p)}(100)(20-2p)$
$= \frac{20-2p}{20-p}$

To solve $\frac{20-2p}{20-p} < -1$, multiply both sides by the positive quantity $20 - p$ to get $20 - 2p < (-1)(20 - p)$, or $20 - 2p < p - 20$, so $40 < 3p$, so $40/3 < p < 20$.

21. Elasticity of demand at price $p = 15$ is, by definition, the relative change in demand divided by the relative change in price, as price increases from 15 to an amount slightly larger than 15. So if (rel change in demand)/(rel change in price) is less than (-1), then rel change in demand is less than (-1)(rel change in price). This means that demand goes down more than price goes up, so revenue should decrease. (See problem 23.)

23. $[pf(p)]' < 0$
if and only if $p'f(p) + pf'(p) < 0$
if and only if $f(p) + pf'(p) < 0$
if and only if $pf'(p) < -f(p)$
if and only if $\frac{pf'(p)}{f(p)} < -1$.

25. $f(x) = 2x(4 - x)$
$f'(x) = 2(4 - x) + 2x(-1) = 8 - 4x$
$= 4(2 - x) = 0$
$x = 2$ is a maximum since $f(x)$ is a downward opening parabola.

27. $2x'(t) = 2x(t)[4 - x(t)] = 0$
$x(t) = 0$, $x(t) = 4$ are critical numbers.
$x'(t) > 0$ for $0 < x(t) < 4$
$x'(t) < 0$ for $x(t) > 4$
So $x(t) = 4$ is the maximum concentration.

$x'(t) = 0.5x(t)[5 - x(t)]$
$x(t) = 0$, $x(t) = 5$ are critical numbers.
$x'(t) > 0$ for $0 < x(t) < 5$
$x'(t) < 0$ for $x(t) > 5$
So $x(t) = 5$ is the maximum concentration.

29.

$$\begin{aligned} y'(t) &= c \cdot y(t)[K - y(t)] \\ y(t) &= Kx(t) \\ y'(t) &= Kx'(t) \\ Kx'(t) &= c \cdot Kx(t)[K - Kx(t)] \\ x'(t) &= c \cdot Kx(t)[1 - x(t)] \\ &= rx(t)[1 - x(t)] \\ r &= cK \end{aligned}$$

31. $x'(t) = [a - x(t)][b - x(t)]$
for $x(t) = a$,
$x'(t) = [a - a][b - a] = 0$
So the concentration of product is staying the same.

If $a < b$ and $x(0) = 0$ then $x'(t) > 0$ for $0 < x < a < b$
$x'(t) < 0$ for $a < x < b$
Thus $x(t) = a$ is a maximum.

33. $$x(0) = \frac{a[1 - e^{-(b-a)\cdot 0}]}{1 - \left(\frac{a}{b}\right) e^{-(b-a)\cdot 0}} = \frac{a[1-1]}{1 - \left(\frac{a}{b}\right)} = 0$$

$$\lim_{t\to\infty} x(t) = \frac{a[1-0]}{1-0} = a$$

For $a = 2$ and $b = 3$ the graph looks like this:

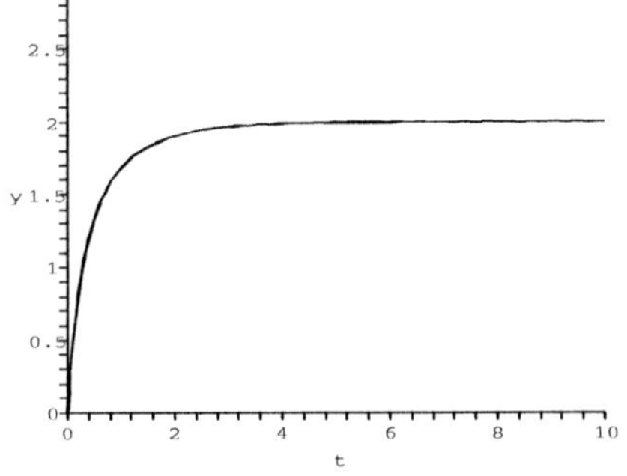

35. The first inflection point occurs around $f = 1/3$, before the step up. The second occurs at the far right of the graph. The equivalence point is presumably more stable. The first inflection point would be hard to measure, since the pH takes drastic leap right after the inflection point occurs.

37. $R(x) = \dfrac{rx}{k+x}$, $x \geq 0$

$$R'(x) = \frac{rk}{(k+x)^2}$$

There are no critical numbers. Any possible maximum would have to be at the endpoint $x = 0$, but in fact R is increasing on $[0, \infty)$, so there is no maximum (although as x goes to infinity, R approaches r).

39.

$$\begin{aligned} PV^{7/5} &= c \\ \frac{d}{dP}\left(PV^{7/5}\right) &= \frac{d}{dP}(c) = 0 \\ V^{7/5} + \frac{7}{5}PV^{2/5}\frac{dV}{dP} &= 0 \\ V + \frac{7}{5}P\frac{dV}{dP} &= 0 \\ \frac{dV}{dP} &= \frac{-5}{7}\frac{V}{P}. \end{aligned}$$

But $V^{7/5} = c/P$, so $V = (c/P)^{5/7}$. Hence

$$\begin{aligned} \frac{dV}{dP} &= \frac{-5}{7}\frac{V}{P} \\ &= \frac{-5}{7}\frac{(c/P)^{5/7}}{P} = \frac{-5c^{5/7}}{7P^{12/7}}. \end{aligned}$$

As pressure increases, volume decreases.

41. $m'(x) = 4 - \cos x$, so the rod is less dense at the ends.

43. $m'(x) = 4$, so the rod is homogeneous.

45. $$\begin{aligned} Q'(t) &= e^{-2t} \cdot (-2)(\cos 3t - 2\sin 3t) \\ &\quad + e^{-2t}((-\sin 3t \cdot 3) - 2\cos 3t \cdot 3) \\ &= e^{-2t}(-8\cos 3t + \sin 3t) \text{ amps} \end{aligned}$$

47. As $t \to \infty$, $Q(t) \to 4\sin 3t$, so $e^{-3t}\cos 2t$ is called the transient term and $4\sin 3t$ is called the steady-state value.

$Q'(t) = e^{-3t} \cdot (-3) \cos 2t$
$\quad + e^{-3t}(-\sin 2t \cdot 2) + 4 \cos 3t \cdot 3$
$\quad = e^{-3t}(-3\cos 2t - 2\sin 2t)$
$\quad + 12 \cos 3t$

The transient term is $e^{-3t}(-3\cos 2t - 2\sin 2t)$ and the steady-state value is $12\cos 3t$.

49. The rate of population growth is given by
$f(p) = 4p(5-p) = 4(5p - p^2)$
$f'(p) = 4(5-2p)$,
so the only critical number is $p = 2.5$. Since the graph of f is a parabola opening down, this must be a max.

51.

$$p'(t) = \frac{-B(1+Ae^{-kt})'}{(1+Ae^{-kt})^2} = \frac{-B(-kAe^{-kt})}{(1+Ae^{-kt})^2} = \frac{kABe^{-kt}}{(1+Ae^{-kt})^2} = \frac{kABe^{-kt}}{1+2Ae^{-kt}+A^2e^{-2kt}} = \frac{kAB}{e^{kt}+2A+A^2e^{-kt}}$$

As t goes to infinity, the exponential term goes to 0, and so the limiting population is

$$\frac{B}{1+A(0)} = B.$$

53. For $a = 70$, $b = 0.2$,

$$f(t) = \frac{70}{1+3e^{-0.2t}} = 70(1+3e^{-0.2t})^{-1}$$
$$f(2) = \frac{7-0}{1+3e^{-0.2\cdot 2}} \approx 23$$
$$f'(t) = -70(1+3e^{-0.2t})^{-2}(3e^{-0.2t})(-0.2) = \frac{42e^{-0.2t}}{(1+3e^{-0.2t})^2}$$
$$f'(2) = \frac{42e^{0.2\cdot 2}}{(1+3e^{-0.2\cdot 2})^2} \approx 3.105$$

This says that at time $t = 2$ hours, the rate at which the spread of the rumor is increasing is about 3% of the population per hour.

$$\lim_{t\to\infty} f(t) = \frac{70}{1+0} = 70$$

so 70% of the population will eventually hear the rumor.

55. $$f'(x) = \frac{-64x^{-1.4}(4x^{-0.4}+15)}{(4x^{-0.4}+15)^2} - \frac{(160x^{-0.4}+90)(-1.6x^{-1.4})}{(4x^{-0.4}+15)^2} = \frac{-816x^{-1.4}}{(4x^{-0.4}+15)^2} < 0$$

So $f(x)$ is decreasing. This shows that pupils shrink as light increases.

57. If for some x marginal revenue equals marginal cost, then
$P'(x) = R'(x) - C'(x) = 0$,
so x is a critical number, but it may not be a maximum.

59. If v is not greater than c, the fish will never make any headway.

$$E'(v) = \frac{v(v-2c)}{(v-c)^2}$$

so the only critical number is $v = 2c$. When v is large, $E(v)$ is large, and when v is just a little bigger than c, $E(v)$ is large, so we must have a minimum.

Ch. 3 Review Exercises

1. $f(x) = e^{3x}$, $x_0 = 0$,
$f'(x) = 3e^{3x}$
$L(x) = f(x_0) + f'(x_0)(x - x_0)$
$\quad = f(0) + f'(0)(x-0)$
$\quad = e^{3\cdot 0} + 3e^{3\cdot 0}x$
$\quad = 1 + 3x$

3. $f(x) = \sqrt[3]{x} = x^{1/3}$, $x_0 = 8$
$f'(x) = \frac{1}{3}x^{-2/3}$
$$\begin{aligned} L(x) &= f(x_0) + f'(x_0)(x - x_0) \\ &= f(8) + f'(8)(x-8) \\ &= \sqrt[3]{8} + \tfrac{1}{3}(8)^{-2/3}(x-8) \\ &= 2 + \tfrac{1}{12}(x-8) \end{aligned}$$
$L(7.96) = 2 + \frac{1}{12}(7.96 - 8) \approx 1.99666$

5. From the graph of $f(x) = x^3 + 5x - 1$, there is one root.
$f'(x) = 3x^2 + 5$
Starting with $x_0 = 0$, Newton's method gives $x_1 = 0.2$, $x_2 = 0.198437$, and $x_3 = 0.198437$.

7. Near an inflection point, the rate of change of the rate of change of $f(x)$ is very small so there aren't any big dropoffs or sharp increases nearby to make the linear approximation inaccurate.

9. $\displaystyle\lim_{x\to 1}\frac{x^3-1}{x^2-1}$ is type $\frac{0}{0}$;

L'Hôpital's Rule gives

$$\lim_{x\to 1}\frac{3x^2}{2x} = \frac{3}{2}.$$

11. $\displaystyle\lim_{x\to 0}\frac{e^{2x}}{x^4+2}$ is type $\frac{\infty}{\infty}$;

applying L'Hôpital's Rule twice gives:

$$\begin{aligned} &\lim_{x\to\infty}\frac{2e^{2x}}{4x^3} \\ &= \lim_{x\to\infty}\frac{4e^{2x}}{12x^2} = \lim_{x\to\infty}\frac{8e^{2x}}{24x} \\ &= \lim_{x\to\infty}\frac{16e^{2x}}{24} = \infty \end{aligned}$$

13.

$$\begin{aligned} L &= \lim_{x\to 2^+}\left|\frac{x+1}{x-2}\right|^{\sqrt{x^2-4}} \\ \ln L &= \lim_{x\to 2^+}\left(\sqrt{x^2-4}\ln\left|\frac{x+1}{x-2}\right|\right) \\ &= \lim_{x\to 2^+}\left(\frac{\ln\left|\frac{x+1}{x-2}\right|}{(x^2-4)^{-1/2}}\right) \\ &= \lim_{x\to 2^+}\left(\frac{\left|\frac{x-2}{x+1}\right|\frac{-3}{(x-2)^2}}{-x(x^2-4)^{-3/2}}\right) \\ &= \lim_{x\to 2^+}\left(\frac{3(x^2-4)^{3/2}}{x(x+1)(x-2)}\right) \\ &= \lim_{x\to 2^+}\left(\frac{3(x-2)^{1/2}(x+2)^{3/2}}{x(x+1)}\right) \\ \ln L &= 0 \\ L &= 1 \end{aligned}$$

15.

$$\begin{aligned} \lim_{x\to 0^+}(\tan x\ln x) &= \lim_{x\to 0^+}\left(\frac{\ln x}{\cot x}\right) \\ &= \lim_{x\to 0^+}\left(\frac{1/x}{-\csc^2 x}\right) \\ &= \lim_{x\to 0^+} -\left(\frac{\sin^2 x}{x}\right) \\ &= -\lim_{x\to 0^+}\left(\frac{\sin x}{x}\sin x\right) \\ &= (-1)(0) = 0 \end{aligned}$$

17. $f'(x) = 3x^2 + 6x - 9 = 3(x^2 + 2x - 3)$
$= 3(x+3)(x-1)$
So the critical numbers are $x = 1$ and $x = -3$.
$f'(x) > 0$ on $(-\infty, -3) \cup (1, \infty)$
$f'(x) < 0$ on $(-3, 1)$
Hence f is increasing on $(-\infty, -3)$ and on $(1, \infty)$ and f is decreasing on $(-3, 1)$. Thus there is a local max at $x = -3$ and a local min at $x = 1$.

$f''(x) = 3(2x + 2) = 6(x + 1)$
$f''(x) > 0$ on $(-1, \infty)$

$f''(x) < 0$ on $(-\infty, -1)$
Hence f is concave up on $(-1, \infty)$ and concave down on $(-\infty, -1)$, and there is an inflection point at $x = -1$.

19. $f'(x) = 4x^3 - 12x^2 = 4x^2(x-3)$
$x = 0, 3$ are critical numbers.
$f'(x) > 0$ on $(3, \infty)$
$f'(x) < 0$ on $(-\infty, 0) \cup (0, 3)$
f increasing on $(3, \infty)$, decreasing on $(-\infty, 3)$ so $x = 3$ is a local min.
$f''(x) = 12x^2 - 24x = 12x(x-2)$
$f''(x) > 0$ on $(-\infty, 0) \cup (2, \infty)$
$f''(x) < 0$ on $(0, 2)$
f is concave up on $(-\infty, 0) \cup (2, \infty)$, concave down on $(0, 2)$ so $x = 0, 2$ are inflection points.

21. $f'(x) = e^{-4x} + xe^{-4x}(-4) = e^{-4x}(1 - 4x)$ $x = 1/4$ is a critical number.
$f'(x) > 0$ on $\left(-\infty, \frac{1}{4}\right)$
$f'(x) < 0$ on $\left(\frac{1}{4}, \infty\right)$
f increasing on $\left(-\infty, \frac{1}{4}\right)$, decreasing on $\left(-\frac{1}{4}, \infty\right)$ so $x = 1/4$ is a local max.
$f''(x) = e^{-4x}(-4)(1-4x) + e^{-4x}(-4)$
$= -4e^{-4x}(2 - 4x)$
$f''(x) > 0$ on $\left(\frac{1}{2}, \infty\right)$
$f''(x) < 0$ on $\left(-\infty, \frac{1}{2}\right)$
f is concave up on $\left(\frac{1}{2}, \infty\right)$, concave down on $\left(-\infty, \frac{1}{2}\right)$ so $x = 1/2$ is inflection point.

23. $f'(x) = \dfrac{x^2 - (x-90)(2x)}{x^4}$
$= \dfrac{-(x-180)}{x^3}$
$x = 180$ is the only critical number.
$f'(x) < 0$ on $(-\infty, 0) \cup (180, \infty)$
$f'(x) > 0$ on $(0, 180)$
$f(x)$ is decreasing on $(-\infty, 0) \cup (180, \infty)$ and increasing on $(0, 180)$ so $f(x)$ has a local maximum at $x = 180$.
$f''(x) = -\dfrac{x^3 - (x-180)(3x^2)}{x^6}$
$= -\dfrac{-2x + 540}{x^4}$
$f''(x) < 0$ on $(-\infty, 0) \cup (0, 270)$
$f''(x) > 0$ on $(270, \infty)$ so $x = 90$ is an inflection point.

25. $f'(x) = \dfrac{x^2 + 4 - x(2x)}{(x^2+4)^2}$
$= \dfrac{4 - x^2}{(x^2+4)^2}$
$x = \pm 2$ are critical numbers.
$f'(x) > 0$ on $(-2, 2)$
$f'(x) < 0$ on $(-\infty, -2) \cup (2, \infty)$
f increasing on $(-2, 2)$, decreasing on $(-\infty, -2)$ and on $(2, \infty)$ so f had a local min at $x = -2$ and a local max at $x = 2$.
$f''(x) =$
$\dfrac{-2x(x^2+4)^2 - (4-x^2)[2(x^2+4) \cdot 2x]}{(x^2+4)^4}$
$= \dfrac{2x^3 - 24x}{(x^2+4)^3}$
$f''(x) > 0$ on $\left(-\sqrt{12}, 0\right) \cup \left(\sqrt{12}, \infty\right)$
$f''(x) < 0$ on $\left(-\infty, -\sqrt{12}\right) \cup \left(0, \sqrt{12}\right)$
f is concave up on $\left(-\sqrt{12}, 0\right) \cup \left(\sqrt{12}, \infty\right)$, concave down on $\left(-\infty, -\sqrt{12}\right) \cup \left(0, \sqrt{12}\right)$ so $x = \pm\sqrt{12}, 0$ are inflection points.

27. $f'(x) = 3x^2 + 6x - 9$
$= 3(x+3)(x-1)$
$x = -3$, $x = 1$ are critical numbers, but $x = -3 \notin [0, 4]$.
$f(0) = 0^3 + 3 \cdot 0^2 - 9 \cdot 0 = 0$
$f(4) = 4^3 + 3 \cdot 4^2 - 9 \cdot 4 = 76$
$f(1) = 1^3 + 3 \cdot 1^2 - 9 \cdot 1 = -5$
So $f(4) = 76$ is absolute max on $[0, 4]$, $f(1) = -5$ is absolute min.

29. $f'(x) = \frac{4}{5}x^{-1/5}$
$x = 0$ is critical number.
$f(-2) = (-2)^{4/5} \approx 1.74$
$f(3) = (3)^{4/5} \approx 2.41$
$f(0) = (0)^{4/5} = 0$
$f(0) = 0$ is absolute min, $f(3) = 3^{4/5}$ is absolute max.

31. $f'(x) = 3x^2 + 8x + 2$

$f'(x) = 0$ when

$$x = \frac{-8 \pm \sqrt{64-24}}{6} = -\frac{4}{3} \pm \frac{\sqrt{10}}{3}$$

$x = -\frac{4}{3} - \frac{\sqrt{10}}{3}$ is local max, $x = -\frac{4}{3} + \frac{\sqrt{10}}{3}$ is local min.

33. $f'(x) = 5x^4 - 4x + 1 = 0$
$x \approx 0.2553,\ 0.8227$
local min at $x \approx 0.8227$,
local max at $x \approx 0.2553$.

35. One possible graph:

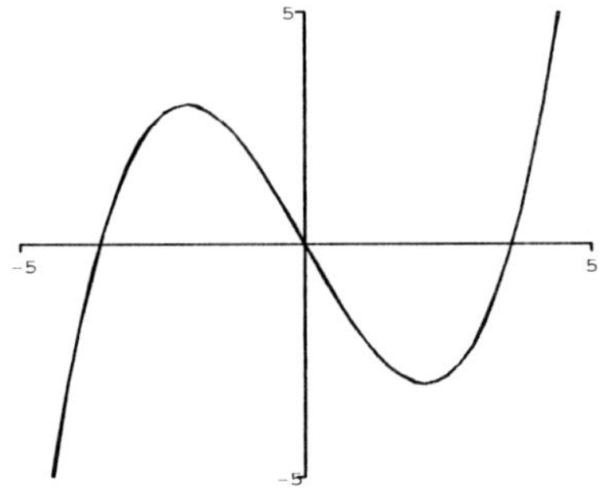

37. $f'(x) = 4x^3 + 12x^2 = 4x^2(4x+3)$
$f''(x) = 12x^2 + 24x = 12x(x+2)$
$f'(x) > 0$ on $(-3, 0) \cup (0, \infty)$
$f'(x) < 0$ on $(-\infty, -3)$
$f''(x) > 0$ on $(-\infty, -2) \cup (0, \infty)$
$f''(x) < 0$ on $(-2, 0)$
f increasing on $(-3, \infty)$, decreasing on $(-\infty, -3)$, concave up on $(-\infty, -2) \cup (0, \infty)$, concave down on $(-2, 0)$, local min at $x = -3$, inflection points at $x = -2,\ 0$.
$f(x) \to \infty$ as $x \to \pm\infty$.

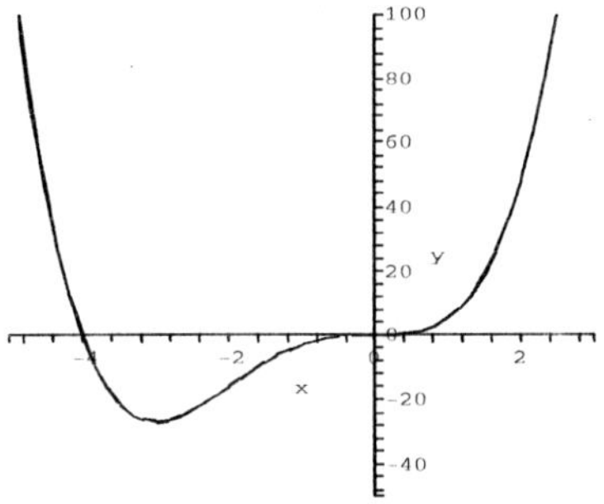

39. $f'(x) = 4x^3 + 4 = 4(x^3+1)$
$f''(x) = 12x^2$
$f'(x) > 0$ on $(-1, \infty)$
$f'(x) < 0$ on $(-\infty, -1)$
$f''(x) > 0$ on $(-\infty, 0) \cup (0, \infty)$
f increasing on $(-1, \infty)$, decreasing on $(-\infty, -1)$, concave up on $(-\infty, \infty)$, local min at $x = -1$.
$f(x) \to \infty$ as $x \to \pm\infty$.

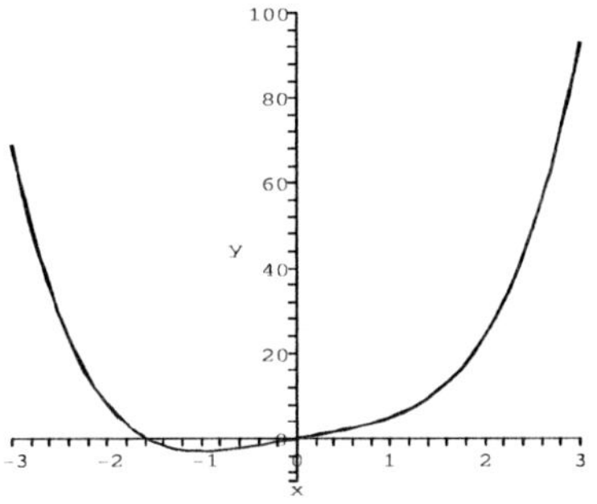

41. $f'(x) = \dfrac{x^2 + 1 - x(2x)}{(x^2+1)^2}$
$= \dfrac{1 - x^2}{(x^2+1)^2}$
$f''(x) =$
$\dfrac{-2x(x^2+1)^2 - (1-x^2)2(x^2+1)2x}{(x^2+1)^4}$
$= \dfrac{2x(x^2-3)}{(x^2+1)^4}$
$f'(x) > 0$ on $(-1, 1)$
$f'(x) < 0$ on $(-\infty, -1) \cup (1, \infty)$
$f''(x) > 0$ on $\left(-\sqrt{3}, 0\right) \cup \left(\sqrt{3}, \infty\right)$
$f''(x) < 0$ on $\left(-\infty, -\sqrt{3}\right) \cup \left(0, \sqrt{3}\right)$
f increasing on $(-1, 1)$, decreasing on $(-\infty, -1)$ and on $(1, \infty)$, concave up

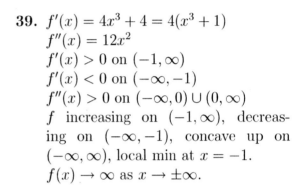

on

$$\left(-\sqrt{3},0\right)\cup\left(\sqrt{3},\infty\right),$$

concave down on

$$\left(-\infty,-\sqrt{3}\right)\cup\left(0,\sqrt{3}\right),$$

local min at $x = -1$, local max at $x = 1$, inflection points at $0, \pm\sqrt{3}$.

$$\lim_{x\to\infty}\frac{x}{x^2+1} = \lim_{x\to-\infty}\frac{x}{x^2+1} = 0$$

So f has a horizontal asymptote at $y = 0$.

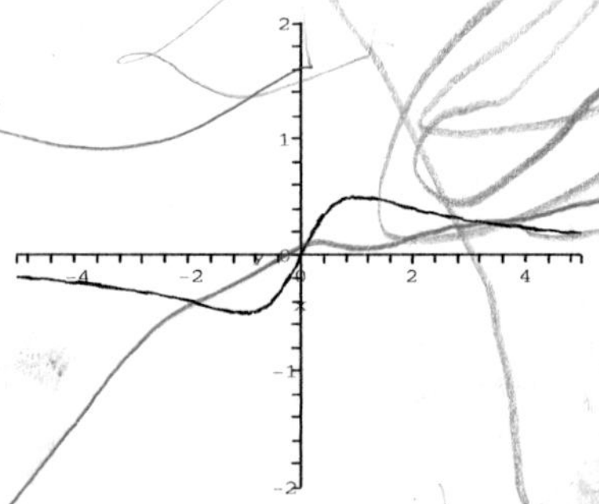

43. $$f'(x) = \frac{(2x)(x^2+1) - x^2(2x)}{(x^2+1)^2} = \frac{2x}{(x^2-1)^2}$$

$$f''(x) = \frac{2(x^2+1)^2 - 2x\cdot 2(x^2+1)2x}{(x^2+1)^4} = \frac{2-6x^2}{(x^2+1)^3}$$

$f'(x) > 0$ on $(0,\infty)$
$f'(x) < 0$ on $(-\infty,0)$
$f''(x) > 0$ on $\left(-\sqrt{\frac{1}{3}},\sqrt{\frac{1}{3}}\right)$
$f''(x) < 0$ on $\left(-\infty,-\sqrt{\frac{1}{3}}\right)\cup\left(\sqrt{\frac{1}{3}},\infty\right)$
f increasing on $(0,\infty)$ decreasing on $(-\infty,0)$, concave up on

$$\left(-\sqrt{\frac{1}{3}},\sqrt{\frac{1}{3}}\right),$$

concave down on

$$\left(-\infty,-\sqrt{\frac{1}{3}}\right)\cup\left(\sqrt{\frac{1}{3}},\infty\right),$$

local min at $x = 0$, inflection points at $x = \pm\sqrt{1/3}$.

$$\lim_{x\to\infty}\frac{x^2}{x^2+1} = \lim_{x\to-\infty}\frac{x^2}{x^2+1} = 1$$

So f has a horizontal asymptote at $y = 1$.

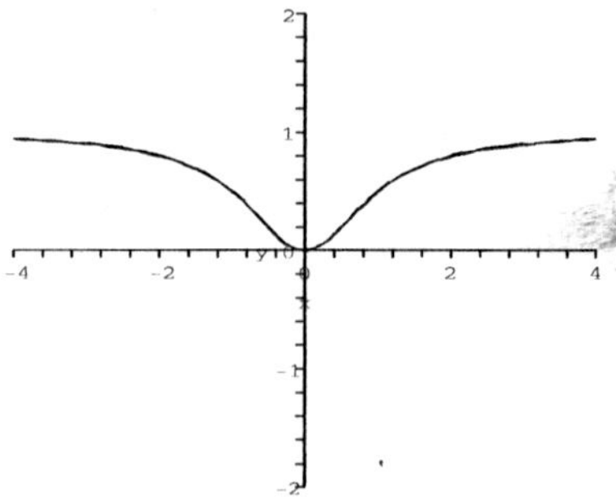

45. $$f'(x) = \frac{3x^2(x^2-1) - x^3(2x)}{(x^2-1)^2} = \frac{x^4-3x^2}{(x^2-1)^2}$$

$$f''(x) = \frac{(4x^3-6x)(x^2-1)^2}{(x^2-1)^4} - \frac{(x^4-3x^2)2(x^2-1)2x}{(x^2-1)^4} = \frac{2x^3+6x}{(x^2-1)^4}$$

$f'(x) > 0$ on $(-\infty,-\sqrt{3})\cup(\sqrt{3},\infty)$
$f'(x) < 0$ on $(-\sqrt{3},-1)\cup(-1,0)\cup(0,1)\cup(1,\sqrt{3})$
$f''(x) > 0$ on $(-1,0)\cup(1,\infty)$
$f''(x) < 0$ on $(-\infty,-1)\cup(0,1)$
f increasing on $(-\infty,-\sqrt{3})$ and on $(\sqrt{3},\infty)$; decreasing on $(-\sqrt{3},-1)$ and on $(-1,1)$ and on $(1,\sqrt{3})$; concave up on $(-1,0)\cup(1,\infty)$, concave down on $(-\infty,-1)\cup(0,1)$; $x = -\sqrt{3}$ local max; $x = \sqrt{3}$ local min; $x = 0$ inflection point. f is undefined at $x = -1$ and $x = 1$.

$\lim_{x\to 1^+} \frac{x^3}{x^2-1} = \infty$, and

$\lim_{x\to 1^-} \frac{x^3}{x^2-1} = -\infty$

So f has vertical asymptotes at $x = 1$ and $x = -1$.

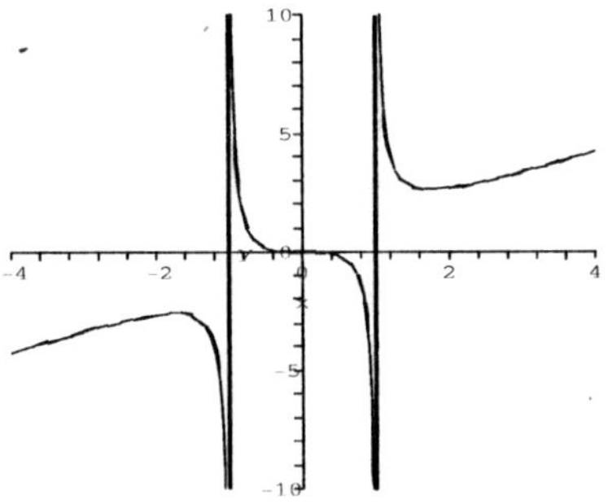

47. $d = \sqrt{(x-2)^2 + (y-1)^2}$
$= \sqrt{(x-2)^2 + (2x^2-1)^2}$
$f(x) = (x-2)^2 + (2x^2-1)^2$
$f'(x) = 2(x-2) + 2(2x^2-1)4x$
$= 16x^3 - 6x - 4$
$f'(x) = 0$ when $x \approx 0.8237$
$f'(x) < 0$ on $(-\infty, 0.8237)$
$f'(x) > 0$ on $(0.8237, \infty)$
So $x \approx 0.8237$ corresponds to the closest point.
$y = 2x^2 = 2(0.8237)^2 = 1.3570$
$(0.8237, 1.3570)$ is closest to $(2, 1)$.

49. $C(x) = 6\sqrt{4^2 + (4-x)^2} + 2\sqrt{2^2 + x^2}$
$C'(x) =$
$6 \cdot \frac{1}{2}[16 + (4-x)^2]^{-1/2} \cdot 2(4-x)(-1)$
$+ 2\frac{1}{2}(4 + x^2)^{-1/2} \cdot 2x$
$= \frac{6(x-4)}{\sqrt{16+(4-x)^2}} + \frac{2x}{\sqrt{4+x^2}}$
$C'(x) = 0$ when $x \approx 2.864$
$C'(x) < 0$ on $(0, 2.864)$
$C'(x) > 0$ on $(2.864, 4)$
So $x \approx 2.864$ gives the minimum cost.
Locate highway corner $4 - 2.864 = 1.136$ miles east of point A.

51. Area: $A = 2\pi r^2 + 2\pi rh$
Convert to in^3:
16 fl oz = 16 fl oz $\cdot$ 1.80469 in^3/fl oz
$= 28.87504\text{in}^3$
Volume: $V = \pi r^2 h$

$$h = \frac{\text{Vol}}{\pi r^2} = \frac{28.87504}{\pi r^2}$$

$A(r) = 2\pi\left(r^2 + \frac{28.87504}{\pi r}\right)$
$A'(r) = 2\pi\left(2r - \frac{28.87504}{\pi r^2}\right)$
$2\pi r^3 = 28.87504$

$$r = \sqrt[3]{\frac{28.87504}{2\pi}} \approx 1.663$$

$A'(r) < 0$ on $(0, 1.663)$
$A'(r) > 0$ on $(1.663, \infty)$
So $r \approx 1.663$ gives the minimum surface area.

$$h = \frac{28.87504}{\pi(1.663)^2} \approx 3.325$$

53. Let θ_1 be the angle from the horizontal to the upper line segment defining θ and let θ_2 be the angle from the horizontal to the lower line segment defining θ. Then the length of the side opposite θ_2 is $\frac{H-P}{2}$ while the length of the side opposite θ_1 is $\frac{H+P}{2}$. Then

$$\theta(x) = \theta_1 - \theta_2$$
$$= \tan^{-1}\left(\frac{H+P}{2x}\right) - \tan^{-1}\left(\frac{H-P}{2x}\right)$$

and so

$$\theta'(x) = \frac{1}{1 + \left(\frac{H+P}{2x}\right)^2}\left(-\frac{H+P}{2x^2}\right) - \frac{1}{1+\left(\frac{H-P}{2x}\right)^2}\left(-\frac{H-P}{2x^2}\right).$$

We set this equal to 0:

$$0 = \frac{-2(H+P)}{4x^2 + (H+P)^2} + \frac{2(H-P)}{4x^2 + (H-P)^2}$$

and solve for x:

$$\frac{2(H+P)}{4x^2+(H+P)^2} = \frac{2(H-P)}{4x^2+(H-P)^2}$$

$$8x^2(H+P) - 8x^2(H-P)$$
$$= 2(H-P)(H+P)^2$$
$$- 2(H+P)(H-P)^2$$
$$8x^2(2P) = 2(H-P)(H+P)(2P)$$
$$x^2 = \frac{H^2-P^2}{4}$$
$$x = \frac{\sqrt{H^2-P^2}}{2}.$$

55. $Q'(t) = -3e^{-3t}\sin 2t + e^{-3t}\cos 2t \cdot 2$
$= e^{-3t}(2\cos 2t - 3\sin 2t)$ amps

57. $\rho(x) = m'(x) = 2x$
As you move along the rod to the right, its density increases.

59. $C'(x) = 0.04x + 20$
$C'(20) = 0.04(20) + 20 = 20.8$
$C(20) - C(19) =$
$0.02(20)^2 + 20(20) + 1800$
$- [0.02(19)^2 + 20(19) + 1800]$
$= 20.78$

Chapter 4

Integration

4.1 Antiderivatives

1. $\dfrac{x^4}{4}, \dfrac{x^4}{4} + 3, \dfrac{x^4}{4} - 2$

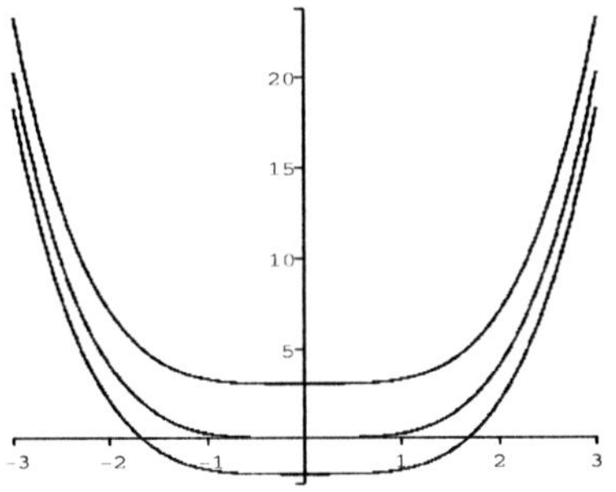

3. $e^x, e^x + 1, e^x - 3$

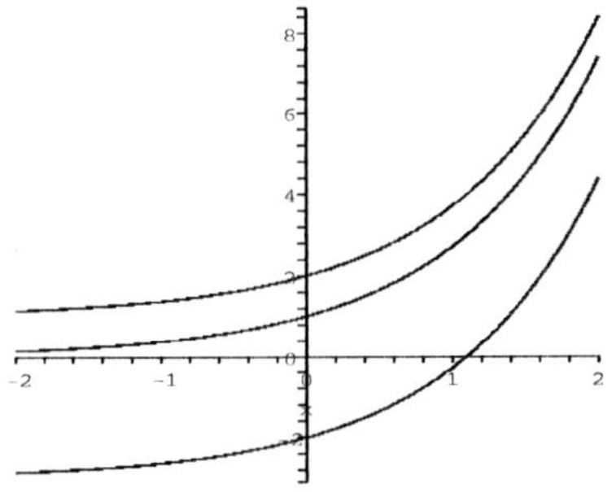

5. $\displaystyle\int (3x^4 - 3x)\,dx = \frac{3}{5}x^5 - \frac{3}{2}x^2 + c$

7. $\displaystyle\int \left(3\sqrt{x} - \frac{1}{x^4}\right) dx = 2x^{3/2} + \frac{x^{-3}}{3} + c$

9. $\displaystyle\int \frac{x^{1/3} - 3}{x^{2/3}}\,dx$

$\displaystyle = \int (x^{-1/3} - 3x^{-2/3})\,dx$

$\displaystyle = \frac{3}{2}x^{2/3} - 9x^{1/3} + c$

11. $\displaystyle\int (2\sin x + \cos x)\,dx$

$= -2\cos x + \sin x + c$

13. $\displaystyle\int 2\sec x \tan x\,dx = 2\sec x + c$

15. $\displaystyle\int 5\sec^2 x\,dx = 5\tan x + c$

17. $\displaystyle\int (3e^x - 2)\,dx = 3e^x - 2x + c$

19. $\displaystyle\int (3\cos x - 1/x)\,dx$

$= 3\sin x - \ln|x| + c$

21. $\displaystyle\int \frac{4x}{x^2+4}\,dx = 2\ln|x^2 + 4| + c$

23. $\displaystyle\int \left(5x - \frac{3}{e^x}\right) dx = \frac{5}{2}x^2 + \frac{3}{e^x} + c$

25. $\displaystyle\int \frac{e^x}{e^x + 3}\,dx = \ln|e^x + 3| + c$

27. $\displaystyle\int \frac{e^x + 3}{e^x}\,dx = \int (1 + 3e^{-x})\,dx$

$= x - 3e^{-x} + c$

29. $\displaystyle\int x^{1/4}(x^{5/4} - 4)\,dx$

$\displaystyle = \int (x^{3/2} - 4x^{1/4})\,dx$

$\displaystyle = \frac{2}{5}x^{5/2} - \frac{16}{5}x^{5/4} + c$

31. a) N/A

b) $\displaystyle\int (\sqrt{x^3} + 4)\,dx = \frac{2}{5}x^{5/2} + 4x + c$

33. a) N/A

b) $\displaystyle\int \sec^2 x\,dx = \tan x + c$

35. Use a CAS to find antiderivatives and verify by computing the derivatives:

1.11(b) $\int \sec x\, dx = \ln|\sec x + \tan x| + c$
Verify:
$$\frac{d}{dx} \ln|\sec x + \tan x|$$
$$= \frac{\sec x \tan x + \sec^2 x}{\sec x + \tan x} = \sec x$$

1.11(f) $\int x \sin 2x\, dx$
$$= \frac{\sin 2x}{4} - \frac{x \cos 2x}{2} + c$$
Verify:
$$\frac{d}{dx}\left(\frac{\sin 2x}{4} - \frac{x \cos 2x}{2}\right)$$
$$= \frac{2\cos 2x}{4} - \frac{\cos 2x - 2x \sin 2x}{2}$$
$$= x \sin 2x$$

37. Use a CAS to find antiderivatives and verify by computing the derivatives:

(a) $\int x^2 e^{-x^3}\, dx = -\frac{1}{3} e^{-x^3} + c$
Verify:
$$\frac{d}{dx}\left(-\frac{1}{3} e^{-x^3}\right)$$
$$= -\frac{1}{3} e^{-x^3} \cdot (-3x^2)$$
$$= x^2 e^{-x^3}$$

(b) $\int \frac{1}{x^2 - x}\, dx = \ln|x-1| - \ln|x| + c$
Verify:
$$\frac{d}{dx}(\ln|x-1| - \ln|x|)$$
$$= \frac{1}{x-1} - \frac{1}{x} = \frac{x - (x-1)}{x(x-1)}$$
$$= \frac{1}{x(x-1)} = \frac{1}{x^2 - x}$$

(c) $\int \sec x dx = \ln|\sec x + \tan x| + c$
Verify:
$$\frac{d}{dx}[\ln|\sec x + \tan x|]$$
$$= \frac{\sec x \tan x + \sec^2 x}{\sec x + \tan x}$$
$$= \frac{\sec x(\sec x + \tan x)}{\sec x + \tan x} = \sec x$$

39. Finding the antiderivative,
$f(x) = 3e^x + \frac{x^2}{2} + c.$

Since $f(0) = 4$, we have
$4 = f(0) = 3 + c$. Therefore,
$f(x) = 4 \sin x + 1.$

41. Finding the antiderivative of f'' gives
$f'(x) = 12x + c_1.$

Since $f'(0) = 2$, we have $2 = f'(0) = c_1$ and therefore
$f'(x) = 12x + 2.$

Finding the antiderivative of $f'(x)$ gives
$f(x) = 6x^2 + 2x + c_2.$

Since $f(0) = 3$, we have $3 = f(0) = c_2$ and
$f(x) = 6x^2 + 2x + 3.$

43. Taking antiderivatives,
$f''(x) = 3 \sin x + 4x^2$
$f'(x) = -3 \cos x + \frac{4}{3}x^3 + c_1$
$f(x) = -3 \sin x + \frac{1}{3}x^4 + c_1 x + c_2.$

45. Taking antiderivatives,
$f'''(x) = 4 - 2/x^3$
$f''(x) = 4x + x^{-2} + c_1$
$f'(x) = 2x^2 - x^{-1} + c_1 x + c_2$
$f(x) = \frac{2}{3}x^3 - \ln|x| + \frac{c_1}{2}x^2 + c_2 x + c_3$

47. Position is the antiderivative of velocity,
$s(t) = 3t - 6t^2 + c.$
Since $s(0) = 3$, we have $c = 3$. Thus,
$s(t) = 3t - 6t^2 + 3.$

49. First we find velocity, which is the antiderivative of acceleration,
$v(t) = -3 \cos t + c_1.$
Since $v(0) = 0$ we have
$-3 + c_1 = 0$, $c_1 = 3$ and
$v(t) = -3 \cos t + 3.$

Position is the antiderivative of velocity,
$s(t) = -3\sin t + 3t + c_2$.
Since $s(0) = 4$, we have $c_2 = 4$. Thus,
$s(t) = -3\sin t + 3t + 4$.

51. The key is to find the velocity and position functions. We start with constant acceleration a, a constant. Then, $v(t) = at + v_0$ where v_0 is the initial velocity. The initial velocity is 30 miles per hour, but since our time is in seconds, it is probably best to work in feet per second ($30\text{mph} = 44\text{ft/s}$). $v(t) = at + 44$.

We know that the car accelerates to 50 mph ($50\text{mph} = 73\text{ft/s}$) in 4 seconds, so $v(4) = 73$. Therefore, $a \cdot 4 + 44 = 73$ and $a = \frac{29}{4}$ ft/s

So,
$v(t) = \frac{29}{4}t + 44$ and
$s(t) = \frac{29}{8}t^2 + 44t + s_0$
where s_0 is the initial position. We can assume the the starting position is $s_0 = 0$.

Then, $s(t) = \frac{29}{8}t^2 + 44t$ and the distance traveled by the car during the 4 seconds is $s(4) = 234$ feet.

53. There are many correct answers, but any correct answer will be a vertical shift of this answer.

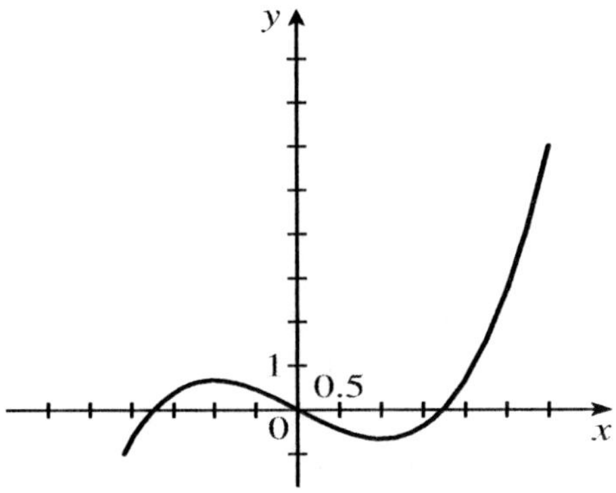

55. All functions that have the derivative shown in Exercise 53 are vertical translations of the graph given as the answer for Exercise 53.

57. To estimate the acceleration over each interval, we estimate $v'(t)$ by computing the slope of the tangent lines. For example, for the interval $[0, 0.5]$:
$a \approx \frac{v(0.5) - v(0)}{0.5 - 0} = -31.6 \text{ m/s}^2$.
Notice, acceleration should be negative since the object is falling.

To estimate the distance traveled over the interval, we estimate the velocity and multiply by the time (distance is rate times time). For an estimate for the velocity, we will use the average of the velocities at the endpoints. For example, for the interval $[0, 0.5]$, the time interval is 0.5 and the velocity is -11.9. Therefore the position changed is $(-11.9)(0.5) = -5.95$ meters. The distance traveled will be 5.95 meters (distance should be positive).

Interval	Accel	Dist
$[0.0, 0.5]$	-31.6	5.95
$[0.5, 1.0]$	-24.2	12.925
$[1.0, 1.5]$	-11.6	17.4
$[1.5, 2.0]$	-3.6	19.3

59. To estimate the speed over the interval, we first approximate the acceleration over the interval by averaging the acceleration at the endpoint of the interval. Then, the velocity will be the acceleration times the length of time. The slope of the tangent lines. For example, for the interval $[0, 0.5]$ the average acceleration is -0.9 and $v(0.5) = 70 + (-0.9)(0.5) = 69.55$.

And, the distance traveled is the speed times the length of time. For the time $t = 0.5$, the distance would be $\dfrac{70 + 69.55}{2} \times 0.5 \approx 34.89$ meters.

Time	Speed	Dist
0	70	0
0.5	69.55	34.89
1.0	70.3	69.85
1.5	70.35	105.01
2.0	70.65	104.26

61. We start by taking antiderivatives:
$f'(x) = x^2/2 - x + c_1$
$f(x) = x^3/6 - x^2/2 + c_1 x + c_2.$

Now, we use the data that we are given. We know that $f(1) = 2$ and $f'(1) = 3$, which gives us
$3 = f'(1) = 1/2 - 1 + c_1,$
and
$1 = f(1) = 1/6 - 1/2 + c_1 + c_2.$

Therefore $c_1 = 7/2$ and $c_2 = -13/6$ and the function is
$$f(x) = \frac{x^3}{6} - \frac{x^2}{2} + \frac{7x}{2} - \frac{13}{6}.$$

63. Let $u = x^2$; then $du = 2xdx$.
$$\int 2x \cos x^2\, dx = \int \cos u\, du$$
$= \sin u + c$
$= \sin x^2 + c$

65. $\dfrac{d}{dx}\left[2x \sin 2x + x^2 2 \cos 2x\right]$
$= 2(x \sin 2x + x^2 \cos 2x)$

Therefore,
$$\int \left(x \sin 2x + x^2 \cos 2x\right) dx$$
$= \dfrac{1}{2} x^2 \sin 2x + c$

67. $\displaystyle\int \frac{x \cos(x^2)}{\sqrt{\sin(x^2)}} dx = \sqrt{\sin(x^2)} + c$

69. $\displaystyle\int \frac{-1}{\sqrt{1 - x^2}} dx = \cos^{-1}(x) + c_1$
$\displaystyle\int \frac{-1}{\sqrt{1 - x^2}} dx = -\sin^{-1}(x) + c_2$

Therefore,
$\cos^{-1} x + c_1 = -\sin^{-1} x + c_2$

Therefore,
$\sin^{-1} x + \cos^{-1} x =$ constant

To find the value of the constant, let x be any convenient value. Suppose $x = 0$; then $\sin^{-1} 0 = 0$ and $\cos^{-1} 0 = \pi/2$, so
$\sin^{-1} x + \cos^{-1} x = \dfrac{\pi}{2}$

71. To derive these formulas, all that needs to be done is to take the derivatives to see that the integrals are correct:
$$\frac{d}{dx}\left(e^x\right) = e^x$$
$$\frac{d}{dx}\left(-e^{-x}\right) = e^{-x}$$

4.2 Sums And Sigma Notation

1. $\displaystyle\sum_{i=1}^{50} i^2 = \frac{(50)(51)(101)}{6} = 42,925$

3. $\displaystyle\sum_{i=1}^{10} \sqrt{i}$
$= 1 + \sqrt{2} + \sqrt{3} + \sqrt{4} + \sqrt{5} + \sqrt{6}$
$+ \sqrt{7} + \sqrt{8} + \sqrt{9} + \sqrt{10}$
≈ 22.47

5. $\displaystyle\sum_{i=1}^{6} 3i^2 = 3 + 12 + 27 + 48 + 75 + 108$
$= 273$

7. $\displaystyle\sum_{i=6}^{10} (4i + 2)$
$= (4(6) + 2) + (4(7) + 2) + (4(8) + 2)$
$+ (4(9) + 2) + (4(10) + 2)$
$= 26 + 30 + 34 + 38 + 42$
$= 170$

9. $\sum_{i=1}^{70}(3i-1)$

$= 3\cdot\sum_{i=1}^{70} i - 70$

$= 3\cdot\frac{70(71)}{2} - 70$

$= 7,385$

11. $\sum_{i=1}^{40}(4-i^2)$

$= 160 - \sum_{i=1}^{40} i^2$

$= 160 - \frac{(40)(41)(81)}{6}$

$= 160 - 22,140$

$= -21,980$

13. $\sum_{i=1}^{100}(i^2-3i+2)$

$= \sum_{i=1}^{100} i^2 - 3\cdot\sum_{i=1}^{100} i + 200$

$= \frac{(100)(101)(201)}{6} - 3\cdot\frac{100(101)}{2} + 200$

$= 338,350 - 15,150 + 200$

$= 323,400$

15. $\sum_{i=1}^{200}(4-3i-i^2)$

$= 800 - 3\cdot\sum_{i=1}^{200} i - \sum_{i=1}^{200} i^2$

$= 800 - 3\cdot\frac{200(201)}{2} - \frac{(200)(201)(401)}{6}$

$= -2,746,200$

17. $\sum_{i=0}^{n}(i^2-3)$

$= \sum_{i=0}^{n} i^2 + \sum_{i=0}^{n}(-3)$

$= 0 + \sum_{i=1}^{n} i^2 + (n+1)(-3)$

$= \frac{n(n+1)(2n+1)}{6} - 3(n+1)$

$= \frac{(n+1)(2n^2+n-18)}{6}$

19. $\sum_{i=1}^{n}\frac{1}{n}\left[\left(\frac{i}{n}\right)^2 + 2\left(\frac{i}{n}\right)\right]$

$= \frac{1}{n}\left[\sum_{i=1}^{n}\frac{i^2}{n^2} + 2\sum_{i=1}^{n}\frac{i}{n}\right]$

$= \frac{1}{n}\left[\frac{1}{n^2}\sum_{i=1}^{n} i^2 + \frac{2}{n}\sum_{i=1}^{n} i\right]$

$= \frac{1}{n}\left[\frac{1}{n^2}\left(\frac{n(n+1)(2n+1)}{6}\right)\right.$
$\left. + \frac{2}{n}\left(\frac{n(n+1)}{2}\right)\right]$

$= \frac{n(n+1)(2n+1)}{6n^3} + \frac{n(n+1)}{n^2}$

$\lim_{n\to\infty}\sum_{i=1}^{n}\frac{1}{n}\left[\left(\frac{i}{n}\right)^2 + 2\left(\frac{i}{n}\right)\right]$

$= \lim_{n\to\infty}\left[\frac{n(n+1)(2n+1)}{6n^3} + \frac{n(n+1)}{n^2}\right]$

$= \frac{2}{6} + 1 = \frac{4}{3}$

21. $\sum_{i=1}^{n}\frac{1}{n}\left[4\left(\frac{2i}{n}\right)^2 - \left(\frac{2i}{n}\right)\right]$

$= \frac{1}{n}\left[16\sum_{i=1}^{n}\frac{i^2}{n^2} - 2\sum_{i=1}^{n}\frac{i}{n}\right]$

$= \frac{1}{n}\left[\frac{16}{n^2}\sum_{i=1}^{n} i^2 - \frac{2}{n}\sum_{i=1}^{n} i\right]$

$= \frac{1}{n}\left[\frac{16}{n^2}\left(\frac{n(n+1)(2n+1)}{6}\right)\right.$
$\left. - \frac{2}{n}\left(\frac{n(n+1)}{2}\right)\right]$

$= \frac{16n(n+1)(2n+1)}{6n^3} - \frac{n(n+1)}{n^2}$

$\lim_{n\to\infty}\sum_{i=1}^{n}\frac{1}{n}\left[4\left(\frac{2i}{n}\right)^2 - \left(\frac{2i}{n}\right)\right]$

$$= \lim_{n\to\infty} \left[\frac{16n(n+1)(2n+1)}{6n^3} - \frac{n(n+1)}{n^2} \right]$$
$$= \frac{16}{3} - 1 = \frac{13}{3}$$

23. $\sum_{i=1}^{n} f(x_i)\Delta x$
$$= \sum_{i=1}^{5} (x_i^2 + 4x_i) \cdot 0.2$$
$$= (0.2^2 + 4(0.2))(0.2) + \dots + (1^2 + 4)(0.2)$$
$$= (0.84)(0.2) + (1.76)(0.2) + (2.76)(0.2) + (3.84)(0.2) + (5)(0.2)$$
$$= 2.84$$

25. $\sum_{i=1}^{n} f(x_i)\Delta x$
$$= \sum_{i=1}^{10} (4x_i^2 - 2) \cdot 0.1$$
$$= (4(2.1)^2 - 2)(0.1) + \dots + (4(3)^2 - 2)(0.1)$$
$$= (15.64)(0.1) + (17.36)(0.1) + (19.16)(0.1) + (21.04)(0.1) + (23)(0.1) + (25.04)(0.1) + (27.16)(0.1) + (29.36)(0.1) + (31.64)(0.1) + (34)(0.1)$$
$$= 24.34$$

27. Distance
$= 50(2) + 60(1) + 70(1/2) + 60(3)$
$= 375$ miles.

29. Remember to convert minutes into hours.
Distance
$$= 15\left(\frac{1}{3}\right) + 18\left(\frac{1}{2}\right) + 16\left(\frac{1}{6}\right) + 12\left(\frac{2}{3}\right)$$
$$= 24\frac{2}{3} \text{ miles.}$$

31. On the time interval $[0, 0.25]$, the estimated velocity is the average velocity $\frac{120+116}{2} = 118$ feet per second. We estimate the distance traveled during the time interval $[0, 0.25]$ to be $(118)(0.25 - 0) = 29.5$ feet.

Altogether, the distance traveled is estimated as
$$= (236/2)(0.25) + (229/2)(0.25) + (223/2)(0.25) + (218/2)(0.25) + (214/2)(0.25) + (210/2)(0.25) + (207/2)(0.25) + (205/2)(0.25)$$
$= 217.75$ feet.

33. Want to prove that
$$\sum_{i=1}^{n} i^3 = \frac{n^2(n+1)^2}{4}$$
is true for all integers $n \geq 1$.

For $n = 1$, we have
$$\sum_{i=1}^{1} i^3 = 1 = \frac{1^2(1+1)^2}{4},$$
as desired. So the proposition is true for $n = 1$.

Next, assume that
$$\sum_{i=1}^{k} i^3 = \frac{k^2(k+1)^2}{4},$$
for some integer $k \geq 1$.

In this case, we have by the induction assumption that for $n = k + 1$,
$$\sum_{i=1}^{n} i^3 = \sum_{i=1}^{k+1} i^3 = \sum_{i=1}^{k} i^3 + (k+1)^3$$
$$= \frac{k^2(k+1)^2}{4} + (k+1)^3$$
$$= \frac{k^2(k+1)^2 + 4(k+1)^3}{4}$$
$$= \frac{(k+1)^2(k^2 + 4k + 4)}{4}$$

$$= \frac{(k+1)^2(k+2)^2}{4}$$
$$= \frac{n^2(n+1)^2}{4}$$
as desired.

35. $\sum_{i=1}^{10}(i^3 - 3i + 1)$
$$= \sum_{i=1}^{10} i^3 - 3\sum_{i=1}^{10} i + 10$$
$$= \frac{100(11)^2}{4} - 3\frac{10(11)}{2} + 10$$
$$= 2,870$$

37. $\sum_{i=1}^{100}(i^5 - 2i^2)$
$$= \sum_{i=1}^{100} i^5 - 2\sum_{i=1}^{100} i^2$$
$$= \frac{(100^2)(101^2)[2(100^2)+2(100)-1]}{12}$$
$$- 2\frac{20(21)(41)}{6}$$
$$= 171,707,655,800$$

39. $\sum_{i=1}^{n}(ca_i + db_i) = \sum_{i=1}^{n} ca_i + \sum_{i=1}^{n} db_i$
$$= c\sum_{i=1}^{n} a_i + d\sum_{i=1}^{n} b_i$$

41. $\sum_{i=1}^{n} e^{6i/n}\left(\frac{6}{n}\right)$
$$= \frac{6}{n}\sum_{i=1}^{n} e^{6i/n}$$
$$= \frac{6}{n}\left(\frac{e^{6/n} - e^6}{1 - e^{6/n}}\right)$$
$$= \frac{6}{n}\left(\frac{1 - e^6}{1 - e^{6/n}} - 1\right)$$
$$= \frac{6}{n}\frac{1 - e^6}{1 - e^{6/n}} - \frac{6}{n}$$
Now $\lim_{n\to\infty} \frac{6}{n} = 0$, and
$$\lim_{n\to\infty} \frac{6}{n}\frac{1 - e^6}{1 - e^{6/n}}$$
$$= 6(1 - e^6)\lim_{n\to\infty} \frac{1/n}{1 - e^{6/n}}$$
$$= 6(1 - e^6)\lim_{n\to\infty} \frac{1}{-6e^{6/n}}$$
$$= e^6 - 1,$$

Thus $\lim_{n\to\infty} \sum_{i=1}^{n} e^{6i/n}\frac{6}{n} = e^6 - 1.$

4.3 Area

1. a) Evaluation points:
0.125, 0.375, 0.625, 0.875.
Notice that $\Delta x = 0.25$.

$$A_4 = [f(0.125) + f(0.375) + f(0.625) + f(0.875)](0.25)$$
$$= [(0.125)^2 + 1 + (0.375)^2 + 1 + (0.625)^2 + 1 + (0.875)^2 + 1](0.5)$$
$$= 1.38125.$$

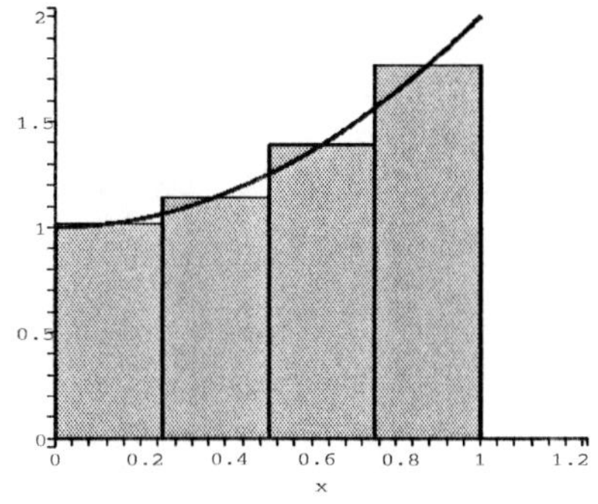

b) Evaluation points:
0.25, 0.75, 1.25, 1.75.
Notice that $\Delta x = 0.5$.

$$A_4 = [f(0.25) + f(0.75) + f(1.25) + f(1.75)](0.5)$$
$$= [(0.25)^2+1+(0.75)^2+1+(1.25)^2+1 + (1.75)^2 + 1](0.5)$$
$$= 4.625.$$

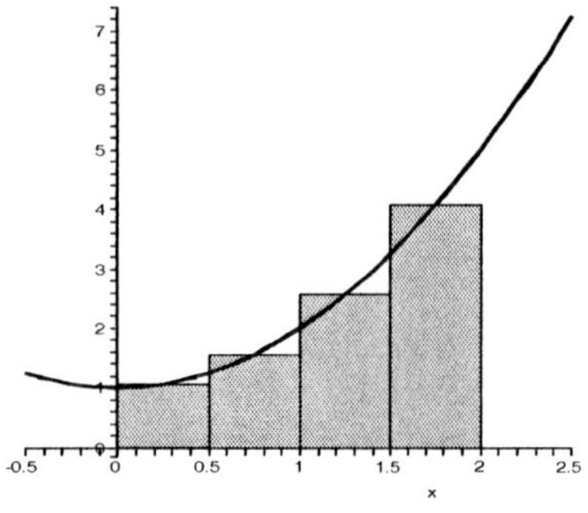

3. a) Evaluation points:
$\pi/8,\ 3\pi/8,\ 5\pi/8,\ 7\pi/8$.
Notice that $\Delta x = \pi/4$.

$$\begin{aligned} A_4 &= [f(\pi/8) + f(3\pi/8) + f(5\pi/8) \\ &\quad + f(7\pi/8)](\pi/4) \\ &= [\sin(\pi/8) + \sin(3\pi/8) + \sin(5\pi/8) \\ &\quad + \sin(7\pi/8)](\pi/4) \\ &= 2.05234. \end{aligned}$$

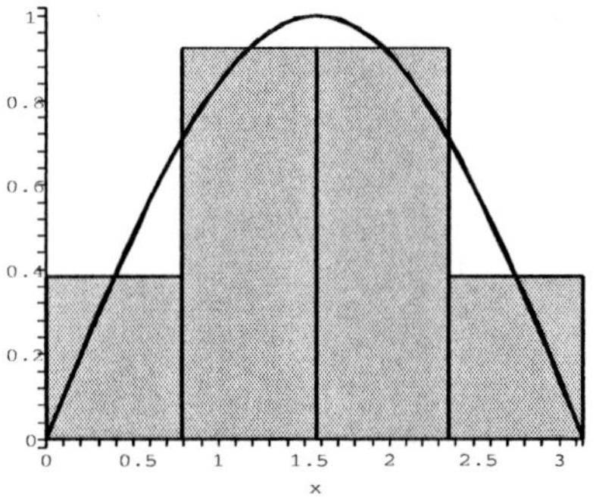

b) Evaluation points:
$\pi/16,\ 3\pi/16,\ 5\pi/16,\ 7\pi/16,\ 9\pi/16,\ 11\pi/16,\ 13\pi/16,\ 15\pi/16$.
Notice that $\Delta x = \pi/8$.

$$\begin{aligned} A_4 &= [f(\pi/16)+f(3\pi/16)+f(5\pi/16) \\ &\quad +f(7\pi/16)+f(9\pi/16)+f(11\pi/16) \\ &\quad + f(13\pi/16) + f(15\pi/16)](\pi/8) \\ &= [\sin(\pi/16)+\sin(3\pi/16)+\sin(5\pi/16) \\ &\quad + \sin(7\pi/16) + \sin(9\pi/16) \\ &\quad + \sin(11\pi/16) + \sin(13\pi/16) \\ &\quad + \sin(15\pi/16)](\pi/8) \\ &= 2.0129. \end{aligned}$$

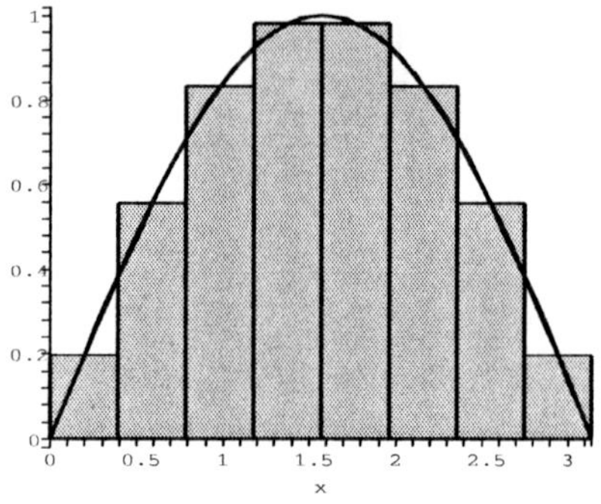

5. a) There are 16 rectangles and the evaluation points are given by $c_i = i\Delta x$ where i is from 0 to 15.

$$\begin{aligned} A_{16} &= \Delta x \sum_{i=0}^{15} f(c_i) \\ &= \frac{1}{16} \sum_{i=0}^{15} \left[\left(\frac{i}{16} \right)^2 + 1 \right] \approx 1.3027 \end{aligned}$$

b) There are 16 rectangles and the evaluation points are given by $c_i = i\Delta x + \frac{\Delta x}{2}$ where i is from 0 to 15.

$$\begin{aligned} A_{16} &= \Delta x \sum_{i=0}^{15} f(c_i) \\ &= \frac{1}{16} \sum_{i=0}^{15} \left[\left(\frac{i}{16} + \frac{1}{32} \right)^2 + 1 \right] \\ &\approx 1.3330 \end{aligned}$$

c) There are 16 rectangles and the evaluation points are given by $c_i = i\Delta x + \Delta x$ where i is from 0 to 15.

$$\begin{aligned} A_{16} &= \Delta x \sum_{i=0}^{15} f(c_i) \\ &= \frac{1}{16} \sum_{i=0}^{15} \left[\left(\frac{i}{16} + \frac{1}{16} \right)^2 + 1 \right] \\ &\approx 1.3652 \end{aligned}$$

7. a) There are 16 rectangles and the evaluation points are the left endpoints which are given by
$c_i = 1 + i\Delta x$ where i is from 0 to 15.

$$A_{16} = \Delta x \sum_{i=0}^{15} f(c_i)$$

$$= \frac{3}{16}\sum_{i=0}^{15}\sqrt{1+\frac{3i}{16}+2} \approx 6.2663$$

b) There are 16 rectangles and the evaluation points are the midpoints which are given by
$c_i = 1 + i\Delta x + \Delta x/2$ where i is from 0 to 15.

$$A_{16} = \Delta x \sum_{i=0}^{15} f(c_i)$$
$$= \frac{3}{16}\sum_{i=0}^{15}\sqrt{1+\frac{3i}{16}+\frac{3}{32}+2}$$
$$\approx 6.3340$$

c) There are 16 rectangles and the evaluation points are the right endpoints which are given by
$c_i = 1 + i\Delta x$ where i is from 1 to 16.

$$A_{16} = \Delta x \sum_{i=1}^{16} f(c_i)$$
$$= \frac{3}{16}\sum_{i=1}^{16}\sqrt{1+\frac{3i}{16}+2} \approx 6.4009$$

9. a) There are 50 rectangles and the evaluation points are given by $c_i = i\Delta x$ where i is from 0 to 49.

$$A_{50} = \Delta x \sum_{i=0}^{50} f(c_i)$$
$$= \frac{\pi}{100}\sum_{i=0}^{50}\cos\left(\frac{\pi i}{100}\right) \approx 1.0156.$$

b) There are 50 rectangles and the evaluation points are given by $c_i = \frac{\Delta x}{2} + i\Delta x$ where i is from 0 to 49.

$$A_{50} = \Delta x \sum_{i=0}^{50} f(c_i)$$
$$= \frac{\pi}{100}\sum_{i=0}^{50}\cos\left(\frac{\pi}{200}+\frac{\pi i}{100}\right)$$
$$\approx 1.00004.$$

c) There are 50 rectangles and the evaluation points are given by $c_i = \Delta x + i\Delta x$ where i is from 0 to 49.

$$A_{50} = \Delta x \sum_{i=0}^{50} f(c_i)$$
$$= \frac{\pi}{100}\sum_{i=0}^{50}\cos\left(\frac{\pi}{100}+\frac{\pi i}{100}\right)$$
$$\approx 0.9842.$$

11.

n	Left Endpoint	Midpoint	Right Endpoint
10	10.56	10.56	10.56
50	10.662	10.669	10.662
100	10.6656	10.6672	10.6656
500	10.6666	10.6667	10.6666
1000	10.6667	10.6667	10.6667
5000	10.6667	10.6667	10.6667

13.

n	Left Endpoint	Midpoint	Right Endpoint
10	15.48000	17.96000	20.68000
50	17.4832	17.9984	18.5232
100	17.7408	17.9996	18.2608
500	17.9480	17.9999	18.0520
1000	17.9740	17.9999	18.0260
5000	17.9948	17.9999	18.0052

15. $\Delta x = \frac{1}{n}$. We will use right endpoints as evaluation points, $x_i = \frac{i}{n}$.

$$A_n = \sum_{i=1}^{n} f(x_i)\Delta x$$
$$= \frac{1}{n}\sum_{i=1}^{n}\left[\left(\frac{i}{n}\right)^2+1\right]$$
$$= \frac{1}{n^3}\sum_{i=1}^{n} i^2 + 1$$
$$= \frac{1}{n^3}\left(\frac{n(n+1)(2n+1)}{6}\right)+1$$
$$= \frac{8n^2+3n+1}{6n^2}$$

Now to compute the exact area, we take the limit as $n \to \infty$:

$$\begin{aligned}A &= \lim_{n\to\infty} A_n \\ &= \lim_{n\to\infty} \frac{8n^2+3n+1}{6n^2} \\ &= \lim_{n\to\infty} \frac{8}{6} + \frac{3}{6n} + \frac{1}{6n^2} \\ &= \frac{4}{3}.\end{aligned}$$

17. $\Delta x = \frac{2}{n}$. We will use right endpoints as evaluation points, $x_i = 1 + \frac{2i}{n}$.

$$\begin{aligned}A_n &= \sum_{i=1}^{n} f(x_i)\Delta x \\ &= \frac{2}{n}\sum_{i=1}^{n} 2\left(1+\frac{2i}{n}\right)^2 + 1 \\ &= \frac{2}{n}\sum_{i=1}^{n}\left(\frac{8i^2}{n^2}+\frac{8i}{n}+3\right) \\ &= \frac{16}{n^3}\sum_{i=1}^{n} i^2 + \frac{16}{n^2}\sum_{i=1}^{n} i + 6 \\ &= \frac{16}{n^3}\left(\frac{n(n+1)(2n+1)}{6}\right) \\ &\quad + \frac{16}{n^2}\left(\frac{n(n+1)}{2}\right) + 6 \\ &= \frac{16n(n+1)(2n+1)}{6n^3} + \frac{16n(n+1)}{2n^2} \\ &\quad + 6.\end{aligned}$$

Now to compute the exact area, we take the limit as $n \to \infty$:

$$\begin{aligned}A &= \lim_{n\to\infty} A_n \\ &= \lim_{n\to\infty}\left(\frac{16n(n+1)(2n+1)}{6n^3} + \frac{16n(n+1)}{2n^2} + 6\right) \\ &= \lim_{n\to\infty} \frac{32}{6} + \frac{16}{2} + 6 \\ &= 19\frac{1}{3}.\end{aligned}$$

19. Using left hand endpoints:

$L_8 = [f(0.0) + f(0.1) + f(0.2) + f(0.3) + f(0.4) + f(0.5) + f(0.6) + f(0.7)](0.1)$
$= (2.0 + 2.4 + 2.6 + 2.7 + 2.6 + 2.4 + 2.0 + 1.4)(0.1) = 1.81.$

Right endpoints:

$R_8 = [f(0.1) + f(0.2) + f(0.3) + f(0.4) + f(0.5) + f(0.6) + f(0.7) + f(0.8)](0.2)$
$= (2.4 + 2.6 + 2.7 + 2.6 + 2.4 + 2.0 + 1.4 + 0.6)(0.1) = 1.67.$

21. Using left hand endpoints:

$L_8 = [f(1.0) + f(1.1) + f(1.2) + f(1.3) + f(1.4) + f(1.5) + f(1.6) + f(1.7)](0.1)$
$= (1.8 + 1.4 + 1.1 + 0.7 + 1.2 + 1.4 + 1.82 + 2.4)(0.1) = 1.182.$

Right endpoints:

$R_8 = [f(1.1) + f(1.2) + f(1.3) + f(1.4) + f(1.5) + f(1.6) + f(1.7) + f(1.8)](0.1)$
$= (1.4 + 1.1 + 0.7 + 1.2 + 1.4 + 1.82 + 2.4 + 2.6)(0.1) = 1.262.$

23. Let L, M, and R be the values of the Riemann sums with left endpoints, midpoints and right endpoints. Let A be the area under the curve. Then: $L < M < A < R$.

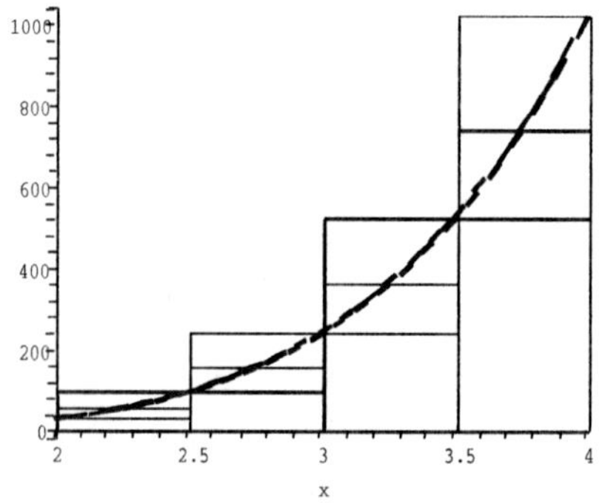

25. Let L, M, and R be the values of the Riemann sums with left endpoints, midpoints and right endpoints. Let A be the area under the curve. Then: $R < A < M < L$.

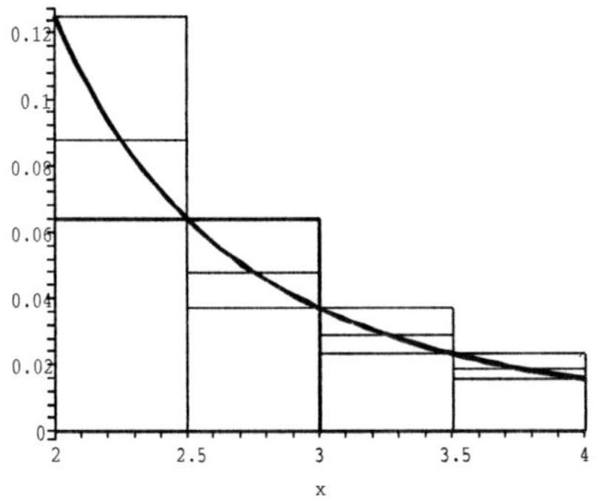

27. There are many possible answers here. One possibility is to use $x = 1/6$ on $[0, 0.5]$ and $x = 1/2$ on $[0.5, 1]$.

29. We subdivide the interval $[a, b]$ into n equal subintervals. If you are located at $a + (b - a)/n$ (the first right endpoint), then each step of distance Δx takes you to a new right endpoint. To arrive at the i-th right endpoint, you have to take $(i - 1)$ steps to the right of distance Δx. Therefore, $c_i = a+(b-a)/n+(i-1)\Delta x = a+i\Delta x$.

31. We subdivide the interval $[a, b]$ into n equal subintervals. The first evaluation point is $a+\Delta x/2$. From this evaluation point, each step of distance Δx takes you to a new evaluation point. To arrive at the i-th evaluation point, you have to take $(i - 1)$ steps to the right of distance Δx. Therefore,
$c_i = a + \Delta x/2 + (i - 1)\Delta x$
$= a + (i - 1/2)\,\Delta x$, for $i = 1, \ldots, n$.

33. $A \approx (0.2 - 0.1)(0.002) + (0.3 - 0.2)(0.004)+(0.4-0.3)(0.008)+(0.5-0.4)(0.014)+(0.6-0.5)(0.026)+(0.7-0.6)(0.048)+(0.8-0.7)(0.085)+(0.9-0.8)(0.144) + (0.95 - 0.9)(0.265) + (0.98 - 0.95)(0.398) + (0.99 - 0.98)(0.568)+(1-0.99)(0.736)+1/2\cdot [(0.1 - 0)(0.002) + (0.2 - 0.1)(0.004 - 0.002) + (0.3 - 0.2)(0.008 - 0.004) + (0.4 - 0.3)(0.014-0.008)+(0.5-0.4)(0.026-0.014) + (0.6 - 0.5)(0.048 - 0.026) + (0.7 - 0.6)(0.085 - 0.048) + (0.8 - 0.7)(0.144-0.085)+(0.9-0.8)(0.265-0.144) + (0.95 - 0.9)(0.398 - 0.265) + (0.98 - 0.95)(0.568 - 0.398) + (0.99 - 0.98)(0.736 - 0.568)\ (1 - 0.99)(1 - 0.736)]$
≈ 0.092615

The Lorentz curve looks like:

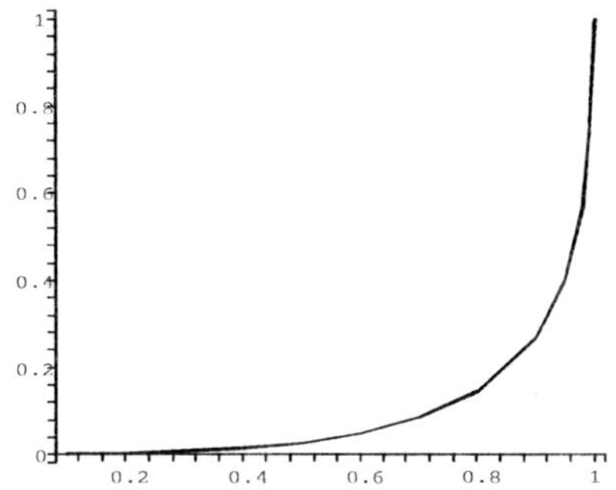

35. $U_4 = \dfrac{2}{4}\displaystyle\sum_{i=1}^{4}\left(\frac{i}{2}\right)^2$

$= \dfrac{1}{8}\displaystyle\sum_{i=1}^{4} i^2 = \dfrac{1}{8}\left[1^2 + 2^2 + 3^2 + 4^2\right]$

$= \dfrac{30}{8} = 3.75$

$L_4 = \dfrac{2}{4}\displaystyle\sum_{i=1}^{4}\left(\frac{i-1}{2}\right)^2$

$= \dfrac{1}{8}\displaystyle\sum_{i=1}^{4} i^2 = \dfrac{1}{8}\left[0^2 + 1^2 + 2^2 + 3^2\right]$

$= \dfrac{14}{8} = 1.75$

37. a) $U_n = \dfrac{2}{n}\displaystyle\sum_{i=1}^{n}\left(\frac{2i}{n}\right)^2$

$= \left(\dfrac{2}{n}\right)^3 \displaystyle\sum_{i=1}^{n} i^2$

$= \left(\dfrac{2}{n}\right)^3 \dfrac{n(n+1)(2n+1)}{6}$

$= \dfrac{4}{3}\dfrac{n(n+1)(2n+1)}{n^3}$

$$= \frac{4}{3}\left(1+\frac{1}{n}\right)\left(2+\frac{1}{n}\right)$$

$$\lim_{n\to\infty} U_n = \frac{4}{3}(2) = \frac{8}{3}$$

b) $$L_n = \frac{2}{n}\sum_{i=1}^{n}\left(\frac{2(i-1)}{n}\right)^2$$

$$= \left(\frac{2}{n}\right)^3 \sum_{i=1}^{n}(i-1)^2$$

$$= \left(\frac{2}{n}\right)^3 \sum_{i=1}^{n-1} i^2$$

$$= \left(\frac{2}{n}\right)^3 \frac{(n-1)(n)(2n-1)}{6}$$

$$= \frac{4}{3}\frac{(n-1)(n)(2n-1)}{n^3}$$

$$= \frac{4}{3}\left(1-\frac{1}{n}\right)\left(2-\frac{1}{n}\right)$$

$$\lim_{n\to\infty} L_n = \frac{4}{3}(2) = \frac{8}{3}$$

39. a) $$U_n = \frac{2}{n}\sum_{i=1}^{n}\left[\left(0+\frac{2}{n}i\right)^3+1\right]$$

$$= \frac{2}{n}\sum_{i=1}^{n}\left[\left(\frac{2i}{n}\right)^3+1\right]$$

$$= \left(\frac{2}{n}\right)^4 \sum_{i=1}^{n} i^3 + \sum_{i=1}^{n} 1$$

$$= \frac{2^4}{n^4}\left[\frac{n^2(n+1)^2}{4} + \frac{2}{n}(n)\right]$$

$$= \frac{4(n+1)^2}{n^2} + 2$$

$$= \frac{4(n^2+2n+1)}{n^2} + 2$$

$$= 4\left(1+\frac{2}{n}+\frac{1}{n^2}\right)+2$$

$$= 6+\frac{8}{n}+\frac{4}{n^2}$$

$$\lim_{n\to\infty} U_n = 6$$

b) $$L_n = \frac{2}{n}\sum_{i=0}^{n-1}\left[\left(0+\frac{2}{n}i\right)^3+1\right]$$

$$= \frac{2}{n}\sum_{i=0}^{n-1}\left[\left(\frac{2i}{n}\right)^3+1\right]$$

$$= \left(\frac{2}{n}\right)^4 \sum_{i=0}^{n-1} i^3 + \sum_{i=1}^{n} 1$$

$$= \frac{2^4}{n^4}\left[\frac{(n-1)^2n^2}{4} + \frac{2}{n}(n)\right]$$

$$= \frac{4(n-1)^2}{n^2} + 2$$

$$= \frac{4(n^2-2n+1)}{n^2} + 2$$

$$= 4\left(1-\frac{2}{n}+\frac{1}{n^2}\right)+2$$

$$= 6-\frac{8}{n}+\frac{4}{n^2}$$

$$\lim_{n\to\infty} L_n = 6$$

4.4 The Definite Integral

1. $$\int_0^3 (x^3+x)\,dx$$

$$= \sum_{i=1}^{n}(c_i^3+c_i)\Delta x = \sum_{i=1}^{n}(c_i^3+c_i)\cdot\frac{3}{n},$$

$$c_i = \frac{x_i+x_{i-1}}{2}, x_i = \frac{3i}{n}$$

$n \geq 20 \implies$ Riemann sum ≈ 24.65

3. $$\int_0^\pi \sin x^2\,dx$$

$$= \sum_{i=1}^{n}\sin c_i^2\Delta x = \sum_{i=1}^{n}\sin c_i^2\left(\frac{\pi}{n}\right),$$

$$c_i = \frac{x_i+x_{i-1}}{2}, x_i = \frac{i\pi}{n}$$

$n \geq 4 \implies$ Riemann sum ≈ 0.80

5. For n rectangles, $\Delta x = \frac{1}{n}$, $x_i = i\Delta x$.

$$R_n = \sum_{i=1}^{n} f(x_i)\,\Delta x$$
$$= \sum_{i=1}^{n} 2x_i\,\Delta x = \frac{1}{n}\sum_{i=1}^{n} 2\left(\frac{i}{n}\right)$$
$$= \frac{2}{n^2}\sum_{i=1}^{n} i = \frac{2}{n^2}\left(\frac{n(n+1)}{2}\right)$$
$$= \frac{(n+1)}{n}$$

To compute the value of the integral, we take the limit as $n \to \infty$,

$$\int_1^2 2x\,dx = \lim_{n\to\infty} R_n$$
$$= \lim_{n\to\infty} \frac{(n+1)}{n} = 1$$

7. For n rectangles, $\Delta x = 2/n$, $x_i = i\Delta x = 2i/n$.

$$R_n = \sum_{i=1}^{n} f(x_i)\,\Delta x$$
$$= \sum_{i=1}^{n} (x_i^2)\,\Delta x = \frac{2}{n}\sum_{i=1}^{n} 2\left(\frac{2i}{n}\right)^2$$
$$= \frac{2}{n}\sum_{i=1}^{n} \frac{4i^2}{n^2} = \frac{8}{n^3}\sum_{i=1}^{n} i^2$$
$$= \frac{8}{n^3}\left(\frac{n(n+1)(2n+1)}{6}\right)$$
$$= \frac{4(n+1)(2n+1)}{3n^2}$$

To compute the value of the integral, we take the limit as $n \to \infty$,

$$\int_0^3 (x^2+1)\,dx = \lim_{n\to\infty} R_n$$
$$= \lim_{n\to\infty} \frac{4(n+1)(2n+1)}{3n^2} = \frac{8}{3}$$

9. For n rectangles, $\Delta x = 2/n$, $x_i = 1 + i\Delta x = 1 + 2i/n$

$$R_n = \sum_{i=1}^{n} f(x_i)\,\Delta x$$
$$= \sum_{i=1}^{n} (x_i^2 - 3)\,\Delta x$$
$$= \frac{2}{n}\sum_{i=1}^{n} \left[\left(1+\frac{2i}{n}\right)^2 - 3\right]$$
$$= \sum_{i=1}^{n} \left(\frac{8i}{n^2} + \frac{8i^2}{n^3} - \frac{4}{n}\right)$$
$$= \frac{8n(n+1)}{2n^2} + \frac{8n(n+1)(2n+1)}{6n^3} - 4$$

To compute the value of the integral, we take the limit as $n \to \infty$,

$$\int_1^3 (x^2-3)\,dx = \lim_{n\to\infty} R_n$$
$$= \frac{8}{2} + \frac{16}{6} - 4 = \frac{8}{3}$$

11. Notice that the graph of $y = 4 - x^2$ is above the x-axis between $x = -2$ and $x = 2$:

$$\int_{-2}^{2} (4 - x^2)\,dx$$

13. Notice that the graph of $y = x^2 - 4$ is below the x-axis between $x = -2$ and $x = 2$. Since we are asked for area and the area in question is below the x-axis, we have to be a bit careful.

$$\int_{-2}^{2} -(x^2 - 4)\,dx$$

15. $\displaystyle\int_0^{\pi} \sin x\,dx$

17.
$$\left|\int_0^1 (x^3 - 3x^2 + 2x)\,dx\right| + \left|\int_1^2 (x^3 - 3x^2 + 2x)\,dx\right|$$
$$= \int_0^1 (x^3 - 3x^2 + 2x)\,dx - \int_1^2 (x^3 - 3x^2 + 2x)\,dx$$

19. The total distance is the total area under the curve whereas the total displacement is the signed area under the curve. In this case, from $t = 0$ to $t = 4$, the function is always positive so the total distance is equal to the total displacement. This means we want to compute the definite integral $\int_0^4 40(1-e^{-2t})\,dt$. We compute various right hand sums for different values of n:

n	R_n
10	146.9489200
20	143.7394984
50	141.5635684
100	140.7957790
500	140.1662293
1000	140.0865751

It looks like these are converging to about 140. So, the total distance traveled is approximately 140 and the final position is

$$s(b) \approx s(0) + 140 = 0 + 140 = 140.$$

21. $\displaystyle\int_0^2 f(x)\,dx + \int_2^3 f(x)\,dx = \int_0^3 f(x)\,dx$

23. $\displaystyle\int_0^2 f(x)\,dx + \int_2^1 f(x)\,dx = \int_0^1 f(x)\,dx$

25.

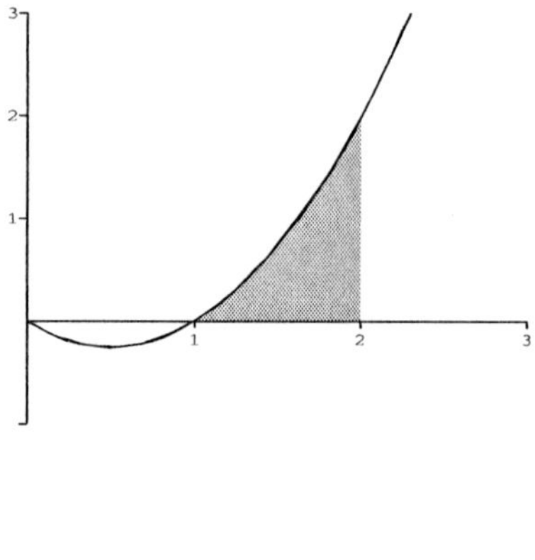

27.

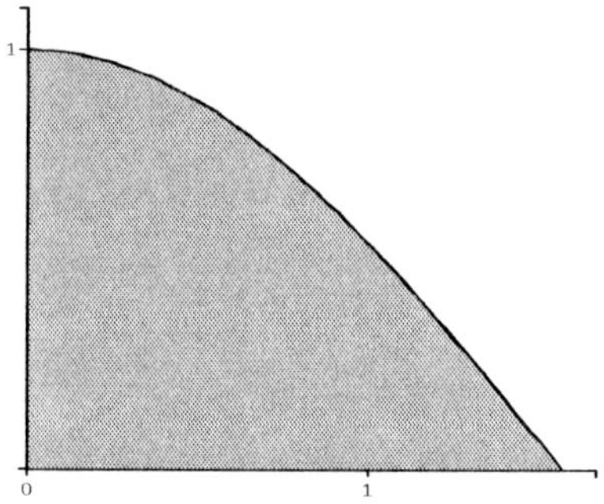

29. The function $f(x) = 3\cos x^2$ is decreasing on $[\pi/3, \pi/2]$. Therefore, on this interval, the maximum occurs at the left endpoint and is $f(\pi/3) = 3\cos(\pi^2/9)$. The minimum occurs at the right endpoint and is $f(\pi/2) = 3\cos(\pi^2/4)$.

Using these to estimate the value of the integral gives the following inequality:

$$\frac{\pi}{6}\cdot\left(3\cos\frac{\pi^2}{4}\right) \le \int_{\pi/3}^{\pi/2} 3\cos x^2\,dx \le \frac{\pi}{6}\cdot\left(3\cos\frac{\pi^2}{9}\right)$$

$$-1.23 \le \int_{\pi/3}^{\pi/2} 3\cos x^2\,dx \le 0.72$$

31. The function $f(x) = \sqrt{x^2+1}$ is increasing on $[0,2]$. Therefore, on this interval, the maximum occurs at the right endpoint and is $f(2) = \sqrt{5}$. The minimum occurs at the left endpoint and is $f(0) = 1$.

Using these to estimate the value of the integral gives the following inequality:

$$(2)(1) \le \int_0^2 \sqrt{x^2+1}\,dx \le (2)(\sqrt{5})$$

$$2 \le \int_0^2 \sqrt{x^2+1}\,dx \le 4.472$$

33. We are looking for a value c, such that

$$f(c) = \frac{1}{2-0}\int_0^2 3x^2\,dx$$

Since $\int_0^2 3x^2\,dx = 8$, we want to find c so that $f(c) = 4$ or, $3c^2 = 4$

Solving this equation using the quadratic formula gives $c = \pm\frac{2}{\sqrt{3}}$

We are interested in the value that is in the interval $[0, 2]$, so $c = \frac{2}{\sqrt{3}}$.

35. $$f_{ave} = \frac{1}{4}\int_0^4 (2x+1)\,dx$$
$$= \frac{1}{4}\lim_{n\to\infty}\sum_{i=1}^{n}\frac{4}{n}\left(\frac{8i}{n}+1\right)$$
$$= \lim_{n\to\infty}\sum_{i=1}^{n}\left(\frac{8n(n+1)}{2n^2}+1\right)$$
$$= 4 + 1 = 5$$

37. $$f_{ave} = \frac{1}{1-0}\int_0^1 (x^2-1)\,dx$$
$$= \lim_{n\to\infty}\sum_{i=1}^{n}\frac{1}{n}\left[\left(1+\frac{2i}{n}\right)^2 - 1\right]$$
$$= \lim_{n\to\infty}\sum_{i=1}^{n}\frac{1}{n}\left(\frac{4i}{n}+\frac{4i^2}{n^2}\right)$$
$$= \lim_{n\to\infty}\left(\frac{4n(n+1)}{2n^2}+\frac{4n(n+1)(2n+1)}{6n^3}\right)$$
$$= 2 + \frac{4}{3} = \frac{10}{3}$$

39. This is just a restatement of the Integral Mean Value Theorem.

41. Between $x = 0$ and $x = 2$, the area below the x-axis is much less than the area above the x-axis. Therefore

$$\int_0^2 f(x)\,dx > 0$$

43. Between $x = 0$ and $x = 2$, the area below the x-axis is slightly greater than the area above the x-axis. Therefore

$$\int_0^2 f(x)\,dx < 0$$

45. Imagine the interval $[0, 2]$ is divided into n subintervals. If n is even, then the point $x = 1$ must be one of the boundary points. If we take the midpoint evaluations to approximate Riemann sums for $\int_0^2 f(x)\,dx$ and $\int_0^2 g(x)\,dx$, all the values $f(c_i)$ and $g(c_i)$ are going to be exactly the same for same index number i, since the only difference between $f(x)$ and $g(x)$ occurs at $x = 1$, and $x = 1$ is never going to be one of the c_i's. Thus the approximated Riemann sums for $\int_0^2 f(x)\,dx$ and $\int_0^2 g(x)\,dx$ are going to be the same.

47. $$\int_0^4 f(x)\,dx$$
$$= \int_0^1 f(x)\,dx + \int_1^4 f(x)\,dx$$
$$= \int_0^1 2x\,dx + \int_1^4 4\,dx$$

$\int_0^1 2x\,dx$ is the area of a triangle with base 1 and height 2 and therefore has area $\frac{1}{2}(1)(2) = 1$.

$\int_1^4 4\,dx$ is the area of a rectangle with base 3 and height 4 and therefore has area $(3)(4) = 12$.

Therefore

$$\int_0^4 f(x)\,dx = 1 + 12 = 13$$

49. Since $b(t)$ represents the birthrate (in births per month), the total number of births from time $t=0$ to $t=12$ is given by the integral $\int_0^{12} b(t)\,dt$.

Similarly, the total number of deaths from time $t=0$ to $t=12$ is given by the integral $\int_0^{12} a(t)\,dt$.

Of course, the net change in population is the number of birth minus the number of deaths:

Population Change
= Births − Deaths

$$= \int_0^{12} b(t)\,dt - \int_0^{12} a(t)\,dt$$

$$= \int_0^{12} [b(t) - a(t)]\,dt.$$

Next we solve the inequality
$410 - 0.3t > 390 + 0.2t$
$20 > 0.5t$ then $t < 40$ months

Therefore $b(t) > a(t)$ when $t < 40$ months. The population is increasing when the birth rate is greater than the death rate, which is during the first 40 month. After 40 months, the population is decreasing. The population would reach a maximum at $t = 40$ months.

51. From $PV = 10$ we get $P(V) = 10/V$. By definition,

$$\int_2^4 P(V)\,dV = \int_2^4 \frac{10}{V}\,dV$$

$$= \lim_{n\to\infty} \sum_{i=1}^{n} \frac{2}{n} \cdot \frac{10}{2 + \frac{2i}{n}}$$

An estimate of the value of this integral is setting $n = 100$, and then the integral ≈ 6.93

53. $$\frac{1}{2-0}\int_0^2 f(x)dx = 5$$

$$\int_0^2 f(x)dx = 10$$

and

$$\frac{1}{6-2}\int_2^6 f(x)dx = 11$$

$$\int_2^6 f(x)dx = 44$$

The average value of f over $[0,6]$ is

$$\frac{1}{6-0}\int_0^6 f(x)dx$$

$$= \frac{1}{6}\left(\int_0^2 f(x)dx + \int_2^6 f(x)dx\right)$$

$$= \frac{1}{6}(10+44) = 9$$

55. $$\int_0^2 3x dx = \frac{1}{2}bh = \frac{1}{2}(2)(6) = 6$$

57. $$\int_0^2 \sqrt{4-x^2} = \frac{1}{4}\pi r^2 = \frac{1}{4}\pi\left(2^2\right) = \pi$$

59. (a) Average temperature

$$= \frac{1}{24}[3(44)+3(52)+3(70)+3(82) +3(86)+3(80)+3(72)+3(56)]$$

$$= \frac{3}{24}[44+52+70+82+86+80 +72+56]$$

$$= \frac{542}{8} = 67.75$$

(b) average temperature

$$= \frac{1}{24}[3(46)+3(44)+3(52)+3(70) +3(82)+3(86)+3(80)+3(72)]$$

$$= \frac{3}{24}[46+44+52+70+82+86 +80+72]$$

$$= \frac{1}{8}[532] = 66.5$$

61. Since r is the rate at which items are shipped, rt is the number of items shipped between time 0 and time t. Therefore, $Q - rt$ is the number of items remaining in inventory at time t. Since $Q - rt = 0$ when $t = Q/r$, the formula is valid for $0 \le t \le Q/r$. The average value of $f(t) = Q - rt$ on the time interval $[0, Q/r]$ is $\dfrac{1}{Q/r - 0}\displaystyle\int_0^{Q/r} f(t)dt$

$$= \frac{r}{Q}\int_0^{Q/r}(Q - rt)dt$$
$$= \frac{r}{Q}\left[Qt - \frac{1}{2}rt^2\right]_0^{Q/r}$$
$$= \frac{r}{Q}\left[\frac{Q^2}{r} - \frac{r}{2}\frac{Q^2}{r^2}\right]$$
$$= \frac{r}{Q}\left[\frac{Q^2}{2r}\right] = \frac{Q}{2}$$

63. Delivery is completed in time Q/p, and since in that time Qr/p items are shipped, the inventory when delivery is completed is

$$Q - \frac{Qr}{p} = Q\left(1 - \frac{r}{p}\right)$$

The inventory at any time is given by

$$g(t) = \begin{cases} (p-r)t & \text{for } t \in \left[0, \frac{Q}{p}\right] \\ Q - rt & \text{for } t \in \left[\frac{Q}{p}, \frac{Q}{r}\right] \end{cases}$$

The graph of g has two linear pieces. The average value of g over the interval $[0, Q/r]$ is the area under the graph (which is the area of a triangle of base Q/r and height $Q(1 - r/p)$) divided by the length of the interval (which is the base of the triangle). Thus the average value of the function is $(1/2)bh$ divided by b, which is

$$(1/2)h = (1/2)Q(1 - r/p)$$

This time the total cost

$$f(Q) = c_0\frac{D}{Q} + c_c\frac{Q}{2}(1 - \frac{r}{p})$$

$$f'(Q) = -\frac{c_0 D}{Q^2} + \frac{c_c(1 - \frac{r}{p})}{2}$$

$f'(Q) = 0$ gives $\dfrac{c_0 D}{Q^2} = \dfrac{c_c}{2}(1 - \dfrac{r}{p})$

$$Q = \sqrt{\frac{2c_0 D}{c_c(1 - r/p)}}$$

The order size to minimize the total cost is

$$Q = \sqrt{\frac{2c_0 D}{c_c(1 - r/p)}}$$

65. The maximum of

$$F(t) = 9 - 10^8(t - 0.0003)^2$$

occurs when $10^8(t - 0.0003)^2$ reaches its minimum, that is, when $t = 0.0003$. At that time

$F(0.0003) = 9$ thousand pounds.

We estimate the value of

$$\int_0^{0.0006}\left[9 - 10^8(t - 0.0003)^2\right]dt$$

using midpoint sum and $n = 20$, and get $m\Delta v \approx 0.00360$ thousand pound-seconds, so $\Delta v \approx 360$ ft per second.

67. Since $f(x) = x^3$ is an odd function, the area under the curve and above the x-axis from 0 to 1 is the same as the area under the x-axis and above the curve for -1 to 0.

You can see that $\int_{-1}^{1} x^3 e^{-x} dx < 0$ by thinking of e^{-x} as a weighting factor. Since the weighting factor is always positive, and is greater for $x < 0$ than it is for $x > 0$, the "negative" area to the left of $x = 0$ counts more than the "positive" area to the right of $x = 0$.

4.5 The Fundamental Theorem Of Calculus

1. $\displaystyle\int_0^2 (2x - 3)\,dx$

$$= (x^2 - 3x)\Big|_0^2 = -2$$

3. $\displaystyle\int_{-1}^{1} (x^3 + 2x)\,dx$
$\displaystyle = \left(\frac{x^4}{4} + x^2\right)\Big|_{-1}^{1} = 0$

5. $\displaystyle\int_0^4 (\sqrt{x} + 3x)\,dx$
$\displaystyle = \left(\frac{2}{3}x^{\frac{3}{2}} + \frac{3x^2}{2}\right)\Big|_0^4 = \frac{88}{3}$

7. $\displaystyle\int_0^1 (x\sqrt{x} + x^{-\frac{1}{2}})\,dx$
$\displaystyle = \left(\frac{2}{5}x^{\frac{5}{2}} + 2x^{\frac{1}{2}}\right)\Big|_0^1 = \frac{12}{5}$

9. $\displaystyle\int_0^{\frac{\pi}{4}} (\sec x \tan x)\,dx$
$\displaystyle = \sec x\,\Big|_0^{\frac{\pi}{4}} = \sqrt{2} - 1$

11. $\displaystyle\int_{\pi/2}^{\pi} (2\sin x - \cos x)\,dx$
$\displaystyle = (-2\cos x - \sin x)\,\Big|_{\pi/2}^{\pi} = 3$

13. $\displaystyle\int_0^{1/2} \frac{3}{\sqrt{1-x^2}}\,dx$
$\displaystyle = 3\sin^{-1} x\,\Big|_0^{1/2}$
$\displaystyle = 3\left(\frac{\pi}{6} - 0\right) = \frac{\pi}{2}$

15. $\displaystyle\int_1^4 \frac{x-3}{x}\,dx$
$\displaystyle = \int_1^4 \left(1 - 3x^{-1}\right)\,dx$
$\displaystyle = (x - 3\ln|x|)\,\Big|_1^4 = 3 - 3\ln 4$

17. $\displaystyle\int_0^4 x(x-2)\,dx$
$\displaystyle = \left(\frac{x^3}{3} - x^2\right)\Big|_0^4 = \frac{16}{3}$

19. $\displaystyle\int_0^{\ln 2} (e^{x/2})^2\,dx$
$\displaystyle = (e^x)\,\Big|_0^{\ln 2} = 2 - 1 = 1$

21. $\displaystyle\int_0^2 \sqrt{x^2+1}\,dx$
$\displaystyle = \lim_{n\to\infty} \sum_{i=1}^{n} \frac{2}{n}\sqrt{\left(\frac{2i}{n} + 1\right)}$

Estimating using $n = 20$, we get the Riemann sum ≈ 2.96

23. $\displaystyle\int_1^4 \frac{x^2}{x^2+4}\,dx$
$\displaystyle = \lim_{n\to\infty} \sum_{i=1}^{n} \frac{3}{n}\frac{(1+(3i/n)^2)}{(1+3i/n)^2+4}$

Estimating using $n = 20$, we get the Riemann sum ≈ 1.71

25. $\displaystyle\int_0^{\pi/4} \frac{\sin x}{\cos^2 x}\,dx$
$\displaystyle = \int_0^{\pi/4} \tan x \sec x\,dx$
$\displaystyle = \sec x\,\Big|_0^{\pi/4} = \sqrt{2} - 1$

27. $f'(x) = x^2 - 3x + 2$

29. $\displaystyle f'(x) = \left(e^{-(x^2)^2} + 1\right)\frac{d}{dx}(x^2)$
$\displaystyle = \left(e^{-x^4} + 1\right)(2x)$

31. $f'(x) = -\ln(x^2+1)$

33. $y'(x) = \sin\sqrt{x^2+\pi^2}$
At the point in question, $y(0) = 0$ and $y'(0) = \sin\pi = 0$

Therefore, the tangent line has slope 0 and passes through the point $(0,0)$. The equation of this line is $y = 0$.

35. $y'(x) = \cos(\pi x^3)$
At the point in question, $y(2) = 0$ and $y'(2) = \cos 8\pi = 1$

Therefore, the tangent line has slope 1 and passes through the point $(2,0)$. The equation of this line is $y = x - 2$.

37. $f'(x) = x^2 - 3x + 2$

Setting $f'(x) = 0$ we get

$(x-1)(x-2) = 0, x = 1, 2.$

$$f'(x)\begin{cases} > 0 & \text{when } t < 1 \text{ or } t > 2 \\ < 0 & \text{when } 1 < t < 2 \end{cases}$$

$$f(1) = \int_0^1 (t^2 - 3t + 2)\,dt$$

$$= (t^3/3 - 3t^2/2 + 2t)\Big|_0^1 = \frac{5}{6}$$

$$f(2) = \int_0^2 (t^2 - 3t + 2)\,dt$$

$$= (t^3/3 - 3t^2/2 + 2t)\Big|_0^2 = \frac{2}{3}$$

Hence $f(x)$ has a local maximum at the point $(1, 5/6)$ and a local minimum at the point $(2, 2/3)$.

39. The graph of $y = 4 - x^2$ is above the x-axis over the interval $[-2, 2]$.

$$\int_{-2}^{2} (4 - x^2)\,dx$$

$$= (4x - \frac{x^3}{3})\Big|_{-2}^{2}$$

$$= \frac{32}{3}$$

41. The graph of $y = x^2$ is above the x-axis over the interval $[0, 2]$.

$$\int_0^2 x^2\,dx = \frac{x^3}{3}\Big|_0^2 = \frac{8}{3}$$

43. The graph of $y = \sin x$ is above the x-axis over the interval $[0, \pi]$.

$$\int_0^\pi \sin x\,dx = -\cos x\Big|_0^\pi = 2$$

45. If you look at the graph of $1/x^2$, it is obvious that there is positive area between the curve and the x-axis over the interval $[-1, 1]$. In addition to this, there is a vertical asymptote in the interval that we are integrating over which should alert us to a possible problem.

The problem is that $1/x^2$ is not continuous on $[-1, 1]$ (the discontinuity occurs at $x = 0$) and that continuity is one of the conditions in the Fundamental Theorem of Calculus, Part I (Theorem 4.1).

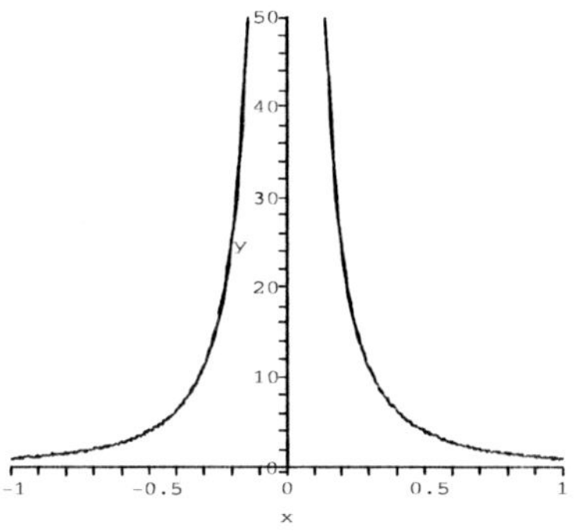

47. $s(t) = 40t + \cos t + c$,
$s(0) = 0 + \cos 0 + c = 2$ so therefore $c = 1$ and
$s(t) = 40t + \cos t + 1$.

49. $v(t) = 4t - \dfrac{t^2}{2} + c_1$,
$v(0) = c_1 = 8$ so therefore $c_1 = 8$ and
$v(t) = 4t - \dfrac{t^2}{2} + 8$.

$s(t) = 2t^2 - \dfrac{t^3}{6} + 8t + c_2$,
$s(0) = c_2 = 0$ so therefore $c_2 = 0$ and
$s(t) = 2t^2 - \dfrac{t^3}{6} + 8t$.

51. $\omega(t) = 10t + c_1$ and $\omega(0) = 0$ gives that $c_1 = 0$ and hence $\omega(t) = 10t$.
$\omega(0.8) = 8$ rad/s.
$v(0.8) = 3(8) = 24$ ft/s.
$\theta(t) = 5t^2 + c_2$ and $\theta(0) = 0$, so $c_2 = 0$ and $\theta(t) = 5t^2$
$\theta(0.8) = 5(0.8^2) = 3.2$ rad.

53. $f_{ave} = \dfrac{1}{3-1}\displaystyle\int_1^3 (x^2 - 1)\,dx$
$= \dfrac{1}{2}\left(\dfrac{x^3}{3} - x\right)\Big|_1^3 = \dfrac{10}{3}.$

55. $f_{ave} = \frac{1}{1-0}\int_0^1 (2x - 2x^2)\,dx$
$= \left(x^2 - \frac{2x^3}{3}\right)\Big|_0^1 = \frac{1}{3}$

57. $f_{ave} = \frac{1}{\pi/2 - 0}\int_0^{\pi/2} \cos x\,dx$
$= \frac{2}{\pi}(\sin x)\Big|_0^{\pi/2} = \frac{2}{\pi}$

59. $\int_0^3 f(x)\,dx < \int_0^2 f(x)\,dx$
$< \int_0^1 f(x)\,dx$

61. Using the Fundamental Theorem of Calculus, it follows that an antiderivative of e^{-x^2} is $\int_a^x e^{-t^2}\,dt$ where a is a constant.

63. $CS = \int_0^Q D(q)\,dq - PQ$
$= \int_0^Q (150 - 2q - 3q^2)\,dq - PQ$
$= (150q - q^2 - q^3)\Big|_0^Q - PQ$
$= 150Q - Q^2 - Q^3$
$- (150 - 2Q - 3Q^2)Q$
$= Q^2 + 2Q^3$

When $Q = 4$,
$CS = 16 + 2(64) = 144$ dollars.

When $Q = 6$,
$CS = 36 + 2(216) = 468$ dollars.
The consumer surplus is higher for $Q = 6$ than that for $Q = 4$.

65. The next shipment must arrive when the inventory is zero. This occurs at time T: $f(t) = Q - r\sqrt{t}$
$f(T) = 0 = Q - r\sqrt{T}$
$r\sqrt{T} = Q$
$T = \frac{Q^2}{r^2}$
The average value of f on $[0, T]$ is:
$\frac{1}{T}\int_0^T f(t)\,dt$
$= \frac{1}{T}\int_0^T (Q - rt^{1/2})\,dt$
$= \frac{1}{T}\left[Qt - \frac{2}{3}rt^{3/2}\right]_0^T$
$= \frac{1}{T}\left[QT - \frac{2}{3}rT^{3/2}\right]$
$= Q - \frac{2}{3}r\sqrt{T}$
$= Q - \frac{2}{3}r\frac{Q}{r}$
$= \frac{Q}{3}$

67. When $a < 2$ or $a > 2$, f is continuous. Using the Fundamental Theorem of Calculus,
$\left[\lim_{x\to a} F(x)\right] - F(a)$
$= \lim_{x\to a}[F(x) - F(a)]$
$= \lim_{x\to a}\left[\int_0^x f(t)\,dt - \int_0^a f(t)\,dt\right]$
$= \lim_{x\to a}\left[\int_a^x f(t)\,dt\right]$
$= 0$

When $a = 2$,

$\lim_{x\to a^-}\left[\int_a^x f(t)\,dt\right]$
$= \lim_{x\to 2^-}\left[\int_2^x t\,dt\right]$
$= \lim_{x\to 2^-}\left[\frac{t^2}{2}\right]_0^x$
$= \lim_{x\to 2^-}\left[\frac{x^2}{2} - \frac{2^2}{2}\right]$
$= 0$ and

$\lim_{x\to a^+}\left[\int_a^x f(t)\,dt\right]$
$= \lim_{x\to 2^+}\left[\int_2^x (t+1)\,dt\right]$
$= \lim_{x\to 2^+}\left[\frac{t^2}{2} + t\right]_0^x$

$$= \lim_{x \to 2^+} \left[\frac{x^2}{2} + x - \frac{2^2}{2} - 2 \right]$$
$$= 0$$

Thus, for all values of a,

$$\left[\lim_{x \to a} F(x) \right] - F(a) = 0$$
$$\lim_{x \to a} F(x) = F(a)$$

Thus, F is continuous for all x. However, $F'(2)$ does not exist, which is shown as follows:

$$F'(2) = \lim_{h \to 0} \frac{F(2+h) - F(2)}{h}$$
$$= \lim_{h \to 0} \frac{1}{h} \left[\int_0^{2+h} f(t)\,dt - \int_0^2 f(t)\,dt \right]$$
$$= \lim_{h \to 0} \frac{1}{h} \int_2^{2+h} f(t)\,dt$$

We'll show that this limit does not exist by showing that the left and right limits are different. The right limit is

$$\lim_{h \to 0^+} \frac{1}{h} \int_2^{2+h} f(t)\,dt$$
$$= \lim_{h \to 0^+} \frac{1}{h} \int_2^{2+h} (t+1)\,dt$$
$$= \lim_{h \to 0^+} \frac{1}{h} \left[\frac{t^2}{2} + t \right]_2^{2+h}$$
$$= \lim_{h \to 0^+} \frac{1}{h} \left[\frac{(2+h)^2}{2} + 2 + h - \frac{2^2}{2} - 2 \right]$$
$$= \lim_{h \to 0^+} \frac{1}{h} \left[\frac{h^2 + 4h + 4}{2} + 2 + h - 4 \right]$$
$$= \lim_{h \to 0^+} \frac{1}{h} \left[\frac{h^2}{2} + 3h \right]$$
$$= \lim_{h \to 0^+} \frac{1}{h} \left[\frac{h}{2} + 3 \right]$$
$$= 3$$

The left limit is $\lim_{h \to 0^-} \frac{1}{h} \int_2^{2+h} f(t)\,dt$

$$= \lim_{h \to 0^-} \frac{1}{h} \int_2^{2+h} t\,dt$$
$$= \lim_{h \to 0^-} \frac{1}{h} \left[\frac{t^2}{2} \right]_2^{2+h}$$
$$= \lim_{h \to 0^-} \frac{1}{h} \left[\frac{(2+h)^2}{2} - \frac{2^2}{2} \right]$$
$$= \lim_{h \to 0^-} \frac{1}{h} \left[\frac{h^2 + 4h + 4}{2} - 2 \right]$$
$$= \lim_{h \to 0^-} \frac{1}{h} \left[\frac{h}{2} + 2 \right]$$
$$= 2$$

Thus, $F'(2)$ does not exist. This result does not contradict the Fundamental Theorem of Calculus, because in this situation, $f(x)$ is not continuous, and thus The Fundamental Theorem of Calculus does not apply.

69. $g(x) = \int_0^x \left[\int_0^u f(t)\,dt \right] du$

$$g'(x) = \int_0^x f(t)\,dt$$
$$g''(x) = f(x)$$

A zero of f corresponds to a zero of the second derivative of g (possibly an inflection point of g).

71. The integrals in parts (a) and (c) are improper, because the integrands have asymptotes at one of the limits of integration. The Fundamental Theorem of Calculus applies to the integral in part (b).

73. (a) $\lim_{n \to \infty} \frac{1}{n} \left[\sin \frac{\pi}{n} + \sin \frac{2\pi}{n} + \cdots + \sin \pi \right]$

$$= \int_0^1 \sin(\pi x)\,dx$$
$$= -\frac{1}{\pi} \cos(\pi x) \Big|_0^1$$
$$= -\frac{1}{\pi} (\cos \pi - \cos 0)$$
$$= -\frac{1}{\pi} (-1 - 1)$$
$$= \frac{2}{\pi}$$

(b) $\lim_{n\to\infty} \frac{2}{n}\left[\frac{1}{1+2/n} + \frac{1}{1+4/n} + \cdots + \frac{1}{3}\right]$

$= \int_0^2 \frac{1}{1+x}\,dx$

$= \ln|1+x|\Big|_0^2$

$= \ln 3 - \ln 1$

$= \ln 3$

75. Let $F(x) = \int_{a(x)}^{b(x)} f(t)\,dt$,

$G(x) = \int_0^{b(x)} f(t)\,dt$,

$H(x) = \int_0^{a(x)} f(t)\,dt$,

Then

$F(x) = G(x) - H(x)$.

$G'(x) = f(b(x))b'(x)$

$H'(x) = f(a(x))a'(x)$.

$F'(x) = G'(x) - H'(x)$

$= f(b(x))b'(x) - f(a(x))a'(x)$.

4.6 Integration by Substitution

1. Let $u = x^3+2$ and then $du = 3x^2\,dx$ and

$\int x^2\sqrt{x^3+2}\,dx = \frac{1}{3}\int u^{-1/2}\,du$

$= \frac{2}{9}u^{3/2} + c = \frac{2}{9}(x^3+2)^{3/2} + c$

3. Let $u = \sqrt{x}+2$ and then $du = \frac{1}{2}x^{-1/2}\,dx$ and

$\int \frac{(\sqrt{x}+2)^3}{\sqrt{x}}\,dx = 2\int u^3\,du$

$= \frac{2}{4}u^4 + c = \frac{1}{2}(\sqrt{x}+2)^4 + c$

5. Let $u = x^4+3$ and then $du = 4x^3\,dx$ and

$\int x^3\sqrt{x^4+3}\,dx = \frac{1}{4}\int u^{1/2}\,du$

$= \frac{1}{6}u^{3/2} + c = \frac{1}{6}(x^4+3)^{3/2} + c$

7. Let $u = \cos x$ and then $du = -\sin x\,dx$ and

$\int \frac{\sin x}{\sqrt{\cos x}}\,dx = -\int \frac{du}{\sqrt{u}}$

$= -2\sqrt{u} + c = -2\sqrt{\cos x} + c$

9. Let $u = x^3$ and then $du = 3x^2\,dx$ and

$\int x^2\cos x^3\,dx = \frac{1}{3}\int \cos u\,du$

$= \frac{1}{3}\sin u + c = \frac{1}{3}\sin x^3 + c$

11. Let $u = x^2+1$ and then $du = 2x\,dx$ and

$\int xe^{x^2+1}\,dx = \frac{1}{2}\int e^u\,du$

$= \frac{1}{2}e^u + c = \frac{1}{2}e^{x^2+1} + c$

13. Let $u = \sqrt{x}$ and then $du = \frac{1}{2\sqrt{x}}\,dx$ and

$\int \frac{e^{\sqrt{x}}}{\sqrt{x}}\,dx = 2\int e^u\,du$

$= 2e^u + c = 2e^{\sqrt{x}} + c$

15. Let $u = \ln x$, and then $du = \frac{1}{x}\,dx$ and

$\int \frac{\sqrt{\ln x}}{x}\,dx = \int \sqrt{u}\,du$

$= \frac{2}{3}u^{3/2} + c = \frac{2}{3}(\ln x)^{3/2} + c$

17. Let $u = \sqrt{x}+1$ and then $du = \frac{1}{2\sqrt{x}}\,dx$ and

$\int \frac{1}{\sqrt{x}\left(\sqrt{x}+1\right)^2}\,dx = 2\int u^{-2}\,du$

$= -2u^{-1} + c = -2\left(\sqrt{x}+1\right)^{-1} + c$

19. Let $u = \ln x + 1$ and then $du = \frac{1}{x}\,dx$ and

$\int \frac{4}{x(\ln x+1)^2}\,dx = 4\int u^{-2}\,du$

$= -4u^{-1} + c = -4(\ln x + 1)^{-1} + c$

21. Let $u = \sin^{-1} x$ and then $du = \frac{1}{\sqrt{1-x^2}}\,dx$ and

$$\int \frac{(\sin^{-1} x)^3}{\sqrt{1-x^2}}\,dx = \int u^3\,du = \frac{u^4}{4} + c = \frac{(\sin^{-1} x)^4}{4} + c$$

23. Let $u = x^2$ and then $du = 2x dx$ and

$$\int \frac{x}{\sqrt{1-x^4}}dx = \frac{1}{2}\int \frac{1}{\sqrt{1-u^2}}du = \frac{1}{2}\sin^{-1} u + c = \frac{1}{2}\sin^{-1} x^2 + c$$

25. Let $u = x^3$ and then $du = 3x^2dx$ and

$$\int \frac{x^2}{1+x^6}dx = \frac{1}{3}\int \frac{1}{1+u^2}du = \frac{1}{3}\tan^{-1} u + c = \frac{1}{3}\tan^{-1} x^3 + c$$

27. Let $u = x + 7$ and then $du = dx$, $x = u - 7$ and

$$\int \frac{2x+3}{x+7}\,dx = \int \frac{2(u-7)+3}{u}\,du = \int \left(2 - \frac{11}{u}\right) du = 2u - 11\ln|u| + c = 2(x+7) - 11\ln|x+7| + c$$

29. Let $u = \sqrt{1+\sqrt{x}}$ and then $(u^2-1)^2 = x$
$2(u^2-1)(2u)du = dx$ and

$$\int \frac{1}{\sqrt{1+\sqrt{x}}}dx = \int \frac{4u(u^2-1)}{u}\,du = 4\int (u^2-1)\,du = 4\left(\frac{u^3}{3} - u\right) + c = \frac{4}{3}(1+\sqrt{x})^{3/2} - 4(1+\sqrt{x})^{1/2} + c$$

31. Let $u = x^2 + 1$ and then $du = 2x\,dx$, $u(0) = 1$, $u(2) = 5$

$$\int_0^2 x\sqrt{x^2+1}\,dx = \frac{1}{2}\int_1^5 \sqrt{u}\,du = \frac{1}{2}\cdot\frac{2}{3}u^{3/2}\Big|_1^5 = \frac{1}{3}\left(\sqrt{125} - 1\right) = \frac{5}{3}\sqrt{5} - \frac{1}{3}$$

33. Let $u = x^2 + 1$ and then $du = 2x\,dx$, $u(-1) = 2 = u(1)$ and

$$\int_{-1}^1 \frac{x}{(x^2+1)^{1/2}}\,dx = \frac{1}{2}\int_2^2 u^{-1/2}\,du = 0$$

35. Let $u = e^x$ and then $du = e^x dx$, $u(0) = 1$, $u(2) = e^2$ and

$$\int_0^2 \frac{e^x}{1+e^{2x}}dx = \int_1^{e^2} \frac{1}{1+u^2}du = \tan^{-1} u\Big|_1^{e^2} = \tan^{-1} e^2 - \tan^{-1} 1 = \tan^{-1} e^2 - \frac{\pi}{4}$$

37. Let $u = \sin x$ and then $du = \cos x\,dx$ $u(\pi/4) = 1/\sqrt{2}$, $u(\pi/2) = 1$ and

$$\int_{\pi/4}^{\pi/2} \cot x\,dx = \int_{1/\sqrt{2}}^1 \frac{1}{u}\,du = \ln|u|\Big|_{1/\sqrt{2}}^1 = \ln\sqrt{2}$$

39.
$$\int_1^4 \frac{x-1}{\sqrt{x}}\,dx = \int_1^4 (x^{1/2} - x^{-1/2})\,dx = \left(\frac{2}{3}x^{3/2} - 2x^{1/2}\right)\Big|_1^4 = \left(\frac{16}{3} - 4\right) - \left(\frac{2}{3} - 2\right) = \frac{8}{3}$$

41. (a) $\int_0^\pi \sin x^2\,dx \approx .77$ using midpoint evaluation with $n \geq 40$

(b) Let $u = x^2$ and then $du = 2x\,dx$ and

$$\int_0^\pi x\sin x^2\,dx = \frac{1}{2}\int_0^{\pi^2} \sin u\,du = \frac{1}{2}(-\cos u)\Big|_0^{\pi^2} = -\frac{1}{2}\cos\pi^2 + \frac{1}{2} \approx 0.95134$$

43. (a) $\int_0^2 \frac{4x^2}{(x^2+1)^2}\,dx \approx 1.414$ using right endpoint evaluation with $n \geq 50$.

(b) Let $u = x^2 + 1$ and then $du = 2x\,dx, x^2 = u - 1$ and

$$\int_0^2 \frac{4x^3}{(x^2+1)^2}\,dx = \int_1^5 2\frac{u-1}{u^2}\,du$$
$$= \int_1^5 (2u^{-1} - 2u^{-2})\,du$$
$$= \left(2\ln|u| + 2u^{-1}\right)\Big|_1^5$$
$$= 2\ln 5 - \frac{8}{5}$$

45. $\frac{1}{2}\int_0^4 f(u)\,du$

47. $\int_0^1 f(u)\,du$

49. $\int_{-a}^{a} f(x)dx$

$$= \int_{-a}^{0} f(x)dx + \int_0^a f(x)dx$$

Let $u = -x$ and $du = -dx$ in the first integral. Then,

$$\int_{-a}^{a} f(x)\,dx$$
$$= -\int_a^0 f(-u)\,du + \int_0^a f(x)\,dx$$
$$= \int_0^a f(-u)\,du + \int_0^a f(x)\,dx$$

If f is even, then $f(-u) = f(u)$, and so

$$\int_{-a}^{a} f(x)dx$$
$$= \int_0^a f(u)\,du + \int_0^a f(x)\,dx$$
$$= \int_0^a f(x)\,dx + \int_0^a f(x)\,dx$$
$$= 2\int_0^a f(x)\,dx$$

If f is odd, then $f(-u) = -f(u)$, and so

$$\int_{-a}^{a} f(x)dx$$
$$= -\int_0^a f(u)du + \int_0^a f(x)dx$$
$$= -\int_0^a f(x)dx + \int_0^a f(x)dx$$
$$= 0$$

51. Let $u = 10 - x$, so that $du = -dx$. Then,

$$I = \int_0^{10} \frac{\sqrt{x}}{\sqrt{x} + \sqrt{10-x}}\,dx$$
$$= -\int_{x=0}^{x=10} \frac{\sqrt{10-u}}{\sqrt{10-u}+\sqrt{u}}\,du$$
$$= -\int_{u=10}^{u=0} \frac{\sqrt{10-u}}{\sqrt{10-u}+\sqrt{u}}\,du$$
$$= \int_{u=0}^{u=10} \frac{\sqrt{10-u}}{\sqrt{10-u}+\sqrt{u}}\,du$$
$$I = \int_{x=0}^{x=10} \frac{\sqrt{10-x}}{\sqrt{10-x}+\sqrt{x}}\,dx$$

The last equation follows from the previous one because u and x are dummy variables of integration. Now note that

$$\frac{\sqrt{x}}{\sqrt{x}+\sqrt{10-x}}$$
$$= \frac{\sqrt{x}+\sqrt{10-x}-\sqrt{10-x}}{\sqrt{x}+\sqrt{10-x}}$$
$$= 1 - \frac{\sqrt{10-x}}{\sqrt{x}+\sqrt{10-x}}$$

Thus,

$$\int_0^{10} \frac{\sqrt{x}}{\sqrt{x}+\sqrt{10-x}}\,dx$$
$$= \int_0^{10} \left[1 - \frac{\sqrt{10-x}}{\sqrt{x}+\sqrt{10-x}}\right] dx$$
$$= \int_0^{10} 1\,dx - \int_0^{10} \frac{\sqrt{10-x}}{\sqrt{x}+\sqrt{10-x}}\,dx$$
$$I = \int_0^{10} 1\,dx - I$$

$2I = 10$
$I = 5$

53. Let $u = 6 - x$, so that $du = -\,dx$.

Then,

$$I = \int_2^4 \frac{\sin^2(9-x)}{\sin^2(9-x)+\sin^2(x+3)}dx$$
$$= -\int_4^2 \frac{\sin^2(u+3)}{\sin^2(u+3)+\sin^2(9-u)}du$$
$$= \int_2^4 \frac{\sin^2(u+3)}{\sin^2(u+3)+\sin^2(9-u)}du$$
$$= \int_2^4 \frac{\sin^2(x+3)}{\sin^2(x+3)+\sin^2(9-x)}dx$$
$$= \int_2^4 \left[1 - \frac{\sin^2(9-x)}{\sin^2(x+3)+\sin^2(9-x)}\right]dx$$
$$I = \int_2^4 1\,dx - I$$
$2I = 2$
$I = 1$

55. Let $6-u = x+4$; that is, let $u = 2-x$, so that $du = -dx$.

Then,

$$I = \int_0^2 \frac{f(x+4)}{f(x+4)+f(6-x)}dx$$
$$= -\int_2^0 \frac{f(6-u)}{f(6-u)+f(u+4)}du$$
$$= \int_0^2 \frac{f(6-u)}{f(6-u)+f(u+4)}du$$
$$= \int_0^2 \frac{f(6-x)}{f(6-x)+f(x+4)}dx$$
$$= \int_0^2 \frac{f(6-x)+f(x+4)-f(x+4)}{f(6-x)+f(x+4)}dx$$
$$= \int_0^2 \left[1 - \frac{f(x+4)}{f(6-x)+f(x+4)}\right]dx$$
$$I = \int_0^2 1dx - I$$
$2I = 2$
$I = 1$

57. Let $u = x^{1/6}$, so that $du = (1/6)x^{-5/6}dx$, which means $6u^5du = dx$.

Thus,

$$\int \frac{1}{\sqrt{x}+\sqrt[3]{x}}dx$$
$$= 6\int \frac{u^5}{u^3+u^2}du$$
$$= 6\int \frac{u^3}{u+1}du$$
$$= 6\int \left[u^2 - u + 1 - \frac{1}{u+1}\right]du$$
$$= 6\left[\frac{u^3}{3} - \frac{u^2}{2} + u - \ln|u+1|\right] + c$$
$$= 2x^{1/2} - 3x^{1/3} + 6x^{1/6} - 6\ln|x^{1/6}+1| + c$$

59. First let $u = \ln\sqrt{x}$, so that $du = x^{-1/2}(1/2)x^{-1/2}dx$, so that $2du = \frac{1}{x}dx$. Then,

$$\int \frac{1}{x\ln\sqrt{x}}dx = 2\int \frac{1}{u}du$$
$$= 2\ln|u| + c$$
$$= 2\ln|\ln\sqrt{x}| + c$$

Now use the substitution $u = \ln x$, so that $du = \frac{1}{x}dx$. Then,

$$\int \frac{1}{x\ln\sqrt{x}}dx = \int \frac{1}{x\ln(x^{1/2})}dx$$
$$= \int \frac{1}{x\left(\frac{1}{2}\right)\ln x}dx$$
$$= 2\int \frac{1}{u}du$$
$$= 2\ln|u| + c_1$$
$$= 2\ln|\ln x| + c_1$$

The two results differ by a constant, and so are equivalent, as can be seen as follows:

$$2\ln|\ln\sqrt{x}| = 2\ln|\ln(x^{1/2})|$$
$$= 2\ln\left|\frac{1}{2}\ln x\right|$$

$$= 2\left[\ln\frac{1}{2} + \ln|\ln x|\right]$$
$$= 2\ln\frac{1}{2} + 2\ln|\ln x|$$
$$= 2\ln|\ln x| + \text{ constant}$$

61. The point is that if we let $u = x^4$, then we get $x = \pm u^{1/4}$, and so we need to pay attention to the sign of u and x. A safe way is to solve the original indefinite integral in terms of x, and then solve the definite integral using boundary points in terms of x.

$$\int_{-2}^{1} 4x^4 dx = \int_{x=-2}^{x=1} u^{1/4} du$$
$$= \frac{4}{5}u^{5/4}\Big|_{x=-2}^{x=1}$$
$$= \frac{4}{5}x^5\Big|_{x=-2}^{x=1}$$
$$= \frac{4}{5}(1^5 - (-2)^5)$$
$$= \frac{4}{5}(1 - (-32))$$
$$= \frac{4(33)}{5} = \frac{132}{5}$$

63. Let $u = 1/x$, so that $du = -1/x^2 dx$, which means that $-1/u^2 du = dx$. Then,

$$\int_a^1 \frac{1}{x^2+1} dx = -\int_{1/a}^{1} \frac{1/u^2}{1/u^2+1} du$$
$$= \int_1^{1/a} \frac{1}{1+u^2} du$$
$$= \int_1^{1/a} \frac{1}{1+x^2} dx$$

The last equation follows from the previous one because u and x are dummy variables of integration. Thus,

$$\tan^{-1} x\Big|_a^1 = \tan^{-1} x\Big|_1^{1/a}$$
$$\tan^{-1} 1 - \tan^{-1} a = \tan^{-1}\frac{1}{a} - \tan^{-1} 1$$
$$\tan^{-1} a + \tan^{-1}\frac{1}{a} = 2\tan^{-1} 1$$
$$\tan^{-1} a + \tan^{-1}\frac{1}{a} = \frac{\pi}{2}$$

65. $\bar{x} = \dfrac{\int_{-2}^{2} x\sqrt{4-x^2}\,dx}{\int_{-2}^{2}\sqrt{4-x^2}\,dx}$

Examine the denominator of $\bar{x}$, the graph of $\sqrt{4-x^2}$, which is indeed a semicircle, is symmetric over the two intervals $[-2, 0]$ and $[0, 2]$, while multiplying by x changes the symmetry into anti-symmetry. In other words,

$$\int_{-2}^{0} x\sqrt{4-x^2}\,dx = -\int_0^2 x\sqrt{4-x^2}\,dx$$

so that

$$\int_{-2}^{2} x\sqrt{4-x^2}\,dx$$
$$= \int_{-2}^{0} x\sqrt{4-x^2}\,dx + \int_0^2 x\sqrt{4-x^2}\,dx$$
$$= 0$$

Hence $\bar{x} = 0$.

Now the integral $\int_{-2}^{2}\sqrt{4-x^2}\,dx$ is the area of a semicircle with radius 2, thus its value $= (1/2)\pi 2^2 = 2\pi$.

Then

$$\bar{y} = \frac{\int_{-2}^{2}(\sqrt{4-x^2})^2\,dx}{2\int_{-2}^{2}\sqrt{4-x^2}\,dx}$$
$$= \frac{\int_{-2}^{2}(4-x^2)\,dx}{2\cdot 2\pi}$$
$$= \frac{\int_{-2}^{0}(4-x^2)\,dx + \int_0^2(4-x^2)\,dx}{4\pi}$$
$$= \frac{2\int_0^2(4-x^2)\,dx}{4\pi}$$
$$= \frac{\int_0^2(4-x^2)\,dx}{2\pi}$$
$$= \frac{1}{2\pi}\left(4x - \frac{x^3}{3}\right)\Big|_0^2$$
$$= \frac{8}{3\pi}$$

67. $V(t) = V_p \sin(2\pi f t) V^2(t)$
$= V_p^2 \sin^2(2\pi f t)$
$= V_p^2 \left(\frac{1}{2} - \frac{1}{2}\cos(4\pi f t)\right)$
$= \frac{V_p^2}{2}(1 - \cos(4\pi f t))$

$$\text{rms} = \sqrt{f \int_0^{1/f} V^2(t)\, dt}$$
$$= \sqrt{f \int_0^{1/f} \frac{V_p^2}{2}(1 - \cos(4\pi f t))\, dt}$$
$$= \frac{V_p\sqrt{f}}{\sqrt{2}} \sqrt{\int_0^{1/f} (1 - \cos(4\pi f t))\, dt}$$
$$= \frac{V_p\sqrt{f}}{\sqrt{2}} \sqrt{\left(t - \frac{\sin(4\pi f t)}{4\pi f}\right)\Big|_0^{1/f}}$$
$$= \frac{V_p\sqrt{f}}{\sqrt{2}} \sqrt{\frac{1}{f}} = \frac{V_p}{\sqrt{2}}$$

4.7 Numerical Integration

1. Midpoint Rule:
$$\int_0^1 (x^2 + 1)\, dx$$
$$\approx \frac{1}{4}\left[f\left(\frac{1}{8}\right) + f\left(\frac{3}{8}\right) + f\left(\frac{5}{8}\right) + f\left(\frac{7}{8}\right)\right]$$
$$= \frac{85}{64}$$

Trapezoidal Rule:
$$\int_0^1 (x^2 + 1)\, dx$$
$$\approx \frac{1-0}{2(4)}\left[f(0) + 2f\left(\frac{1}{4}\right) + 2f\left(\frac{1}{2}\right) + 2f\left(\frac{3}{4}\right) + f(1)\right]$$
$$= \frac{43}{32}$$

Simpson's Rule:
$$\int_0^1 (x^2 + 1)\, dx$$
$$= \frac{1-0}{3(4)}\left[f(0) + 4f\left(\frac{1}{4}\right) + 2f\left(\frac{1}{2}\right) + 4f\left(\frac{3}{4}\right) + f(1)\right]$$
$$= \frac{4}{3}$$

3. Midpoint Rule:
$$\int_1^3 \frac{1}{x}\, dx$$
$$\approx \frac{3-1}{4}\left[f\left(\frac{5}{4}\right) + f\left(\frac{7}{4}\right) + f\left(\frac{9}{4}\right) + f\left(\frac{11}{4}\right)\right]$$
$$= \frac{1}{2}\left(\frac{4}{5} + \frac{4}{7} + \frac{4}{9} + \frac{4}{11}\right)$$
$$= \frac{3776}{3465}$$

Trapezoidal Rule:
$$\int_1^3 \frac{1}{x}\, dx$$
$$\approx \frac{3-1}{2(4)}\left[f(1) + 2f\left(\frac{3}{2}\right) + 2f(2) + 2f\left(\frac{5}{2}\right) + f(3)\right]$$
$$= \frac{1}{4}\left(1 + \frac{4}{3} + 1 + \frac{4}{5} + \frac{1}{3}\right)$$
$$= \frac{67}{60}$$

Simpson's Rule:
$$\int_1^3 \frac{1}{x}\, dx$$
$$\approx \frac{3-1}{3(4)}\left[f(1) + 4f\left(\frac{3}{2}\right) + 2f(2) + 4f\left(\frac{5}{2}\right) + f(3)\right]$$
$$= \frac{1}{6}\left(1 + \frac{8}{3} + 1 + \frac{8}{5} + \frac{1}{3}\right)$$

$= \dfrac{11}{10}$

5. (a) Left Endpoints:

$$\int_0^2 f(x)\,dx \approx \frac{2-0}{4}[f(0)+f(.5)+f(1)+f(1.5)] = \frac{1}{2}(1+.25+0+.25) = .75$$

(b) Midpoint Rule:

$$\int_0^2 f(x)\,dx \approx \frac{2-0}{4}[f(.25)+f(.75)+f(1.25)+f(1.75)] = \frac{1}{2}(.65+.15+.15+.65) = .7$$

(c) Trapezoidal Rule:

$$\int_0^2 f(x)\,dx \approx \frac{2-0}{2(4)}[f(0)+2f(.5)+2f(1)+2f(1.5)+f(2)] = \frac{1}{4}(1+.5+0+.5+1) = .75$$

7.

n	Midpoint	Trapezoidal	Simpson
10	0.5538	0.5889	0.5660
20	0.5629	0.5713	0.5655
50	0.5652	0.5666	0.5657

9.

n	Midpoint	Trapezoidal	Simpson
10	0.88220	0.88184	0.88207
20	0.88211	0.88202	0.88208
50	0.88209	0.88207	0.88208

11.

n	Midpoint	Trapezoidal	Simpson
10	3.9775	3.9775	3.9775
20	3.9775	3.9775	3.9775
50	3.9775	3.9775	3.9775

13. The exact value of this integral is

$$\int_0^1 5x^4\,dx = x^5\Big|_0^1 = 1-0 = 1$$

n	Midpoint	EM_n
10	1.00832	8.3×10^{-3}
20	1.00208	2.1×10^{-3}
40	1.00052	5.2×10^{-4}
80	1.00013	1.3×10^{-4}

n	Trapezoidal	ET_n
10	0.98335	1.6×10^{-2}
20	0.99583	4.1×10^{-3}
40	0.99869	1.0×10^{-3}
80	0.99974	2.6×10^{-4}

n	Simpson	ES_n
10	1.000066	6.6×10^{-5}
20	1.0000041	4.2×10^{-6}
40	1.00000026	2.6×10^{-7}
80	1.00000016	1.6×10^{-8}

15. The exact value of this integral is

$$\int_0^\pi \cos x\,dx = \sin x\Big|_0^\pi = 0$$

n	Midpoint	EM_n
10	0	0
20	0	0
40	0	0
80	0	0

n	Trapezoidal	ET_n
10	0	0
20	0	0
40	0	0
80	0	0

n	Simpson	ES_n
10	0	0
20	0	0
40	0	0
80	0	0

17. If you double n, the error in the Midpoint Rule is divided by 4, the error in the Trapezoidal Rule is divided by 4 and the error in the Simpson's Rule is divided by 16.

19. Midpoint Rule:
$\ln 4 \approx 1.366162$

Trapezoidal Rule:
$\ln 4 \approx 1.428091$

Simpson's Rule:
$\ln 4 \approx 1.391621$

21. Midpoint Rule:
$\sin 1 \approx 0.843666$

Trapezoidal Rule:
$\sin 1 \approx 0.837084$

Simpson's Rule:
$\sin 1 \approx 0.841489$

23. $f(x) = \frac{1}{x}, f''(x) = \frac{2}{x^3}, f^{(4)}(x) = \frac{24}{x^5}$

Then $K = 2, L = 24$

Hence according to Theorems 9.1 and 9.2

$$|ET_4| \le 2\frac{(4-1)^3}{12 \cdot 4^2} \approx 0.281$$

$$|EM_4| \le 2\frac{(4-1)^3}{24 \cdot 4^2} \approx 0.141$$

$$|ES_4| \le 24\frac{(4-1)^5}{180 \cdot 4^4} \approx 0.127$$

25. Using Theorems 9.1 and 9.2, and the calculation in Example 9.10, we find the following lower bounds for the number of steps needed to guarantee accuracy of 10^{-7} in Exercise 19:

Midpoint: $\sqrt{\frac{2 \cdot 3^3}{24 \cdot 10^{-7}}} \approx 4745$

Trapezoidal: $\sqrt{\frac{2 \cdot 3^3}{14 \cdot 10^{-7}}} \approx 6709$

Simpson's: $\sqrt[4]{\frac{24 \cdot 3^5}{180 \cdot 10^{-7}}} \approx 135$

27. We use $K = 60, L = 120$.

n	$\lvert EM_n \rvert$	Error bound
10	8.3×10^{-3}	2.5×10^{-2}

n	$\lvert ET_n \rvert$	Error bound
10	1.6×10^{-2}	5×10^{-2}

n	$\lvert ES_n \rvert$	Error bound
10	7.0×10^{-5}	6.6×10^{-3}

29. Trapezoidal Rule:

$$\int_0^2 f(x)\,dx$$
$$\approx \frac{2-0}{2(8)}[f(0) + 2f(0.25) + 2f(0.5) + 2f(.75) + 2f(1) + 2f(1.25) + 2f(1.5) + 2f(1.75) + f(2)]$$
$$= \frac{1}{8}[4.0 + 9.2 + 10.4 + 9.6 + 10 + 9.2 + 8.8 + 7.6 + 4.0]$$
$$= 9.1$$

Simpson's Rule:

$$\int_0^2 f(x)\,dx$$
$$\approx \frac{2-0}{3(8)}[f(0) + 4f(.25) + 2f(.5) + 4f(.75) + 2f(1) + 4f(1.25) + 2f(1.5) + 4f(1.75) + f(2)]$$
$$= \frac{1}{12}(4.0 + 18.4 + 10.4 + 19.2 + 10 + 18.4 + 8.8 + 15.2 + 4.0)$$
$$\approx 9.033$$

31. Simpson's Rule:

$$\int_0^{120} f(x)\,dx$$
$$\approx \frac{120-0}{3(12)}[f(0) + 4f(10) + 2f(20) + 4f(30) + 2f(40) + 4f(50) + 2f(60) + 4f(70) + 2f(80) + 4f(90) + 2f(100) + 4f(110) + f(120)]$$

$$= \frac{10}{3}(56+216+116+248+116+232 +124+224+104+192+80+128+22) \approx 6193$$

33. Simpson's Rule:

$$\int_0^{24} v(t)\,dt \approx \frac{12-0}{3(12)}[f(0)+4f(1)+2f(2) + 4f(3)+2f(4)+4f(5)+2f(6) + 4f(7)+2f(8)+4f(9)+2f(10) + 4f(11)+f(12)]$$
$$= \frac{1}{3}(40+168+80+176+96+200 + 92+184+84+176+80+168 + 42) = 529$$

35. Simpson's Rule:

$$\int_0^{2.4} f(x)\,dx \approx \frac{2.4-0}{3(12)}[f(0)+4f(.2)+2f(.4) + 4f(.6)+2f(.8)+4f(1)+2f(1.2) +4f(1.4)+2f(1.6)+4f(1.8)+2f(2) + 4f(2.2)+f(2.4)]$$
$$= \frac{1}{15}(0+.8+.8+4+3.2+8+4.4 +8+3.2+4.8+1.2+.8+0) \approx 2.6$$

37. a) Midpoint Rule:

$$M_n < \int_a^b f(x)\,dx$$

b) Trapezoidal Rule:

$$T_n > \int_a^b f(x)\,dx$$

c) Simpson's Rule: not enough information.

39. a) Midpoint Rule:

$$M_n > \int_a^b f(x)\,dx$$

b) Trapezoidal Rule:

$$T_n < \int_a^b f(x)\,dx$$

c) Simpson's Rule: not enough information.

41. a) Midpoint Rule:

$$M_n < \int_a^b f(x)\,dx$$

b) Trapezoidal Rule:

$$T_n > \int_a^b f(x)\,dx$$

c) Simpson's Rule:

$$S_n = \int_a^b f(x)\,dx$$

43. $\frac{1}{2}(R_L + R_R)$

$$= \sum_{i=0}^{n-1} f(x_i) + \sum_{i=1}^{n} f(x_i)$$
$$= f(x_0) + \sum_{i=1}^{n-1} f(x_i) + \sum_{i=1}^{n-1} f(x_i) + f(x_n)$$
$$= f(x_0) + 2\sum_{i=1}^{n-1} f(x_i) + f(x_n) = T_n$$

45. $I_1 = \int_0^1 \sqrt{1-x^2}\,dx$ is one fourth of the area of a circle with radius 1, so

$$\int_0^1 \sqrt{1-x^2}\,dx = \frac{\pi}{4}$$

$$I_2 = \int_0^1 \frac{1}{1+x^2}\,dx = \arctan x \Big|_0^1 = \arctan 1 - \arctan 0 = \frac{\pi}{4}$$

n	$S_n(\sqrt{1-x^2})$	$S_n(\frac{1}{1+x^2})$
4	0.65652	0.78539
8	0.66307	0.78539

The second integral $\int \frac{1}{1+x^2}\,dx$ provides a better algorithm for estimating π.

47. (a) $\int_{-1}^{1} x\,dx = 0$

$\left(-\frac{1}{\sqrt{3}}\right) + \left(\frac{1}{\sqrt{3}}\right) = 0$

(b) $\int_{-1}^{1} x^2\,dx = \frac{2}{3}$

$\left(-\frac{1}{\sqrt{3}}\right)^2 + \left(\frac{1}{\sqrt{3}}\right)^2 = \frac{2}{3}$

(c) $\int_{-1}^{1} x^3\,dx = 0$

$\left(-\frac{1}{\sqrt{3}}\right)^3 + \left(\frac{1}{\sqrt{3}}\right)^3 = 0$

49. Simpson's Rule is not applicable because $\frac{\sin x}{x}$ is not defined at $x = 0$.

$L = \lim_{x\to 0} \frac{\sin x}{x}$

$= \lim_{x\to 0} \frac{\cos x}{1} = \cos 0 = 1$

The two functions $f(x)$ and $\frac{\sin x}{x}$ differ only at one point $x = 0$, so

$\int_0^{\pi} f(x)\,dx = \int_0^{\pi} \frac{\sin x}{x}\,dx$

We can now apply Simpson's Rule with $n = 2$:

$\int_0^{\pi} f(x)\,dx$

$\approx \frac{\pi}{6}\left(1 + \frac{4\sin\pi}{\pi/2} + \frac{\sin\pi}{\pi}\right)$

$= \frac{\pi}{2}\left(\frac{1}{3} + \frac{8}{3\pi}\right)$

$\approx \frac{\pi}{2}\cdot 1.18$

51. Let I be the exact integral. Then we have

$T_n - I \approx -2(M_n - I)$

$T_n - I \approx 2I - 2M_n$

$T_n + 2M_n \approx 3I$

$\frac{T_n}{3} + \frac{2}{3}M_n \approx I$

4.8 The Natural Logarithm As An Integral

1. $\ln 4 = \ln 4 - \ln 1 = \ln x\Big|_1^4 = \int_1^4 \frac{dx}{x}$

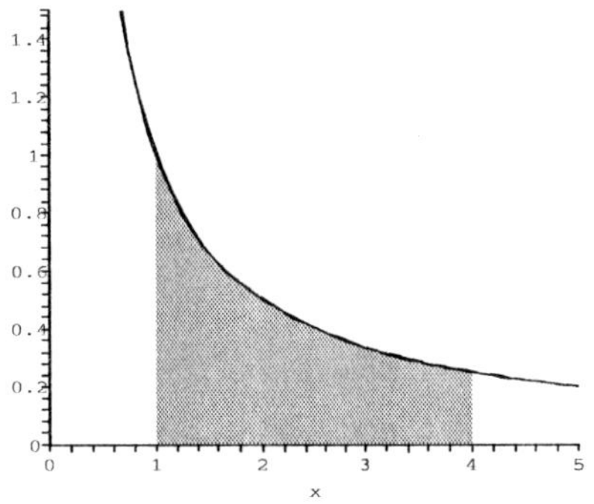

3. $\ln 8.2 = \int_1^{8.2} \frac{dx}{x}$

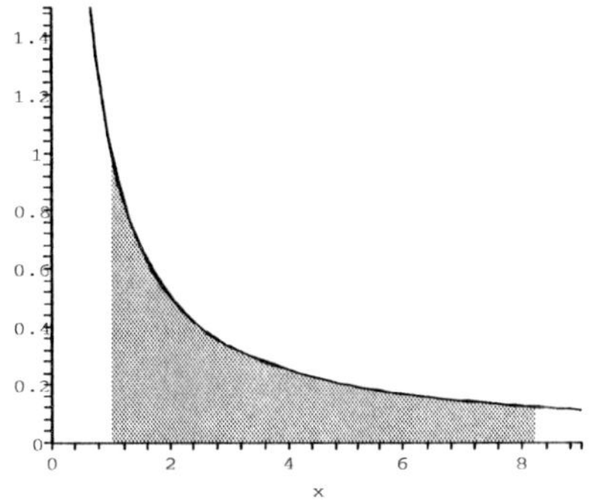

5. $\ln 4 = \int_1^4 \frac{dx}{x}$

$\approx \frac{3}{12}\left(\frac{1}{1} + 4\frac{1}{1.75} + 2\frac{1}{1.5} + 4\frac{1}{3.25} + \frac{1}{4}\right)$

≈ 1.3868

7. (a) Simpson's Rule with $n = 32$:

$\ln 4 = \int_1^4 \frac{dx}{x} \approx 1.386296874$

(b) Simpson's Rule with $n = 64$:

$\ln 4 = \int_1^4 \frac{dx}{x} \approx 1.386294521$

9. $\frac{7}{2}\ln 2$

11. $\ln\left(\dfrac{3^2\cdot\sqrt{3}}{9}\right) = \dfrac{1}{2}\ln 3$

13. $\dfrac{1}{\sqrt{x^2+1}}\cdot\dfrac{1}{2}(x^2+1)^{-\frac{1}{2}}\cdot 2x$

15. $\dfrac{x^5+1}{x^4}\cdot\dfrac{4x^3(x^5+1)-x^4(5x^4)}{(x^5+1)^2}$

17. $\displaystyle\int \frac{3x^3}{x^4+5}\,dx = \frac{3}{4}\ln|x^4+5|+c$

$= \dfrac{3}{4}\ln(x^4+5)+c$

19. $\displaystyle\int \frac{1}{x\ln x}\,dx = \ln|\ln x|+c$

21. $\displaystyle\int \frac{e^{2x}}{1+e^{2x}}\,dx = \frac{1}{2}\ln|1+e^{2x})|+c$

$= \dfrac{1}{2}\ln(1+e^{2x})+c$

23. Let $u = 2/x, du = (-2/x^2)\,dx$

$\displaystyle\int \frac{e^{2/x}}{x^2}\,dx = -\frac{1}{2}\int e^u\,du$

$= -\dfrac{1}{2e^u}+c = -\dfrac{1}{2}e^{2/x}+c$

25. $\displaystyle\int_0^1 \frac{x^2}{x^3-4}\,dx = \frac{1}{3}\ln|x^3-4|\Big|_0^1$

$= \dfrac{1}{3}\ln 3 - \dfrac{1}{3}\ln 4 = \dfrac{1}{3}\ln\dfrac{3}{4}$

27. $\displaystyle\int_0^1 \tan x\,dx = \int_0^1 \frac{\sin x}{\cos x}\,dx$

$= -\ln|\cos x|\Big|_0^1$

$= -\ln|\cos 1| - \ln|\cos 0|$

$= -\ln(\cos 1)$

29. $\displaystyle\int_0^1 \frac{e^x-1}{e^{2x}}\,dx = \int_0^1 (e^{-x}-e^{-2x})\,dx$

$= \left(-e^{-x}+\dfrac{1}{2}e^{-2x}\right)\Big|_0^1$

$= -e^{-1}+\dfrac{1}{2}e^{-2}+\dfrac{1}{2}$

31.

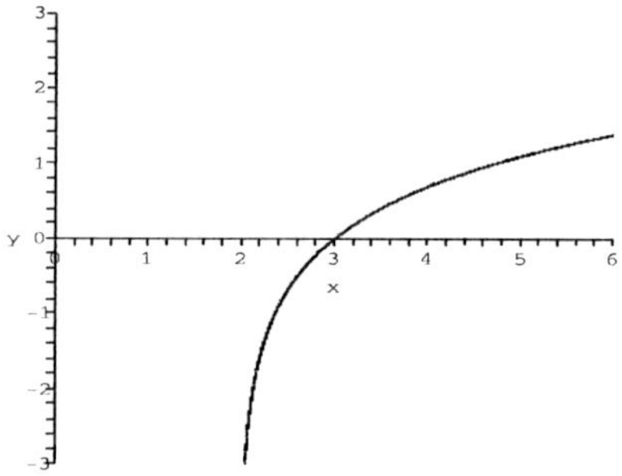

33.

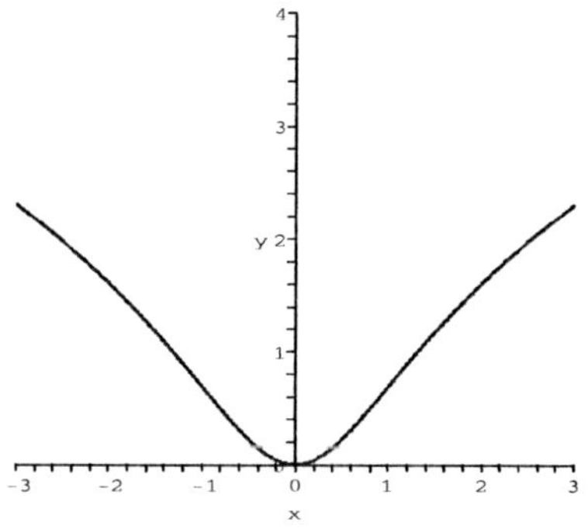

35.

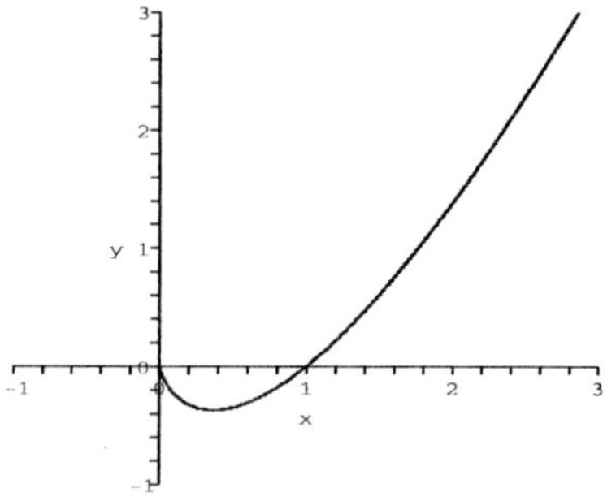

37. $\ln\left(\dfrac{a}{b}\right) = \ln\left(a\cdot\dfrac{1}{b}\right) = \ln a + \ln\left(\dfrac{1}{b}\right)$

$= \ln a - \ln b$

39. $f(x) = \dfrac{1}{1+e^{-x}}$

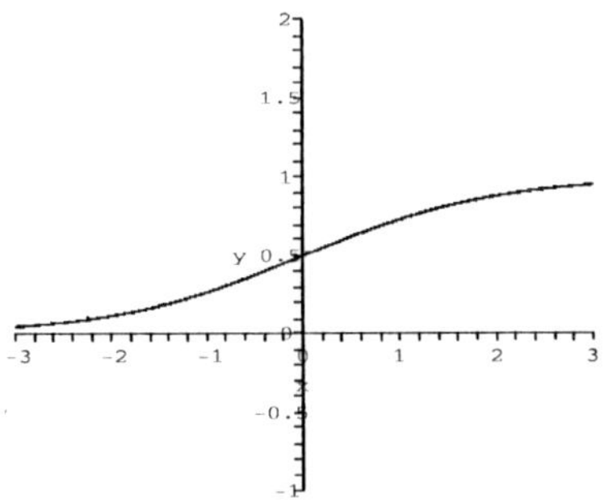

Using $\lim\limits_{x\to\infty} e^{-x} = 0$ we get

$$\lim_{x\to\infty} \frac{1}{1+e^{-x}} = 1$$

Using $\lim\limits_{x\to-\infty} e^{-x} = \infty$ we get

$$\lim_{x\to\infty} \frac{1}{1+e^{-x}} = 0$$

The function $f(x)$ is increasing over $(-\infty, \infty)$ and when $x = 0$,

$$f(0) = \frac{1}{1+1} = \frac{1}{2}.$$

So $g(x) = \begin{cases} 0 & \text{if } x < 0 \\ 1 & \text{if } x \geq 0 \end{cases}$

The threshold value for $g(x)$ to switch is $x = 0$.

One way of modifying the function to move the threshold to $x = 4$ is to let

$$f(x) = \frac{1}{1+e^{-(x-4)}}$$

41. $h = \ln e^h = \displaystyle\int_1^{e^h} \frac{1}{x}\,dx = \frac{e^h - 1}{\bar{x}},$

for some $\bar{x}$ in $(0, h)$

$$\frac{e^h - 1}{h} = \bar{x}$$

as $h \to 0^+, \bar{x} \to 0,$ then

$$\lim_{h\to 0^+} \frac{e^h - 1}{h} = 0$$

$$-h = \ln e^{-h}$$

$$= \int_1^{e^{-h}} \frac{1}{x}\,dx = \frac{e^{-h} - 1}{\bar{x}},$$

for some $\bar{x}$ in $(-h, 0)$

$$\frac{e^{-h} - 1}{-h} = \bar{x}$$

as $h \to 0^+, -h \to 0^-, \bar{x} \to 0,$ then

$$\lim_{h\to 0^+} \frac{e^{-h} - 1}{-h} = 0$$

43. $\ln\left[\lim\limits_{n\to\infty}\left(1+\dfrac{1}{n}\right)^n\right]$

$$= \lim_{n\to\infty} \ln\left(1+\frac{1}{n}\right)^n$$

$$= \lim_{n\to\infty} n\ln\left(1+\frac{1}{n}\right)$$

$$= \lim_{n\to\infty} \frac{\ln(1+1/n}{)}1/n$$

$$= \lim_{n\to\infty} \frac{-1/n^2}{-1/n^2(1+1/n)}$$

$$= \lim_{n\to\infty} \frac{1}{1+1/n}$$

$$= 1$$

45. $s(x) = x^2 \ln(1/x)$

$s'(x) = 2x\ln 1/x + x^2 \cdot x \cdot (-1/x^2)$

$= 2x\ln(1/x) - x = x(2\ln(1/x) - 1)$

$s'(x) = 0$ gives

$x = 0$ (which is impossible) or

$\ln(1/x) = 1/2, x = e^{-1/2}$

Since $s'(x) \begin{cases} < 0 & \text{if } x < e^{-1/2} \\ > 0 & \text{if } x > e^{-1/2} \end{cases}$

The value $x = e^{-1/2}$ maximizes the transmission speed.

Ch. 4 Review Exercises

1. $\displaystyle\int (4x^2 - 3)\,dx = \frac{4}{3}x^3 - 3x + c$

3. $\displaystyle\int \frac{4}{x}\,dx = 4\ln|x| + c$

5. $\displaystyle\int 2\sin 4x\,dx = -\frac{1}{2}\cos 4x + c$

7. $\displaystyle\int (x - e^{4x})\,dx = \frac{x^2}{2} - \frac{1}{4}e^{4x} + c$

9. $\int \frac{x^2+4}{x}\,dx = \int (x + 4x^{-1})\,dx$
$= \frac{x^2}{2} + 4\ln|x| + c$

11. $\int e^x(1-e^{-x})\,dx = \int (e^x - 1)\,dx$
$= e^x - x + c$

13. Let $u = x^2+4$, then $du = 2x\,dx$ and
$\int x\sqrt{x^2+4}\,dx$
$= \frac{1}{2}\int u^{1/2}\,du = \frac{1}{3}\,u^{3/2} + c$
$= \frac{1}{3}\,(x^2+4)^{3/2} + c$

15. Let $u = x^3$, $du = 3x^2\,dx$
$\int 6x^2 \cos x^3\,dx = 2\int \cos u\,du$
$= 2\sin u + c = 2\sin x^3 + c$

17. Let $u = 1/x$, $du = -1/x^2\,dx$
$\int \frac{e^{1/x}}{x^2}\,dx = -\int e^u\,du$
$= -e^u + c = -e^{1/x} + c$

19. $\int \tan x\,dx = \int \frac{\sin x}{\cos x}\,dx$
$= -\ln|\cos x| + c$

21. $f(x) = \int (3x^2+1)\,dx = x^3 + x + c$
$f(0) = c = 2$
$f(x) = x^3 + x + 2$

23. $s(t) = \int (-32t + 10)\,dt$
$= -16t^2 + 10t + c$
$s(0) = c = 2$
$s(t) = -16t^2 + 10t + 2$

25. $\sum_{i=1}^{6}(i^2 + 3i)$
$= (1^2+3\cdot 1) + (2^2+3\cdot 2) + (3^2+3\cdot 3)$
$+ (4^2+3\cdot 4) + (5^2+3\cdot 5) + (6^2+3\cdot 6)$
$= 4 + 10 + 18 + 28 + 40 + 54$
$= 154$

27. $\sum_{i=1}^{100}(i^2 - 1)$
$= \sum_{i=1}^{100} i^2 - \sum_{i=1}^{100} 1$
$= \frac{100(101)(201)}{6} - 100$
$= 338,250$

29. $\frac{1}{n^3}\sum_{i=1}^{n}(i^2 - i)$
$= \frac{1}{n^3}\left(\sum_{i=1}^{n} i^2 - \cdot \sum_{i=1}^{n} i\right)$
$= \frac{1}{n^3}\left(\frac{n(n+1)(2n+1)}{6} - \frac{n(n+1)}{2}\right)$
$= \frac{(n+1)(2n+1)}{6n^2} - \frac{n+1}{2n^2}$

$\lim_{n\to\infty}\frac{1}{n^3}\sum_{i=1}^{n}(i^2 - i)$
$= \lim_{n\to\infty}\left(\frac{(n+1)(2n+1)}{6n^2} - \frac{n+1}{2n^2}\right)$
$= \frac{2}{6} - 0 = \frac{1}{3}$

31. Riemann sum $= \frac{2}{8}\sum_{i=1}^{8} c_i^2 = 2.65625$

33. Riemann sum $= \frac{3}{8}\sum_{i=1}^{8} c_i^2 \approx 4.668$

35. **(a)** Left-endpoints:
$\int_0^{1.6} f(x)\,dx$
$\approx \frac{1.6-0}{8}(f(0) + f(.2) + f(.4)$
$+ f(.6) + f(.8) + f(1) + f(1.2)$
$+ f(1.4))$
$= \frac{1}{5}(1 + 1.4 + 1.6 + 2 + 2.2 + 2.4$
$+ 2 + 1.6)$
$= 2.84$

(b) Right-endpoints:

$$\int_0^{1.6} f(x)\,dx$$
$$\approx \frac{1.6-0}{8}(f(.2)+f(.4)+f(.6)$$
$$+f(.8)+f(1)+f(1.2)+f(1.4)$$
$$+f(1.6))$$
$$=\frac{1}{5}(1.4+1.6+2+2.2+2.4$$
$$+2+1.6+1.4)$$
$$=2.92$$

(c) Trapezoidal Rule:
$$\int_0^{1.6} f(x)\,dx$$
$$\approx \frac{1.6-0}{2(8)}[f(0)+2f(.2)+2f(.4)$$
$$+2f(.6)+2f(.8)+2f(1)$$
$$+2f(1.2)+2f(1.4)+f(1.6)]$$
$$=2.88$$

(d) Simpson's Rule:
$$\int_0^{1.6} f(x)\,dx$$
$$\approx \frac{1.6-0}{3(8)}[f(0)+4f(.2)+2f(.4)$$
$$+4f(.6)+2f(.8)+4f(1)$$
$$+2f(1.2)+4f(1.4)+f(1.6)]$$
$$\approx 2.907$$

37. See Example 7.10.

Simpson's Rule is expected to be most accurate.

39. We will compute the area A_n of n rectangles using right endpoints. In this case $\Delta x = \frac{1}{n}$ and $x_i = \frac{i}{n}$
$$A_n = \sum_{i=1}^{n} f(x_i)\Delta x = \frac{1}{n}\sum_{i=1}^{n} f\left(\frac{i}{n}\right)$$
$$=\frac{1}{n}\sum_{i=1}^{n} 2\cdot\left(\frac{i}{n}\right)^2$$
$$=\frac{2}{n^3}\sum_{i=1}^{n} i^2$$
$$=\left(\frac{2}{n^3}\right)\frac{n(n+1)(2n+1)}{6}$$
$$=\frac{(n+1)(2n+1)}{3n^2}$$
Now, to find the integral, we take the limit:
$$\int_0^1 x^2\,dx = \lim_{n\to\infty} A_n$$
$$=\lim_{n\to\infty}\frac{(n+1)(2n+1)}{3n^2}$$
$$=\frac{2}{3}$$

41. Area $= \int_0^3 (3x-x^2)\,dx$
$$=\left(\frac{3x^2}{2}-\frac{x^3}{3}\right)\Big|_0^3 = \frac{9}{2}$$

43. The velocity is always positive, so distance traveled is equal to change in position.
$$\text{Dist} = \int_1^2 (40-10t)\,dt$$
$$=(40t-5t^2)\Big|_1^2 = 25$$

45. $f_{ave} = \frac{1}{2}\int_0^2 e^x\,dx = \frac{e^2-1}{2} \approx 3.19$

47. $\int_0^2 (x^2-2)\,dx = \left(\frac{x^3}{3}-2x\right)\Big|_0^2 = -\frac{4}{3}$

49. $\int_0^{\pi/2} \sin 2x\,dx = -\frac{1}{2}\cos 2x\Big|_0^{\pi/2} = 1$

51. $\int_0^{10} (1-e^{-t/4})\,dt$
$$=\left(t+4e^{-t/4}\right)\Big|_0^{10} = 6+4e^{-5/2}$$

53. $\int_0^2 \frac{x}{x^2+1}\,dx = \frac{1}{2}\ln|x^2+1|\Big|_0^2$
$$=\frac{\ln 5}{2}$$

55. $\int_0^2 x\sqrt{x^2+4}\,dx$
$$=\left(\frac{1}{2}\cdot\frac{2}{3}\cdot(x^2+4)^{3/2}\right)\Big|_0^2$$

$$= \frac{16\sqrt{2} - 8}{3}$$

57. $\int_0^1 (e^x - 2)^2\, dx = \int_0^1 (e^{2x - 4e^x + 4})\, dx$

$$= \left(\frac{1}{2}e^{2x} - 4e^x + 4x\right)\Big|_0^2$$

$$= \left(\frac{e^2}{2} - 4e + 4\right) - \left(\frac{1}{2} - 4\right)$$

$$= \frac{e^2}{2} - 4e + \frac{15}{2}$$

59. $f'(x) = \sin x^2 - 2$

61. a) Midpoint Rule:

$$\int_0^1 \sqrt{x^2+4}\, dx$$

$$\approx \frac{1-0}{4}\left[f\left(\frac{1}{8}\right) + f\left(\frac{3}{8}\right) + f\left(\frac{5}{8}\right) + f\left(\frac{7}{8}\right)\right]$$

$$\approx 2.079$$

b) Trapezoidal Rule:

$$\int_0^1 \sqrt{x^2+4}\, dx$$

$$\approx \frac{1-0}{2(4)}\left[f(0) + 2f\left(\frac{1}{4}\right) + 2f\left(\frac{1}{2}\right) + 2f\left(\frac{3}{4}\right) + f(1)\right]$$

$$\approx 2.083$$

c) Simpson's Rule:

$$\int_0^1 \sqrt{x^2+4}\, dx$$

$$\approx \frac{1-0}{3(4)}\left[f(0) + 4f\left(\frac{1}{4}\right) + 2f\left(\frac{1}{2}\right) + 4f\left(\frac{3}{4}\right) + f(1)\right]$$

$$\approx 2.080$$

63.

n	Midpoint	Trapezoid	Simpson's
20	2.08041	2.08055	2.08046
40	2.08045	2.08048	2.08046

Chapter 5

Applications of the Definite Integral

5.1 Area Between Curves

1. $\text{Area} = \int_1^3 \left[x^3 - (x^2 - 1)\right] dx$

$= \left(\frac{x^4}{4} - \frac{x^3}{3} + x\right)\Big|_1^3$

$= \left(\frac{81}{4} - \frac{27}{3} + 3\right) - \left(\frac{1}{4} - \frac{1}{3} + 1\right)$

$= \frac{160}{12} = \frac{40}{3}$

3. $\text{Area} = \int_{-2}^0 [e^x - (x - 1)]\, dx$

$= \left(e^x - \frac{x^2}{2} + x\right)\Big|_{-2}^0$

$= (1 - 0 + 0) - \left(e^{-2} - \frac{4}{2} + (-2)\right)$

$= 5 - e^{-2}$

5. $\text{Area} = \int_{-2}^2 \left[7 - x^2 - (x^2 - 1)\right] dx$

$= \left(8x - \frac{2x^3}{3}\right)\Big|_{-2}^2$

$= \left(16 - \frac{16}{3}\right) - \left(-16 + \frac{16}{3}\right)$

$= \frac{64}{3}$

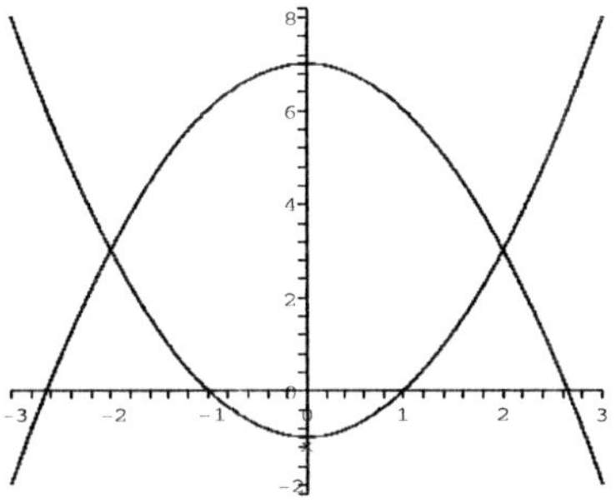

7. $\text{Area} = \int_{-1}^2 (3x + 2 - x^3)\, dx = \frac{27}{4}$

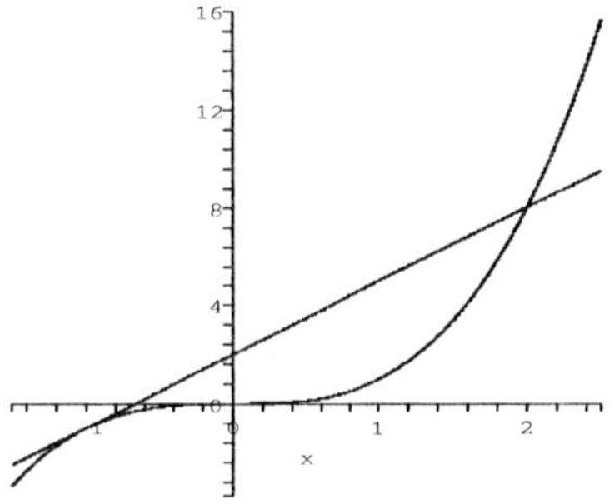

9. $\text{Area} = \int_0^{\sqrt{\ln 4}} \left(4xe^{-x^2} - x\right) dx$

$= -2e^{-x^2} - \frac{x^2}{2}\Big|_0^{\sqrt{\ln 4}}$

$= -2\left[\frac{1}{4} - 1\right] - \frac{\ln 4}{2}$

$= \frac{3 - \ln 4}{2}$

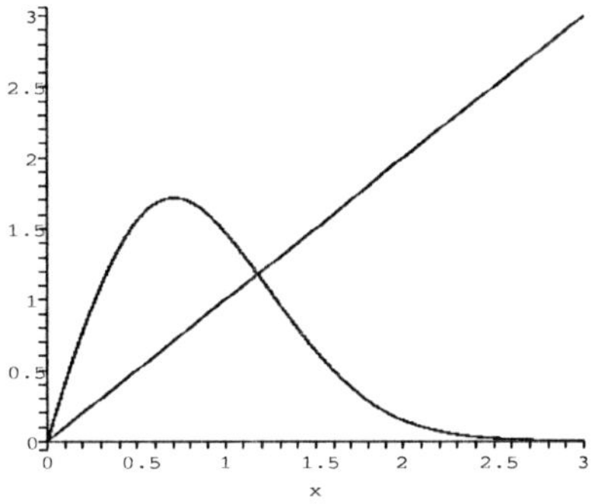

11. $\text{Area} = \int_{-2}^0 \left[x - \frac{5x}{x^2 + 1}\right] dx$

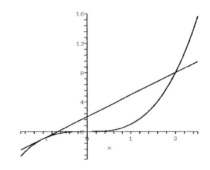

$$+\int_0^2 \left[\frac{5x}{x^2+1} - x\right] dx$$
$$= 2\int_0^2 \left[\frac{5x}{x^2+1} - x\right] dx$$
$$= 2\,\frac{5}{2}\ln\left|x^2+1\right| - \frac{x^2}{2}\,\Big|_0^2$$
$$= 5[\ln 5 - \ln 1] - [4 - 0]$$
$$= 5\ln 5 - 4$$

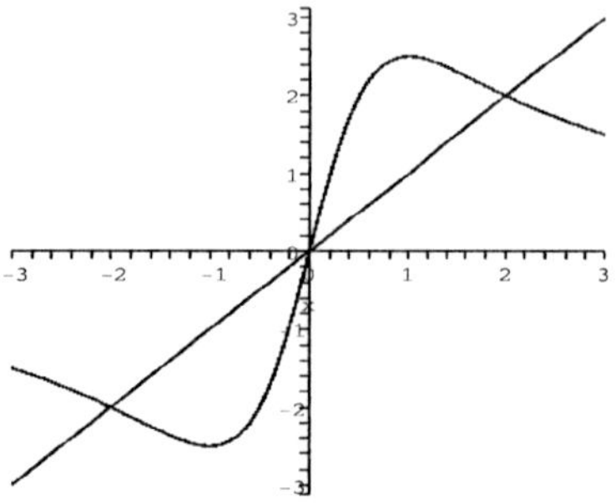

13. Area $= \displaystyle\int_{-.7145}^{0} (1 - x^2) - e^x\, dx$

$$= \left(-e^x + x - \frac{x^3}{3}\right)\Big|_{-.7145}^{0}$$
$$= (-1 + 0 - 0) - (-1.08235)$$
$$= .08235$$

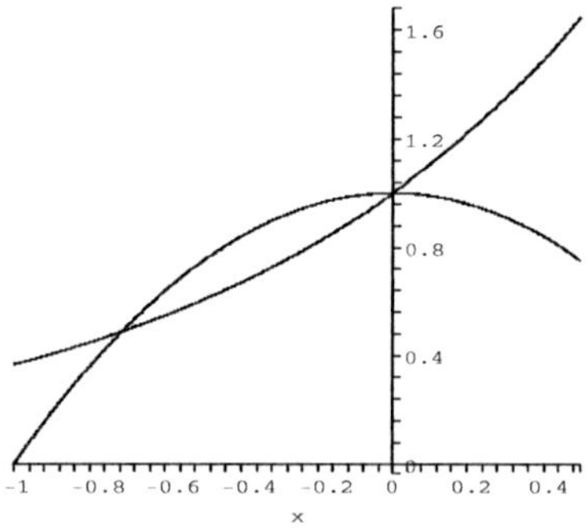

15. Area $= \displaystyle\int_0^{.8767} \left(\sin x - x^2\right) dx$

$$= \left(-\cos x - \frac{x^3}{3}\right)\Big|_0^{.8767}$$
$$\approx .135697$$

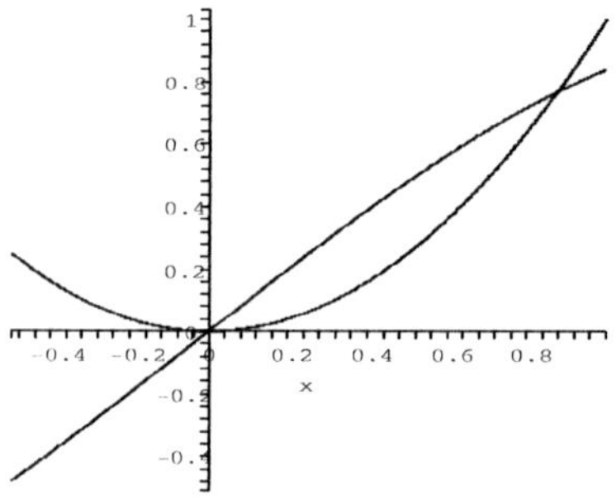

17. Area $= \displaystyle\int_{-1}^{1.3532} \left(2 + x - x^4\right) dx$

$$= \left(2x + \frac{x^2}{2} - \frac{x^5}{5}\right)\Big|_{-1}^{1.3532}$$
$$= 4.01449$$

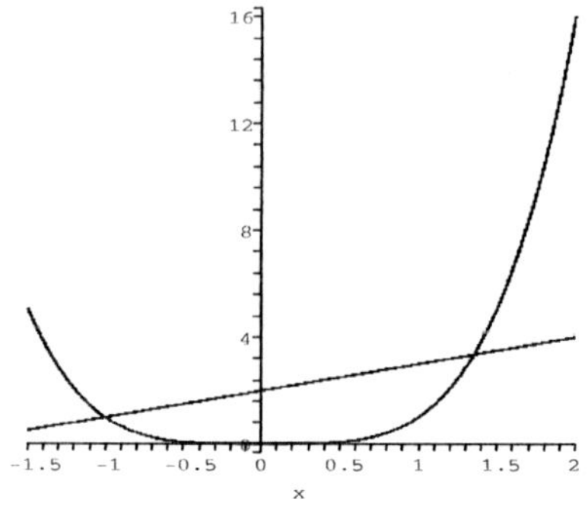

19. Area $= \displaystyle\int_0^1 [(2 - y) - y]dy$

$$= \int_0^1 [2 - 2y]dy$$
$$= \left(2y - y^2\right)\Big|_0^1$$
$$= 1 - 0 = 1$$

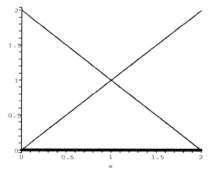

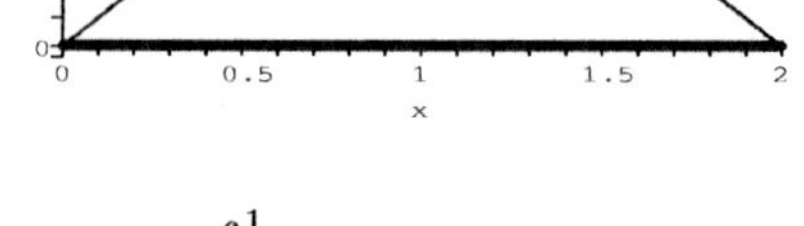

21. Area $= \displaystyle\int_0^1 [x - (-x)]\, dx$

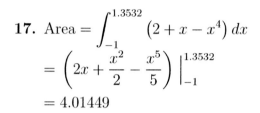

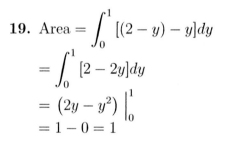

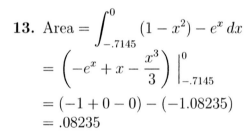

$$= 2\int_0^1 x\,dx = x^2\Big|_0^1$$
$$= 1 - 0 = 1$$

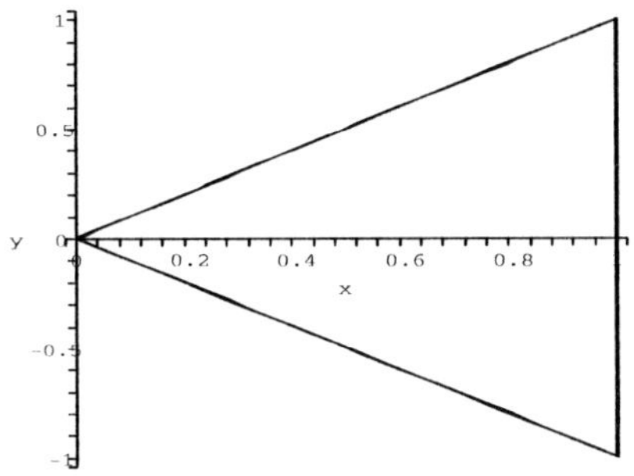

23. $\text{Area} = \int_0^2 [(6-y) - y]dy$
$$= \int_0^2 (6-2y)dy$$
$$= \left(6y - y^2\right)\Big|_0^2$$
$$= (12-4) - (0-0)$$
$$= 8$$

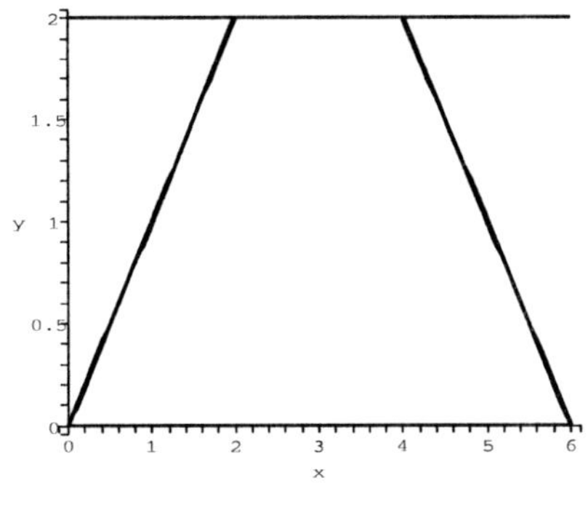

25. $\text{Area} = \int_0^{\ln 2} [e^x - 1]\,dx$
$$+ \int_{\ln 2}^{\ln 4} \left[4e^{-x} - 1\right]\,dx$$
$$= [e^x - x]_0^{\ln 2} + \left[-4e^{-x} - x\right]_{\ln 2}^{\ln 4}$$
$$= [2 - \ln 2 - (1-0)]$$
$$+ \left[-4\left(\frac{1}{4}\right) - \ln 4 - \left(-4\left(\frac{1}{2}\right) - \ln 2\right)\right]$$
$$= 2 - \ln 2 - 1 - 1 - \ln 4 + 2 + \ln 2$$
$$= 2 - \ln 4$$
$$= 2 - 2\ln 2$$

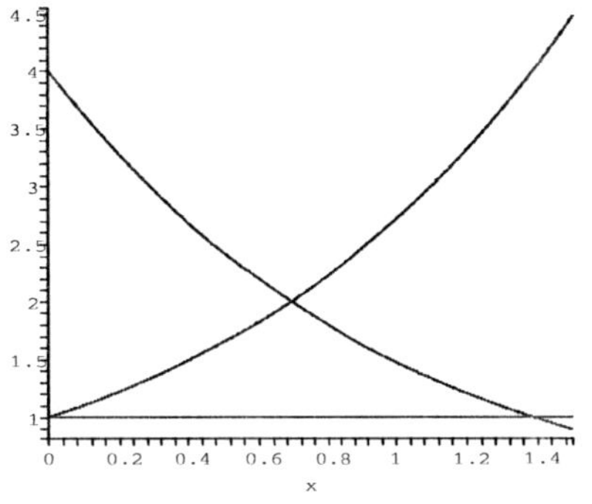

27. $A = \frac{1}{b-a}\int_a^b f(x)\,dx,$
$$= \frac{1}{3-0}\int_0^3 x^2\,dx = \left(\frac{1}{3}\cdot\frac{x^3}{3}\right)\Big|_0^3$$
$$= \frac{27}{9} - 0 = 3$$

Relative to the interval $[0,3]$, the inequality $x^2 < 3$ holds only on the subinterval $[0,\sqrt{3})$. We find

$$\int_0^{\sqrt{3}} \left(3 - x^2\right)dx$$
$$= \left(3x - \frac{x^3}{3}\right)\Big|_0^{\sqrt{3}}$$
$$= (3\sqrt{3} - \sqrt{3}) - (0-0)$$
$$= 2\sqrt{3}, \text{ whereas}$$

$$\int_{\sqrt{3}}^3 \left(x^2 - 3\right)dx$$
$$= \left(\frac{x^3}{3} - 3x\right)\Big|_{\sqrt{3}}^3$$
$$= (9-9) - (\sqrt{3} - 3\sqrt{3})$$
$$= 2\sqrt{3}, \text{ the same.}$$

29. $f(4) = 16.1e^{.07(4)} = 21.3$
$g(4) = 21.3e^{.04(4-4)} = 21.3$

21.3 represents the consumption rate (million barrels per year) at time $t = 4$ (1/1/74).

$$\int_4^{10} \left(16.1e^{.07t} - 21.3e^{.04(t-4)}\right)dt$$

$= \left(230e^{.07t} - 532.5e^{.04(t-4)}\right)\Big|_4^{10}$
$= 14.4$ million barrels saved.

31. For $t \geq 0$,
$b(t) = 2e^{.04t} \geq 2e^{.02t} = d(t)$

$$\int_0^{10} \left(2e^{.04t} - 2e^{.02t}\right) dt$$

$= \left(50e^{.04t} - 100e^{.02t}\right)\Big|_0^{10}$
$= 2.45$ million people.
This number represents births minus deaths, hence population growth over the ten-year interval.

33. $\int_0^{.4} f_c(x) \approx \frac{.4}{3(4)}\{f_c(0) + 4f_c(.1)$
$+2f_c(.2)+4f_c(.3)+f_c(.4)\} = 291.67$

$\int_0^{.4} f_e(x) \approx \frac{.4}{3(4)}\{f_e(0) + 4f_e(.1)$
$+2f_e(.2)+4f_e(.3)+f_e(.4)\} = 102.33$

$$\frac{\int_0^{.4} f_c(x) - \int_0^{.4} f_e(x)}{\int_0^{.4} f_c(x)} \approx \frac{291.67 - 102.33}{291.67}$$
$= .6491\ldots.$

$1 - .6491 = .3508$, so the proportion of energy retained is about 35.08%.

35. $\int_0^{3} f_s(x) \approx \frac{3}{3(4)}\{f_s(0) + 4f_s(.75)$
$+2f_s(1.5)+4f_s(2.25)+f_s(3)\} = 860.$

$\int_0^{3} f_r(x) \approx \frac{3}{3(4)}\{f_r(0) + 4f_r(.75)$
$+2f_r(1.5)+4f_r(2.25)+f_r(3)\} = 800$

$$1 - \left(\frac{860 - 800}{860}\right) = .9302.$$

Energy returned by the tendon is 93.02%.

37. These integrals represent the difference in the distances traveled by the two runners over the time periods in question. If it was a race, the first runner would be ahead by 2 miles after a time of π and the second runner would have caught up to the first runner by the time $t = 2\pi$.

$$\int_0^{\pi} [f(t) - g(t)]\, dt$$
$$= \int_0^{\pi} [10 - (10 - \sin t)]\, dt = 2$$
$$\int_{\pi}^{2\pi} [f(t) - g(t)]\, dt$$
$$= \int_{\pi}^{2\pi} [10 - (10 - \sin t)]\, dt = 0.$$

39. Without formulae or tables, only rough or qualitative estimates are possible.

time	1	2	3	4	5
amount	397	403	401	412	455

$V(3) \approx 374$, $V(4) \approx 374$, $V(5) \approx 404$.

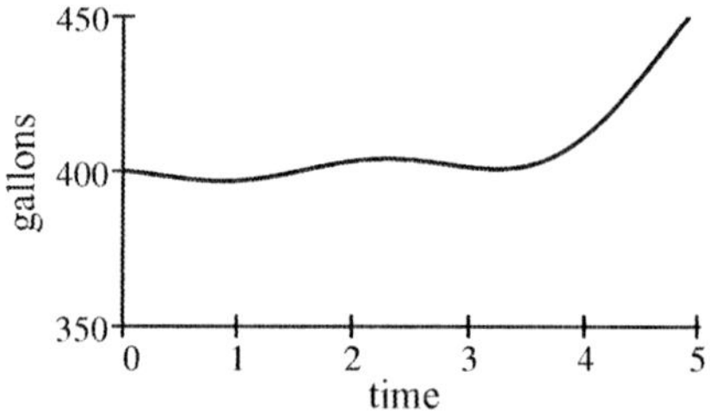

41. In this set-up, p is price and q is quantity. We find that $D(q) = S(q)$ only if $D(q) = S(q)$

$$10 - \frac{q}{40} = 2 + \frac{q}{120} + \frac{q^2}{1200}$$
$12000 - 30q = 2400 + 10q + q^2$
$q^2 + 40q - 9600 = 0$
$(q - 80)(q + 120) = 0$

within the range of the picture only at $q = 80$. Thus $q^* = 80$ and $p^* = D(q^*) = S(q^*) = 8$.

Consumer surplus, as an area, is that part of the picture below the D curve, above $p = p^*$, and to the left of $Q = q^*$.

Numerically in this case the consumer surplus is

$$\int_0^{q^*} [D(q) - p^*]\, dq$$

$$= \int_0^{80} \left(2 - \frac{q}{40}\right) dq$$

$$= 2q - \frac{q^2}{80}\Big|_0^{80} = 160 - 80 = 80$$

The units are dollars (q counting items, p in dollars per item).

43. The curves, meeting as they do at 2 and 5, represent the derivatives C' and R'. The area (a) between the curves over the interval $[0, 2]$ is the loss resulting from the production of the first 2000 items. The area (b) between the curves over the interval $[2, 5]$ is the profit resulting from the production of the next 3000 items. The area (c), as the *sum* of the two previous (call it $(a) + (b)$), is *without meaning*. However, the *difference* $(b) - (a)$ would be the total profit on the first 5000 items, or, if negative, would represent the loss. The area (d) between the curves over the interval $[5, 6]$ represents the loss attributable to the (unprofitable) production of the next thousand items after the first 5000.

45. Let $y_1 = ax^2 + bx + c$, $y_2 = mx + n$, and $u = y_1 - y_2$. If we assume that $a < 0$, then $y_1 > y_2$ on (A, B) and the area between the curves is given by the integral

$$\int_A^B (y_1 - y_2)\, dx$$

$$= \int_A^B u\, dx = ux\Big|_A^B - \int_A^B x du.$$

By assumption, u is zero ($y_1 = y_2$) at both A and B, so the first part of the last expression is zero. We must now show that

$$-\int_A^B x du = -\int_A^B x[2ax+(b-m)]\, dx$$

is the same as

$|a|(B - A)^3/6$

$= |a|(B^3 - 3B^2A + 3BA^2 - A^3)/6.$

But again because $u = 0$ at both A and B, we know that

$aA^2 + bA + c = mA + n$ and

$aB^2 + bB + c = mB + n$.

By subtraction of the first from second, factoring out (and canceling) $B - A$, we learn $a(B + A) = m - b$, so that our target integral is also given by

$$-2a \int_A^B x(x - \frac{A+B}{2})\, dx$$

$$= |a|\{2(B^3 - A^3)/3 \\ - (A + B)(B^2 - A^2)/2\}$$

and the student who cares enough can finish the details.

The case in which $a > 0 (y_2 > y_1)$ is not essentially different.

47. Let the upper parabola be $y = y_1 = qx^2 + v + h$ and let the lower be
$y = y_2 = px^2 + v$. They are to meet at $x = w/2$, so we must have
$qw^2/4 + h = pw^2/4$, hence
$h = (p - q)w^2/4$ or $(q - p)w^2 = -4h$.

Using symmetry, the area between the curves is given by the integral

$$2 \int_0^{w/2} (y_1 - y_2)\, dx$$

$$= 2 \int_0^{w/2} [h + (q - p)x^2]\, dx$$

$= 2[hw/2 + (q - p)w^3/24]$

$= w[h + (q - p)w^2/12]$

$= w[h - 4h/12] = (2/3)wh.$

49. Solve for x in $x - x^2 = L$ we get
$$x = \frac{1 \pm \sqrt{1-4L}}{2}$$
$$A_1 = \int_0^{(1-\sqrt{1-4L})/2} [L - (x - x^2)]\,dx$$
$$= \left(Lx - \frac{x^2}{2} + \frac{x^3}{3}\right)\Big|_0^{(1-\sqrt{1-4L})/2}$$
$$A_2 = \int_{(1-\sqrt{1-4L})/2}^{(1+\sqrt{1-4L})/2} [(x - x^2) - L]\,dx$$
$$= \left(\frac{x^2}{2} - \frac{x^3}{3} - Lx\right)\Big|_{(1-\sqrt{1-4L})/2}^{(1+\sqrt{1-4L})/2}$$
By setting $A_1 = A_2$, we get the final answer
$$L = \frac{16}{3}.$$

5.2 Volume: Slicing, Disks and Washers

1. $V = \int_{-1}^{3} A(x)\,dx = \int_{-1}^{3} (x+2)\,dx$
$$= \left(\frac{x^2}{2} + 2x\right)\Big|_{-1}^{3}$$
$$= \left(\frac{9}{2} + 6\right) - \left(\frac{1}{2} - 2\right)$$
$$= 12$$

3. $V = \pi \int_0^2 (4-x)^2\,dx$
$$= -\frac{\pi}{3}(4-x)^3\Big|_0^2$$
$$= -\frac{\pi}{3}(8 - 64) = \frac{56\pi}{3}$$

5. $V = \int_0^{60} \pi x^2 dy$
$$= \pi \int_0^{60} 60[60 - y]dy$$
$$= 60\pi\ 60y - \frac{y^2}{2}\Big|_0^{60}$$
$$= 60\pi\left[60^2 - \frac{60^2}{2}\right]$$
$$= \frac{60^3\pi}{2} = 108{,}000\pi \text{ ft}^3$$

7. Replacing the 300 by 750 and the 160 by 500 in the previous solution will produce the answer
$$\frac{750^2 500}{3} = 93,750,500 \text{ ft}^3$$
which is one third the area of the base times the height.

9. The key observation in this problem is that by simple proportions, had the steeple continued to a point it would have had height 36, hence 6 extra feet. One can copy the integration method, integrating only to 30, or one can subtract the volume of the missing "point" from the full pyramid. Either way the answer is
$$\frac{3^2 36}{3} - \left(\frac{1}{2}\right)^2 \cdot \frac{6}{3} = \frac{215}{2} \text{ ft}^3$$

11. $V = \pi \int_0^{2\pi} \left(4 + \sin\frac{x}{2}\right)^2 dx$
$$= \pi \int_0^{2\pi} \left(16 + 8\sin\frac{x}{2} + \sin^2\frac{x}{2}\right) dx$$
$$= \pi\left(16x - 16\cos\frac{x}{2} + \frac{1}{2}x - \frac{1}{2}\sin x\right)\Big|_0^{2\pi}$$
$$= 33\pi^2 + 32\pi \text{ in}^3$$

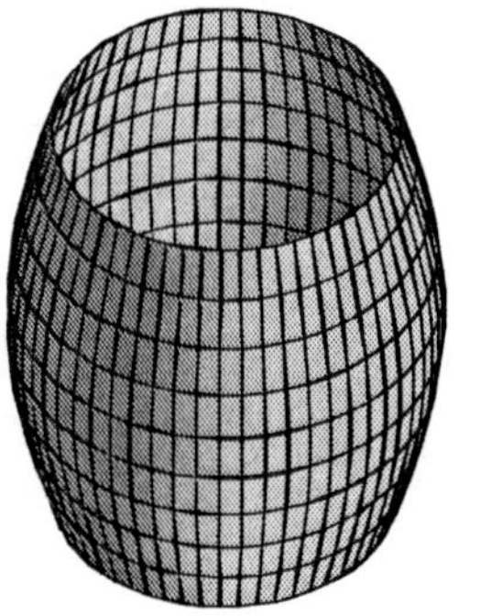

13. $V = \int_0^1 A(x)\,dx$

$\approx \frac{1}{3(10)} [A(0) + 4A(.1) + 2A(.2)$
$+ 4A(.3) + 2A(.4) + 4A(.5)$
$+ 2A(.6) + 4A(.7) + 2A(.8)$
$+ 4A(.9) + A(1.0)]$
$= \frac{7.4}{30} \approx 0.2467\text{cm}^3$

15. $V = \int_0^2 A(x)\,dx$

$\approx \frac{2}{3(4)} [A(0) + 4A(.5) + 2A(1)$
$+4A(1.5) + A(2)]$
$= 2.5 \text{ ft}^3$

17. (a) $V = \pi \int_0^2 (2-x)^2\,dx$

$= -\pi \left(\frac{(2-x)^3}{3} \right) \Big|_0^2$

$= \frac{8\pi}{3}$

(b) $V = \pi \int_0^2 \left[3^2 - \{3 - (2-x)\}^2\right] dx$

$= \pi \int_0^2 \left[9 - \{1+x\}^2\right] dx$

$= \pi\ 9x \Big|_0^2 - \frac{(1+x)^3}{3} \Big|_0^2\Big]$

$= \pi \left[18 - \frac{3^3 - 1^3}{3}\right] = \frac{28\pi}{3}$

19. (a) $V = \pi \int_0^2 (y^2)^2 dy = \pi \int_0^2 y^4 dy$

$= \pi \left(\frac{y^5}{5} \right) \Big|_0^2 = \frac{32\pi}{5}$

(b) $V = \pi \int_0^2 (4)^2 dy$

$- \pi \int_0^2 (4 - y^2)^2 dy$

$= \pi \int_0^2 (-y^4 + 8y^2) dy$

$= \pi \left(-\frac{y^5}{5} + \frac{8y^3}{3} \right) \Big|_0^2$

$= \pi \left[\left(-\frac{32}{5} + \frac{64}{3} \right) - (0 + 0) \right]$

$= \frac{224\pi}{15}$

21. (a) $V = 4\pi e^2 - \pi \int_1^{e^2} (\ln y)^2 dy$

$= 4\pi e^2$
$- \left[y(\ln y)^2 - 2y \ln y + 2y\right] \Big|_1^{e^2}$
$= 4\pi e^2 - (2e^2 - 2)$
$= 2\pi(e^2 + 1).$

(b) $V = \pi \int_0^2 (e^x + 2)^2\,dx$

$- \pi \int_0^2 (2)^2\,dx$

$= \pi \int_0^2 \left(e^{2x} + 4e^x\right) dx$

$= \pi \left(\frac{e^{2x}}{2} + 4e^x \right) \Big|_0^2$

$= \pi \left[\left(\frac{e^4}{2} + 4e^2 \right) - \left(\frac{1}{2} + 4 \right) \right]$

$= \pi \left(\frac{e^4}{2} + 4e^2 - \frac{9}{2} \right)$

23. (a) $V = \pi \int_0^1 \left(\sqrt{\frac{x}{x^2+2}} \right)^2 dx$

$= \frac{\pi}{2} \ln |x^2 + 2| \Big|_0^1$
$= \frac{\pi}{2} \ln \frac{3}{2} \approx 0.637$

(b) $V = \pi \int_0^1 \left[3^2 - \left(3 - \sqrt{\frac{x}{x^2+2}} \right)^2 \right] dx$

$= \pi \int_0^1 \left(6\sqrt{\frac{x}{x^2+2}} - \frac{3x}{x^2+2} \right) dx$

$= \left(6\pi \int_0^1 \sqrt{\frac{x}{x^2+2}}\,dx \right)$

$$-\left(\frac{3\pi}{2}\ln|x^2+2|\Big|_0^1\right)$$
$$\approx 7.4721$$

25. (a) $V = \pi\int_0^3 (3-y)^2 dy$
$$= \pi\int_0^3 (9-6y+y^2)dy$$
$$= \pi\left(9y-3y^2+\frac{y^3}{3}\right)\Big|_0^3$$
$$= 9\pi$$

(b) $V = \pi\int_0^3 (3-x)^2\,dx$
$$= \pi\int_0^3 (9-6x+x^2)\,dx$$
$$= \pi\left(9x-3x^2+\frac{x^3}{3}\right)\Big|_0^3$$
$$= 9\pi$$

(c) $V = \pi\int_0^3 \left[(3)^2-(3-(3-x))^2\right]dx$
$$= \pi\int_0^3 \left[9-x^2\right]dx$$
$$= \pi\left(9x-\frac{x^3}{3}\right)\Big|_0^3$$
$$= 18\pi$$

(d) $V = \pi\int_0^3 \left[((3-x)+3)^2-3^2\right]dx$
$$= \pi\int_0^3 \left[(6-x)^2-9\right]dx$$
$$= \pi\int_0^3 (27-12x+x^2)\,dx$$
$$= \pi\left(27x-6x^2+\frac{x^3}{3}\right)\Big|_0^3$$
$$= 36\pi$$

(e) $V = \pi\int_0^3 [(3)^2-(3-(3-y))^2]dy$
$$= \pi\int_0^3 (9-y^2)dy$$
$$= \pi\left(9y-\frac{y^3}{3}\right)\Big|_0^3$$
$$= 18\pi$$

(f) $V = \pi\int_0^3 [((3-y)+3)^2-(3)^2]dy$
$$= \pi\int_0^3 [(6-y)^2-9]dy$$
$$= \pi\int_0^3 (27-12y+y^2)dy$$
$$= \pi\left(27y-6y^2+\frac{y^3}{3}\right)\Big|_0^3$$
$$= 36\pi$$

27. (a) $V = \int_0^1 \pi(1)^2 dy - \int_0^1 \pi\left(\sqrt{y}\right)^2 dy$
$$= \pi\int_0^1 (1-y)dy$$
$$= \pi\left(y-\frac{y^2}{2}\right)\Big|_0^1$$
$$= \frac{\pi}{2}$$

(b) $V = \int_0^1 \pi\left(x^2\right)^2 dx$
$$= \pi\frac{x^5}{5}\Big|_0^1 = \frac{\pi}{5}$$

(c) $V = \int_0^1 \pi\left(1-\sqrt{y}\right)^2 dy$
$$= \pi\int_0^1 \left(1-2y^{1/2}+y\right)dy$$
$$= \pi\left(y-\frac{4}{3}y^{3/2}+\frac{y^2}{2}\right)\Big|_0^1$$
$$= \frac{\pi}{6}$$

(d) V
$$= \int_0^1 \pi(1)^2\,dx - \int_0^1 \pi\left(1-x^2\right)^2 dx$$
$$= \pi\int_0^1 \left(2x^2-x^4\right)dx$$
$$= \pi\left(\frac{2}{3}x^3-\frac{x^5}{5}\right)\Big|_0^1$$
$$= \frac{7\pi}{15}$$

(e) $V = \int_0^1 \pi(2)^2 dy -$

$$\int \pi (1+\sqrt{y})^2\,dy$$
$$= \pi \int_0^1 (3 - 2y^{1/2} - y)\,dy$$
$$= \pi \left(3y - \frac{4}{3}y^{3/2} - \frac{y^2}{2}\right)\Big|_0^1$$
$$= \frac{7\pi}{6}$$

(f) V
$$= \int_0^1 \pi \left(x^2+1\right)^2\,dx$$
$$- \int_0^1 \pi (1)^2\,dx$$
$$= \pi \int_0^1 \left(x^4 + 2x^2\right)\,dx$$
$$= \pi \left(\frac{x^5}{5} + \frac{2}{3}x^3\right)\Big|_0^1$$
$$= \frac{13\pi}{15}$$

29. $V = \pi \int_0^h \left(\sqrt{\frac{y}{a}}\right)^2\,dy$
$$= \frac{\pi}{a}\int_0^h y\,dy = \frac{\pi h^2}{2a}$$

The volume of a cylinder of height h and radius $\sqrt{h/a}$ is
$$h \cdot \pi(\sqrt{h/a})^2 = \frac{\pi h^2}{a}$$

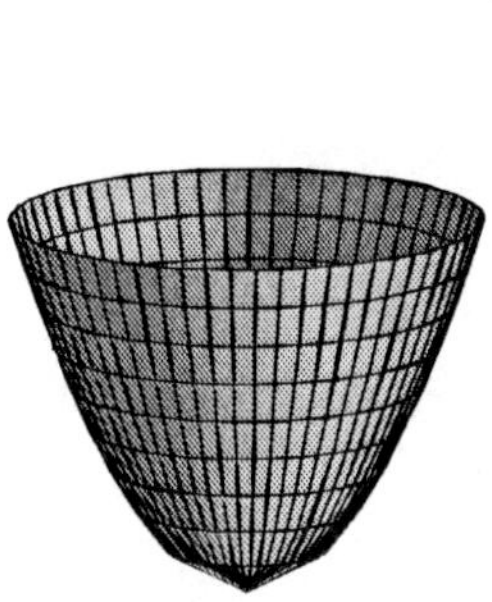

31. We can choose either x or y to be our integration variable,
$$V = \pi \int_{-1}^1 1\,dx = \pi x\Big|_{-1}^1 = 2\pi$$

33. The line connecting the two points $(0,1)$ and $(1,-1)$ has equation $y = -2x+1$, or $x = \frac{1-y}{2}$

$$V = \int_{-1}^1 \pi \left(\frac{1-y}{2}\right)^2\,dy$$
$$= \pi \left(\frac{y}{4} - \frac{y^2}{4} + \frac{y^3}{12}\right)\Big|_{-1}^1$$
$$= \frac{2\pi}{3}$$

35. $V = \pi \int_{-r}^r \left(\sqrt{r^2-y^2}\right)^2\,dy$
$$= \pi \int_{-r}^r (r^2 - y^2)dy$$
$$= \pi \left(r^2 y - \frac{y^3}{3}\right)\Big|_{-r}^r$$
$$= \frac{4}{3}\pi r^3$$

37. If we compute the two volumes using disks parallel to the base, we have identical cross sections, so the volumes are the same.

39. (a) If each of these line segments is the base of square, then the cross-sectional area is evidently

$$A(x) = 4(1-x^2).$$

The volume would be
$$V_a = \int_{-1}^1 A(x)\,dx$$
$$= 2\int_0^1 A(x)\,dx$$
$$= 8\left(x - \frac{x^3}{3}\right)\Big|_0^1 = \frac{16}{3}$$

(b) These segments I_x cannot be the literal "bases" of circles, because circles "sit" on a single point of tangency. They could however be diameters. Assuming so, the cross sectional area would be "$\pi/2$ times radius-squared" or $\pi(1-x^2)/2$. The resulting volume would be $\pi/8$ times the previous case, or $2\pi/3$.

41. Reasoning as in Exercise 39, the line segment I_x is $[x^2, 2-x^2]$, $(1 \le x \le 1)$. The length of this segment is

$$(2-x^2) - x^2 = 2(1-x^2),$$

hence in case (a)

$$A(x) = 4(1-x^2)^2 = 4(1-2x^2+x^4)$$

The volume would again be

$$V = 2\int_0^1 A(x)\,dx$$

$$= 8\left(x - \frac{2x^3}{3} + \frac{x^5}{5}\right)\Big|_0^1$$

$$= 8\left(1 - \frac{2}{3} + \frac{1}{5}\right) = \frac{64}{15}$$

With the same provisos as in Exercise 39, the answer to (b) would be $\pi/8$ times the (a)-case, or $8\pi/15$.

For (c), the volume would be $\sqrt{3}/4$ times the (a)-case, or $16\sqrt{3}/15$.

43. This time the line segment I_x is $[0, e^{-2x}]$, $(0 \le x \le \ln 5)$. If (a) this is the base of a square, the cross-sectional area is $A(x) = (e^{-2x})^2 = e^{-4x}$. The volume V_a would be the integral

$$\int_0^{\ln 5} A(x)\,dx$$

$$= \int_0^{\ln 5} e^{-4x}\,dx = \frac{-e^{-4x}}{4}\Big|_0^{\ln 5}$$

$$= \frac{1-\left(\frac{1}{5}\right)^4}{4} = \frac{156}{625} = .2496$$

In the (b)-case, the segment I_x is the base of a *semi*circle, so the cross-sectional area would be

$$\left(\frac{1}{2}\right)\pi\left(\frac{e^{-2x}}{2}\right)^2 = \left(\frac{\pi}{8}\right)e^{-4x}$$

The resulting volume V_b would be

$$(\pi/8)V_a = \frac{39\pi}{1250} \approx .09802$$

45. We must estimate $\pi\int_0^3 [f(x)]^2\,dx$.

The given table can be extended to give these respective values for $f(x)2$: $4, 1.44, .81, .16, 1.0, 1.96,\ 2.56$. Simpson's approximation to the integral would be

$$\frac{3}{(3)(6)}\{4 + 4(1.44) + 2(.81)$$
$$+ 4(.16) + 2(1.0)$$
$$+4(1.96) + 2.56\}$$

The sum in the braces is 24.42, and this must be multiplied by $\pi/6$ giving a final answer of 12.786.

47. In this problem, let $x = g(y)$ be the equation of the given curve describing the shape of the container. For each height y, let $V(y)$ be the volume of fluid in the container when the depth is y. Later we will estimate $V(y)$. For now, one knows that $V(y)$ is the integral of $\pi[g(y)]^2$, or by the fundamental theorem of calculus, that

$$\frac{dV}{dy} = \pi[g(y)]^2.$$

In actual practice, y and hence V are functions of t (time). Our primary interest is in y as a function of t, but we will obtain this information indirectly, first finding V as a function of y. It appears that $g(y)$ is about $2y$ for $0 < y < 1$, which leads to $[g(y)]^2 = 4y^2$, $V(y) = 4\pi y^3/3$ (on $0 < y < 1$), and $V(1) = 4\pi/3 = 4.2$. We'll keep the *formula* in mind for later, but for now will use the value at $y = 1$ and the crude trapezoidal estimate

$$V(y+1) = V(y)+\pi[g^2(y)+g^2(y+1)]/2$$

to compile the following table:

y	$g(y)$	$g^2(y)$	$V(y)$
1	2	4	4.2
2	2	9	24.6
3	3	9	52.9
4	3	9	81.2
5	4	16	120.4

The assumption of uniform flow rate amounts to $dV/dt =$ constant, and if we start the clock ($t = 0$) as we begin the flow, we get $V = kt$ for some k. The above table, supplemented by the formula when $y < 1$, can be read to give y (vertical) as a function of V (horizontal). But because $V = kt$, the graph looks exactly the same if the horizontal units are time. In the following picture, we have scaled it on the assumption of a flow rate of 120.4 cubic units per minute, a rate which requires one minute to fill the container. The previous formula $4\pi y^3/3 = V(= kt = (120.4)t)$ (on $0 < y < 1$), becomes $y = (3.06)t^{1/3}$ for very small t, and accounts for the (barely discernible) vertical tangent at $t = 0$.

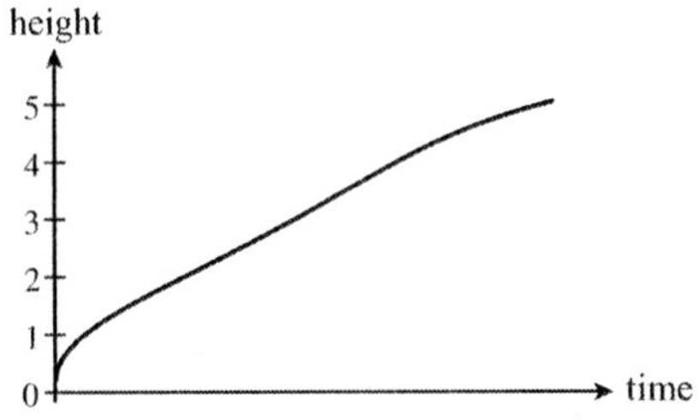

5.3 Volumes by Cylindrical Shells

1. Radius of a shell: $r = 2 - x$
Height of a shell: $h = x^2$

$$V = \int_{-1}^{1} 2\pi(2-x)x^2\,dx = 2\pi\left(\frac{2x^3}{3} - \frac{x^4}{4}\right)\Big|_{-1}^{1} = \frac{8\pi}{3}$$

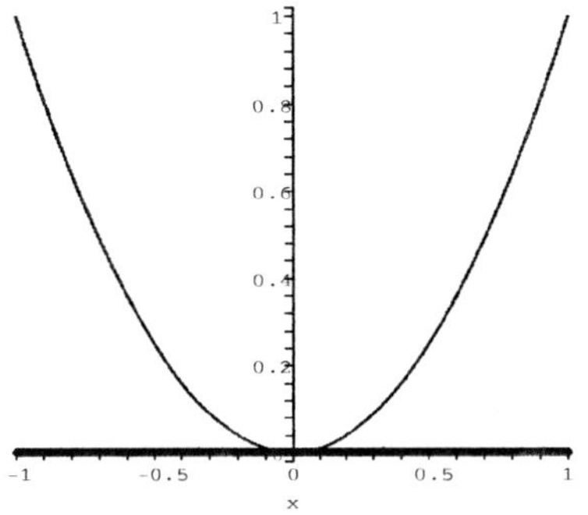

3. Radius of a shell: $r = x$
Height of a shell: $h = 2x$

$$V = \int_0^1 2\pi x(2x)\,dx = \frac{4\pi}{3}x^3\Big|_0^1 = \frac{4\pi}{3}$$

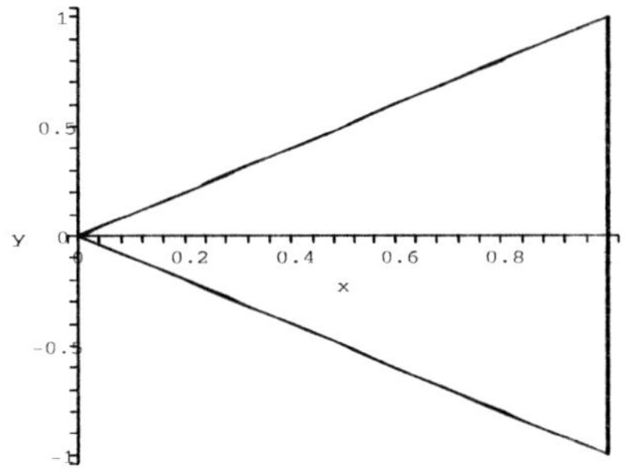

5. Radius of a shell: $r = 3 - y$
Height of a shell: $h = 2y$

$$V = \int_0^2 2\pi(3-y)(2y)\,dy = \frac{40\pi}{3}$$

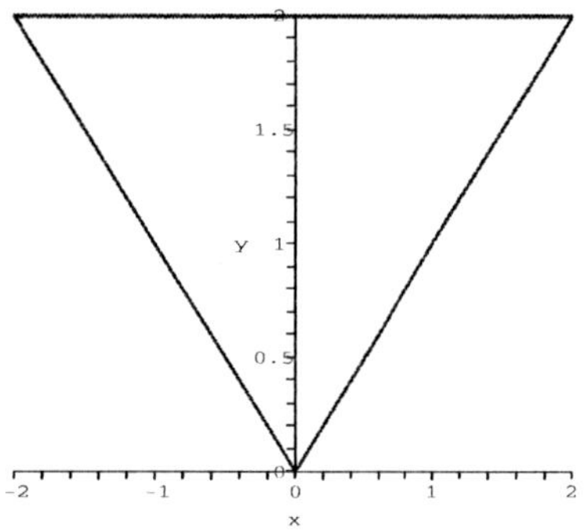

7. Radius of a shell: $r = y$
Height of a shell: $h = \sqrt{1-(y-1)^2}$

$$V = \int_0^2 2\pi y\sqrt{1-(y-1)^2}\,dy$$
$$= \int_{-1}^1 2\pi(u+1)\sqrt{1-u^2}\,du$$
$$= \int_{-1}^1 2\pi u\sqrt{1-u^2}\,du$$
$$+ \int_{-1}^1 2\pi\sqrt{1-u^2}\,du$$
$$= 0 + \frac{1}{2}\pi\cdot 1^1\cdot 2\pi$$
$$= \pi^2$$

9. $V = \int_{-1}^1 2\pi(x+2)\left((2-x^2)-x^2\right)dx$
$$= 2\pi\int_{-1}^1 \left(4+2x-4x^2-2x^3\right)dx$$
$$= 2\pi\left(4x+x^2-\frac{4x^3}{3}-\frac{x^4}{2}\right)\Big|_{-1}^1$$
$$= \frac{32\pi}{3}$$

11. $V = \int_{-2}^2 2\pi(2+y)y^2\,dy$

$$= 2\pi\left(\frac{2}{3}y^3+\frac{y^4}{4}\right)\Big|_{-2}^2$$
$$= \frac{64\pi}{3}$$

13. $V = \int_{-1}^2 2\pi(2-x)(x-(x^2-2))\,dx$
$$= 2\pi\int_{-1}^2 \left(4-3x^2+x^3\right)dx$$
$$= 2\pi\left(4x-x^3+\frac{x^4}{4}\right)\Big|_{-1}^2$$
$$= \frac{27\pi}{2}$$

15. $V = \int_{-2}^4 2\pi(5-y)[9-(y-1)^2]\,dy$
$$= \int_{-2}^4 (y^3-7y^2+2y+40)\,dy$$
$$= \left(\frac{y^4}{4}-\frac{7y^3}{3}+y^2+40y\right)\Big|_{-2}^4$$
$$= 288\pi$$

17. (a) $V = \int_2^4 2\pi(y)(y-(4-y))\,dy$
$$= 2\pi\int_2^4 \left(2y^2-4y\right)dy$$
$$= 2\pi\left(\frac{2y^3}{3}-2y^2\right)\Big|_2^4$$
$$= \frac{80\pi}{3}$$

(b) $V = \int_0^2 2\pi(x)(4-(4-x))\,dx$
$$= \int_2^4 2\pi(x)(4-x)\,dx$$
$$= 2\pi\left(\frac{x^3}{3}\right)\Big|_0^2 + 2\pi\left(2x^2-\frac{x^3}{3}\right)\Big|_2^4$$
$$= 2\pi\left(\frac{8}{3}+\frac{16}{3}\right) = 16\pi$$

(c) $V = \int_2^4 \pi(4-(4-y))^2\,dy$
$$= \int_2^4 \pi(4-y)^2\,dy$$

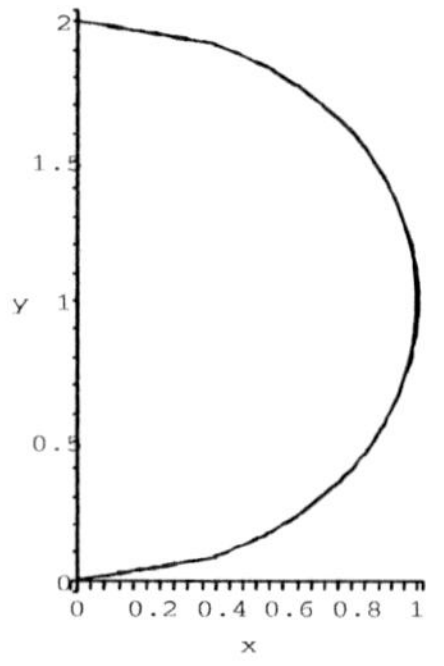

$$= \pi \int_2^4 y^2 dy - \pi \int_2^4 (16 - 8y + y^2) dy$$
$$= \pi \int_2^4 (-16 + 8y) dy$$
$$= \pi \left(-16y + 4y^2\right)\Big|_2^4 = 16\pi$$

(d) $V = \int_2^4 2\pi(4-y)\left(y - (4-y)\right) dy$
$$= 2\pi \int_2^4 \left(-2y^2 + 12y - 16\right) dy$$
$$= 2\pi \left(-\frac{2y^3}{3} + 6y^2 - 16y\right)\Bigg|_2^4$$
$$= \frac{16\pi}{3}$$

19. (a) Method of shells.

$V = \int_{-2}^3 2\pi(3 - x)[x - (x^2 - 6)]\, dx$
$$= \int_{-2}^3 2\pi(-x^3 - 4x^2 - 3x + 18)\, dx$$
$$= \frac{625\pi}{6}$$

(b) Method of washers.

$V = \int_{-2}^3 \pi[(x^2 - 6)^2 - x^2]\, dx$
$$= \int_{-2}^3 \pi(x^4 - 13x^2 + 36)\, dx$$
$$= \frac{250\pi}{3}$$

(c) Method of shells.

$V = \int_{-2}^3 2\pi(3 + x)[x - (x^2 - 6)]\, dx$
$$= \int_{-2}^3 2\pi(x^3 - 2x^2 + 9x + 18)\, dx$$
$$= \frac{875\pi}{6}$$

(d) Method of washers.

$V = \int_{-2}^3 \pi[(6 + x)^2 - (x^2)^2]\, dx$
$$= \int_{-2}^3 \pi(-x^4 + x^2 + 12x + 36)\, dx$$
$$= \frac{500\pi}{3}$$

21. (a) $V \approx 2\pi \int_{-0.89}^{0.89} (2 - x) \cdot (\cos x - x^4)\, dx \approx 16.72$

(b) $V \approx \pi \int_{-0.89}^{0.89} [(2 - x^4)^2 - (2 - \cos x)^2]\, dx \approx 12.64$

(c) $V \approx \pi \int_{-0.89}^{0.89} [(\cos x)^2 - (x^4)^2]\, dx \approx 4.09$

(d) $V \approx 2 \cdot 2\pi \int_0^{0.89} x(\cos x - x^4)\, dx \approx 2.99$

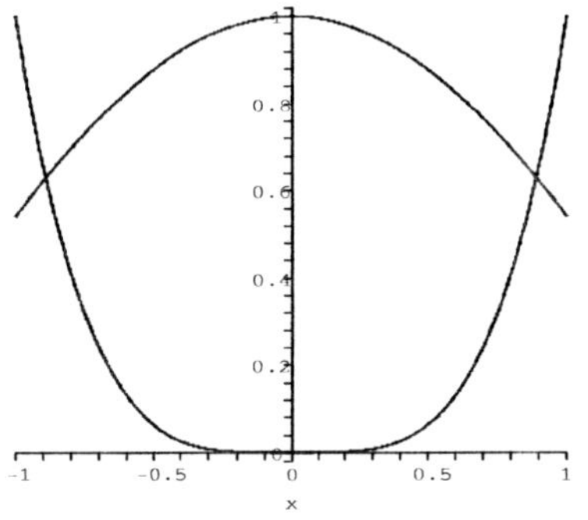

23. (a) $V = \int_0^1 \pi(2 - x)^2\, dx - \int_0^1 \pi\left(x^2\right)^2 dx$
$$= \pi \int_0^1 (x^2 - 4x + 4)\, dx - \pi \int_0^1 x^4\, dx$$
$$= \pi \int_0^1 (-x^4 + x^2 - 4x + 4)\, dx$$

$$= \pi \left(\frac{x^5}{5} + \frac{x^3}{3} - 2x^2 + 4x \right) \Big|_0^1$$
$$= \frac{32\pi}{15}$$

(b) $$V = \int_0^1 2\pi x \left(2 - x - x^2\right)\, dx$$
$$= 2\pi \int_0^1 \left(2x - x^2 - x^3\right)\, dx$$
$$= 2\pi \left(x^2 - \frac{x^3}{3} - \frac{x^4}{4} \right) \Big|_0^1$$
$$= \frac{5\pi}{6}$$

(c) $$V = \int_0^1 2\pi(1-x)(2-x-x^2)\, dx$$
$$= 2\pi \int_0^1 \left(x^3 - 3x + 2\right)\, dx$$
$$= 2\pi \left(\frac{x^4}{4} - \frac{3x^2}{2} + 2x \right) \Big|_0^1$$
$$= \frac{3\pi}{2}$$

(d) $$V = \int_0^1 \pi \left(2 - 2x^2\right)^2\, dx$$
$$= \int_0^1 \pi \left(2 - (2 - x)\right)^2\, dx$$
$$= \pi \int_0^1 (x^4 - 4x^2 + 4)\, dx$$
$$- \pi \int_0^1 x^2\, dx$$
$$= \pi \int_0^1 (x^4 - 5x^2 + 4)\, dx$$
$$= \pi \left(\frac{x^5}{5} - \frac{5x^3}{3} + 4x \right) \Big|_0^1$$
$$= \frac{38\pi}{15}$$

25. (a) $$V = 2\pi \int_0^1 y(2 - y - y^2)\, dy$$
$$= 2\pi \int_0^1 (-y^3 - y^2 + 2y)\, dy$$
$$= 2\pi \left(-\frac{y^4}{4} - \frac{y^3}{3} + y^2 \right) \Big|_0^1$$
$$= \frac{5\pi}{6}$$

(b) $$V = 2\pi \int_0^1 [(2-y)^2 - (y^2)^2]\, dy$$
$$= 2\pi \int_0^1 (-y^4 + y^2 - 4y + 4)\, dy$$
$$= 2\pi \left(-\frac{y^5}{5} + \frac{y^3}{3} - 2y^2 + 4y \right) \Big|_0^1$$
$$= \frac{64\pi}{15}$$

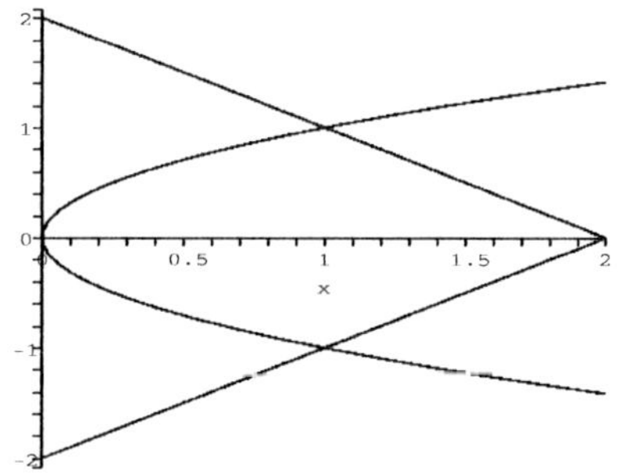

27. Axis of revolution: x-axis
Region bounded by:
$y = 2x - x^2, y = 0$

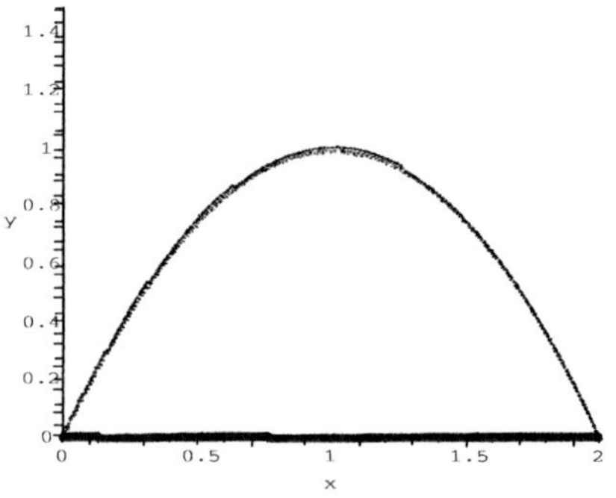

29. Axis of revolution: y-axis
Region bounded by: $x = \sqrt{y}, x = y$

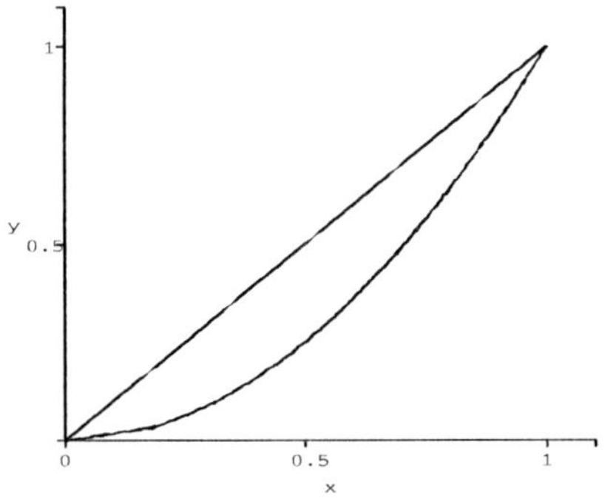

31. Axis of revolution: y-axis
Region bounded by: $y = x, y = x^2$

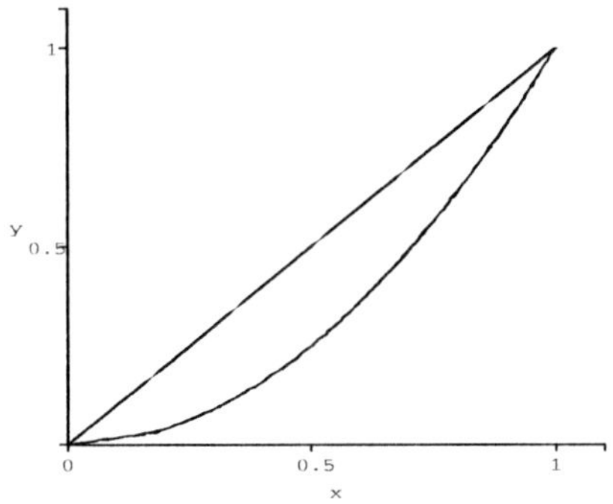

33. If the r-interval $[0, R]$ is partitioned by points r_i, the circular band

$$\{r_i^2 \le x^2 + y^2 \le r_{i+1}^2\}$$

has approximate area $c(r_i)\Delta r_i$ (length times thickness). The limit of the sum of these areas is

$$A = \lim \sum_{i=1}^{n} c(r_i)\Delta r_i = \int_0^R c(r)dr$$

Because we know that $c(r) = 2\pi r$, we can evaluate the integral, getting

$$2\pi \frac{r^2}{2}\bigg|_0^R = \pi R^2.$$

35. The volume that we are looking for is twice the volume of a shell with radius x and height $\sqrt{1-x^2}$. In other words, The bead is mathematically the solid formed up from revolving the region bounded by $y = \sqrt{1-x^2}, x = 1/2$ and the x-axis around the y-axis. Therefore

$$V = 2 \cdot \int_{1/2}^{1} 2\pi x\sqrt{1-x^2}\, dx.$$

Let $u = 1 - x^2, du = -2xdx$, and

$$V = 4\pi \int_{1/2}^{1} x\sqrt{1-x^2}\, dx$$
$$= -\frac{1}{2} 4\pi \int_{3/4}^{0} u^{1/2}\, du$$
$$= 2\pi \cdot \frac{2}{3} u^{3/2}\bigg|_0^{3/4}$$
$$= \frac{\sqrt{3}\pi}{2} \text{ cm}^3.$$

37. $V = \int_0^1 x(1-x^2)\, dx$

$$= \int_0^1 (x - x^3)\, dx$$
$$= \left(\frac{x^2}{2} - \frac{x^4}{4}\right)\bigg|_0^1 = \frac{1}{4}$$

$$V_1 = \int_c^1 x(1-x^2)\, dx$$
$$= \left(\frac{x^2}{2} - \frac{x^4}{4}\right)\bigg|_c^1 = \frac{1}{4} - \frac{c^2}{2} + \frac{c^4}{4}$$

We want
$$V - V_1 = \frac{1}{10}V$$

Then
$$\frac{c^2}{2} - \frac{c^4}{4} = \frac{1}{40}$$
$c \approx 0.2265$

5.4 Arc Length and Surface Area

1. For $n = 2$, the evaluations points are $0, 0.5, 1$

$$s \approx s_1 + s_2$$
$$= \sqrt{(0-0.5)^2 + [f(0) - f(0.5)]^2}$$
$$+ \sqrt{(1-0.5)^2 + [f(1) - f(0.5)]^2}$$

$= \sqrt{0.5^2 + 0.5^4} + \sqrt{0.5^2 + 0.75^2}$

≈ 1.460

For $n = 4$, the evaluations points: $0, 0, 25, 0.5, 0.75, 1$

$s \approx \sum_{i=1}^{4} s_i \approx 1.474$

3. For $n = 2$, the evaluations points are $0, \pi/2, \pi$

$s \approx s_1 + s_2$

$= \sqrt{(\pi/2)^2 + [\cos(\pi/2) - \cos 0]^2}$

$+ \sqrt{(\pi/2)^2 + [\cos \pi - \cos(\pi/2)]^2}$

$= \sqrt{\pi^2 + 4} \approx 3.724$

For $n = 4$, the evaluations points: $0, \pi/4, \pi/2, 3\pi/4, \pi$

$s \approx \sum_{i=1}^{4} s_i \approx 3.790$

5. This is a straight line segment from $(0, 1)$ to $(2, 5)$. As such, its length is

$s = \sqrt{(5-1)^2 + (2-0)^2}$

$= \sqrt{20} = 2\sqrt{5}$

7. $y'(x) = 6x^{1/2}$, the arc length integrand is $\sqrt{1 + (y')^2} = \sqrt{1 + 36x}$.

Let $u = 1 + 36x$ then

$s = \int_1^2 \sqrt{1 + 36x}\, dx$

$= \int_{37}^{73} \sqrt{u} \left(\frac{du}{36}\right)$

$= \frac{2}{3(36)} u^{3/2} \Big|_{37}^{73}$

$= \frac{1}{54}(73\sqrt{73} - 37\sqrt{37})$

≈ 7.3824

9. $y'(x) = \frac{2x}{4} - \frac{1}{2x} = \frac{1}{2}\left(x - \frac{1}{x}\right)$

$1 + (y')^2 = 1 + \frac{1}{4}\left(x^2 - 2 + \frac{1}{x^2}\right)$

$= \frac{1}{4}\left(x^2 + 2 + \frac{1}{x^2}\right)$

$= \left[\frac{1}{2}\left(x + \frac{1}{x}\right)\right]^2$

$s = \frac{1}{2}\int_1^2 \left(x + \frac{1}{x}\right) dx$

$= \frac{1}{2}\left(\frac{x^2}{2} + \ln x\right)\Big|_1^2$

$= \frac{1}{2}\left(\frac{3}{2} + \ln 2\right)$

≈ 1.0965

11. $y'(x) = \frac{x^3}{2} - \frac{1}{2x^3} = \frac{1}{2}\left(x^3 - \frac{1}{x^3}\right)$

$1 + (y')^2 = 1 + \frac{1}{4}\left(x^6 - 2 + \frac{1}{x^6}\right)$

$= \frac{1}{4}\left(x^6 + 2 + \frac{1}{x^6}\right)$

$= \left[\frac{1}{2}(x^3 + \frac{1}{x^3})\right]^2$

$s = \int_{-2}^{-1} \sqrt{1 + (y')^2}\, dx$

$= -\frac{-1}{-2}\int_{-2}^{-1} \left(x^3 + \frac{1}{x^3}\right) dx$

$= \frac{1}{2}\left(-\frac{x^4}{4}\Big|_{-2}^{-1} + \frac{1}{2x^2}\Big|_{-2}^{-1}\right)$

$= \frac{1}{2}\left(\frac{15}{4} + \frac{3}{8}\right) = \frac{33}{16}$

13. $y'(x) = \frac{x^{1/2}}{2} - \frac{x^{-1/2}}{2}$

$= \frac{1}{2}\left(\sqrt{x} - \frac{1}{\sqrt{x}}\right),$

$1 + (y')^2 = 1 + \frac{1}{4}\left(x - 2 + \frac{1}{x}\right)$

$= \frac{1}{4}\left(x + 2 + \frac{1}{x}\right)$

$= \left[\frac{1}{2}\left(\sqrt{x} + \frac{1}{\sqrt{x}}\right)\right]^2.$

$$s = \int_1^4 \sqrt{1+(y')^2}$$
$$= \frac{1}{2}\int_1^4 \left(\sqrt{x} + \frac{1}{\sqrt{x}}\right) dx$$
$$= \frac{x^{3/2}}{3}\bigg|_1^4 + \sqrt{x}\Big|_1^4$$
$$= \frac{7}{3} + 1 = \frac{10}{3}$$

15. $s = \int_{-1}^1 \sqrt{1+(3x^2)^2}\, dx$
$$= \int_{-1}^1 \sqrt{1+9x^4}\, dx \approx 3.0957$$

17. $s = \int_0^2 \sqrt{1+(2-2x)^2}\, dx \approx 2.9578$

19. $s = \int_0^\pi \sqrt{1+(-\sin x)^2}\, dx$
$$= \int_0^\pi \sqrt{1+\sin^2 x}\, dx \approx 3.8201$$

21. $s = \int_0^\pi \sqrt{1+(x\sin x)^2}\, dx = 4.6984$

23. $s = \int_{-10}^{10} \sqrt{1+\left(\frac{1}{2}\left(e^{x/10} - e^{-x/10}\right)\right)^2}\, dx$
$$= \int_{-10}^{10} \frac{1}{2}\left(e^{x/10} + e^{-x/10}\right) dx$$
$$= \left(5e^{x/10} - 5e^{-x/10}\right)\Big|_{10}^{10}$$
$$= 10e - 10e^{-1}$$
$$\approx 23.50402387 \text{ ft.}$$

25. In Example 4.4,

$y(x) = 5(e^{x/10} + e^{-x/10})$
$y(0) = 5(e^0 + e^0) = 10$
$y(-10) = y(10)$
$= 5(e^1 + e^{-1}) = 15.43$

sag $= 15.43 - 10 = 5.43$ ft

A lower estimate for the arc length given the sag would be

$2\sqrt{(10)^2 + (sag)^2}$
$= 2\sqrt{100 + 29.4849} \approx 22.76$

This looks good against the calculated arc length of 23.504.

27. $y = 0$ when $x = 0$ and when $x = 60$, so the punt traveled 60 yards horizontally.

$$y'(x) = 4 - \frac{2}{15}x = \frac{2}{15}(30 - x).$$

This is zero only when $x = 30$, at which point the punt was $(30)^2/15 = 60$ yards high.

$$s = \int_0^{60} \sqrt{1 + \left(4 - \frac{2}{15}x\right)^2}\, dx$$
≈ 139.4 yards

$$v = \frac{s}{4 \text{ sec}} = \frac{139.4 \text{ yards}}{4 \text{ sec}} \cdot \frac{3 \text{ feet}}{1 \text{ yard}}$$
$= 104.55$ ft/s

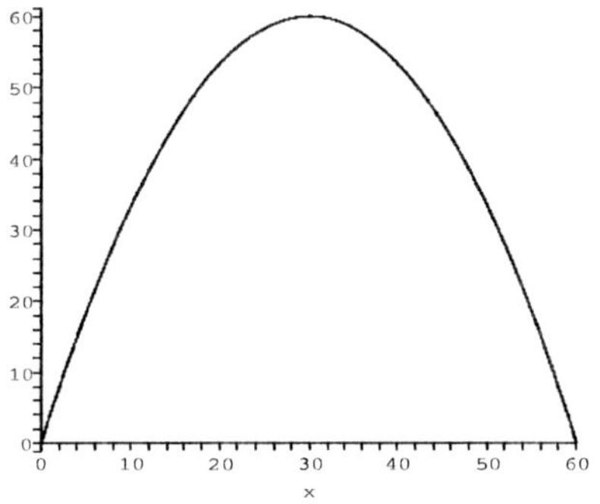

29. All details are provided in the text.

31. $\frac{d}{dx}\sqrt{2}\int_0^x \sqrt{1 - \frac{\sin^2 u}{3}} du$
$$= \frac{1}{2}\sqrt{2} \cdot \sqrt{4 - 2\sin^2 x}$$
$$= \sqrt{1 + \cos^2 x}$$

33. The antiderivatives returned by some CAS still include an integral, indicating that the CAS can't find an antiderivative in closed form (see Exercise 34). In this case numerical integration is the method of choice.

35. $s = 2\pi \int_0^1 y\,ds$
$= 2\pi \int_0^1 x^2\sqrt{1+(2x)^2}\,dx$
≈ 3.8097

37. $s = 2\pi \int_2^2 y\,ds$
$= 2\pi \int_0^2 (2x - x^2)\sqrt{1+(2-2x)^2}\,dx$
≈ 10.9654

39. $s = 2\pi \int_0^1 y\,ds$
$= 2\pi \int_0^1 e^x\sqrt{1+e^{2x}}\,dx \approx 22.9430$

41. $s = 2\pi \int_0^{\pi/2} y\,ds$
$= 2\pi \int_0^{\pi/2} \cos x\sqrt{1+\sin^2 x}\,dx$
≈ 7.2117

43. $L_1 = \int_{-\pi/6}^{\pi/6} \sqrt{1+\cos^2 x}\,dx \approx 1.44829$

L_2
$= \sqrt{\left(\sin\frac{\pi}{6} - \sin\left(-\frac{\pi}{6}\right)\right)^2 + \left(\frac{2\pi}{6}\right)^2}$
≈ 1.44797

Hence

$$\frac{L_2}{L_1} = \frac{1.44797}{1.44829} \approx .9998$$

45. $L_1 = \int_3^5 \sqrt{1+(e^x)^2}\,dx \approx 128.3491$

$L_2 = \sqrt{2^2 + (e^5 - e^3)^2} \approx 128.3432$

Hence

$$\frac{L_2}{L_1} \approx 0.9999$$

47. $s_1 = \int_0^1 \sqrt{1+(6x^5)^2}\,dx$
$= \int_0^1 \sqrt{1+36x^{10}}\,dx \approx 1.672$

$s_2 = \int_0^1 \sqrt{1+(8x^7)^2}\,dx$
$= \int_0^1 \sqrt{1+64x^{14}}\,dx \approx 1.720$

$s_3 = \int_0^1 \sqrt{1+(10x^9)^2}\,dx$
$= \int_0^1 \sqrt{1+100x^{18}}\,dx \approx 1.75$

As $n \to \infty$, the length approaches 2, since one can see that the graph of $y = x^n$ on $[0,1]$ approaches a path consisting of the horizontal line segment from $(0,0)$ to $(1,0)$ followed by the vertical line segment from $(1,0)$ to $(1,1)$.

49. $y_1 = x^4, y_1' = 4x^3$
$y_2 = x^2, y_2' = 2x$

Since both are increasing for positive x, y_1 is "steeper" (y_2 is "flatter") if and only if $y_1' > y_2'$, i.e.,

$$4x^3 > 2x,\ x^2 > \frac{1}{2},\ x > \sqrt{\frac{1}{2}}$$

51. $S = \int_{-1}^1 2\pi\sqrt{1-y^2}\sqrt{1+\left(\frac{y}{\sqrt{1-y^2}}\right)^2}\,dy$
$= \int_{-1}^1 2\pi\sqrt{1-y^2}\sqrt{\frac{1}{\sqrt{1-y^2}}}\,dy$
$= \int_{-1}^1 2\pi\,dy = 4\pi$

53. $6\pi : 4\pi : (\sqrt{5}+1)\pi = 3 : 2 : \tau$

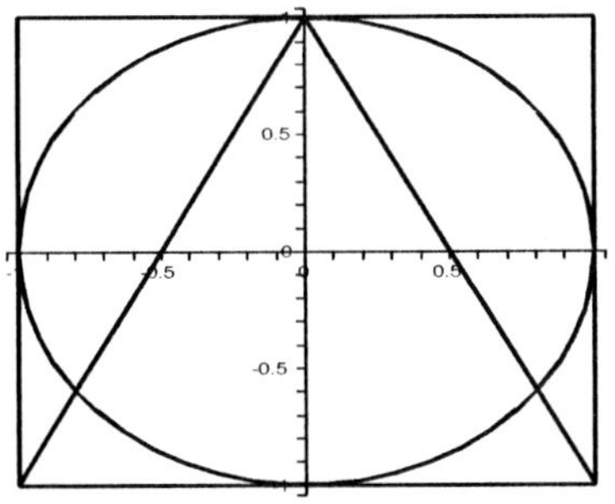

5.5 Projectile Motion

1. $y(0) = 80$, $y(0) = 0$

3. $y(0) = 60$, $y(0) = 10$

5. The initial conditions are
$y(0) = 30$ and $y'(0) = 0$
We want to find $y'(t)$ when $y(t) = 0$.

We start with the equation $y''(t) = -32$. Integrating gives $y'(t) = -32t + c_1$. From the initial velocity, we have $0 = y'(0) = -32(0) + c_1$, and so $y'(t) = -32t$.

Integrating again gives
$y(t) = -16t^2 + c_2$. From the initial position, we have
$30 = y(0) = -16(0) + c_2$ and so
$y(t) = -16t^2 + 30$.

Solving $y(t) = 0$ gives $t = \pm\sqrt{\frac{15}{8}}$. The positive solution is the solution we are interested in. This is the time when the diver hits the water. The diver's velocity is therefore

$$y'\left(\sqrt{\frac{15}{8}}\right) = -32\sqrt{\frac{15}{8}} \approx -43.8 \text{ ft/sec}$$

7. If an object is dropped (time zero, zero initial velocity) from an initial height of y_0, then the impact moment is $t_0 = \sqrt{y_0}/4$ and the impact velocity (ignoring possible negative sign) is

$$v_{\text{impact}} = 32t_0 = 8\sqrt{y_0}.$$

Therefore if the object is dropped from 30 ft, the impact velocity is

$$8\sqrt{30} \approx 43.8178 \text{ feet per second.}$$

If dropped from 120 ft, impact velocity is

$$8\sqrt{120} \approx 87.6356 \text{ feet per second.}$$

From 3000 ft, impact velocity is

$$8\sqrt{3000} \approx 438.178 \text{ feet per second.}$$

From a height of $h\,y_0$, the impact velocity is

$$8\sqrt{hy_0} = 8\sqrt{h}\sqrt{y_0} = \sqrt{h}\,(8\sqrt{y_0}),$$

which is to say that impact velocity increases by a factor of $\sqrt{h}$ when initial height increases by a factor of h.

9. See the solution to Exercise 7. We are told that $t_0 = 4$, hence
$v_{\text{impact}} = (32)4 = 128$ feet per second, and $y_0 = (v_0/8)^2 = 16^2 = 256$ feet.

11. With y the height at time t, we go back to $y'' = -32$,
$v(t) = y' = -32t + v(0)$, and
$y = -16t^2 + tv(0) + y(0)$.

In this case, with launch at 64 feet per second from ground level, we have
$v(0) = 64$ and $y(0) = 0$, hence
$y = -16t^2 + 64t = -16(t-2)^2 + 64$ (completing the square), and
$v(t) = -32(t-2)$.

This is zero at $t = 2$, which is when he peaks. At time 2, $y = 64$. He remains in the air until the next time $y = 0$, which is at $t = 4$. We see that $v(4) = -64$, which is, except for sign, the same as his launch velocity.

13. Reviewing the solution to Exercise 11, the difference is that $v(0)$ is unknown. However, we still see that $y = -16t^2 + tv(0) = -t[16t - v(0)]$ (factoring, rather than completing the square). The second time that $y = 0$ can be seen to occur at time $t_2 = v(0)/16$, at which time
$v(t_2) = -32t_2 + v(0) = v(0)(-2+1)$
$= -v(0)$

Now we see
$v(t) = -32t + v(0) = -32t + 16t_2$
$= -16(2t - t_2)$

The peak was therefore at time $t_2/2$, at which time the height was
$-(t_2/2)[16t2/2 - v(0)]$
$= -(t2/2)[(v(0)/2) - v(0)]$
$= -(v(0)/32)[-v(0)/2] = v(0)^2/64.$

In summary, $y_{\text{max}} = [v(0)/8]^2$ in this problem (and more generally, $y_{\text{max}} = [v(0)/8]^2 + y(0))$.
If $y_{\text{max}} = 20$ inches $= 5/3$ feet, then $v(0)/8 = \sqrt{5/3}$, and
$v(0) = 8\sqrt{5/3} \approx 10.33$ feet per second.

This is considerably less than Michael Jordan's initial velocity of about 17 feet per second, but the difference in velocity is not as dramatic as in height (20 inches to 54 inches).

15. If the initial conditions are
$y(0) = H$ and $y'(0) = 0$

Integrating $y''(t) = -32$ gives
$y'(t) = -32t + c_1$.

The initial condition gives
$y'(t) = -32t + v_0 = -32t$.

Integrating gives
$y(t) = -16t^2 + c_2$.

The initial condition gives
$y(t) = -16t^2 + H$.

The impact occurs when $y(t_0) = 0$ or when $t_0 = \sqrt{y_0}/4 = \sqrt{H}/4$.

Therefore the impact velocity is

$y'(t_0) = -32t_0 = -8\sqrt{H}$

17. The initial conditions are
$s(0) = 0, s'(0) = 0$.

Integrating $s''(t) = -32$ gives
$s'(t) = -32t + c_1$.

The initial condition gives
$s'(t) = -32t$.

Integrating gives
$s(t) = -16t^2 + c_2$.

The initial condition gives
$s(t) = -16t^2$.

Realizing that -32 was given in feet per second^2, and we are using centimeters now, we use
1 foot $= 30.48$ cms
and get

$s(t) = -487.68t^2$ cm

The yardstick is grabbed when $s(t_0) = -d$, that is when

$$t_0 = \frac{\sqrt{d}}{487.68} \approx 0.045\sqrt{d}$$

19. The time t_0 when the lead ball hits the ground satisfies

$$179 = 12800 \ln\left(\cosh\left(\frac{t_0}{20}\right)\right)$$

$$\cosh\left(\frac{t_0}{20}\right) = e^{179/12800}$$

$$t_0 \approx 3.3526$$

At time t_0, the hight of the wood ball is

$$179 - \frac{7225}{8}\ln\left(\cosh\left(\frac{16}{85}t_0\right)\right)$$
$$\approx 179 - 169.0337 = 9.9663 \text{ ft}$$

21. The starting point is

$y'' = -9.8$

$y'(0) = 98\sin(\pi/3) = 49\sqrt{3}.$

We get

$$y(t) = -4.9t^2 + ty'(0)$$
$$= -4.9t(t - [v(0)/4.9])$$
$$= -4.9t(t - 10\sqrt{3})$$

The flight time is $10\sqrt{3}$.

As to the horizontal range, we have $x'(t)$ constant and forever equal to $98\cos(\pi/3) = 49$. Therefore $x(t) = 49t$ and in this case, the horizontal range is $49(10\sqrt{3})$ (meters).

23. This problem modifies Example 5.5 by using a service angle of 6° (where the Example 5.5 used 7°) and no other changes. Here the serve hits the net.

Next we want to find the range for which the serve will be in.

If θ is the angle, then the initial conditions are

$x'(0) = 176\cos\theta$, $x(0) = 0$

$y'(0) = 176\sin\theta$, $y(0) = 10$

Integrating $x''(t) = 0$ and $y''(t) = -32$, then using the initial conditions gives

$x'(t) = 176\cos\theta$

$x(t) = 176(\cos\theta)t$

$y'(t) = -32t + 176\sin\theta$

$y(t) = -16t^2 + 176(\sin\theta)t + 10$

To make sure the serve is in, we see what happens at the net and then when the ball hits the ground. First, the ball passes the net when $x = 39$ or when $39 = 176(\cos\theta)t$. Solving gives

$$t = \frac{39}{176\cos\theta}$$

Plugging this in for the function $y(t)$ gives

$$y\left(\frac{39}{176\cos\theta}\right)$$
$$= -16\left(\frac{39}{176\cos\theta}\right)^2$$
$$+ 176(\sin\theta)\left(\frac{39}{176\cos\theta}\right) + 10$$
$$= -\frac{1521}{1936}\sec^2\theta + 39\tan\theta + 10$$

We want to ensure that this value is greater than 3 so we determine the values of θ that give $y > 3$ (using a graphing calculator or CAS). This restriction means that we must have

$$-0.15752 < \theta < 1.5507$$

Next, we want to determine when the ball hits the ground. This is when

$$0 = y(t) = -16t^2 + 176(\sin\theta)t + 10$$

We solve this equation using the quadratic formula to get

$$t = \frac{-176\sin\theta \pm \sqrt{176^2\sin^2\theta + 640}}{-32}$$

We are interested in the positive solution, so

$$t = \frac{176\sin\theta + \sqrt{176^2\sin^2\theta + 640}}{32}$$

Substituting this in to $x(t) = 176(\cos\theta)t$ gives

$$x = 44\cos\theta\left(22\sin\theta + \sqrt{484\sin^2\theta + 10}\right)$$

We want to determine the values of θ that ensure that $x < 60$. Using a graphing calculator or a CAS gives $\theta < -0.13429$

Putting together our two conditions on θ now gives the possible range of angles for which the serve will be in:

$$-0.15752 < \theta < -0.13429$$

25. Let $(x(t), y(t))$ be the trajectory. In this case
$y(0) = 6, x(0) = 0$
$y'(0) = 0, x'(0) = 130$
$x''(t) \equiv 0, x'(t) \equiv 130$
$x(t) = 130t$

This is 60 at time $t = 6/13$. Meanwhile,
$y''(t) = -32, y'(t) = -32t$
$y(t) = -16t^2 + 6$
$$y\left(\frac{6}{13}\right) = -16\left(\frac{6}{13}\right)^2 + 6 = \frac{438}{169}$$
$$y\left(\frac{6}{13}\right) \approx 2.59 \text{ ft}$$

27. Let $(x(t), y(t))$ be the trajectory. In this case 5° is converted to $\pi/36$ radians.

$y(0) = 5, x(0) = 0$
$y'(0) = 120 \sin\frac{\pi}{36} \approx 10.46$
$x'(0) = 120 \cos\frac{\pi}{36} \approx 119.54$
$x''(0) \equiv 0$
$x'(t) \equiv 119.54$
$x(t) = 119.54t$
This is 120 when
$t = 120/119.54 = 1.00385\ldots$

Meanwhile,
$y''(t) = -32$
$y'(t) = -32t + 10.46$
$y(t) = -16t^2 + 10.46t + 5$
$y(1.00385) = -16(1.00385)^2$
$y(1.00385) + 10.46(1.00385) + 5$
$y(1.00385) \approx -.62$ ft

29. (a) Assuming that the ramp height h is the same as the height of the cars, this problem seems to be asking for the initial speed v_0 required to achieve a horizontal flight distance of 125 feet from a launch angle of 30° above the horizontal. We may assume $x(0) = 0$, $y(0) = h$, and we find

$y'(0) = v_0 \sin\frac{\pi}{6} = \frac{v_0}{2}$
$x'(0) = v_0 \cos\frac{\pi}{6} = \frac{\sqrt{3}}{2}v_0$
$y''(t) \equiv -32,\ x''(t) \equiv 0$
$y'(t) = -32t + \frac{v_0}{2},\ x'(t) = \frac{\sqrt{3}}{2}v_0$
$y(t) = -16t^2 + \frac{v_0}{2}t + h,$
$x(t) = \frac{\sqrt{3}}{2}v_0 t.$

$x(t)$ will be 125 if $t = 250/\left(\sqrt{3}\nu_0\right)$ at which time we require that y be h. Therefore
$$-16\left(\frac{250}{\sqrt{3}v_0}\right)^2 + \frac{v_0}{2}\left(\frac{250}{\sqrt{3}v_0}\right) = 0$$

$$v_0 = \sqrt{\frac{8000}{\sqrt{3}}} \approx 68 \text{ ft/s}$$

(b) With an angle of $45° = \pi/4$, the equations become
$y'(0) = v_0 \sin\frac{\pi}{4} = \frac{v_0}{\sqrt{2}}$
$x'(0) = v_0 \cos\frac{\pi}{4} = \frac{v_0}{\sqrt{2}}$
$y''(t) = -32, \quad x''(t) = 0$
$y'(t) = -32t + \frac{v_0}{\sqrt{2}}, \quad x'(t) = \frac{v_0}{\sqrt{2}}$
$y(t) = -16t^2 + \frac{v_0 t}{\sqrt{2}} + h,$
$x(t) = \frac{v_0 t}{\sqrt{2}}$
where h is the height of the ramp.

We now solve $x(t) = 125$ which gives
$$t_0 = t = \frac{125\sqrt{2}}{v_0}$$
At this distance, we want the car to be at a height h to clear the cars. This gives the equation $y(t_0) = h$, or

$$-16\left(\frac{125\sqrt{2}}{v_0}\right)^2$$
$$+\frac{125v_0\sqrt{2}}{v_0\sqrt{2}}+h=h$$
Solving for v_0 gives

$$v_0 = 20\sqrt{10} \approx 63.24 \text{ ft/s}.$$

31. (a) In this case with $\theta_0 = 0$ and $\omega = 1$

$$x''(t) = -25\sin(4t)$$
$$x'(0) = x(0) = 0$$
$$x'(t) = \frac{25}{4}\cos 4t - \frac{25}{4}$$
$$x(t) = \frac{25}{16}\sin 4t - \frac{25}{4}t$$

(b) With $\theta_0 = \frac{\pi}{2}$ and $\omega = 1$

$$x''(t) = -25\sin\left(4t+\frac{\pi}{2}\right)$$
$$x'(0) = x(0) = 0$$
$$x'(t) = \frac{25}{4}\cos\left(4t+\frac{\pi}{2}\right)$$
$$x(t) = \frac{25}{16}\sin\left(4t+\frac{\pi}{2}\right) - \frac{25}{16}$$

33. From Exercise 5, time of impact is

$$t = \frac{\sqrt{30}}{4} \text{ seconds.}$$

$2\frac{1}{2}$ somersaults corresponds to 5π radians of revolution. Therefore the average angular velocity is

$$\frac{5\pi}{\sqrt{30}/4} = \frac{20\pi}{\sqrt{30}} \approx 11.47 \text{ rad/sec}$$

35. Let $(x(t), y(t))$ be the trajectory of the center of the basketball. We are assuming that $y(0) = 6$, $x(0) = 0$, the angle of launch θ of the shot is $52°$ ($\theta = \frac{13\pi}{45}$ in radians) and the initial speed is 25 feet per second. Therefore

$$y'(0) = 25\sin\frac{13\pi}{45} \approx 19.70$$
$$x'(0) = 25\cos\frac{13\pi}{45} \approx 15.39$$
$$y''(t) \equiv -32, x''(t) \equiv 0$$
$$y'(t) = -32t + 19.70, x'(t) \equiv 15.39$$
$$y(t) = -16t^2 + 19.70t + 6,$$
$$x(t) = 15.39t.$$

x will be 15 when t is about $15/15.39 = .9746\ldots$, at which time y will be about

$$-16(.9746\ldots)^2 + 19.70(.9746\ldots) + 6 \approx 10$$

In other words, the center of the ball is at position $(15, 10)$ and the shot is good.

More generally, with unknown θ, the number 19.70 is replaced by $25\sin\theta$, while the number 15.39 is replaced by $25\cos\theta$. y will be exactly 10 if

$$-16t^2 + 25t\sin\theta + 6 = 10$$
$$t = \frac{25\sin\theta + \sqrt{625\sin^2\theta - 256}}{32}$$
$$x = 25t\cos\theta.$$

As a function of θ, this last expression is too complicated to use calculus (easily) to maximize and minimize it on the θ-interval $(48°, 57°)$, but quick spreadsheet calculations give these values:

(Observe that x is not a monotonic function of θ in this range. It takes its maximum when θ is between 52.4 and 52.5 degrees. The evidence is overwhelming that all the shots will be good.)

θ	t	x
degrees	seconds	feet
48.0	0.8757	14.6484
49.0	0.9021	14.7958
50.0	0.9274	14.9024
51.0	0.9516	14.9710
52.0	0.9748	15.0038
52.1	0.9771	15.0051
52.2	0.9793	15.0062
52.3	0.9816	15.0069
52.4	0.9838	15.0073
52.5	0.9861	15.0073
52.6	0.9883	15.0070
52.7	0.9905	15.0064
52.8	0.9928	15.0054
52.9	0.9950	15.0042
53.0	0.9972	15.0026
54.0	1.0187	14.9690
55.0	1.0394	14.9044
56.0	1.0594	14.8100
57.0	1.0787	14.6869

37. $85^\circ = \dfrac{17}{36}\pi$ radiance.

$$x'(0) = 100 \cdot \cos\left(\frac{17}{36}\pi\right) \approx 8.72$$

$$y'(0) = 100 \cdot \sin\left(\frac{17}{36}\pi\right) \approx 99.62$$

$x''(0) = -20$
$y''(0) = 0$

$y(t) = 99.62t$
$x(t) = -10t^2 + 8.72t$
$y(t_0) = 90$ when $t_0 = 0.903$
$x(t_0) = x(0.903) \approx -0.29$

The ball just barely gets into the goal.

39. Let $(x(t), y(t))$ be the trajectory of the ship. Some of our data is in feet, so we will take $g = -32$ in this problem. We have
$y''(t) - 32$
$y'(t) = -32t + y'(0)$
$y(t) = -16t^2 + y'(0)t + y(0)$

$x'(t) \equiv c$
$x(t) = ct + x(0)$

Solving for t, we have

$$\frac{1}{c}(x - x(0)) = t.$$

Substituting this expression for t in $y(t)$, we have

$$y - y(0) = -16\left[\frac{1}{c}(x - x(0))\right]^2 + y'(0)\left[\frac{1}{c}(x - x(0))\right]$$

Hence the path is a parabola.

Turning to the question of the duration of weightlessness, we can assume $x(0) = 0$, and we know that $y'(t) = 0$ when $y - y(0) = 2500$. For this unknown time t_1 (the moment when y' is zero), we have $0 = -32t_1 + y'(0)$. Therefore $t_1 = y'(0)/32$, and

$$2500 = y(t_1) - y(0) = -16\left[\frac{y'(0)}{32}\right]^2 + y'(0)\left[\frac{y'(0)}{32}\right] = \frac{y'(0)^2}{64},$$

hence $y'(0)2 = 64(2500)$,
$y'(0) = 8(50) = 400$, and
$t_1 = 400/32 = 25/2$.

We now know that

$$y - y(0) = -16t^2 + 400t$$

for all t. The second time (t_2) that $y(t) = y(0)$ (after time zero) occurs when

$$t = 400/16 = 25 \text{ seconds.}$$

This is the duration of the weightless experience. Note that $t_2 = 2t_1$. The plane must pull out of the dive soon after this time.

41. Let $y(t)$ be the height of the first ball at time t, and let v_{0y} be the initial velocity. We can assume $y(0) = 0$.

As usual, we have

$y'' = -32,\ y' = -32t + v_{0y}$,
$y = -16t2 + tv_{0y}$.

The second return to height zero is at time $t = 16/v_{0y}$. If this is to be $5/2$, then $v_{0y} = 40$. But the maximum occurs at time

$$v_{0y}/32 = 5/4,,$$

at which time the height $(y(5/4))$ is

$$-16(25/16) + 40(5/4) = 25 \text{ feet}.$$

For eleven balls, the difference is that the second return to zero is to be at time $11/4$, hence $v_{0y} = 44$, and the maximum height is 30.25.

43. The student must first study the solution to Exercise 51. Here we have the additional x-component of the motion, which as in so many problems is $x(t) = tv_{0x}$. With initial *speed* of v_0, and initial angle α from the *vertical,* we have $v_{0y} = v_0 \cos\alpha$ and $v_{0x} = v_0 \sin\alpha$. The horizontal distance at elapsed time $v_{0y}/16$ (time of return to initial height) is by formula $x(v_{0y}/16) = (v_{0y}/16)v_{0x}$ which defines ω. As in Exercise 41, the maximum height occurs at time $v_{0y}/32$, and at this time the height h is

$-16(v_{0y}/32)^2 + v_{0y}(v_{0y}/32) = v_{0y}^2/64$
$= (v0y/64)(16\omega/v_{0x})$
$= (\omega/4)(\cos\alpha/\sin\alpha) = \omega/(4\tan\alpha)$.

Thus $\omega = 4h\tan\alpha$.

45. We must use the result
$\Delta\alpha \approx \dfrac{\Delta\omega}{4h}$ from Exercise 44.

With $h = 25$ from Exercise 51 (10 balls) and $\omega = 1$, we get
$\Delta\alpha$ about $1/100 = .01$ radians
or about $.6°$.

47. The initial conditions are
$x'(0) = 220\cos(9.3°) \approx 217.1,\ x(0) = 0$
$y'(0) = 220\sin(9.3°) \approx 35.55,\ y(0) = 0$

Integrating $x''(t) = 0$ and $y''(t) = -32$ and using the initial conditions gives
$x'(t) = 217.1$
$x(t) = 217.1t$
$y'(t) = -32t + 35.55$
$y(t) = -16t + 35.55t$

Solving for $y(t) = 0$ gives
$t = 0$ and $t = 2.222$. Plugging this into $x(t)$ gives
$x(2.222) = 482.4$ ft

49. With trajectory (x, y), and assuming $x(0) = 0$ and $y(0) = 0$, we have by now seen many times the conclusion

$y = -gt^2 + tv\sin\theta$.

The return to ground level occurs at time
$t = 2v\sin\theta/g$, at which time the horizontal range is
$x = tv\cos\theta = v^2\sin(2\theta)/g$.

With $v = 60$ ft per second and $\theta = 25°$, and on earth with $g = 32$, this is about 86 feet, a short chip shot. On the moon with $g = 5.2$, it is about 530.34 ft.

51. Here we have initial conditions
$y'(0) = 41.07$ ft/sec, $y(0) = 0$

If we start with the equation $y''(t) = -g$, then using the initial conditions gives
$y'(t) = -gt + 41.07$
$y(t) = -\dfrac{g}{2}t^2 + 41.07t$

The maximum height occurs with $y'(t) = 0$ or $t = \frac{41.07}{g}$. Plugging this in to $y(t)$ gives

$$H = y\left(\frac{41.07}{g}\right)$$
$$= -\frac{g}{2}\left(\frac{41.07}{g}\right)^2 + 41.07\left(\frac{41.07}{g}\right)$$
$$= \frac{41.07^2}{2g}$$

So, on Earth ($g = 32$) gives a maximum height of

$$H \approx 26.36 \text{ feet}$$

53. Let $(x(t), y(t))$ be the trajectory of the paint ball, and let $z(t)$ be the height of the target at time t.

We do assume that $y(0) = z(0)$ (target opposite shooter at time of shot) and $y'(0) = 0$ (aiming directly at the target, hence using an initially horizontal trajectory), and as a result $y - z$ has second derivative 0, and initial value 0.

However, this only tells us that

$$y - z = [y'(0) - z'(0)]t = -z'(0)t$$

and if the target is already in motion ($z'(0)$ not zero), the shot may miss at 20 feet or any distance.

If on the other hand, the target is stationary at the moment of the shot, then the shot hits at20 feet or any other distance.

55. (a) The speed at the bottom is given by $\frac{1}{2}mv^2 = mgH, v = \sqrt{2gH}$

(b) Use the result from (a) $v = \sqrt{2gH} = \sqrt{2 \cdot 16g} = 4\sqrt{2g}$
$= 4\sqrt{2 \cdot 32} = 32$ ft/s

(c) At half way down,
$\frac{1}{2}mv^2 + mh8 = mh16,$
$v = \sqrt{2 \cdot (16 - 8)g} = 4\sqrt{g}$
$= 4\sqrt{32} \approx 22.63$ ft/s

(d) At half way down, the slope of the line tangent to $y = x^2$ is $2 \cdot \sqrt{8} = 4\sqrt{2}$

Hence we know that

$$\frac{v_y}{v_x} = 4\sqrt{2}$$

At the same time,

$$(v_y)^2 + (v_x)^2 = (4\sqrt{g})^2$$
$$v_x^2 = \frac{16g}{33}$$
$$v_x = 4\sqrt{\frac{g}{33}} \approx 3.939 \text{ ft/s}$$
$$v_y = 16\sqrt{\frac{2g}{33}} \approx 22.282 \text{ ft/s}$$

5.6 Applications of Integration to Physics and Engineering

1. We first determine the value of the spring constant k. We convert to feet so that our units of work is in foot-pounds.

$$5 = F(1/3) = \frac{k}{3} \text{ and so } k = 15.$$

$$W = \int_0^6 F(x)\,dx$$
$$= \int_0^{1/2} 15x\,dx = \frac{15}{8} \text{ foot-pounds.}$$

3. The force is constant (250 pounds) and the distance is 20/12 feet, so the work is

$$W = Fd = (250)(20/12)$$
$$= 1250/3 \text{ foot-pounds.}$$

5. If x is between 0 and 30,000 feet, then the weight of the rocket at altitude x is $10000 - \frac{1}{15}x$.

Therefore the work is

$$\int_0^{30,000} \left(10,000 - \frac{x}{15}\right) dx$$
$$= \left(10,000x - \frac{x^2}{30}\right)\Big|_0^{30,000}$$
$$= 270,000,000 \text{ ft-lb}$$

7. $W = \int_0^1 800x(10x)\,dx$

$$= \left(400x^2 - \frac{800}{3}x^3\right)\Big|_0^1$$
$$= \frac{400}{3} \text{ mile-lb}$$
$$= 704,000 \text{ ft-lb}$$

9. $W = \int_0^{100} 62.4\pi$

$$\cdot (100x - x^2)(200 + x)\,dx$$
$$= 62.4\pi \int_0^{100} \left(20,000x - 100x^2 - x^3\right) dx$$
$$= 8,168,140,899 \text{ ft-lb}$$

11. The difference between this Exercise and the Example 6.3 is that here the integral runs from 10 to 20 rather than from 0 to 20. In order to compile the number, however, it is easier to evaluate the integral from 0 to 10, getting

$$62.4\pi \left[400\frac{10^2}{2} - 40\frac{10^3}{3} + \frac{10^4}{4}\right]$$
$$\approx 1.797 \cdot 10^6$$

This number is then subtracted from the Example 6.3 solution $2.61 \cdot 10^6$, leaving a final answer of about $8.1 \cdot 10^5$ foot-pounds.

13. $W = \int_0^{10} ax\,dx = \frac{100a}{2}$

$$W_1 = \int_0^c ax\,dx = \frac{ac^2}{2}$$
$$W_1 = \frac{W}{2} \text{ gives } \frac{ac^2}{2} = \frac{1}{2}\frac{100a}{2}$$
$$c = \sqrt{50} \approx 7.1 \text{ feet}$$

The answer is greater than 5 feet because the deeper the laborer digs, the more distance it is required for him to lift the dirt out of the hole.

15. We estimate the integral using Simpson's Rule:

$$J = \int_0^{.0008} F(t)dt$$
$$\approx \frac{.0008}{3(8)}[0 + 4(1000) + 2(2100)$$
$$+ 4(4000) + 2(5000) + 4(5200)$$
$$+ 2(2500) + 4(1000) + 0]$$
$$\approx 2.133$$

$$2.13 = J = m\Delta v = .01\Delta v$$
$$\Delta v = 213$$

The velocity after impact is therefore $213 - 100 = 113$ ft/sec.

17. We compute the impulse using Simpson's rule.

$$J \approx \frac{.6}{3(6)}[0 + 4(8000) + 2(16,000)$$
$$+ 4(24,000) + 2(15,000) + 4(9000)$$
$$+ 0]$$
$$\approx 7533.3$$

$$7533.3 = J = m\Delta v = 200\Delta v$$
$$\Delta v = 37.7 \text{ ft/sec}$$

Since the velocity after the crash is zero, this number is the estimated original velocity.

19. $F'(t)$ is zero at $t = 3$, and the maximum thrust is $F(3) = 30/e \approx 11.0364$

It is implicit in the drawing that the thrust is zero after time 6 . Therefore the impulse is

$$\int_0^6 10te^{-t/3}dt$$
$$= 90 - 270e^{-2} \approx 53.55.$$

21. $m = \int_0^6 \left(\frac{x}{6} + 2\right) dx = 15$

$$M = \int_0^6 x\left(\frac{x}{6} + 2\right) dx = 48$$

Therefore,

$$\bar{x} = \frac{M}{m} = \frac{48}{15} = \frac{16}{5} = 3.2$$

So the center of mass is to the right of $x = 3$.

23. $m = \int_{-3}^{27} \left(\frac{1}{46} + \frac{x+3}{690}\right)^2 dx$

$$= \frac{690}{3}\left(\frac{1}{46} + \frac{x+3}{690}\right)^3\Bigg|_{-3}^{27}$$

$\approx .0614$ slugs

≈ 31.5 oz

25. $M = \int_{-3}^{27} x\left(\frac{1}{46} + \frac{x+3}{690}\right)^2 dx$

≈ 1.0208

$$\bar{x} = \frac{M}{m} = \frac{1.0208}{.0614} \approx 16.6 \text{ in.}$$

This is 3 inches less than the bat of Example 6.5, a reflection of the translation three inches to the left on the number line.

27. $m = \int_0^{30} .00468\left(\frac{3}{16} + \frac{x}{60}\right) dx$

$\approx .0614$ slugs

$$M = \int_0^{30} .00468x\left(\frac{3}{16} + \frac{x}{60}\right) dx$$

≈ 1.0969

weight $= m(32)(16) = 31.4$ oz

$$\bar{x} = \frac{M}{m} = \frac{1.0969}{.0614} \approx 17.8 \text{ in.}$$

29. Area of the base is $\frac{1}{2}(3+1) = 2.$

Area of the body is $1 \times 4 = 4$.

Area of the tip is $\frac{1}{2}(1 \times 1) = \frac{1}{2}.$

Base:

$$m = \int_0^1 \rho(3 - 2x)\, dx = \frac{5}{12} \approx .4167.$$

Body:

$$m = \int_1^5 \rho\, dx = 12\rho$$
$$\bar{x} = \frac{M}{m} = 3$$

Tip:

$$m = \int_5^6 \rho(6 - x)\, dx \approx 2.67\rho$$
$$\bar{x} = \frac{M}{m} = \frac{16}{3} \approx 5.33$$

31. The x-coordinate of the centroid is the same as the center of mass from $x = 0$ to $x = 4$ with density $\rho(x) = \frac{3}{2}x$, hence

$$\bar{x} = \frac{M}{m} = \frac{\int_0^4 3/2 \cdot x^2\, dx}{\int_0^4 3/2 \cdot x\, dx} = \frac{8}{3}$$

The y-coordinate of the centroid is the same as the center of mass from $y = 0$ to $y = 6$ with density $\rho(y) =$

$6 - \frac{2}{3}y$, hence

$$\bar{y} = \frac{M}{m} = \frac{\int_0^6 2/3 \cdot \left(6y - \frac{2}{3}y^2\right)\,dy}{\int_0^6 2/3 \cdot \left(6 - \frac{2}{3}y\right)\,dy} = 2$$

So the center of the given triangle is the point $(8/3, 2)$.

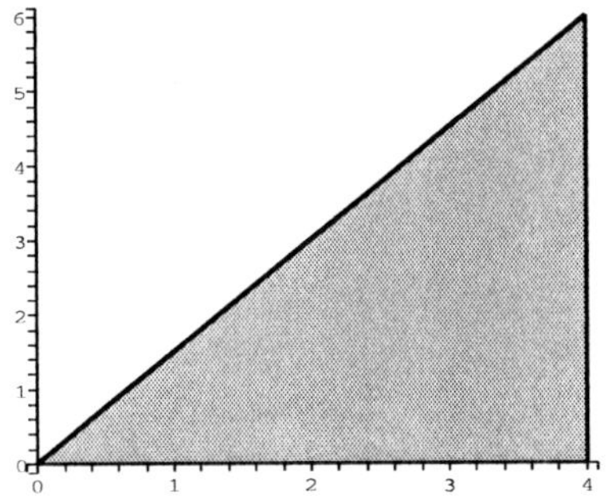

33. This time the x-coordinate of the centroid is obviously $x = 0$, so the question remains to find the y-coordinate.

This is the same as finding the center of mass from $y = 0$ to $y = 4$ with density $\rho(y) = \sqrt{4-y}$, hence

$$\bar{y} = \frac{M}{m} = \frac{\int_0^4 y\sqrt{4-y}\,dy}{\int_0^4 \sqrt{4-y}\,dy}$$

$$= \frac{-\int_4^0 (4u^{1/2} - u^{3/2})\,du}{-\int_4^0 u^{1/2}\,du}$$

$$= \frac{(8/3 \cdot u^{3/2} - 2/5 \cdot u^{5/2})\Big|_0^4}{2/3 \cdot u^{3/2}\Big|_0^4}$$

$$= \frac{8}{5}$$

So the centroid is the pint $(0, 8/5)$.

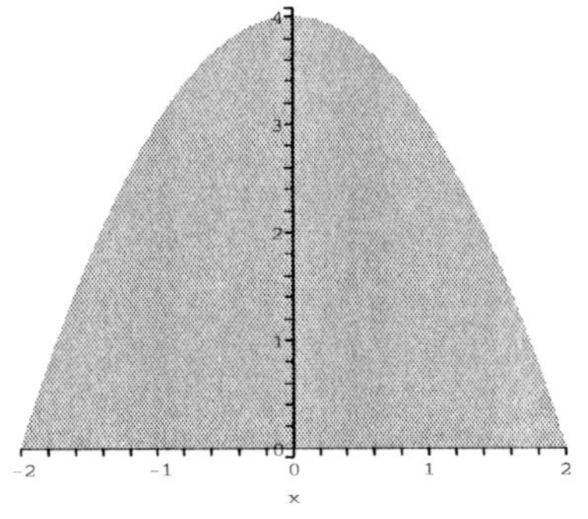

35. With x the depth, the horizontal width is a linear function of x, given by $x + 40$. Hence

$$F = \int_0^{60} 62.4x(x+40)\,dx$$

$$= 62.4\left(\frac{x^3}{3} + 20x^2\right)\Big|_0^{60}$$

$$= 8{,}985{,}600 \text{ lb}$$

37. Let x be the vertical deviation above the center of the window, the horizontal width of the window is given by $2\sqrt{25-x^2}$, depth of water $40 + x$, and hydrostatic force

$$62.4\int_{-5}^{5} (x+40)2\sqrt{25-x^2}\,dx$$

$$= 62.4\int_{-5}^{5} 2x\sqrt{25-x^2}\,dx$$

$$+ 62.4(40)\int_{-5}^{5} 2\sqrt{25-x^2}\,dx$$

$$\approx 196{,}035 \text{ pounds.}$$

39. Assuming that the center of the circular window descends to 1000 feet, then by the previous principle, after converting the three inch radius to 1/4 feet, we get $F = 12{,}252$ pounds. An alternate calculation in which x is the deviation downward from the top edge of the window, would be

$$F = \int_0^{0.5} 62.4(999.75 + x)$$

$$\cdot\, 2\sqrt{(0.25)^2 - (0.25 - x)^2}\, dx$$
$$= \int_0^{0.5} 124.8(999.75+x)\sqrt{0.5x - x^2}\, dx$$
$$\approx 12,252 \text{ lb}$$

41. With $h(t)$ the height of the vaulter at time t (time measured from the peak of his vault)

$h(0) = 20, h'(0) = 0, h''(0) = -32$
$h'(t) = -32t, h(t) = -16t^2 + 20$

$$h(t) = 0 \text{ for } t = \sqrt{\frac{20}{16}} = \frac{\sqrt{5}}{2}$$

$$v = h'\left(\frac{\sqrt{5}}{2}\right) = -16\sqrt{5}$$
$$\approx -35.7 \text{ ft/s}$$

$$\frac{1}{2}mv^2 = \frac{1}{2}\left(\frac{200}{32}\right)(16^2 \cdot 5)$$
$$= 4000 \text{ ft-lb}$$

43. (100 tons)(20 miles/hr)
$$= \frac{(100 \cdot 2000 \text{ lbs})(20 \cdot 5280 \text{ ft})}{3600 \text{ sec}}$$
$$\approx 5{,}866{,}667 \text{ ft-lb/s}$$
$$= \frac{5{,}866{,}667}{550} \text{ hp}$$
$$\approx 10,667 \text{ hp}$$

45. The bat in Exercise 23 models the bat of Example 6.5 choked up 3 in.

From Example 6.5:
$$f(x) = \left(\frac{1}{46} + \frac{x}{690}\right)^2;$$
$$\int_{-3}^{27} f(x) \cdot x^2\, dx \approx 27.22.$$

From Exercise 23:
$$f(x) = \left(\frac{1}{46} + \frac{x+3}{690}\right)^2;$$
$$\int_{-3}^{27} f(x) \cdot x^2\, dx \approx 20.54.$$

Reduction in moment:
$$\frac{27.22 - 20.54}{27.22} \approx 24.5\%$$

47. $\displaystyle\int_{-a}^{a} 2\rho x^2 b\sqrt{1 - \frac{x^2}{a^2}}\, dx = \frac{1}{4}\rho\pi a^3 b$

49. Using the formula in Exercise 46, we find that the moments are 1323.8 for the wooden racket, 1792.9 for the mid-sized racket, and 2361.0 for the oversized racket. The ratios are
$$\frac{\text{mid}}{\text{wood}} \approx 1.35, \quad \frac{\text{over}}{\text{wood}} \approx 1.78$$

5.7 Probability

1. $f(x) = 4x^3 \geq 0$ for $0 \leq x \leq 1$ and
$$\int_0^1 4x^3\, dx = x^4\Big|_0^1 = 1 - 0 = 1$$

3. $f(x) = x + 2x^3 \geq 0$ for $0 \leq x \leq 1$ and
$$\int_0^1 (x + 2x^3)\, dx = \frac{x^2}{2} + \frac{x^4}{2}\Big|_0^1 = 1$$

5. $f(x) = \dfrac{1}{2}\sin x \geq 0$ over $[0, \pi]$ and
$$\int_0^{\pi} \frac{1}{2}\sin x\, dx = \frac{1}{2} - \cos x\Big|_0^{\pi} = 1.$$

7. We solve for c:
$$1 = \int_0^1 cx^3\, dx = \frac{c}{4}$$
which gives $c = 4$.

9. We solve for c:
$$1 = \int_0^1 ce^{-4x}\, dx = -\frac{c}{4}(e^{-4} - 1)$$
which gives $c = \dfrac{4}{1 - e^{-4}}$.

11. We solve for c:
$$1 = \int_0^2 2ce^{-cx}\, dx = 2 - 2e^{-2c}$$
which gives $c = \dfrac{1}{2}\ln 2$.

13. $P(70 \le x \le 72)$

$$= \int_{70}^{72} \frac{.4}{\sqrt{2\pi}} e^{-.08(x-68)^2}\, dx$$
$$\approx 0.157$$

15. $P(84 \le x \le 120)$

$$= \int_{84}^{120} \frac{.4}{\sqrt{2\pi}} e^{-.08(x-68)^2}\, dx$$
$$\approx 7.76 \times 10^{-11}$$

17. $P\left(0 \le x \le \frac{1}{4}\right) = \int_0^{1/4} 6e^{-6x}\, dx$

$$= -e^{-6x}\Big|_0^{1/4}$$
$$= (-e^{-3/2} + 1) \approx .77687$$

19. $P(1 \le x \le 2) = \int_1^2 6e^{-6x}\, dx$

$$= -e^{-6x}\Big|_1^2$$
$$= (-e^{-12} + e^{-6}) \approx .00247$$

21. $P(0 \le x \le 1) = \int_0^1 4xe^{-2x}\, dx$

$$= 1 - 3e^{-2} \approx .594$$

23. Mean:

$$\int_0^{10} x(4xe^{-2x})\, dx \approx 0.9999995$$

25. (a) Mean:

$$\mu = \int_a^b xf(x)\, dx = \int_0^1 3x^3\, dx$$
$$= \frac{3}{4} = 0.75$$

(b) Median, we must solve for m:

$$\frac{1}{2} = \int_a^m f(x)\, dx$$
$$= \int_0^m 3x^2\, dx = m^3$$

which gives $m = \frac{1}{\sqrt[3]{2}} \approx 0.7937.$

27. (a) Mean:

$$\mu = \int_a^b xf(x)\, dx$$
$$= \int_0^\pi \frac{1}{2} x \sin x\, dx$$
$$= \frac{1}{2}(\sin x - x\cos x)\Big|_0^\pi$$
$$= \frac{\pi}{2}$$

(b) Median, we must solve for m:

$$\frac{1}{2} = \int_a^m f(x)\, dx$$
$$= \int_0^m \sin x\, dx = \frac{1}{2}(1 - \cos m)$$

which gives

$$m = \cos^{-1}(0) = \frac{\pi}{2} \approx 1.57.$$

29. (a) Mean:

$$\mu = \int_a^b xf(x)\, dx$$
$$= \frac{\ln 3}{2} \int_0^3 xe^{-kx}\, dx$$
$$= -\frac{3}{2}\left(\frac{-2 + \ln 3}{\ln 3}\right) \approx 1.2307$$

(b) Median, we must solve for m:

$$\frac{1}{2} = \int_a^m f(x)\, dx$$
$$= \frac{\ln 3}{2} \int_0^m e^{-kx}\, dx$$

which gives

$$m = 3\,\frac{\ln(3) - \ln(2)}{\ln(3)} \approx 1.1072.$$

31. Density $f(x) = ce^{-4x}, [0, b], b > 0$

$$1 = \int_0^b ce^{-4x}\, dx$$
$$= -\frac{c}{4}e^{-4x}\Big|_0^b = -\frac{c}{4}\left(e^{-4b} - 1\right)$$
$$c = \frac{4}{1 - e^{-4b}}$$

As $b \to \infty$, $c \to 4$

33. Density $f(x) = ce^{-6x}, [0, b], b > 0$

$$1 = \int_0^b ce^{-6x}\,dx$$
$$= \frac{-c}{6}e^{-6x}\Big|_0^b = -\frac{c}{6}\left(e^{-6b} - 1\right)$$

$$c = \frac{6}{1 - e^{-6b}}$$

As $b \to \infty$, $c \to 6$

$$\mu = \int_0^b xce^{-6x}\,dx$$
$$= \frac{ce^{-6c}}{36}(-6x - 1)\Big|_0^b$$
$$= \frac{ce^{-6b}}{36}(-6b - 1) + \frac{c}{36}$$

As $b \to \infty$, $\mu \to \dfrac{1}{6}$

35. (a) $P(h \le 3)$
$$= P(0) + P(1) + P(2) + P(3)$$
$$= \frac{1}{256} + \frac{8}{256} + \frac{28}{256} + \frac{56}{256}$$
$$= \frac{93}{256}$$

(b) $P(h > 4)$
$$= P(5) + P(6) + P(7) + P(8)$$
$$= \frac{56}{256} + \frac{28}{256} + \frac{8}{256} + \frac{1}{256}$$
$$= \frac{93}{256}$$

(c) $P(0)+P(8) = \dfrac{1}{256} + \dfrac{1}{256} = \dfrac{1}{128}$

(d) $P(1) + P(3) + P(5) + P(7)$
$$= \frac{8}{256} + \frac{56}{256} + \frac{56}{256} + \frac{8}{256}$$
$$= \frac{1}{2}$$

37. (a) $P(4/3) + P(4/2) + P(4/1) + P(4/0) = 0.1659 + 0.2073 + 0.2073 + 0.1296 = 0.7101$

(b) $P(0/4) + P(1/4) + P(2/4) + P(3/4) = 0.0256 + 0.0615 + 0.0922 + 0.1106 = 0.2899$

(c) $P(0/4) + P(4/0) = 0.0256 + 0.1296 = 0.1552$

(d) $P(2/4) + P(3/4) + P(4/3) + P(4/2) = 0.0922 + 0.1106 + 0.1659 + 0.2073 = 0.576$

39. About c:

$$1 = \int_1^{100} cp^{-2}dp = -cp^{-1}\Big|_1^{100}$$
$$= -c\left(\frac{1}{100} - 1\right).$$

therefore, $c = \dfrac{100}{99}$

The requested probability is

$$\int_{60}^{70} cp^{-2}\,dp = -cp^{-1}\Big|_{60}^{70}$$
$$= \frac{100}{99}\left(\frac{1}{70}\quad\frac{1}{60}\right) \approx .0024$$

41. $f(x) = \dfrac{.4}{\sqrt{2\pi}}e^{-.08(x-68)^2}$

$$f'(x) = \frac{-.064}{\sqrt{2\pi}}(x - 68)e^{-.08(x-68)^2}$$

$$f''(x) = \frac{-.064}{\sqrt{2\pi}}e^{-.08(x-68)^2}$$
$$\cdot \left(1 - .16(x - 68)^2\right)$$

The second derivative is zero when

$x-68 = \pm 1/\sqrt{0.16} = \pm 1/0.4 = \pm 5/2$

Thus the standard deviation is $\dfrac{5}{2}$.

43. $f'(p) = mp^{m-1}(1 - p)^{n-m}$
$$- (n - m)p^m(1 - p)^{n-m-1}$$

$f'(p) = 0$ when $p = \dfrac{m}{n}$ and

$$f'(p)\begin{cases} < 0 & \text{if } p < m/n \\ > 0 & \text{if } p > m/n \end{cases}$$

Hence $f(p)$ is maximized when $p = \dfrac{m}{n}$.

In common senses, in order for an event to happen m times in n tries, the probability of the event itself should be about m/n.

45. Suppose the statement is not true. Then there must be a game before which the player's winning percentage is smaller than 75% and after which the player's winning percentage is greater than 75%. Then there are integers a and b (note that $a \geq m, b \geq n$ and $a - b = m - n$), such that

$$\frac{a}{b} < \frac{3}{4} \text{ and } \frac{a+1}{b+1} > \frac{3}{4}.$$

Then

$$4a < 3b, \text{ and } 4a + 4 > 3b + 3$$
$$3b + 4 > 4a + 4 > 3b + 3.$$

But there is no integer between the two numbers $3b + 4$ and $3b + 3$, and thus such situation will never happen. Thus there must be a game after which the player's winning percentage is exactly 75%.

47. The probability of a $2k$-goal game ending in a $k - k$ tie is

$$f(2k) = \frac{(2k)\cdots(k+1)}{(k)\cdots(1)} p^k (1-p)^k$$

$f(2k) < f(2k - 2)$ for general k.

$$\frac{f(2k)}{f(2k-2)} = 2\frac{2k-1}{k} p(1-p)$$

Here $\dfrac{2k-1}{k} = 2 - \dfrac{1}{k} < 2.$

On the other hand,

$$\left(p - \frac{1}{2}\right)^2 \geq 0, p^2 - p + \frac{1}{4} \geq 0$$
$$p - p^2 \leq \frac{1}{4}, p(1-p) \leq \frac{1}{4}$$

Now we get

$$\frac{f(2k)}{f(2k-2)} = 2\frac{2k-1}{k} p(1-p)$$
$$< 2 \cdot 2 \cdot \frac{1}{4} = 1$$

So $f(2k) < f(2k-2)$. In other words, the probability of a tie is decreasing as the number of goals increases.

49. To find the maximum, we take the derivative and set it equal to zero:

$$f'(x) = -2ax(bx-1)(bx+1)e^{-b^2x^2} = 0.$$

This gives critical numbers

$$x = 0, \pm\frac{1}{b}.$$

Since this will be a pdf for the interval $[0, 4m]$, we only have to check that there is a maximum at $\frac{1}{b}$. An easy check shows that $f'(x) > 0$ on the interval $\left[0, \frac{1}{b}\right]$ and $f'(x) < 0$ for $x > \frac{1}{b}$. Therefore there is a maximum at $x = m = \frac{1}{b}$ (the most common speed).

To find a in terms of m, we want the total probability equal to 1. Since $m = \frac{1}{b}$, we also make the substitution $b = \frac{1}{m}$.

$$1 = \int_0^{4m} ax^2 e^{-x^2/m^2}\, dx$$

Solving for a gives

$$a = \left(\int_0^{4m} x^2 e^{-x^2/m^2}\, dx\right)^{-1}$$

Note: this integral is not expressible in terms of elementary functions, so we will leave it like this. Using a CAS, one can find that

$$a \approx 2.2568m^{-3}$$

Ch. 5 Review Exercises

1. $\text{Area} = \int_0^{\pi} (x^2 + 2 - \sin x)\, dx$

$= \left(\frac{x^3}{3} + 2x + \cos x\right)\Big|_0^{\pi}$

$= \frac{\pi^3}{3} + 2\pi - 2$

3. $\text{Area} = \int_0^1 x^3 - (2x^2 - x)\, dx$

$= \left(\frac{x^4}{4} - \frac{2}{3}x^3 + \frac{x^2}{2}\right)\Big|_0^1 = \frac{1}{12}$

5. Solving $e^{-x} = 2 - x^2$ we get $x \approx -0.537, 1.316$

$\text{Area} \approx \int_{-.537}^{1.316} (2 - x^2 - e^{x})\, dx$

$= \left(2x - \frac{x^3}{3} + e^{-x}\right)\Big|_{-.537}^{1.316} \approx 1.452$

7. $\text{Area} = \int_0^1 x^2\, dx + \int_1^2 (2 - x)\, dx$

$= \frac{x^3}{3}\Big|_0^1 + \left(2x - \frac{x^2}{2}\right)\Big|_1^2$

$= \frac{1}{3} + (4 - 2) - \left(2 - \frac{1}{2}\right) = \frac{5}{6}$

9. If P is the population at time t, the equation is

$P'(t) = \text{birth rate} - \text{death rate}$
$= (10 + 2t) - (4 + t) = 6 + t$

Thus $P = 6t + t^2/2 + P(0)$, so at time $t = 6$,

$$P(6) = 36 + 18 + 10{,}000 = 10{,}054.$$

Alternatively,

$A = \int_0^6 [(10 + 2t) - (4 + t)]dt$

$= \int_0^6 (6 + t)dt = \left(6t + \frac{t^2}{2}\right)\Big|_0^6 = 54$

population $= 10{,}000 + 54 = 10{,}054$

11. $V = \int_0^2 \pi(3 + x)^2\, dx$

$= \pi \int_0^2 (9 + 6x + x^2)\, dx$

$= \pi \left(9x + 3x^2 + \frac{x^3}{3}\right)\Big|_0^2$

$= \frac{98\pi}{3}$

13. Use trapezoidal estimate:

$V = 0.4\left(\frac{0.4}{2} + 1.4 + 1.8 + 2.0 + 2.1 + 1.8 + 1.1 + \frac{0.4}{2}\right)$

≈ 4.2

15. (a) V

$= \int_{-2}^2 \pi(4)^2\, dx - \int_{-2}^2 \pi(x^2)^2\, dx$

$= \pi \int_{-2}^2 (16 - x^4)\, dx$

$= \pi \left(16x - \frac{x^5}{5}\right)\Big|_{-2}^2$

$= \frac{256\pi}{5}$

(b) $V = \int_0^4 \pi(\sqrt{y})^2 dy = \pi \int_0^4 y dy$

$= \frac{\pi y^2}{2}\Big|_0^4 = 8\pi$

(c) $V = \int_0^4 \pi(2 + \sqrt{y})^2 dy$

$- \int_0^4 \pi(2 - \sqrt{y})^2 dy$

$= \pi \int_0^4 (4 + 4y^{1/2} + y)dy$

$- \pi \int_0^4 (4 - 4y^{1/2} + y)dy$

$= \pi \int_0^4 (8y^{1/2})dy$

$$= 8\pi \cdot \frac{2}{3} y^{3/2}\Big|_0^4 = \frac{128\pi}{3}$$

(d) $$V = \int_{-2}^{2} \pi(6)^2\,dx - \int_{-2}^{2} \pi(x^2+2)^2\,dx$$
$$= \pi \int_{-2}^{2} (-x^4 - 4x^2 + 32)\,dx$$
$$= \pi \left(-\frac{x^5}{5} - \frac{4x^3}{3} + 32x \right)\Big|_{-2}^{2}$$
$$= \frac{1408\pi}{15}$$

17. (a) $$V = \int_0^1 2\pi y((2-y)-y)dy$$
$$= 2\pi \int_0^1 (2y - 2y^2)dy$$
$$= 2\pi \left(y^2 - \frac{2y^3}{3} \right)\Big|_0^1$$
$$= \frac{2\pi}{3}$$

(b) $$V = \int_0^1 \pi(2-y)^2 dy - \int_0^1 \pi(y)^2 dy$$
$$= \pi \int_0^1 (4-4y)dy$$
$$= \pi\,(4y - 2y^2)\big|_0^1 = 2\pi$$

(c) $$V = \int_0^1 \pi((2-y)+1)^2 dy - \int_0^1 \pi(y+1)^2 dy$$
$$= \pi \int_0^1 (9 - 6y + y^2)dy - \pi \int_0^1 (y^2 + 2y + 1)dy$$
$$= \pi \int_0^1 (8 - 8y)dy$$
$$= \pi\,(8y - 4y^2)\big|_0^1 = 4\pi$$

(d) $$V = \int_0^1 2\pi(4-y)((2-y)-y)dy$$
$$= 2\pi \int_0^1 (8 - 10y + 2y^2)dy$$
$$= 2\pi \left(8y - 5y^2 + \frac{2y^3}{3} \right)\Big|_0^1$$
$$= \frac{22\pi}{3}$$

19. $s = \int_{-1}^{1} \sqrt{1 + (4x^3)^2}\,dx \approx 3.2$

21. $s \int_{-2}^{2} \sqrt{1 + \left(\frac{e^{x/2}}{2}\right)^2}\,dx \approx 4.767$

23. $S = \int_0^1 2\pi(1-x^2)\sqrt{1+4x^2}\,dx$
≈ 5.483

25. $h''(t) = -32$
$h(0) = 64, h'(0) = 0$
$h'(t) = -32t$
$h(t) = -16t^2 + 64$

This is zero when $t = 2$, at which time $h'(2) = -32(2) = -64$. The speed at impact is reported as 64 feet per second.

27. $y''(t) = -32, x''(t) = 0,$
$y(0) = 0, x(0) = 0$
$y'(0) = 48 \sin\left(\frac{\pi}{9}\right)$
$x'(0) = 48 \cos\left(\frac{\pi}{9}\right)$
$y'(0) \approx 16.42, x'(0) \approx 45.11$
$y'(t) = -32t + 16.42$
$y(t) = -16t^2 + 16.42t$

This is zero at $t = 1.026$. Meanwhile,

$x'(t) \equiv 45.11$
$x(t) = 45.11t$
$x(1.026) = 45.11(1.026) \approx 46.3$ ft
This is the horizontal range.

29. $y(0) = 6, x(0) = 0$
$y'(0) = 80 \sin\left(\frac{2\pi}{45}\right) \approx 11.13,$

$x'(0) = 80\cos\left(\frac{2\pi}{45}\right) \approx 79.22$
$y''(t) = -32, x''(t) = 0$
$y'(t) = -32t + 11.13$
$y(t) = -16t^2 + 11.13t + 6$
$x'(t) = 79.22$
$x(t) = 79.22t$

This is 120 (40 yards) when t is about 1.51. At this time, the vertical height (if still in flight) would be $y(1.51) = -16(1.51)^2 + 11.13(1.51) + 6$ $= -13.6753$,

Since this is negative, we conclude the ball is not still in flight, has hit the ground, and was not catchable.

31. $h''(t) = -32$
$h'(0) = v_0$
$h(0) = 0$
$h'(t) = -32t + v_0$
This is zero at $t = v_0/32$.
$h\left(\frac{v_0}{32}\right) = -16\left(\frac{v_0^2}{32^2}\right) + \frac{v_0^2}{32} = \frac{v_0^2}{64}$
If this is to be 128, then clearly v_0 must be

$$\sqrt{(64)(128)} = 64\sqrt{2} \text{ ft/sec.}$$

Impact speed from ground to ground is the same as launch speed, which can be verified by first finding the time t of return to ground:
$-16t^2 + v_0 t = 0$
$t = v_0/16$
and then compiling

$$h'(v_0/16) = -32(v_0/16) + v_0 = -v_0$$

33. $F = kx,\ 60 = k \cdot 1,\ k = 60$

$$W = \int_0^{2/3} 60x\,dx = 30x^2\Big|_0^{2/3}$$

$$= \frac{30 \cdot 4}{9} = \frac{40}{3} \text{ ft-lb}$$

35. $m = \int_0^4 (x^2 - 2x + 8)\,dx$

$$= \left(\frac{x^3}{3} - x^2 + 8x\right)\Big|_0^4 = \frac{112}{3}$$

$$M = \int_0^4 x\left(x^2 - 2x + 8\right)\,dx$$

$$= \int_0^4 (x^3 - 2x^2 + 8x)\,dx$$

$$= \left(\frac{x^4}{4} - \frac{2x^3}{3} + 4x^2\right)\Big|_0^4 = \frac{256}{3}$$

$$\bar{x} = \frac{M}{m} = \frac{\frac{256}{3}}{\frac{112}{3}} = \frac{256}{112} = \frac{16}{7}$$

Center of mass is greater than 2 because the object has greater density on the right side of the interval $[0, 4]$.

37. $F = \int_0^{80} 62.4x(140 - x)\,dx$

$$= 62.4 \int_0^{80} (140x - x^2)\,dx$$

$$= 62.4\left(70x^2 - \frac{x^3}{3}\right)\Big|_0^{80}$$

$$= 62.4(80)^2(130/3)$$

$$\approx 17{,}305{,}600 \text{ lb}$$

39. $J \approx \frac{.0008}{3(8)}\{0 + 4(800) + 2(1600)$
$+ 4(2400) + 2(3000) + 4(3600)$
$+2(2200) + 4(1200) + 0\}$
$= 1.52$

$J = m\Delta v$

$1.52 = .01\Delta v$
$\Delta v = 152$ ft/s
$152 - 120 = 32$ ft/s

41. $f(x) = x + 2x^3$ on $[0, 1]$
$f(x) \geq 0$ for $0 \leq x \leq 1$ and

$$\int_0^1 (x + 2x^3)\,dx = \left(\frac{x^2}{2} + \frac{x^4}{2}\right)\Big|_0^1 = 1$$

43.

$$1 = \int_1^2 \frac{c}{x^2}\,dx = \left.\frac{-c}{x}\right|_1^2 = \frac{-c}{2} + c = \frac{c}{2}$$

Therefore $c = 2$

45. (a) $P(x < .5) = \int_0^{.5} 4e^{-4x}\,dx$

$= -e^{-4x}\Big|_0^{.5} = 1 - e^{-2} \approx .864$

(b) $P(.5 \leq x \leq 1) = \int_{.5}^{1} 4e^{-4x}\,dx$

$= -e^{-4x}\Big|_{.5}^{1} = -e^{-4} + e^{-2} \approx .117$

47. (a) $\mu = \int_0^1 x\left(x + 2x^3\right)\,dx$

$= \left.\frac{x^3}{3} + \frac{2x^5}{5}\right|_0^1 = \frac{11}{15} \approx 0.7333$

(b) $\frac{1}{2} = \int_0^c \left(x + 2x^3\right)\,dx$

$= \left.\frac{x^2}{2} + \frac{x^4}{2}\right|_0^c = \frac{c^2}{2} + \frac{c^4}{2}$

Therefore $c^2 + c^4 = 1$,

$$c = \sqrt{\frac{-1 + \sqrt{5}}{2}} \approx 0.786$$

Chapter 6

Integration Techniques

6.1 Review of Formulas and Techniques

1. $\displaystyle\int \sin 6x\,dx = -\frac{1}{6}\cos x + c$

3. $\displaystyle\int \sec 2x \tan 2x\,dx = \frac{1}{2}\sec 2x + c$

5. $\displaystyle\int e^{3-2x}\,dx = -\frac{1}{2}e^{3-2x} + c$

7. Let $u = 1 + x^{2/3}$, $du = \frac{2}{3}x^{-1/3}\,dx$

$$\int \frac{4}{x^{1/3}(1+x^{2/3})}\,dx = 4\left(\frac{3}{2}\right)\int u^{-1}\,du$$
$$= 6\ln|u| + C = 6\ln|1+x^{2/3}| + c$$

9. Let $u = \sqrt{x}, du = \frac{1}{2\sqrt{x}}dx$

$$\int \frac{\sin\sqrt{x}}{\sqrt{x}}\,dx = 2\int \sin u\,du$$
$$= -2\cos u + C = -2\cos\sqrt{x} + c$$

11. Let $u = \sin x$, $du = \cos x\,dx$

$$\int_0^{\pi} \cos x e^{\sin x}\,dx = \int_0^0 e^u\,du = 0$$

13. $\displaystyle\int_{-\pi/4}^{0} \sec x \tan x\,dx$
$$= \sec x\Big|_{-\pi/4}^{0} = 1 - \sqrt{2}$$

15. $\displaystyle\frac{3}{16+x^2}\,dx = \frac{3}{4}\tan^{-1}\frac{x}{4} + c$

17. Let $u = x^3$, $du = 3x^2\,dx$

$$\frac{x^2}{1+x^6}\,dx = \frac{1}{3}\int \frac{1}{1+u^2}\,du$$
$$= \frac{1}{3}\tan^{-1}u + C = \frac{1}{3}\tan^{-1}x^3 + c$$

19. $\displaystyle\frac{1}{\sqrt{4-x^2}}\,dx = \sin^{-1}\frac{x}{2} + c$

21. Let $u = x^2$, $du = 2x\,dx$

$$\int \frac{x}{\sqrt{1-x^4}}\,dx = \frac{1}{2}\int \frac{1}{\sqrt{1-u^2}}\,du$$
$$= \frac{1}{2}\sin^{-1}u + C = \frac{1}{2}\sin^{-1}x^2 + c$$

23. $\displaystyle\int \frac{4}{5+2x+x^2}\,dx$
$$= 4\int \frac{1}{4+(x+1)^2}\,dx$$
$$= 2\tan^{-1}\left(\frac{x+1}{2}\right) + c$$

25. $\displaystyle\int \frac{4x}{5+2x+x^2}\,dx$
$$= \int \frac{4x+4}{5+2x+x^2}\,dx - \int \frac{4}{5+2x+x^2}\,dx$$
$$= 2\ln|4+(x+1)^2|$$
$$- 2\tan^{-1}\left(\frac{x+1}{2}\right) + c$$

27. $\displaystyle\int (x^2+4)^2\,dx = \int (x^4+8x^2+16)\,dx$
$$= \frac{x^5}{5} + \frac{8}{3}x^3 + 16x + c$$

29. $\displaystyle\int \frac{1}{\sqrt{3-2x-x^2}}\,dx$
$$= \int \frac{1}{\sqrt{4-(x+1)^2}}\,dx$$

$= \arcsin\left(\frac{x+1}{2}\right) + c$

31. $\int \frac{1+x}{1+x^2}\,dx$
$= \int \frac{1}{1+x^2}\,dx + \frac{1}{2}\int \frac{2x}{1+x^2}\,dx$
$= \tan^{-1} x + \frac{1}{2}\ln|1+x^2| + c$

33. $\int_{-2}^{-1} e^{\ln(x^2+1)}\,dx = \int_{-2}^{-1}(x^2+1)\,dx$
$= \left(\frac{x^3}{3}+x\right)\Big|_{-2}^{-1} = \frac{10}{3}$

35. $\int_3^4 x\sqrt{x-3}\,dx$
$= \int_3^4 (x-3+3)\sqrt{x-3}\,dx$
$= \int_3^4 (x-3)^{3/2}\,dx + 3\int_3^4 (x-3)^{1/2}\,dx$
$= \frac{2}{5}(x-3)^{5/2}\Big|_3^4 + 3\cdot\frac{2}{3}(x-3)^{3/2}\Big|_3^4$
$= \frac{12}{5}$

37. Substituting $u = e^x$

$\int_0^2 \frac{e^x}{1+e^{2x}}\,dx = \tan^{-1} e^x\Big|_0^2$
$= \tan^{-1} e^2 - \frac{\pi}{4}$

39. $\int_1^4 \frac{x^2+1}{\sqrt{x}}\,dx$
$= \int_1^4 x^{3/2}\,dx + \int_1^4 x^{-1/2}\,dx$
$= \frac{2}{5}x^{5/2}\Big|_1^4 + 2x^{1/2}\Big|_1^4 = \frac{72}{5}$

41. $\int \frac{5}{3+x^2}\,dx = \frac{5}{\sqrt{3}}\arctan\frac{x}{\sqrt{3}} + c$

$\int \frac{5}{3+x^3}\,dx$: N/A

43. $\int \ln x\,dx$: N/A

Substituting $u = \ln x$,
$\int \frac{\ln x}{2x}\,dx = \frac{1}{4}\ln^2 x + c$

45. $\int e^{-x^2}\,dx$: N/A

Substituting $u = -x^2$
$\int xe^{-x^2}\,dx = -\frac{1}{2}e^{-x^2} + c$

47. $\int_0^2 f(x)\,dx$
$= \int_0^1 \frac{x}{x^2+1}\,dx + \int_1^2 \frac{x^2}{x^2+1}\,dx$
$= \frac{1}{2}\ln|x^2+1|\Big|_0^1 + \int_1^2\left(1-\frac{1}{x^2+1}\right)dx$
$= \frac{1}{2}\ln 2 + (x - \arctan x)\Big|_1^2$
$= \frac{\ln 2}{2} + 1 + \frac{\pi}{4} - \arctan 2$

49. $\int \frac{4x+1}{2x^2+4x+10}\,dx$
$= \int \frac{4x+4}{2x^2+4x+10}\,dx$
$- \int \frac{3}{2x^2+4x+10}\,dx$
$= \ln|2x^2+4x+10|$
$- \frac{3}{2}\int \frac{1}{(x+1)^2+4}\,dx$
$= \ln|2x^2+4x+10|$
$- \frac{3}{4}\tan^{-1}\left(\frac{x+1}{2}\right) + c$

6.2 Integration by Parts

1. Let $u = x$, $dv = \cos x\,dx$ so that $du = dx, v = \sin x$

$\int x\cos x\,dx$
$= x\sin x - \int \sin x\,dx$
$= x\sin x + \cos x + c$

3. Let $u = x$, $dv = e^{2x}\,dx$ so that $du = dx,\ v = \frac{1}{2}e^{2x}$

$$\int xe^{2x}\,dx$$
$$= \frac{1}{2}xe^{2x} - \int \frac{1}{2}e^{2x}\,dx$$
$$= \frac{1}{2}xe^{2x} - \frac{1}{4}e^{2x} + c$$

5. Let $u = \ln x,\ dv = x^2\,dx$
$du = \frac{1}{x}\,dx,\ v = \frac{1}{3}x^3$

$$\int x^2 \ln x\,dx$$
$$= \frac{1}{3}x^3 \ln x - \int \frac{1}{3}x^3 \cdot \frac{1}{x}\,dx$$
$$= \frac{1}{3}x^3 \ln x - \frac{1}{3}\int x^2\,dx$$
$$= \frac{1}{3}x^3 \ln x - \frac{1}{9}x^3 + c$$

7. Let $u = x^2,\ dv = e^{-3x}\,dx$
$du = 2x\,dx,\ v = -\frac{1}{3}e^{-3x}$

$$I = \int x^2 e^{-3x}\,dx$$
$$= -\frac{1}{3}x^2e^{-3x} - \int \left(-\frac{1}{3}e^{-3x}\right) \cdot 2x\,dx$$
$$= -\frac{1}{3}x^2e^{-3x} + \frac{2}{3}\int xe^{-3x}\,dx$$

Let $u = x,\ dv = e^{-3x}\,dx$
$du = dx,\ v = -\frac{1}{3}e^{-3x}$

$$I = -\frac{1}{3}x^2e^{-3x} + \frac{2}{3}\left[-\frac{1}{3}xe^{-3x} - \int \left(-\frac{1}{3}e^{-3x}\right)dx\right]$$
$$= -\frac{1}{3}x^2e^{-3x} - \frac{2}{9}xe^{-3x} + \frac{2}{9}\int e^{-3x}\,dx$$
$$= -\frac{1}{3}x^2e^{-3x} - \frac{2}{9}xe^{-3x} - \frac{2}{27}e^{-3x} + c$$

9. Let $I = \int e^x \sin 4x\,dx$
$u = e^x,\ dv = \sin 4x\,dx$
$du = e^x\,dx,\ v = -\frac{1}{4}\cos 4x$

$$I = -\frac{1}{4}e^x \cos 4x - \int \left(-\frac{1}{4}\cos 4x\right) e^x\,dx$$
$$= -\frac{1}{4}e^x \cos 4x + \frac{1}{4}\int e^x \cos 4x\,dx$$

Use integration by parts again, this time let $u = e^x,\ dv = \cos 4x\,dx$
$du = e^x\,dx,\ v = \frac{1}{4}\sin 4x$

$$I = -\frac{1}{4}e^x \cos 4x + \frac{1}{4}\left(\frac{1}{4}e^x \sin 4x - \int \frac{1}{4}(\sin 4x)e^x\,dx\right)$$
$$I = -\frac{1}{4}e^x \cos 4x + \frac{1}{16}e^x \sin 4x - \frac{1}{16}I$$
so
$$\frac{17}{16}I = -\frac{1}{4}e^x \cos 4x + \frac{1}{16}e^x \sin 4x + c_1$$
$$I = -\frac{4}{17}e^x \cos 4x + \frac{1}{17}e^x \sin 4x + c$$

11. Let $I = \int \cos x \cos 2x\,dx$
and $u = \cos x,\ dv = \cos 2x\,dx$
$du = \sin x\,dx,\ v = \frac{1}{2}\sin 2x$

$$I = \frac{1}{2}\cos x \sin 2x - \int \frac{1}{2}\sin 2x(-\sin x)\,dx$$
$$= \frac{1}{2}\cos x \sin 2x + \frac{1}{2}\int \sin x \sin 2x\,dx$$

Let $u = \sin x,\ dv = \sin 2x\,dx$
$du = \cos x,\ dx\ v = -\frac{1}{2}\cos 2x$

$$I = \frac{1}{2}\cos x \sin 2x + \frac{1}{2}\left[-\frac{1}{2}\cos 2x \sin x - \int \left(-\frac{1}{2}\cos 2x\right)\cos x\,dx\right]$$
$$= \frac{1}{2}\cos x \sin 2x - \frac{1}{4}\cos 2x \sin x + \frac{1}{4}I\,dx, \text{ so}$$
$$\frac{3}{4}I = \frac{1}{2}\cos x \sin 2x - \frac{1}{4}\cos 2x \sin x + c_1$$
$$I = \frac{2}{3}\cos x \sin 2x - \frac{1}{3}\cos 2x \sin x + c$$

13. Let $u = x,\ dv = \sec^2 x\,dx$
$du = dx,\ v = \tan x$

$$\int x \sec^2 x\,dx = x \tan x - \int \tan x\,dx$$
$$= x \tan x - \int \frac{\sin x}{\cos x}\,dx$$

Let $u = \cos x$, $du = -\sin x\,dx$

$$\int x \sec^2 x\,dx = x \tan x + \int \frac{1}{u}\,du$$
$$= x \tan x + \ln|u| + c$$
$$= x \tan x + \ln|\cos x| + c$$

15. Let $u = (\ln x)^2$, $dv = dx$

$du = 2\dfrac{\ln x}{x}\,dx$, $v = x$

$$I = \int (\ln x)^2\,dx$$
$$= x(\ln x)^2 - \int x \cdot 2\frac{\ln x}{x}\,dx$$
$$= x(\ln x)^2 - 2\int \ln x\,dx$$

Integration by parts again,

$u = \ln x$, $dv = dx$

$du = \dfrac{1}{x}\,dx$, $v = x$

$$I = x(\ln x)^2 - 2\left[x \ln x - \int x \cdot \frac{1}{x}\,dx\right]$$
$$= x(\ln x)^2 - 2x \ln x + 2\int dx$$
$$= x(\ln x)^2 - 2x \ln x + 2x + c$$

17. Let $u = \ln(\sin x)$, $dv = \cos x\,dx$

$du = \dfrac{1}{\sin x} \cdot \cos x\,dx$, $v = \sin x$

$$I = \int \cos x \ln(\sin x)\,dx$$
$$= \sin x \ln(\sin x) - \int \sin x \cdot \frac{1}{\sin x} \cdot \cos x\,dx$$
$$= \sin x \ln(\sin x) - \int \cos x\,dx$$
$$= \sin x \ln(\sin x) - \sin x + c$$

19. Let $u = x$, $dv = \sin 2x\,dx$

$du = dx$, $v = -\dfrac{1}{2}\cos 2x$

$$\int_0^1 x \sin 2x\,dx$$
$$= -\frac{1}{2}x\cos 2x\Big|_0^1 - \int_0^1 \left(-\frac{1}{2}\cos 2x\right)dx$$
$$= -\frac{1}{2}(1\cos 2 - 0\cos 0)$$
$$+ \frac{1}{2}\int_0^1 \cos 2x\,dx$$
$$= -\frac{1}{2}\cos 2 + \frac{1}{2}\left(\frac{1}{2}\sin 2x\right)\Big|_0^1$$
$$= -\frac{1}{2}\cos 2 + \frac{1}{4}(\sin 2 - \sin 0)$$
$$= -\frac{1}{2}\cos 2 + \frac{1}{4}\sin 2$$

21. Let $u = x$, $dv = \cos \pi x\,dx$

$du = dx$, $v = \dfrac{1}{\pi}\sin \pi x$

$$\int_0^1 x \cos \pi x\,dx$$
$$= \frac{1}{\pi}x \sin \pi x\Big|_0^1 - \int_0^1 \frac{1}{\pi}\sin \pi x\,dx$$
$$= \frac{1}{\pi}\sin \pi - 0 + \frac{1}{\pi^2}\cos \pi x\Big|_0^1$$
$$= 0 + \frac{1}{\pi^2}(\cos \pi - \cos 0) = -\frac{2}{\pi^2}$$

23. Let $u = x$, $dv = \sin \pi x\,dx$

$du = dx$, $v = -\dfrac{1}{\pi}\cos \pi x$

$$\int_0^1 x \sin \pi x\,dx$$
$$= -\frac{1}{\pi}x \cos \pi x\Big|_0^1 - \int_0^1 \left(-\frac{1}{\pi}\cos \pi x\right)dx$$
$$= -\frac{1}{\pi}\cos \pi - 0 + \frac{1}{\pi^2}\sin \pi x\Big|_0^1$$
$$= -\frac{1}{\pi}(-1) + \frac{1}{\pi^2}(\sin \pi - \sin 0) = \frac{1}{\pi}$$

25. Let $u = \ln x$, $dv = dx$

$du = \dfrac{1}{x}\,dx$, $v = x$

$$\int_1^{10} \ln x\,dx = x \ln x\Big|_1^{10} - \int_1^{10} x \cdot \frac{1}{x}\,dx$$
$$= 10 \ln 10 - 1 \ln 1 - \int_1^{10} dx$$
$$= 10 \ln 10 - 0 - x\Big|_1^{10}$$
$$= 10 \ln 10 - 9$$

27. Let $u = \cos^{-1} x$, $dv = dx$

$du = -\dfrac{1}{\sqrt{1 - x^2}}\,dx$, $v = x$

$$I = \int \cos^{-1} x\,dx$$
$$= x\cos^{-1} x - \int x\left(-\frac{1}{\sqrt{1-x^2}}\right)dx$$
$$= x\cos^{-1} x + \int \frac{x}{\sqrt{1-x^2}}\,dx$$

Substituting $u = 1 - x^2$, $du = -2x\,dx$

$$I = x\cos^{-1} x + \int \frac{1}{\sqrt{u}}\left(-\frac{1}{2}\,du\right)$$
$$= x\cos^{1} x - \frac{1}{2}\int u^{-1/2}\,du$$
$$= x\cos^{-1} x - \frac{1}{2}\cdot 2u^{1/2} + c$$
$$= x\cos^{-1} x - \sqrt{1-x^2} + c$$

29. Substituting $u = \sqrt{x}, du = \dfrac{1}{2\sqrt{x}}\,dx$

$$I = \int \sin\sqrt{x}\,dx = 2\int u\sin u\,du$$
$$= 2(-u\cos u + \sin u) + c$$
$$= 2(-\sqrt{x}\cos\sqrt{x} + \sin\sqrt{x}) + c$$

31. Let $u = \sin(\ln x)$, $dv = dx$

$du = \cos(\ln x)\dfrac{dx}{x}$, $v = x$

$$I = \int \sin(\ln x)\,dx$$
$$= x\sin(\ln x) - \int \cos(\ln x)\,dx$$

Integration by parts again,

$u = \cos(\ln x)$, $dv = dx$

$du = -\sin(\ln x)\dfrac{dx}{x}$, $v = x$

$$\int \cos(\ln x)\,dx$$
$$= x\cos(\ln x) + \int \sin(\ln x)\,dx$$
$$I = x\sin(\ln x) - x\cos(\ln x) - I$$
$$2I = x\sin(\ln x) - x\cos(\ln x) + c_1$$
$$I = \frac{1}{2}x\sin(\ln x) - \frac{1}{2}x\cos(\ln x) + c$$

33. Let $u = e^{2x}$, $du = 2e^{2x}\,dx$

$$I = \int e^{6x}\sin(e^{2x})\,dx$$
$$= \frac{1}{2}\int u^2\sin u\,du$$

Let $v = u^2$, $dw = \sin u\,du$

$dv = 2u\,du$, $w = -\cos u$

$$I = \frac{1}{2}\left(-u^2\cos u + 2\int u\cos u\,du\right)$$
$$= -\frac{1}{2}u^2\cos u + \int u\cos u\,du$$
$$= -\frac{1}{2}u^2\cos u + (u\sin u + \cos u) + c$$
$$= -\frac{1}{2}e^{4x}\cos(e^{2x}) + e^{2x}\sin(e^{2x})$$
$$+ \cos(e^{2x}) + c$$

35. Let $u = \sqrt[3]{x} = x^{1/3}$, $du = \dfrac{1}{3}x^{-2/3}\,dx$,

$3u^2\,du = dx$

$$I = \int e^{\sqrt[3]{x}}\,dx = 3\int u^2e^u\,du$$
$$= 3\left(u^2e^u - 2\int ue^u\,du\right)$$
$$= 3u^2e^u - 6\left(ue^u - \int e^u\,du\right)$$
$$= 3u^2e^u - 6ue^u + 6e^u + c$$

Hence $\displaystyle\int_0^8 e^{\sqrt[3]{x}}\,dx = \int_0^2 3u^2e^u\,du$

$$= \left(3u^2e^u - 6ue^u + 6e^u\right)\Big|_0^2$$
$$= 6e^2 - 6$$

37. n times. Each integration reduces the power of x by 1.

39. Let$u = \cos^{n-1} x, dv = \cos x\,dx$

$du = (n-1)(\cos^{n-2} x)(-\sin x)\,dx$,

$v = \sin x$

$$\int \cos^n x\,dx$$
$$= \sin x\cos^{n-1} x$$
$$-\int (\sin x)(n-1)(\cos^{n-2} x)(-\sin x)\,dx$$
$$= \sin x\cos^{n-1} x$$
$$+\int (n-1)(\cos^{n-2} x)(\sin^2 x)\,dx$$
$$= \sin x\cos^{n-1} x$$
$$+\int (n-1)(\cos^{n-2} x)(1-\cos^2 x)\,dx$$
$$= \sin x\cos^{n-1} x$$
$$+\int (n-1)(\cos^{n-2} x - \cos^n x)\,dx$$

Thus $\int \cos^n x\,dx$

$= \sin x \cos^{n-1} x + \int (n-1)\cos^{n-2} x\,dx$

$- (n-1)\int \cos^n x\,dx.$

$n\int \cos^n x\,dx = \sin x \cos^{n-1} x$

$+ (n-1)\int \cos^{n-2} x\,dx$

$\int \cos^n x\,dx$

$= \frac{1}{n}\sin x \cos^{n-1} x + \frac{n-1}{n}\int \cos^{n-2} x\,dx$

41. $\int x^3 e^x\,dx = e^x(x^3 - 3x^2 + 6x - 6) + c$

43. $\int \cos^3 x\,dx$

$= \frac{1}{3}\cos^2 x \sin x + \frac{2}{3}\int \cos x\,dx$

$= \frac{1}{3}\cos^2 x \sin x + \frac{2}{3}\sin x + c$

45. $\int_0^1 x^4 e^x\,dx$

$= e^x(x^4 - 4x^3 + 12x^2 - 24x + 24)\Big|_0^1$

$= 9e - 24$

47. $\int_0^{\pi/2} \sin^5 x\,dx$

$= -\frac{1}{5}\sin^4 x \cos x\Big|_0^{\pi/2} + \frac{4}{5}\int_0^{\pi/2} \sin^3 x\,dx$

$= -\frac{1}{5}\sin^4 x \cos x\Big|_0^{\pi/2}$

$+ \frac{4}{5}\left(-\frac{1}{3}\sin^2 x \cos x - \frac{2}{3}\cos x\right)\Big|_0^{\pi/2}$

(Using Exercise 40)

$= -\frac{1}{5}\left(\sin^4\left(\frac{\pi}{2}\right)\cos\frac{\pi}{2} - \sin^4 0 \cos 0\right)$

$+ \frac{4}{5}\left(-\frac{1}{3}\sin^2\left(\frac{\pi}{2}\right)\cos\frac{\pi}{2} - \frac{2}{3}\cos\frac{\pi}{2}\right)$

$= \frac{8}{15}$

49. m even :

$\int_0^{\pi/2} \sin^m x\,dx$

$= \frac{(m-1)(m-3)\ldots 1}{m(m-2)\ldots 2} \cdot \frac{\pi}{2}$

m odd:

$\int_0^{\pi/2} \sin^m x\,dx$

$= \frac{(m-1)(m-3)\ldots 2}{m(m-2)\ldots 3}$

51. First column: each row is the derivative of the previous row;

Second column: each row is the antiderivative of the previous row.

53.

	$\cos x$	
x^4	$\sin x$	+
$4x^3$	$-\cos x$	−
$12x^2$	$-\sin x$	+
$24x$	$\cos x$	−
24	$\sin x$	+

$\int x^4 \cos x\,dx$

$= x^4 \sin x + 4x^3 \cos x - 12x^2 \sin x$

$- 24x \cos x + 24 \sin x + c$

55.

	e^{2x}	
x^4	$e^{2x}/2$	+
$4x^3$	$e^{2x}/4$	−
$12x^2$	$e^{2x}/8$	+
$24x$	$e^{2x}/16$	−
24	$e^{2x}/32$	+

$\int x^4 e^{2x}\,dx$

$= \left(\frac{x^4}{2} - x^3 + \frac{3x^2}{2} - \frac{3x}{2} + \frac{3}{4}\right)e^{2x} + c$

57.

	e^{-3x}	
x^3	$-e^{-3x}/3$	+
$3x^2$	$e^{-3x}/9$	−
$6x$	$-e^{-3x}/27$	+
6	$e^{-3x}/81$	−

$$\int x^3e^{-3x}\,dx$$
$$=\left(-\frac{x^3}{3}-\frac{x^2}{3}-\frac{2x}{9}-\frac{2}{27}\right)e^{-3x}+c$$

59. (a) Use the identity

$$\cos A\cos B$$
$$=\frac{1}{2}\left[\cos(A-B)+\cos(A+B)\right]$$

This identity gives

$$\int_{-\pi}^{\pi}\cos(mx)\cos(nx)\,dx$$
$$=\int_{-\pi}^{\pi}\frac{1}{2}[\cos((m-n)x)$$
$$+\cos((m+n)x)]\,dx$$
$$=\frac{1}{2}\left[\frac{\sin((m-n)x)}{m-n}\right.$$
$$\left.+\frac{\sin((m+n)x)}{m+n}\right]\Bigg|_{-\pi}^{\pi}$$
$$=0$$

It is important that $m\neq n$ because otherwise

$$\cos((m-n)x)=\cos 0=1.$$

(b) Use the identity

$$\sin A\sin B$$
$$=\frac{1}{2}[\cos(A-B)-\cos(A+B)]$$

This identity gives

$$\int_{-\pi}^{\pi}\sin(mx)\sin(nx)\,dx$$
$$=\int_{-\pi}^{\pi}\frac{1}{2}[\cos((m-n)x)$$
$$-\cos((m+n)x)]\,dx$$
$$=\frac{1}{2}\left[\frac{\sin((m-n)x)}{m-n}\right.$$
$$\left.-\frac{\sin((m+n)x)}{m+n}\right]\Bigg|_{-\pi}^{\pi}$$
$$=0$$

It is important that $m\neq n$ because otherwise

$$\cos((m-n)x)=\cos 0=1.$$

61. The only mistake is the misunderstanding of antiderivatives.

In this problem,

$$\int e^xe^{-x}\,dx$$

is understood as a group of antiderivatives of e^xe^{-x}, not a fixed function.

So the subtraction by $\int e^xe^{-x}\,dx$ on both sides of

$$\int e^xe^{-x}\,dx=-1+\int e^xe^{-x}\,dx$$

does not make sense.

63. Let $u=\ln x,\ dv=e^x\,dx$

$du=\dfrac{dx}{x},\ v=e^x$

$$\int e^x \ln x dx = e^x \ln x - \int \frac{e^x}{x} dx$$
$$\int e^x \ln x dx + \int \frac{e^x}{x} dx = e^x \ln x + C$$

Hence
$$\int e^x \left(\ln x + \frac{1}{x}\right) dx = e^x \ln x + c$$

65. The quotient rule says:
$$\frac{d}{dx}\left(\frac{f(x)}{g(x)}\right) = \frac{f'(x)g(x) - f(x)g'(x)}{g^2(x)}$$
$$= \frac{f'(x)}{g(x)} - \frac{f(x)g'(x)}{g^2(x)}$$
In the form of antiderivatives, we get
$$\int \left[\frac{f'(x)}{g(x)} - \frac{f(x)g'(x)}{g^2(x)}\right] dx = \frac{f(x)}{g(x)}$$
$$\int \frac{f'(x)}{g(x)} dx - \int \frac{f(x)g'(x)}{g^2(x)} dx = \frac{f(x)}{g(x)}$$
$$\int \frac{f'(x)}{g(x)} dx = \frac{f(x)}{g(x)} + \int \frac{f(x)g'(x)}{g^2(x)} dx$$

6.3 Trigonometric Techniques of Integration

1. Let $u = \sin x,\ du = \cos x\, dx$
$$\int \cos x \sin^4 x\, dx = \int u^4\, du$$
$$= \frac{1}{5}u^5 + c = \frac{1}{5}\sin^5 x + c$$

3. Let $u = \sin x,\ du = \cos x\, dx$
$$\int_0^{\pi/4} \cos x \sin^3 x\, dx = \int_0^{1/\sqrt{2}} u^3\, du$$
$$= \frac{1}{4}u^4\Big|_0^{1/\sqrt{2}} = \frac{1}{4\cdot(\sqrt{2})^4} = \frac{1}{16}$$

5. Let $u = cosx,\ du = -sinx\, dx$
$$\int_0^{\pi/2} \cos^2 x \sin x\, dx = \int_1^0 u^2(-du)$$
$$= \left(-\frac{1}{3}u^3\right)\Big|_1^0 = \frac{1}{3}$$

7. $$\int \cos^2 x\, dx = \frac{1}{2}\int (1 + \cos 2x)\, dx$$
$$= \frac{1}{2}x + \frac{1}{4}\sin 2x + c$$

9. Let $u = \sec x,\ du = \sec x \tan x\, dx$
$$\int \tan x \sec^3 x\, dx$$
$$= \int \tan x \sec x \sec^2 x\, dx = \int u^2\, du$$
$$= \frac{1}{3}u^3 + c = \frac{1}{3}\sec^3 x + c$$

11. Let $u = \tan x,\ du = \sec^2 x\, dx$
$$\int_0^{\pi/4} \tan^4 x \sec^4 x\, dx$$
$$= \int_0^{\pi/4} \tan^4 x \sec^2 x \sec^2 x\, dx$$
$$= \int_0^{\pi/4} \tan^4 x(1 + \tan^2 x)\sec^2 x\, dx$$
$$= \int_0^1 u^4(1 + u^2)\, du$$
$$= \int_0^1 u^4 + u^6\, du \frac{u^5}{5} + \frac{u^7}{7}\Big|_0^1 = \frac{12}{35}$$

13. $$\int \cos^2 x \sin^2 x\, dx$$
$$= \int \frac{1}{2}(1 + \cos 2x)\cdot\frac{1}{2}(1 - \cos 2x)\, dx$$
$$= \frac{1}{4}\int (1 - \cos^2 2x)\, dx$$
$$= \frac{1}{4}\int \left[1 - \frac{1}{2}(1 + \cos 4x)\right] dx$$
$$= \frac{1}{4}\left(\frac{1}{2}x - \frac{1}{8}\sin 4x\right) + c$$
$$= \frac{1}{8}x - \frac{1}{32}\sin 4x + c$$

15. Let $u = cosx,\ du = -\sin x\, dx$
$$\int_{-\pi/3}^0 \sqrt{\cos x}\sin^3 x\, dx$$
$$= \int_{-\pi/3}^0 \sqrt{\cos x}(1 - \cos^2 x)\sin x\, dx$$
$$= \int_{1/2}^1 \sqrt{u}(1 - u^2)(-du)$$

$$= \int_{1/2}^{1} (u^{5/2} - u^{1/2}) du$$
$$= \frac{2}{7}u^{7/2} - \frac{2}{3}u^{3/2}\Big|_{1/2}^{1}$$
$$= \frac{25}{168}\sqrt{2} - \frac{8}{21}$$

17. Let $x = 3\sin\theta, -\frac{\pi}{2} < \theta < \frac{\pi}{2}$

$dx = 3\cos\theta\, d\theta$

$$\int \frac{1}{x^2\sqrt{9-x^2}}\,dx = \int \frac{3\cos\theta}{9\sin^2\theta \cdot 3\cos\theta} d\theta$$
$$= \frac{1}{9}\int \csc^2\theta d\theta = -\frac{1}{9}\cot\theta + C$$

By drawing a diagram, we see that if $x = \sin\theta$, then $\cot\theta = \frac{\sqrt{9-x^2}}{x}$.

Thus the integral $= -\frac{\sqrt{9-x^2}}{9x} + c$

19. This is the area of a quarter of a circle of radius 2.

$$\int_0^2 \sqrt{4-x^2}\,dx = \pi$$

21. Let $x = 3\sec\theta, dx = 3\sec\theta\tan\theta d\theta$.

$$I = \int \frac{x^2}{\sqrt{x^2-9}}\,dx$$
$$= \int \frac{27\sec^2\theta\sec\theta\tan\theta}{\sqrt{9\sec^2\theta - 9}}\,d\theta$$
$$= \int 9\sec^3\theta\,d\theta$$

Use integration by parts. Let $u = \sec\theta$ and $dv = \sec^2\theta\,d\theta$. This gives

$$\int \sec^3\theta\,d\theta$$
$$= \sec\theta\tan\theta - \int \sec\theta\tan^2\theta\,d\theta$$
$$= \sec\theta\tan\theta - \int \sec\theta(\sec^2\theta - 1)\,d\theta$$
$$= \sec\theta\tan\theta + \int \sec\theta\,d\theta - \int \sec^3\theta\,d\theta$$

$$2\int \sec^3\theta\,d\theta$$
$$= \sec\theta\tan\theta + \int \sec\theta\,d\theta$$

$$\int \sec^3\theta\,d\theta$$
$$= \frac{1}{2}\sec\theta\tan\theta + \frac{1}{2}\int \sec\theta\,d\theta$$

This leaves us to compute $\int \sec\theta\,d\theta$. For this notice if $u = \sec\theta + \tan\theta$ then $du = \sec\theta\tan\theta + \sec^2\theta$.

$$\int \sec\theta\,d\theta$$
$$= \int \frac{\sec\theta(\sec\theta + \tan\theta)}{\sec\theta + \tan\theta}\,d\theta$$
$$= \int \frac{1}{u}\,du = \ln|u| + c$$
$$= \ln|\sec\theta + \tan\theta| + c$$

Putting all these together and using $\sec\theta = \frac{x}{3}$, $\tan\theta = \frac{\sqrt{x^2-9}}{3}$:

$$\int \frac{x^2}{\sqrt{x^2-9}}\,dx = \int 9\sec^3\theta\,d\theta$$
$$= \frac{9}{2}\sec\theta\tan\theta + \frac{9}{2}\int \sec\theta\,d\theta$$
$$= \frac{9}{2}\sec\theta\tan\theta + \frac{9}{2}\ln|\sec\theta + \tan\theta| + c$$
$$= \frac{9}{2}\left(\frac{x}{3}\right)\left(\frac{\sqrt{x^2-9}}{3}\right)$$
$$+ \frac{9}{2}\ln\left|\frac{x}{3} + \frac{\sqrt{x^2-9}}{3}\right| + c$$
$$= \frac{x\sqrt{x^2-9}}{2} + \frac{9}{2}\ln\left|\frac{x+\sqrt{x^2-9}}{3}\right| + c$$

23. Let $x = 2\sec\theta, dx = 2\sec\theta\tan\theta d\theta$

$$\int \frac{2}{\sqrt{x^2-4}}\,dx = \int \frac{4\sec\theta\tan\theta}{2\tan\theta}\,d\theta$$
$$= 2\int \sec\theta\,d\theta$$
$$= 2\ln|2\sec\theta + 2\tan\theta| + c$$
$$= 2\ln\left|x + \sqrt{x^2-4}\right| + c$$

25. Let $x = 3\tan\theta, dx = 3\sec^2\theta d\theta$

$$\int \frac{x^2}{\sqrt{x^2+9}}\,dx$$
$$= \int \frac{27\tan^2\theta\sec^2\theta}{\sqrt{9\tan^2\theta + 9}}\,d\theta$$

$= \int 9\tan^2\theta \sec\theta \, d\theta$
$= 9\int (\sec^2\theta - 1)\sec\theta \, d\theta$
$= 9\int \sec^3\theta \, d\theta - 9\int \sec\theta \, d\theta$
$= \frac{9}{2}\sec\theta\tan\theta - \frac{9}{2}\ln|\sec\theta + \tan\theta| + c$
$= \frac{9}{2}\left(\frac{\sqrt{x^2+9}}{3}\right)\left(\frac{x}{3}\right)$
$- \frac{9}{2}\ln\left|\frac{\sqrt{x^2+9}}{3} + \frac{x}{3}\right| + c$

$= \frac{x\sqrt{x^2+9}}{2} - \frac{9}{2}\ln\left|\frac{x+\sqrt{x^2+9}}{3}\right| + c$

27. Let $x = 4\tan\theta$, $dx = 4\sec^2\theta d\theta$

$\int \sqrt{x^2+16}\,dx$
$= \int \sqrt{16\tan^2\theta + 16} \cdot 4\sec^2\theta d\theta$
$= 16\int \sec^3\theta d\theta$
$= 16\left(\frac{1}{2}\sec\theta\tan\theta + \frac{1}{2}\int \sec\theta d\theta\right)$
$= 8\sec\theta\tan\theta + 8\int \sec\theta d\theta$
$= 8\sec\theta\tan\theta + 8\ln|\sec\theta + \tan\theta| + c$
$= \frac{1}{2}x\sqrt{16+x^2}$
$+ 8\ln\left|\frac{1}{4}\sqrt{16+x^2} + \frac{x}{4}\right| + c$

29. Let $u = x^2 + 8$, $du = 2x\,dx$

$\int_0^1 x\sqrt{x^2+8}\,dx = \frac{1}{2}\int_8^9 u^{1/2}\,du$
$= \frac{1}{3}u^{3/2}\Big|_8^9 = \frac{27 - 16\sqrt{2}}{3}$

31. Using $u = \tan x$ gives
$\int \tan x \sec^4 x\,dx$
$= \int \tan x(1+\tan^2 x)\sec^2 x\,dx$
$= \int u(1+u^2)\,du$
$= \int (u+u^3)\,du = \frac{1}{2}u^2 + \frac{1}{4}u^4 + c$
$= \frac{1}{2}\tan^2 x + \frac{1}{4}\tan^4 x + c$
Using $u = \sec x$ gives
$\int \tan x \sec^4 x\,dx$
$= \int \tan x \sec x \sec^3 x\,dx$
$= \int u^3\,du = \frac{1}{4}u^4 + c$
$= \frac{1}{4}\sec^4 x + c$

33. This is using integration by parts followed by substitution
$u = \sec^{n-2} x$, $dv = \sec^2 x\,dx$
$du = (n-2)\sec^{n-2} x\tan x\,dx$, $v = \tan x$

$I = \int \sec^n x\,dx = \sec^{n-2} x\tan x$
$- (n-2)\int \sec^{n-2}(\sec^2 x - 1)\,dx$
$= \sec^{n-2} x\tan x$
$- (n-2)\int (\sec^n x - \sec^{n-2} x)\,dx$
$= \sec^{n-2} x\tan x - (n-2)I$
$+ (n-2)\int \sec^{n-2} x\,dx$

$(n-1)I$
$= \sec^{n-2} x\tan x + (n-2)\int \sec^{n-2} x\,dx$

$I = \frac{\sec^{n-2} x\tan x}{n-1} + \frac{n-2}{n-1}\int \sec^{n-2} x\,dx$

35. The average power
$= \frac{1}{\frac{2\pi}{\omega}}\int_0^{2\pi/\omega} RI^2\cos^2(\omega t)\,dt$
$= \frac{\omega RI^2}{2\pi}\int_0^{2\pi/\omega} \frac{1}{2}[1+\cos(2\omega t)]\,dt$
$= \frac{\omega RI^2}{4\pi}\left[t + \frac{1}{2\omega}\sin(2\omega t)\right]\Big|_0^{2\pi/\omega}$
$= \frac{\omega RI^2}{4\pi}\left[\frac{2\pi}{\omega} + \frac{1}{2\omega}\sin\left(\frac{4\omega\pi}{\omega}\right) - 0\right]$
$= \frac{1}{2}RI^2$

37. Using a CAS we get

(Ex 3.2) $\int \cos^4 x \sin^3 x\, dx$
$= -\frac{1}{7}\sin x^2 \cos x^5 - \frac{2}{35}\cos x^5 +$
c

(Ex 3.3) $\int \sqrt{\sin x}\cos^5 x\, dx$
$= \frac{2}{11}\sin x^{11/2} - \frac{4}{7}\sin x^{7/2}$
$+\frac{2}{3}\sin x^{3/2} + c$

(Ex 3.5) $\int \cos^4 x\, dx$
$= \frac{1}{4}\cos x^3 \sin x + \frac{3}{8}\cos x \sin x +$
$\frac{3}{8}x + C$

(Ex 3.6) $\int \tan^3 x \sec^3 x\, dx$
$= 1/5\,\frac{\sin x^4}{\cos x^5} + 1/15\,\frac{\sin x^4}{\cos x^3}$
$-1/15\,\frac{\sin x^4}{\cos x} - 1/15\,\sin x^2 \cos x$
$-\,2/15\,\cos x + c$

Obviously my CAS used different techniques. The answers given by the book are simpler.

39. $-\frac{1}{7}\sin^2 x \cos^5 x - \frac{2}{35}\cos^5 x$
$= -\frac{1}{7}(1-\cos^2 x)\cos^5 x - \frac{2}{35}\cos^5 x$
$= \frac{1}{7}\cos^7 x - \frac{1}{5}\cos^5 x$

The conclusion is $c = 0$

6.4 Integration of Rational Functions Using Partial Fractions

1. $\frac{x-5}{x^2-1} = \frac{x-5}{(x+1)(x-1)}$
$= \frac{A}{x+1} + \frac{B}{x-1}$

$x - 5 = A(x-1) + B(x+1)$
$x = -1 : -6 = -2A; A = 3$
$x = 1 : -4 = 2B; B = -2$

$\frac{x-5}{x^2-1} = \frac{3}{x+1} - \frac{2}{x-1}$
$\int \frac{x-5}{x^2-1}\,dx = \int \left(\frac{3}{x+1} - \frac{2}{x-1}\right) dx$
$= 3\ln|x+1| - 2\ln|x-1| + c$

3. $\frac{6x}{x^2-x-2} = \frac{6x}{(x-2)(x+1)}$
$= \frac{A}{x-2} + \frac{B}{x+1}$

$6x = A(x+1) + B(x-2)$
$x = 2 : 12 = 3A; A = 4$
$x = -1 : -6 = -3B; B = 2$
$\frac{6x}{x^2-x-2} = \frac{4}{x-2} + \frac{2}{x+1}$
$\int \frac{6x}{x^2-x-2}\,dx$
$= \int \left(\frac{4}{x-2} + \frac{2}{x+1}\right) dx$
$= 4\ln|x-2| + 2\ln|x+1| + c$

5. $\frac{-x+5}{x^3-x^2-2x} = \frac{-x+5}{x(x-2)(x+1)}$
$= \frac{A}{x} + \frac{B}{x2} + \frac{C}{x+1}$

$-x+5 = A(x-2)(x+1) + Bx(x+1)$
$+\, cx(x-2)$
$x = 0 : 5 = -2A : A = -\frac{5}{2}$
$x = 2 : 3 = 6B : B = \frac{1}{2}$

$x = -1 : 6 = 3C : C = 2$

$$\frac{-x+5}{x^3-x^2-2x} = -\frac{5/2}{x} + \frac{1/2}{x-2} + \frac{2}{x+1}$$

$$\int \frac{-x+5}{x^3-x^2-2x}\,dx = \int\left(-\frac{5/2}{x} + \frac{1/2}{x-2} + \frac{2}{x+1}\right)dx = -\frac{5}{2}\ln|x| + \frac{1}{2}\ln|x-2| + 2\ln|x+1| + c$$

7. $$\frac{x^3+x+2}{x^2+2x-8} = x-2+\frac{13x-14}{x^2+2x-8} = x-2+\frac{13x-14}{(x+4)(x-2)} = x-2+\frac{A}{x+4}+\frac{B}{x-2}$$

$13x-14 = A(x-2)+B(x+4)$
$x=-4 : -66 = -6A; A = 11$
$x = 2 : 12 = 6B; B = 2$

$$\frac{x^3+x+2}{x^2+2x-8} = x-2+\frac{11}{x+4}+\frac{2}{x-2}$$

$$\int \frac{x^3+x+2}{x^2+2x-8}\,dx = \int\left(x-2+\frac{11}{x+4}+\frac{2}{x-2}\right)dx = \frac{x^2}{2} - 2x + 11\ln|x+4| + 2\ln|x-2| + c$$

9. $$\frac{5x-23}{6x^2-11x-7} = \frac{5x-23}{(2x+1)(3x-7)} = \frac{A}{2x+1}+\frac{B}{3x-7}$$

$5x-23 = A(3x-7)+B(2x+1)$
$x = -\frac{1}{2} : -\frac{51}{2} = -\frac{17}{2}A; A = 3$
$x = \frac{7}{3} : -\frac{34}{3} = \frac{17}{3}B; B = -2$

$$\frac{5x-23}{6x^2-11x-7} = \frac{3}{2x+1} - \frac{2}{3x-7}$$

$$\int \frac{5x-23}{6x^2-11x-7}\,dx = \int\left(\frac{3}{2x+1}-\frac{2}{3x-7}\right)dx = \frac{3}{2}\ln|2x+1| - \frac{2}{3}\ln|3x-7| + c$$

11. $$\frac{x-1}{x^3+4x^2+4x} = \frac{x-1}{x(x+2)^2} = \frac{A}{x}+\frac{B}{x+2}+\frac{C}{(x+2)^2}$$

$x-1 = A(x+2)^2 + Bx(x+2) + cx$
$x = 0 : -1 = 4A; A = -\frac{1}{4}$
$x = -2 : -3 = -2C; C = \frac{3}{2}$
$x = 1 : 0 = 9A + 3B + c; B = \frac{1}{4}$

$$\frac{x-1}{x^3+4x^2+4x} = -\frac{1/4}{x} + \frac{1/4}{x+2} + \frac{3/2}{(x+2)^2}$$

$$\int \frac{x-1}{x^3+4x^2+4x}\,dx = \int\left(-\frac{1/4}{x}+\frac{1/4}{x+2}+\frac{3/2}{(x+2)^2}\right)dx = -\frac{1}{4}\ln|x| + \frac{1}{4}\ln|x+2| - \frac{3}{2(x+2)} + c$$

13. $$\frac{x+4}{x^3+3x^2+2x} = \frac{x+4}{x(x+2)(x+1)} = \frac{A}{x}+\frac{B}{x+2}+\frac{C}{x+1}$$

$x+4 = A(x+2)(x+1) + Bx(x+1) + cx(x+2)$
$x = 0 : 4 = 2A; A = 2$
$x = -2 : 2 = 2B; B = 1$
$x = -1 : 3 = -C; C = -3$

$$\frac{x+4}{x^3+3x^2+2x} = \frac{2}{x} + \frac{1}{x+2} - \frac{3}{x+1}$$

$$\int \frac{x+4}{x^3+3x^2+2x}\,dx = \int\left(\frac{2}{x}+\frac{1}{x+2}-\frac{3}{x+1}\right)dx = 2\ln|x| + \ln|x+2| - 3\ln|x+1| + c$$

15. $\dfrac{x+2}{x^3+x} = \dfrac{x+2}{x(x^2+1)}$
$= \dfrac{A}{x} + \dfrac{Bx+c}{x^2+1}$

$x+2 = A(x^2+1) + (Bx+c)x$
$= Ax^2 + A + Bx^2 + cx$
$= (A+B)x^2 + cx + A$
$A = 2; C = 1; B = -2$

$\dfrac{x+2}{x^3+x} = \dfrac{2}{x} + \dfrac{-2x+1}{x^2+1}$
$\displaystyle\int \frac{x+2}{x^3+x}\,dx = \int \left(\frac{2}{x} + \frac{-2x+1}{x^2+1}\right) dx$
$\displaystyle= \int \left(\frac{2}{x} - \frac{2x}{x^2+1} + \frac{1}{x^2+1}\right) dx$
$= 2\ln|x| - \ln(x^2+1) + \tan^{-1}x + c$

17. $\dfrac{4x-2}{16x^4-1} = \dfrac{4x-2}{(4x^2+1)(2x+1)(2x-1)}$
$= \dfrac{Ax+B}{4x^2+1} + \dfrac{C}{2x+1} + \dfrac{D}{2x-1}$

$4x-2 = (Ax+B)(2x+1)(2x-1) + c(2x-1)(4x^2+1) + D(2x+1)(4x^2+1)$
$A = -4; B = 1; C = 1; D = 0$

$\dfrac{4x-2}{16x^4-1} = \dfrac{-4x+1}{4x^2+1} + \dfrac{1}{2x+1}$
$\displaystyle\int \frac{4x-2}{16x^4-1}\,dx$
$\displaystyle= \int \left(\frac{-4x+1}{4x^2+1} + \frac{1}{2x+1}\right) dx$
$\displaystyle= \int \left(-\frac{1}{2}\frac{8x}{4x^2+1} + \frac{1}{4x^2+1} + \frac{1}{2x+1}\right) dx$
$= -\frac{1}{2}\ln|4x^2+1| + \frac{1}{2}\tan^{-1}(2x)$
$+ \frac{1}{2}\ln|2x+1| + c$

19. $\dfrac{4x^2-7x-17}{6x^2-11x-10}$
$= \dfrac{2}{3} + \dfrac{1}{3}\dfrac{x-31}{(2x-5)(3x+2)}$
$= \dfrac{2}{3} + \dfrac{1}{3}\left[\dfrac{A}{2x-5}\right]$

$x-31 = A(3x+2) + B(2x-5)$

$x = \frac{5}{2} : -\frac{57}{2} = \frac{19}{2}A, A = -3;$
$x = -\frac{2}{3} : -\frac{95}{3} = -\frac{19}{3}B, B = 5;$
$\dfrac{4x^2-7x-17}{6x^2-11x-10}$
$= \dfrac{2}{3} + \dfrac{1}{3}\left[\dfrac{-3}{2x-5} + \dfrac{5}{3x+2}\right]$
$\displaystyle\int \frac{4x^2-7x-17}{6x^2-11x-10}\,dx$
$\displaystyle= \int \left(\frac{2}{3} - \frac{1}{2x-5} + \frac{5/3}{3x+2}\right) dx$
$= \frac{2}{3}x - \frac{1}{2}\ln|2x-5| + \frac{5}{9}\ln|3x+2| + c$

21. $\dfrac{2x+3}{x^2+2x+1} = \dfrac{2x+3}{(x+1)^2}$
$= \dfrac{A}{x+1} + \dfrac{B}{(x+1)^2} 2x+3 = A(x+1) + B$

$x = -1 : B = 1; A = 2$

$\dfrac{2x+3}{x^2+2x+1} = \dfrac{2}{x+1} + \dfrac{1}{(x+1)^2}$
$\displaystyle\int \frac{2x+3}{x^2+2x+1}\,dx$
$\displaystyle= \int \left(\frac{2}{x+1} + \frac{1}{(x+1)^2}\right) dx$
$= 2\ln|x+1| - \dfrac{1}{x+1} + c$

23. $\dfrac{x^3-4}{x^3+2x^2+2x} = 1 + \dfrac{-2x^2-2x-4}{x(x^2+2x+2)}$
$= 1 + \dfrac{A}{x} + \dfrac{Bx+c}{x^2+2x+2}$

$-2x^2 - 2x - 4 = A(x^2+2x+2) + (Bx+c)x$
$= (A+B)x^2 + (2A+c)x + 2A$
$A = -2; B = 0; C = 2$

$\dfrac{x^3-4}{x^3+2x^2+2x}$
$= 1 + \dfrac{-2}{x} + \dfrac{2}{x^2+2x+2}$

$$\int \frac{x^3-4}{x^3+2x^2+2x}\,dx$$
$$= \int \left(1+\frac{-2}{x}+\frac{2}{(x+1)^2+1}\right)dx$$
$$= x - 2\ln|x| + 2\tan^{-1}(x+1)+c$$

25. $\dfrac{x^3+x}{3x^2+2x+1}$
$= \dfrac{x}{3}-\dfrac{2}{9}+\dfrac{1}{9}\dfrac{10x+2}{3x^2+2x+1}$

This is already the PDF, since $3x^2+2x+1$ has no real roots

$$\int \frac{x^3+x}{3x^2+2x+1}\,dx$$
$$= \int \left(\frac{x}{3}-\frac{2}{9}+\frac{1}{9}\frac{10x+2}{3x^2+2x+1}\right)dx$$
$$= \int \left(\frac{x}{3}-\frac{2}{9}+\frac{1}{9}\frac{5}{3}\frac{6x+2}{3x^2+2x+1}\right.$$
$$\left.-\frac{1}{9}\frac{4}{3}\frac{1}{3(x+1/3)^2+2/3}\right)dx$$
$$= \frac{x^2}{6}-\frac{2}{9}x+\frac{5}{27}\ln(3x^2+2x+1)$$
$$-\frac{2\sqrt{2}}{27}\tan^{-1}\left(\frac{3x+1}{\sqrt{2}}\right)+c$$

27. $\dfrac{4x^2+3}{x^3+x^2+x}=\dfrac{A}{x}+\dfrac{Bx+c}{x^2+x+1}$

$4x^2+3 = A(x^2+x+1)+(Bx+c)x$
$= Ax^2+Ax+A+Bx^2+cx$
$A=3; C=-3; B=1$

$$\frac{4x^2+3}{x^3+x^2+x}=\frac{3}{x}+\frac{x-3}{x^2+x+1}$$
$$\int \frac{4x^2+3}{x^3+x^2+x}\,dx$$
$$= \int \left(\frac{3}{x}+\frac{x-3}{x^2+x+1}\right)dx$$
$$= \int \left(\frac{3}{x}+\frac{x+1/2}{x^2+x+1}-\frac{7/2}{x^2+x+1}\right)dx$$
$$= 3\ln|x|+\frac{1}{2}\ln|x^2+x+1|$$
$$-\frac{7}{\sqrt{3}}\tan^{-1}\left(\frac{2x+1}{\sqrt{3}}\right)+c$$

29. $\dfrac{3x^3+1}{x^3-x^2+x-1}$
$= 3+\dfrac{3x^2-3x+4}{x^3-x^2+x-1}$
$= 3+\dfrac{3x^2-3x+4}{(x^2+1)(x-1)}$
$= 3+\dfrac{Ax+B}{x^2+1}+\dfrac{C}{x-1}$

$3x^2-3x+4 = (Ax+B)(x-1)+c(x^2+1)$
$= Ax^2-Ax+Bx-B+cx^2+c$
$x=1: 4=2C; C=2$
$A+c=3: A=1$
$-A+B=-3: B=-2$

$$\frac{3x^3+1}{x^3-x^2+x-1}=3+\frac{x-2}{x^2+1}+\frac{2}{x-1}$$
$$\int \frac{3x^3+1}{x^3-x^2+x-1}\,dx$$
$$= \int \left(3+\frac{x-2}{x^2+1}+\frac{2}{x-1}\right)dx$$
$$= \int \left(3+\frac{x}{x^2+1}-\frac{2}{x^2+1}+\frac{2}{x-1}\right)dx$$
$$= 3x+\frac{1}{2}\ln(x^2+1)-2\tan^{-1}x$$
$$+2\ln|x-1|+c$$

31. $\dfrac{4x^2+2}{(x^2+1)^2}=\dfrac{Ax+B}{x^2+1}+\dfrac{Cx+D}{(x^2+1)^2}$

$4x^2+2=(Ax+B)(x^2+1)+(Cx+D)$
$= Ax^3+Bx^2+(A+c)x+(B+D)$
$A=0; B=4; C=0; D=-2$

$$\frac{4x^2+2}{(x^2+1)^2}=\frac{4}{x^2+1}+\frac{-2}{(x^2+1)^2}$$

33. $\dfrac{2x^2+4}{(x^2+4)^2}=\dfrac{Ax+B}{x^2+4}+\dfrac{Cx+D}{(x^2+4)^2}$

$2x^2+4=(Ax+B)(x^2+4)+cx+D=$
$Ax^3+4Ax+Bx^2+4B+cx+D$
$A=0; B=2$
$4A+c=0: C=0$
$4B+D=4: D=-4$

$$\frac{2x^2+4}{(x^2+4)^2} = \frac{2}{x^2+4} - \frac{4}{(x^2+4)^2}$$

35. $\dfrac{4x^2+3}{(x^2+x+1)^2}$
$= \dfrac{Ax+B}{x^2+x+1} + \dfrac{Cx+D}{(x^2+x+1)^2}$

$4x^2+3 = (Ax+B)(x^2+x+1)+cx+D = Ax^3+Ax^2+Ax+Bx^2+Bx+B+cx+D$
$A=0$
$A+B=4 : B=4$
$A+B+c=0 : C=-4$
$B+D=3 : D=-1$

$\dfrac{4x^2+3}{(x^2+x+1)^2}$
$= \dfrac{4}{x^2+x+1} - \dfrac{4x+1}{(x^2+x+1)^2}$

37. Let $u = x^3+1$, $du = 3x^2\,dx$

$$\int \frac{3}{x^4+x}\,dx = \int \frac{3x^2}{x^3(x^3+1)}\,dx$$
$= \int \dfrac{1}{(u-1)u}\,du$
$= \int \left(\dfrac{1}{u-1} - \dfrac{1}{u}\right) du$
$= \ln|u-1| - \ln|u| + c$
$= \ln\left|\dfrac{u-1}{u}\right| + c$
$= \ln\left|\dfrac{x^3}{x^3+1}\right| + c$

On the other hand, we can let
$u = \dfrac{1}{x}$, $du = -\dfrac{1}{x^2}\,dx$

$$\int \frac{3}{x^4+x}\,dx = -\int \frac{3u^2}{1+u^3}\,du$$
$= -\ln|1+u^3| + c$
$= -\ln|1+1/x^3| + c$

To see that the two answers are equivalent, note that
$\ln\left|\dfrac{x^3}{x^3+1}\right| = -\ln\left|\dfrac{x^3+1}{x^3}\right|$
$= -\ln|1+1/x^3|$

6.5 Integration Table and Computer Algebra Systems

1. $\displaystyle\int \frac{x}{(2+4x)^2}\,dx$
$= \dfrac{2}{16(2+4x)} + \dfrac{1}{16}\ln|2+4x| + c$
$= \dfrac{1}{8(2+4x)} + \dfrac{1}{16}\ln|2+4x| + c$

3. Substitute $u = 1+e^x$

$$\int e^{2x}\sqrt{1+e^x}\,dx = \int (u-1)\sqrt{u}\,du$$
$= \int (u^{3/2} - u^{1/2})\,du$
$= \dfrac{2}{5}u^{5/2} - \dfrac{2}{3}u^{3/2} + c$
$= \dfrac{2}{5}(1+e^x)^{5/2} - \dfrac{2}{3}(1+e^x)^{3/2} + c$

5. Substitute $u = 2x$

$$\int \frac{x^2}{\sqrt{1+4x^2}}\,dx$$
$= \dfrac{1}{8}\displaystyle\int \frac{u^2}{\sqrt{1+u^2}}\,du$
$= \dfrac{1}{8}\left[\dfrac{u}{2} - \sqrt{1+u^2}\right.$
$\left. - \dfrac{1}{2}\ln(u+\sqrt{1+u^2})\right] + c$
$= \dfrac{1}{8}x\sqrt{1+4x^2}$
$- \dfrac{1}{16}\ln(2x+\sqrt{1+4x^2}) + c$

7. Substitute $u = x^3$

$$\int x^8\sqrt{4-x^6}\,dx = \int u^2\sqrt{4-u^2}\cdot\frac{1}{3}\,du$$
$= \dfrac{1}{3}\displaystyle\int u^2\sqrt{4-u^2}\,du$
$= \dfrac{1}{3}\left[\dfrac{u}{8}(2u^2-4)\sqrt{4-u^2} + \dfrac{16}{8}\sin^{-1}\dfrac{u}{2}\right]$

$$+c$$
$$= \frac{1}{24}x^3(2x^6 - 4)\sqrt{4 - x^6}$$
$$+ \frac{2}{3}\sin^{-1}\frac{x^3}{2} + c$$

$$\int_0^1 x^8\sqrt{4 - x^6}\,dx = \frac{\pi}{9} - \frac{\sqrt{3}}{12}$$

9. Substitute $u = e^x$

$$\int \frac{e^x}{\sqrt{e^{2x}+4}}\,dx = \int \frac{1}{\sqrt{u^2+4}}\,du$$
$$= \ln(u + \sqrt{4+u^2}) + c$$
$$= \ln(e^x + \sqrt{4+e^{2x}}) + c$$

$$\int_0^{\ln 2} \frac{x^x}{\sqrt{e^{2x}+4}}\,dx = \ln\left(\frac{2\sqrt{2}+2}{1+\sqrt{5}}\right).$$

11. Substitute $u = x - 3$

$$\int \frac{\sqrt{6x - x^2}}{(x-3)^2}\,dx$$
$$= \int \frac{\sqrt{(u+3)(6-(u+3))}}{u^2}\,du$$
$$= \int \frac{\sqrt{9-u^2}}{u^2}\,du$$
$$= -\frac{1}{u}\sqrt{9-u^2} - \sin^{-1}\frac{u}{3} + c$$
$$= -\frac{1}{x-3}\sqrt{9-(x-3)^2}$$
$$- \sin^{-1}\left(\frac{x-3}{3}\right) + c$$

13. $\displaystyle\int \tan^6 x\,dx = \frac{1}{5}\tan^5 x - \int \tan^4 x\,dx$

$$= \frac{1}{5}\tan^5 x - \left[\frac{1}{3}\tan^3 x - \int \tan^2 x\,dx\right]$$
$$= \frac{1}{5}\tan^5 x - \frac{1}{3}\tan^3 x + \tan x - x + c$$

15. Substitute $u = \sin x$

$$\int \frac{\cos x}{\sin x\sqrt{4+\sin x}}\,dx = \int \frac{1}{u\sqrt{4+u}}\,du$$
$$= \frac{1}{\sqrt{4}}\ln\left|\frac{\sqrt{4+u}-2}{\sqrt{4+u}+2}\right| + c$$
$$= \frac{1}{2}\ln\left|\frac{\sqrt{4+\sin x}-2}{\sqrt{4+\sin x}+2}\right| + c$$

17. Substitute $u = x^2$

$$\int x^3\cos x^2\,dx = \frac{1}{2}\int u\cos u\,du$$
$$= \frac{1}{2}(\cos u + u\sin u) + c$$
$$= \frac{1}{2}\cos x^2 + \frac{1}{2}x^2\sin x^2 + c$$

19. Substitute $u = \cos x$

$$\int \frac{\sin x\cos x}{\sqrt{1+\cos x}}\,dx = -\int \frac{u}{\sqrt{1+u}}\,du$$
$$= -\frac{2}{3}(u-2)\sqrt{1+u} + c$$
$$= -\frac{2}{3}(\cos x - 2)\sqrt{1+\cos x} + c$$

21. Substitute $u = \sin x$

$$\int \frac{\sin^2 x\cos x}{\sqrt{\sin^2 x + 4}}\,dx = \int \frac{u^2}{\sqrt{u^2+4}}\,du$$
$$= \frac{u}{2}\sqrt{4+u^2} - \frac{4}{2}\ln(u + \sqrt{4+u^2}) + c$$
$$= \frac{1}{2}\sin x\sqrt{4+\sin^2 x}$$
$$- 2\ln(\sin x + \sqrt{4+\sin^2 x}) + c$$

23. Substitute $u = -\frac{2}{x^2}$

$$\int \frac{e^{-2/x^2}}{x^3}\,dx = \frac{1}{4}\int e^u\,du$$
$$= \frac{1}{4}e^u + c = \frac{1}{4}e^{-2/x^2} + c$$

25. $\displaystyle\int \frac{x}{\sqrt{4x - x^2}}\,dx$

$$= -\sqrt{4x - x^2} + 2\cos^{-1}\left(\frac{2-x}{2}\right) + c$$

27. Substitute $u = e^x$

$$\int e^x\tan^{-1}(e^x)\,dx = \int \tan^{-1} u\,du$$
$$= u\tan^{-1} u - \frac{1}{2}\ln(1+u^2) + c$$
$$= e^x\tan^{-1} e^x - \frac{1}{2}\ln(1+e^{2x}) + c$$

29. Answer depends on CAS used.

31. Any answer is wrong because the integrand is undefined for all $x \neq 1$.

33. Answer depends on CAS used.

35. Answer depends on CAS used.

6.6 Improper Integrals

1. improper, function not defined at $x = 0$

3. not improper, function continuous on entire interval

5. improper, function not defined at $x = 0$

7. $\int_0^1 x^{-1/3}\,dx = \lim_{R\to 0^+} \int_R^1 x^{-1/3}\,dx$
$= \lim_{R\to 0^+} \frac{3}{2}x^{2/3}\Big|_R^1$
$= \lim_{R\to 0^+} \frac{3}{2}(1 - R^{2/3}) = \frac{3}{2}$

9. $\int_1^\infty x^{-4/5}\,dx = \lim_{R\to\infty} \int_1^R x^{-4/5}\,dx$
$= \lim_{R\to\infty} 5x^{1/5}\Big|_1^R = \lim_{R\to\infty} 5R^{1/5} - 5 = \infty$
So the original integral diverges.

11. $\int_{10}^1 \frac{1}{\sqrt{x-1}}\,dx = \lim_{R\to 1^+} \int_{10}^R \frac{1}{\sqrt{x-1}}\,dx$
$= \lim_{R\to 1^+} 2\sqrt{x-1}\Big|_{10}^R$
$= \lim_{R\to 1^+} 2(\sqrt{R-1} - 3) = -6$

13. $\int_0^1 \ln x\,dx = \lim_{R\to 0^+} \int_R^1 \ln x\,dx$
$= \lim_{R\to 0^+} (x\ln x - x)\Big|_R^1$
$= \lim_{R\to 0^+} (-1 - R\ln R + R)$
$= -1 - \lim_{R\to 0^+} \frac{\ln R}{1/R}$
$= -1 - \lim_{R\to 0^+} \frac{1/R}{-1/R^2}$
$= -1 + \lim_{R\to 0^+} R = -1$

15. $\int_0^3 \frac{2}{x^2-1}\,dx$
$= \int_0^3 \left(-\frac{1}{x+1} + \frac{1}{x-1}\right)dx$
$= \lim_{R\to 1^-} \int_0^R \left(-\frac{1}{x+1} + \frac{1}{x-1}\right)dx +$
$\lim_{R\to 1^+} \int_R^3 \left(-\frac{1}{x+1} + \frac{1}{x-1}\right)dx$
Both of these integrals behave like
$\lim_{R\to 0^+} \int_R^1 \frac{1}{x}\,dx = \lim_{R\to 0^+} (\ln 1 - \ln R)$
$= \lim_{R\to 0^+} \ln\left(\frac{1}{R}\right) = \infty$
So the original integral diverges.

17. $\int_0^\infty xe^x\,dx = \lim_{R\to\infty} \int_0^R xe^x\,dx$
$= \lim_{R\to\infty} (xe^x - e^x)\Big|_0^R$
$= \lim_{R\to\infty} e^R(R-1) + 1 = \infty$
So the original integral diverges.

19. Substitute $u = 3x$
$I = \int_{-\infty}^1 x^2 e^{3x}\,dx = \frac{1}{27}\int_{-\infty}^3 u^2 e^u\,du$
$= \frac{1}{27}\lim_{R\to-\infty} (u^2e^u - 2ue^u + 2e^u)\Big|_R^3$
$= \frac{5}{27}e^3 - \frac{1}{27}\lim_{R\to-\infty} e^R(R^2 - 2R + 2)$
But $\lim_{R\to-\infty} e^R(R^2 - 2R + 2)$
$= \lim_{R\to\infty} e^{-R}(R^2 + 2R + 2)$
$= \lim_{R\to\infty} \frac{R^2 + 2R + 2}{e^R} = 0$
Hence $I = \frac{5}{27}e^3$

21. $\int_{-\infty}^{-1} \frac{1}{x^2}\,dx = \lim_{R\to-\infty} \int_R^{-1} \frac{1}{x^2}\,dx$
$= \lim_{R\to-\infty} -\frac{1}{x}\Big|_R^{-1} = 1 + \lim_{R\to-\infty} \frac{1}{R} = 1$
$\int_{-1}^0 \frac{1}{x^2}\,dx = \lim_{R\to 0^+} \int_{-1}^R \frac{1}{x^2}\,dx$
$= \lim_{R\to 0^+} -\frac{1}{x}\Big|_{-1}^R = -1 - \lim_{R\to 0^+} \frac{1}{R}$

$= \infty$

So the original integral diverges.

23. $\displaystyle\int_{-\infty}^{\infty} \frac{1}{1+x^2}\,dx$

$\displaystyle = \int_{-\infty}^{0} \frac{1}{1+x^2}\,dx + \int_{0}^{\infty} \frac{1}{1+x^2}\,dx$

$\displaystyle = \lim_{R\to-\infty} \int_{R}^{0} \frac{1}{1+x^2}\,dx$

$\displaystyle + \lim_{R\to\infty} \int_{0}^{R} \frac{1}{1+x^2}\,dx$

$\displaystyle = \lim_{R\to\infty} \tan^{-1} x|_R^0 + \lim_{R\to\infty} \tan^{-1} x|_0^R$

$\displaystyle = \lim_{R\to-\infty} (\tan^{-1} 0 - \tan^{-1} R)$

$\displaystyle + \lim_{R\to\infty} (\tan^{-1} R - \tan^{-1} 0)$

$\displaystyle = 0 - \left(-\frac{\pi}{2}\right) + \frac{\pi}{2} - 0 = \pi$

25. $\displaystyle\int_{0}^{\pi/2} \cot x\,dx = \lim_{R\to0^+} \int_{R}^{\pi/2} \frac{\cos x}{\sin x}\,dx$

$\displaystyle = \lim_{R\to0^+} \ln|\sin x|\Big|_R^{\pi/2}$

$\displaystyle = \ln|\sin(\pi/2)| - \lim_{R\to0^+} \ln|\sin R|$

$= \infty$

So the original integral diverges.

27. $\displaystyle\int_{0}^{2} \frac{x}{x^2-1}\,dx$

$\displaystyle = \int_{0}^{1} \frac{x}{x^2-1}\,dx + \int_{1}^{2} \frac{x}{x^2-1}\,dx$

$\displaystyle = \lim_{R\to1^-} \int_{0}^{R} \frac{x}{x^2-1}\,dx$

$\displaystyle + \lim_{R\to1^+} \int_{R}^{2} \frac{x}{x^2-1}\,dx$

$\displaystyle = \lim_{R\to1^-} \frac{1}{2}\ln|x^2-1|\Big|_0^R$

$\displaystyle + \lim_{R\to1^+} \frac{1}{2}\ln|x^2-1|\Big|_R^2$

$\displaystyle = \lim_{R\to1^-} \left(\frac{1}{2}\ln|R^2-1| - \frac{1}{2}\ln|-1|\right)$

$= -\infty$

So the original integral diverges.

29. $\displaystyle\int_{0}^{1} \frac{2}{\sqrt{1-x^2}}\,dx = \lim_{R\to1^-} \int_{0}^{R} \frac{2}{\sqrt{1-x^2}}\,dx$

$\displaystyle = F \lim_{R\to1^-} 2\sin^{-1} x\Big|_0^R$

$\displaystyle = \lim_{R\to1^-} 2(\sin^{-1} R - \sin^{-1} 0)$

$\displaystyle = 2\left(\frac{\pi}{2} - 0\right) = \pi$

31. Substitute $u = \sqrt{x}$

$$\int \frac{1}{\sqrt{x}e^{\sqrt{x}}}\,dx = \int 2e^{-u}\,du$$

Hence $\displaystyle\int_{0}^{\infty} \frac{1}{\sqrt{x}e^{\sqrt{x}}}\,dx$

$\displaystyle = \lim_{R\to0^+} \int_{R}^{1} \frac{1}{\sqrt{x}e^{\sqrt{x}}}\,dx$

$\displaystyle + \lim_{R\to\infty} \int_{1}^{R} \frac{1}{\sqrt{x}e^{\sqrt{x}}}\,dx$

$\displaystyle = \lim_{R\to0^+} \left(\frac{-2}{e^{\sqrt{x}}}\Big|_R^1\right) + \lim_{R\to\infty} \left(\frac{-2}{e^{\sqrt{x}}}\Big|_1^R\right)$

$\displaystyle = \lim_{R\to0^+} \left(\frac{-2}{e} + \frac{2}{e^R}\right) + \lim_{R\to\infty} \left(\frac{-2}{e} + \frac{2}{e^R}\right)$

$= 1 + 1 = 2$

33. $\displaystyle\int_{0}^{\infty} \cos x\,dx = \lim_{R\to\infty} \int_{0}^{R} \cos x\,dx$

$\displaystyle = \lim_{R\to\infty} \sin x\Big|_0^R = \lim_{R\to\infty} (\sin R - \sin 0)$

Therefore the original integral diverges.

35. Converges if and only if $n < 1$.

$\displaystyle\int_{0}^{1} x^{-n}\,dx = \lim_{R\to0^+} \int_{R}^{1} x^{-n}\,dx$

$\displaystyle = \lim_{R\to0^+} \frac{x^{-n+1}}{-n+1}\Big|_R^1$

$\displaystyle = \lim_{R\to0^+} \left(\frac{1}{-n+1} - \frac{R^{-n+1}}{-n+1}\right)$

This limit exists only if $-n+1 \geq 0$ or $n \leq 1$. But, note that if $n = 1$ then the integral is $\displaystyle\int_{1}^{\infty} \frac{1}{x}\,dx$ which diverges. Therefore the integral converges if and only if $n < 1$.

37. $\displaystyle\int_{0}^{\infty} xe^{cx}\,dx$ converges for $c < 0$

$\int_{-\infty}^{0} xe^{cx}\,dx$ converges for $c > 0$

39. $0 < \frac{x}{1+x^3} < \frac{x}{x^3} = \frac{1}{x^2}$

$\int_1^\infty \frac{1}{x^2}\,dx = \lim_{R\to\infty} \int_1^R \frac{1}{x^2}\,dx$

$= \lim_{R\to\infty} \left(-\frac{1}{x}\right)\Big|_1^R$

$= \lim_{R\to\infty} \left(-\frac{1}{R}+1\right) = 1$

So $\int_1^\infty \frac{x}{1+x^3}\,dx$ converges.

41. $\frac{x}{x^{3/2}-1} > \frac{x}{x^{3/2}} = \frac{1}{x^{1/2}} > 0$

$\int_2^\infty x^{-1/2}\,dx = \lim_{R\to\infty} \int_2^R x^{-1/2}\,dx$

$= \lim_{R\to\infty} 2\sqrt{x}\Big|_2^R$

$= \lim_{R\to\infty} (2\sqrt{R} - 2\sqrt{2}) = \infty$

So $\int_2^\infty \frac{x}{x^{3/2}-1}\,dx$ diverges

43. $0 < \frac{3}{x+e^x} < \frac{3}{e^x}$

$\int_0^\infty \frac{3}{e^x}\,dx = \lim_{R\to\infty} \int_0^R \frac{3}{e^x}\,dx$

$= \lim_{R\to\infty} \left(-\frac{3}{e^x}\right)\Big|_0^R$

$= \lim_{R\to\infty} \left(-\frac{3}{e^R}+3\right) = 3$

So $\int_0^\infty \frac{3}{x+e^x}\,dx$ converges

45. $\frac{\sin^2 x}{1+e^x} \le \frac{1}{1+e^x} < \frac{1}{e^x}$

$\int_0^\infty \frac{1}{e^x}\,dx = \lim_{R\to\infty} \int_0^R \frac{1}{e^x}\,dx$

$= \lim_{R\to\infty} \int (-e^{-x})\Big|_0^R$

$= \lim_{R\to\infty} (-e^{-R}+1) = 1$

So $\int_0^\infty \frac{\sin^2 x}{1+e^x}\,dx$ converges.

47. $\frac{x^2e^x}{\ln x} > e^x$

$\int_2^\infty e^x\,dx = \lim_{R\to\infty} \int_2^R e^x\,dx$

$= \lim_{R\to\infty} e^x\Big|_2^R$

$= \lim_{R\to\infty} (e^R - e^2) = \infty$

So $\int_2^\infty \frac{x^2e^x}{\ln x}\,dx$ diverges

49. Let $u = \ln 4x,\ dv = x\,dx$

$du = \frac{dx}{x},\ v = \frac{x^2}{2}$

$\int x\ln 4x\,dx = \frac{1}{2}x^2\ln 4x - \frac{1}{2}\int x\,dx$

$= \frac{1}{2}x^2\ln 4x - \frac{x^2}{4} + c$

$I = \int_0^1 x\ln 4x\,dx = \lim_{R\to 0^+} \int_R^1 x\ln 4x\,dx$

$= \lim_{R\to 0^+} \left(\frac{1}{2}x^2\ln 4x - \frac{x^2}{4}\right)\Big|_R^1$

$= -\frac{1}{4} - \lim_{R\to 0^+} \left(\frac{1}{2}R^2\ln 4R - \frac{R^2}{4}\right)$

$= -\frac{1}{4} - \frac{1}{2}\lim_{R\to 0^+} R^2\ln 4R$

$\lim_{R\to 0^+} R^2\ln 4R = \lim_{R\to 0^+} \frac{\ln 4R}{R^{-2}}$

$= \lim_{R\to 0^+} \frac{R^{-1}}{-2R^{-3}} = \lim_{R\to 0^+} \frac{R^2}{-2} = 0$

Hence $I = -\frac{1}{4}$

51. The volume is finite:

$V = \pi\int_1^\infty \frac{1}{x^2}\,dx = \pi\lim_{R\to\infty} \int_1^R \frac{1}{x^2}\,dx$

$= \pi\lim_{R\to\infty} -\frac{1}{x}\Big|_1^R$

$= \pi\lim_{R\to\infty} \left(-\frac{1}{R}+1\right) = \pi$

The surface area is infinite:

$S = 2\pi\int_1^\infty \frac{1}{x}\sqrt{1+\frac{1}{x^4}}\,dx$

$$\frac{1}{x}\sqrt{1+\frac{1}{x^4}} > \frac{1}{x}$$

and $\displaystyle\int_1^{\infty} \frac{1}{x}\,dx = \lim_{R\to\infty}\int_1^R \frac{1}{x}\,dx$

$$= \lim_{R\to\infty} \ln|x|\Big|_1^R = \lim_{R\to\infty} \ln R = \infty$$

53. True, this statement can be proved using the integration by parts:
$\displaystyle\int f(x)\,dx = xf(x) - \int g(x)\,dx,$
where $g(x)$ is some function related to $f(x)$.

55. False, consider $f(x) = \ln x$

57. $\displaystyle I_p = \int_0^1 x^{-p}\,dx = \lim_{R\to 0^+}\int_R^1 x^{-p}\,dx$

$$= \lim_{R\to 0^+} \frac{x^{-p+1}}{-p+1}\Big|_R^1$$

$$= \lim_{R\to 0^+} \frac{1-R^{-p+1}}{-p+1}$$

We need $p < 1$ for the above limit to converge. If this is the case,
$\displaystyle I_p = \frac{1}{-p+1},$

and for such p

$$\int_0^1 (1-x)^{-p}\,dx$$

$$= \lim_{R\to 1^-}\int_0^R (1-x)^{-p}\,dx$$

$$= \lim_{R\to 1^-} \frac{-(1-x)^{-p+1}}{-p+1}\Big|_0^R$$

$$= \lim_{R\to 1^-} \frac{1-(1-R)^{-p+1}}{-p+1}$$

$$= \frac{1}{-p+1} = I_p$$

59. Substitute $u = \sqrt{k}x$

$$\int_{-\infty}^{\infty} e^{-kx^2}\,dx = \frac{1}{\sqrt{k}}\int_{-\infty}^{\infty} e^{-u^2}\,du$$

$$= \frac{\sqrt{\pi}}{\sqrt{k}}$$

61. (a) $\displaystyle\int_0^{\infty} ke^{-2x}\,dx = \lim_{R\to\infty}\int_0^R ke^{-2x}\,dx$

$$= -\frac{k}{2}\lim_{R\to\infty} e^{-2x}\Big|_0^R$$

$$= -\frac{k}{2}\lim_{R\to\infty}(e^{-2R}-1) = \frac{k}{2} = 1$$

So $k = 2$

(b) $\displaystyle\int_0^{\infty} ke^{-4x}\,dx = \lim_{R\to\infty}\int_0^R ke^{-4x}\,dx$

$$= -\frac{k}{4}\lim_{R\to\infty} e^{-4x}\Big|_0^R$$

$$= -\frac{k}{4}\lim_{R\to\infty}(e^{-4R}-1) = \frac{k}{4} = 1$$

So $k = 4$

(c) If $r > 0$:

$$\int_0^{\infty} ke^{-rx}\,dx = \lim_{R\to\infty}\int_0^R ke^{-rx}\,dx$$

$$= -\frac{k}{r}\lim_{R\to\infty} e^{-rx}\Big|_0^R$$

$$= -\frac{k}{r2}\lim_{R\to\infty}(e^{-rR}-1) = \frac{k}{r} = 1$$

So $k = r$

If ≤ 0:

The integral $\displaystyle\int_0^{\infty} ke^{-rx}\,dx$ diverges for any value of k, so there is no value of k to make the function $f(x) = k$ a pdf.

63. From Exercise 61 (c) we know that this r has to be positive.
Substitute $u = rx$

$$\mu = \int_0^{\infty} xf(x)\,dx = \int_0^{\infty} rxe^{-rx}\,dx$$

$$= \lim_{R\to\infty}\int_0^R rxe^{-rx}\,dx$$

$$= \frac{1}{r}\lim_{R\to\infty}\int_0^R ue^{-u}\,du$$

$$= \frac{1}{r}\lim_{R\to\infty} e^{-u}(-u-1)\Big|_0^R$$

$$= \lim_{R\to\infty} \frac{-R-1}{e^R} + \frac{1}{r}$$

$$= 0 + \frac{1}{r} = \frac{1}{r}$$

65. The probability is given by the formula

$$\int_{\mu}^{\infty} re^{-rx}\,dx = \int_{\frac{1}{r}}^{\infty} re^{-rx}\,dx$$
$$= \lim_{R\to\infty}\int_{\frac{1}{r}}^{R} re^{-rx}\,dx = \lim_{R\to\infty}\left(-e^{-rx}\right)\Big|_{\frac{1}{r}}^{R}$$
$$= \lim_{R\to\infty}\left(e^{-rR} + e^{-1}\right) = e^{-1} < \frac{1}{2}$$

The mean is different from the median, so it is not odd that the probability we just found is not equal to 1/2.

67. Improper because $\tan(\pi/2)$ is not defined.

The two integrals
$$\int_0^{\pi/2} \frac{1}{1+\tan x}\,dx = \int_0^{\pi/2} f(x)\,dx$$
because the two integrand only differ at one value of x, and that except for this value, everything is proper.

$$g(x) = \begin{cases} \dfrac{\tan x}{1+\tan x} & \text{if } 0 \le x < \dfrac{\pi}{2} \\ 0 & \text{if } x = \dfrac{\pi}{2} \end{cases}$$

69. Substitute $u = \frac{\pi}{2} - x$

$$\int_0^{\pi/2} \ln(\sin x)\,dx$$
$$= -\int_{\pi/2}^{0} \ln(\sin(\pi/2 - u))\,du$$
$$= \int_0^{\pi/2} \ln(\cos u)\,du = \int_0^{\pi/2} \ln(\cos x)\,dx$$

Moreover,
$$2\int_0^{\pi/2} \ln(\sin x)\,dx$$
$$= \int_0^{\pi/2} \ln(\cos x)\,dx + \int_0^{\pi/2} \ln(\sin x)\,dx$$
$$= \int_0^{\pi/2} [\ln(\cos x) + \ln(\sin x)]\,dx$$
$$= \int_0^{\pi/2} \ln(\sin x\cos x)\,dx$$
$$= \int_0^{\pi/2} [\ln(\sin(2x)) - \ln 2]\,dx$$
$$= \int_0^{\pi/2} \ln(\sin(2x))\,dx - \frac{\pi}{2}\ln 2$$
$$= \frac{1}{2}\int_0^{\pi} \ln(\sin x)\,dx - \frac{\pi}{2}\ln 2$$

Hence,
$$2\int_0^{\pi/2} \ln(\sin x)\,dx$$
$$= \frac{1}{2}\int_0^{\pi} \ln(\sin x)\,dx - \frac{\pi}{2}\ln 2$$

On the other hand, we notice that the graph of $\sin x$ is symmetric over the interval $[0, \pi]$ across the line $x = \pi/2$, hence
$$\int_0^{\pi} \ln(\sin x)\,dx = 2\int_0^{\pi/2} \ln(\sin x)\,dx$$
and then
$$\frac{1}{2}\int_0^{\pi} \ln(\sin x)\,dx = \int_0^{\pi/2} \ln(\sin x)\,dx$$
So we get
$$\int_0^{\pi/2} \ln(\sin x)\,dx = -\frac{\pi}{2}\ln 2$$

71. Use integration by parts twice, first time
let $u = -\frac{1}{2}x^3, dv = -2xe^{-x^2}\,dx$
second time
let $u = -\frac{1}{2}x, dv = -2xe^{-x^2}\,dx$

$$\int x^4 e^{-x^2}\,dx$$
$$= -\frac{1}{2}x^3e^{-x^2} + \int \frac{3}{2}x^2e^{-x^2}\,dx$$
$$= -\frac{1}{2}x^3e^{-x^2}$$
$$+ \frac{3}{2}\left(-\frac{1}{2}xe^{-x^2} + \frac{1}{2}\int e^{-x^2}\,dx\right)$$
$$= -\frac{1}{2}x^3e^{-x^2} - \frac{3}{4}xe^{-x^2} + \frac{3}{4}\int e^{-x^2}\,dx$$

Putting integration limits to all the above, and realizing that when taking limits to $\pm\infty$, all multiples of e^{-x^2} as shown in above will go to 0 (we have seen this a lot of times before). Then we get

$$\int_{-\infty}^{\infty} x^4 e^{-x^2}\,dx = \frac{3}{4}\int_{-\infty}^{\infty} e^{-x^2}\,dx = \frac{3}{4}\sqrt{\pi}$$

This means when $n = 2$, the statement

$$\int_{-\infty}^{\infty} x^{2n} e^{-x^2}\,dx = \frac{(2n-1)\cdots 3\cdot 1}{2^n}\sqrt{\pi}$$

is true. (We can also check that the case for $n = 1$ is correct.)

For general n, supposing that the statement is true for all $m < n$, then integration by parts gives

$$\int x^{2n} e^{-x^2}\,dx$$
$$= -\frac{1}{2}x^{2n-1}e^{-x^2} + \frac{2n-1}{2}\int x^{2n-2}e^{-x^2}\,dx$$

and hence

$$\int_{-\infty}^{\infty} x^{2n} e^{-x^2}\,dx$$
$$= \frac{2n-1}{2}\int_{-\infty}^{\infty} x^{2n-2}e^{-x^2}\,dx$$
$$= \frac{2n-1}{2}\cdot\frac{(2n-3)\cdots 3\cdot 1}{2^{n-1}}\sqrt{\pi}$$
$$= \frac{(2n-1)\cdots 3\cdot 1}{2^n}\sqrt{\pi}$$

73. Substitute $u = 0.001x$

$$\mu = \lim_{R\to\infty}\int_0^R 0.001xe^{-0.001x}\,dx$$
$$= \lim_{R\to\infty}\int_0^R 1000ue^{-u}\,du$$
$$= \lim_{R\to\infty} 1000e^{-u}(-u-1)\Big|_0^R$$
$$= \lim_{R\to\infty} 1000\left[e^{-R}(-R-1)+1\right]$$
$$= 1000$$

75. $$\int_0^{35}\frac{1}{40}e^{-x/40}\,dx = -e^{-x/40}\Big|_0^{35}$$
$$= 1 - e^{-35/40}$$

$$P(x > 35) = 1 - \text{ above } = e^{-35/40}$$

$$\int_0^{40}\frac{1}{40}e^{-x/40}\,dx = -e^{-x/40}\Big|_0^{40}$$
$$= 1 - e^{-40/40}$$

$$P(x > 40) = 1 - \text{ above } = e^{-40/40}$$

$$\int_0^{45}\frac{1}{40}e^{-x/40}\,dx = -e^{-x/40}\Big|_0^{45}$$
$$= 1 - e^{-45/40}$$

$$P(x > 45) = 1 - \text{ above } = e^{-45/40}$$

Hence,

$$P(x > 40|x > 35) = \frac{P(x > 40)}{P(x > 35)}$$
$$= \frac{e^{-40/40}}{e^{-35/40}} = e^{-5/40} \approx 0.8825$$

$$P(x > 45|x > 40) = \frac{P(x > 45)}{P(x > 40)}$$
$$= \frac{e^{-45/40}}{e^{-40/40}} = e^{-5/40} \approx 0.8825$$

77. Since

$$\int_0^A ce^{-cx}\,dx = -e^{-cx}\Big|_0^A = 1 - e^{-cA}$$

$$P(x > m+n|x > n) = \frac{P(x > m+n)}{P(x > n)}$$
$$= \frac{1 - \int_0^{m+n} ce^{-cx}\,dx}{1 - \int_0^m ce^{-cx}\,dx\,dx} = \frac{e^{-c(m+n)}}{e^{-cn}}$$
$$= e^{-cm}$$

79. (a) For $x \ge 0$,

$$F_1(x) = \int_{-\infty}^x f_1(t)\,dt$$
$$= \int_0^x f_1(t)\,dt$$
$$= \int_0^x 2e^{-2t}\,dt = -e^{-2t}\Big|_0^x$$
$$= 1 - e^{-2x}$$

$$\Omega_1(r) = \frac{\int_r^\infty [1 - F_1(x)]\,dx}{\int_{-\infty}^r F_1(x)\,dx}$$
$$= \frac{\int_r^\infty e^{-2x}\,dx}{\int_0^r (1 - e^{-2x})\,dx}$$
$$= \frac{\frac{1}{2}e^{-2r}}{r + \frac{1}{2}e^{-2r} - \frac{1}{2}} = \frac{e^{-2r}}{2r + e^{-2r} - 1}$$

(b) For $0 \le x \le 1$,

$$F_2(x) = \int_{-\infty}^{x} f_2(t)\,dt$$
$$= \int_0^x f_2(t)\,dt$$
$$= \int_0^x 1\,dt = t\Big|_0^x = x$$

$$\Omega_2(r) = \frac{\int_r^\infty [1 - F_2(x)]\,dx}{\int_{-\infty}^r F_2(x)\,dx}$$
$$= \frac{\int_r^1 (1-x)\,dx}{\int_0^r x\,dx}$$
$$= \frac{\frac{1}{2} - r + \frac{r^2}{2}}{\frac{r^2}{2}} = \frac{1 - 2r + r^2}{r^2}$$

(c) $$\mu_1 = \int_{-\infty}^{\infty} x f_1(x)\,dx$$
$$= \int_0^\infty 2xe^{-2x}\,dx$$
$$= \lim_{R\to\infty} \int_0^R 2xe^{-2x}\,dx$$
$$= \lim_{R\to\infty} e^{-2x}(-x - 1/2)\Big|_0^R$$
$$= \lim_{R\to\infty} e^{-2R}(R + 1/2) + \frac{1}{2} = \frac{1}{2}$$

$$\mu_2 = \int_{-\infty}^{\infty} x f_2(x)\,dx$$
$$= \int_0^1 x\,dx = \frac{x^2}{2}\Big|_0^1 = \frac{1}{2}$$

$\mu_1 = \mu_2$ and when $r = 1/2$

$$\Omega_1(1/2) = \frac{e^{-2\cdot 1/2}}{2\cdot 1/2 + e^{-2\cdot 1/2} - 1}$$
$$= 1$$

$$\Omega_2(1/2) = \frac{1 - 2\cdot 1/2 + (1/2)^2}{(1/2)^2}$$
$$= 1$$

(d) The graph of $f_2(x)$ is more stable than that of $f_1(x)$.

$f_1(x) > f_2(x)$ for $0 < x < 0.34$ and $f_1(x) < f_2(x)$ for $x > 1$.

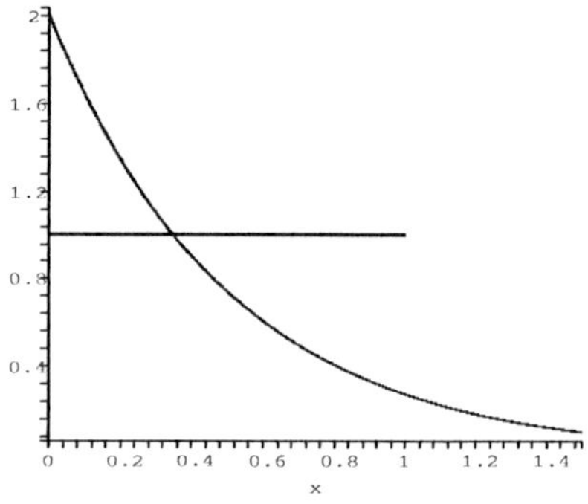

(e) $$\Omega_1(r) = 1 - \frac{2r - 1}{e^{-2r} + 2r - 1}$$
$$\Omega_2 r = 1 - \frac{2r - 1}{r^2}$$

and $r^2 - (e^{-2r} + 2r - 1$
$= e^{-2r} + (r-1)^2 > 0$

This means
when $r < 1/2$, $\Omega_1(r) < \Omega_2(r)$
when $r > 1/2$, $\Omega_1(r) > \Omega_2(r)$

In terms of this example, we see that the riskier investment is only disadvantageous when r small, and will be better when r large.

Ch. 6 Review Exercises

1. Substitute $u = \sqrt{x}$

$$\int \frac{e^{\sqrt{x}}}{\sqrt{x}}\,dx = 2\int e^u\,du$$
$$= 2e^u + c = 2e^{\sqrt{x}} + c$$

3. Use the table of integrals,

$$\int \frac{x^2}{\sqrt{1 - x^2}}\,dx$$
$$= -\frac{1}{2}x\sqrt{1 - x^2} + \frac{1}{2}\sin^{-1} x + c$$

5. Use integration by parts, twice:

$$\int x^2 e^{-3x}\,dx$$
$$= -\frac{1}{3}x^2 e^{-3x} + \frac{2}{3}\int xe^{-3x}\,dx$$
$$= -\frac{1}{3}x^2 e^{-3x}$$

$$+\frac{2}{3}\left(-\frac{1}{3}xe^{-3x}+\frac{1}{3}\int e^{-3x}\,dx\right)$$
$$=-\frac{1}{3}x^2e^{-3x}-\frac{2}{9}xe^{-3x}-\frac{2}{27}e^{-3x}+c$$

7. Substitute $u = x^2$
$$\int \frac{x}{1+x^4}\,dx = \frac{1}{2}\int \frac{du}{1+u^2}$$
$$=\frac{1}{2}\tan^{-1}u+c=\frac{1}{2}\tan^{-1}x^2+c$$

9. $\displaystyle \frac{x^3}{4+x^4}\,dx = \frac{1}{4}\ln(4+x^4)+c$

11. $\displaystyle \int e^{2\ln x}\,dx = \int x^2\,dx = \frac{x^3}{3}+c$

13. Integration by parts,
$$\int_0^1 x\sin 3x\,dx$$
$$=-\frac{1}{3}x\cos 3x\Big|_0^1+\frac{1}{3}\int_0^1 \cos 3x\,dx$$
$$=-\frac{1}{3}\cos 3+\frac{1}{9}\sin 3x\Big|_0^1$$
$$=-\frac{1}{3}\cos 3+\frac{1}{9}\sin 3$$

15. Use the table of integrals
$$\int_0^{\pi/2}\sin^4 x\,dx$$
$$=-\frac{1}{4}\sin^3 x\cos x\Big|_0^{\pi/2}$$
$$+\frac{3}{4}\left(\frac{x}{2}-\frac{1}{2}\sin x\cos x\right)\Big|_0^{\pi/2}$$
$$=\frac{3\pi}{16}$$

17. Use integration by parts,
$$\int_{-1}^1 x\sin \pi x\,dx$$
$$=-\frac{1}{\pi}x\cos \pi x\Big|_{-1}^1+\frac{1}{\pi}\int_{-1}^1 \cos \pi x\,dx$$
$$=\frac{2}{\pi}+\frac{1}{\pi^2}\sin \pi x\Big|_{-1}^1=\frac{2}{\pi}$$

19. Use integration by parts
$$\int_1^2 x^3\ln x\,dx=\frac{x^4}{4}\ln x\Big|_1^2-\frac{1}{4}\int_1^2 x^3\,dx$$
$$=4\ln 2-\frac{x^4}{16}\Big|_1^2=4\ln 2-\frac{15}{16}$$

21. Substitute $u = \sin x$
$$\int \cos x\sin^2 x\,dx=\int u^2\,du$$
$$=\frac{u^3}{3}+c=\frac{\sin^3 x}{3}+c$$

23. Substitute $u = \sin x$
$$\int \cos^3 x\sin^3 x\,dx=\int (1-u^2)u^3\,du$$
$$=\frac{u^4}{4}-\frac{u^6}{6}+c=\frac{3\sin^4 x-2\sin^6 x}{12}+c$$

25. Substitute $u = \tan x$
$$\int \tan^2 x\sec^4 x\,dx=\int u^2(1+u^2)\,du$$
$$=\frac{u^3}{3}+\frac{u^5}{5}+c=\frac{5\tan^3 x+3\tan^5 x}{15}+c$$

27. Substitute $u = \sin x$
$$\int \sqrt{\sin x}\cos^3 x\,dx$$
$$=\int u^{1/2}(1-u^2)\,du$$
$$=\frac{2}{3}u^{3/2}-\frac{2}{7}u^{7/2}+c$$
$$=\frac{2}{3}\sin^{3/2}x-\frac{2}{7}\sin^{7/2}x+c$$

29. Complete the square,
$$\int \frac{2}{8+4x+x^2}\,dx$$
$$=\int \frac{2}{(x+2)^2+2^2}\,dx$$
$$=\tan^{-1}\left(\frac{x+2}{2}\right)+c$$

31. Use the table of integrals,
$$\int \frac{2}{x^2\sqrt{4-x^2}}\,dx=-\frac{\sqrt{4-x^2}}{2x}+c$$

33. Substitute $u = 9-x^2$
$$\int \frac{x^3}{\sqrt{9-x^2}}\,dx=-\frac{1}{2}\int \frac{(9-u)}{u^{1/2}}\,du$$
$$=-\frac{9}{2}\int u^{-1/2}\,du+\frac{1}{2}\int u^{1/2}\,du$$
$$=-9u^{1/2}+\frac{1}{3}u^{3/2}+c$$
$$=-9(9-x^2)^{1/2}+\frac{1}{3}(9-x^2)^{3/2}+c$$

35. Substitute $u = x^2 + 9$

$$\int \frac{x^3}{\sqrt{x^2+9}}\,dx = \frac{1}{2}\int (u-9)u^{-1/2}\,du$$
$$= \frac{1}{3}u^{3/2} - 9u^{1/2} + c$$
$$= \frac{1}{3}(x^2+9)^{3/2} - 9(x^2+9)^{1/2} + c$$

37. Use the method of PFD

$$\int \frac{x+4}{x^2+3x+2}\,dx$$
$$= \int \left(\frac{3}{x+1} + \frac{-2}{x+2}\right) dx$$
$$= 3\ln|x+1| - 2\ln|x+2| + c$$

39. Use the method of PFD

$$\int \frac{4x^2+6x-12}{x^3-4x}\,dx$$
$$= \int \left(\frac{3}{x} + \frac{-1}{x+2} + \frac{2}{x-2}\right) dx$$
$$= 3\ln|x| - \ln|x+2| + 2\ln|x-2| + c$$

41. Use the table of integrals,

$$\int e^x \cos 2x\,dx$$
$$= \frac{(\cos 2x + 2\sin 2x)e^x}{5} + c$$

43. Substitute $u = x^2 + 1$

$$\int x\sqrt{x^2+1}\,dx = \frac{1}{2}\int u^{1/2}\,du$$
$$= \frac{1}{3}u^{3/2} + c = \frac{1}{3}(x^2+1)^{3/2} + c$$

45. $$\frac{4}{x^2-3x-4} = \frac{A}{x+1} + \frac{B}{x-4}$$

$$4 = A(x-4) + B(x+1)$$
$$= (A+B)x + (-4A+B)$$
$$A = -\frac{4}{5};\ B = \frac{4}{5}$$

$$\frac{4}{x^2-3x-4} = \frac{-4/5}{x+1} + \frac{4/5}{x-4}$$

47. $$\frac{-6}{x^3+x^2-2x} = \frac{A}{x} + \frac{B}{x-1} + \frac{C}{x+2}$$

$$-6 = A(x-1)(x+2) + Bx(x+2) + cx(x-1)$$
$$A = -3;\ B = -2;\ C = -1$$

$$\frac{-6}{x^3+x^2-2x} = \frac{-3}{x} + \frac{-2}{x-1} + \frac{-1}{x+2}$$

49. $$\frac{x-2}{x^2+4x+4} = \frac{A}{x+2} + \frac{B}{(x+2)^2}$$

$$x - 2 = A(x+2) + B$$
$$A = 1;\ B = -4$$

$$\frac{x-2}{x^2+4x+4} = \frac{1}{x+2} + \frac{-4}{(x+2)^2}$$

51. Substitute $u = e^{2x}$

$$\int e^{3x}\sqrt{4+e^{2x}}\,dx$$
$$= \int e^{2x}\sqrt{4e^{2x}+e^{4x}}\,dx$$
$$= \frac{1}{2}\int \sqrt{4u+u^2}\,du$$
$$= \frac{1}{2}\int \sqrt{(u+2)^2-4}\,du$$
$$= \frac{1}{4}(u+2)\sqrt{(u+2)^2-4}$$
$$- \ln|(u+2) + \sqrt{(u+2)^2-4}| + c$$
$$= \frac{(e^{2x}+2)\sqrt{4e^{2x}+e^{4x}}}{4}$$
$$- \ln\left|(e^{2x}+2) + \sqrt{4e^{2x}+e^{4x}}\right| + c$$

53. $$\int \sec^4 x\,dx$$

$$= \frac{1}{3}\sec^2 x \tan x + \frac{2}{3}\int \sec^2 x\,dx$$
$$= \frac{1}{3}\sec^2 x \tan x + \frac{2}{3}\tan x + c$$

55. Substitute $u = 3 - x$

$$\int \frac{4}{x(3-x)^2}\,dx = -4\int \frac{1}{(3-u)u^2}\,du$$
$$= \frac{4}{9}\ln\left|\frac{3-u}{u}\right| + \frac{4}{3u} + c$$
$$= \frac{4}{9}\ln\left|\frac{x}{3-x}\right| + \frac{4}{3(3-x)} + c$$

57. $$\int \frac{\sqrt{9+4x^2}}{x^2}\,dx = \int \frac{2\sqrt{\frac{9}{4}+x^2}}{x^2}\,dx$$

$$= 2\left(\frac{-\sqrt{\frac{9}{4}+x^2}}{x} + \ln\left|x+\sqrt{9/4+x^2}\right|\right) + c$$

$$= -\frac{\sqrt{9+4x^2}}{x} + 2\ln\left|x+\sqrt{\frac{9}{4}+x^2}\right| + c$$

59. $\int \frac{\sqrt{4-x^2}}{x}\,dx$

$$= \sqrt{4-x^2} - 2\ln\left|\frac{2+\sqrt{4-x^2}}{x}\right| + c$$

61. Substitute $u = x^2 - 1$

$$\int_0^1 \frac{x}{x^2-1}\,dx = \int_{-1}^0 \frac{du}{2u}$$

$$= \lim_{R\to 0^-}\int_{-1}^R \frac{du}{2u} = \lim_{R\to 0^-} \ln|u|\Big|_{-1}^0$$

This limit does not exist, so the integral diverges.

63. $\int_1^\infty \frac{3}{x^2}\,dx = \lim_{R\to\infty}\int_1^R \frac{3}{x^2}\,dx$

$$= \lim_{R\to\infty} -\frac{3}{x}\Big|_1^R = \lim_{R\to\infty} -\frac{3}{R} + 3 = 3$$

65. $\int_0^\infty \frac{4}{4+x^2}\,dx = \lim_{R\to\infty}\int_0^R \frac{4}{4+x^2}\,dx$

$$= \lim_{R\to\infty} 2\tan^{-1}\frac{x}{2}\Big|_0^R$$

$$= \lim_{R\to\infty} 2\tan^{-1} R = \pi$$

67. $\int_0^2 \frac{3}{x^2}\,dx = \lim_{R\to 0^+}\int_R^2 \frac{3}{x^2}\,dx$

$$= \lim_{R\to 0^+} -\frac{3}{x}\Big|_R^2 = \infty$$

So the original integral diverges.

69. If $c(t) = R$, then the total amount of dye is

$$\int_0^T c(t)\,dt = \int_0^T R\,dt = RT$$

If $c(t) = 3te^{2Tt}$, then we can use integration by parts to get

$$\int_0^T 3te^{2Tt}\,dt$$

$$= \frac{3t}{2T}e^{2Tt}\Big|_0^T - \int_0^T \frac{3}{2T}e^{2Tt}\,dt$$

$$= \frac{3}{2}e^{2T^2} - \frac{3}{4T^2}e^{2Tt}\Big|_0^T$$

$$= \frac{3}{2}e^{2T^2} - \frac{3}{4T^2}e^{2T^2} + \frac{3}{4T^2}$$

Since $R = c(T) = 3Te^{2T^2}$

The cardiac output is

$$\frac{RT}{\int_0^T c(t)\,dt} = \frac{3T^2e^{2T^2}}{\frac{3}{2}e^{2T^2} - \frac{3}{4T^2}e^{2T^2} + \frac{3}{4T^2}}$$

$$= \frac{RT^3}{3T^2e^{2T^2}/2 - 3e^{2T^2}/4 + 3/4}$$

71. $f_{n,\text{ave}} = \frac{1}{e^n}\int_0^{e^n} \ln x\,dx$

$$= \frac{1}{e^n}\lim_{R\to 0}\int_R^{e^n} \ln x\,dx$$

$$= \frac{1}{e^n}\lim_{R\to 0}(x\ln x - x)\Big|_R^{e^n}$$

$$= \frac{1}{e^n}\lim_{R\to 0}(ne^n - e^n - R\ln R + R)$$

$$= n - 1$$

73. $R(t) = P(x > t) = 1 - \int_0^t ce^{-cx}\,dx$

$$= 1 - (-e^{-cx})\Big|_0^t = 1 - (1 - e^{-ct}) = e^{-ct}$$

Hence

$$\frac{f(t)}{R(t)} = \frac{ce^{-ct}}{e^{-ct}} = c$$

75. We use a CAS to see that

$$\int_{90}^{100} \frac{1}{\sqrt{450\pi}}e^{-(x-100)^2/450}\,dx$$

$$\approx 24.75\%$$

We can use substitution to get

$$\frac{1}{\sqrt{450\pi}}\int_a^\infty e^{-(x-100)^2/450}\,dx$$

$$= \frac{1}{\sqrt{\pi}}\int_{\frac{a-90}{\sqrt{450}}}^\infty e^{-u^2}\,du$$

Since $\int_{-\infty}^\infty e^{-x^2}\,dx = \sqrt{\pi}$,

$$\int_0^{\infty} e^{-x^2}\, dx = \frac{\sqrt{\pi}}{2}$$

So we want to find the value of a so that

$$\int_0^{\frac{a-90}{\sqrt{450}}} e^{-u^2}\, du = 0.49\sqrt{\pi}$$

Using a CAS we find

$$\frac{a-90}{\sqrt{450}} \approx 1.645, a \approx 125$$

Some body being called a genius need to have a IQ score of at least 125.

Chapter 7

First-Order Differential Equations

7.1 Modeling with Differential Equations

1. Exponential growth with $k = 4$ so we can use Equation (1.4) to arrive at the general solution of $y = Ae^{4t}$. The initial condition gives $2 = A$ so the solution is $y = 2e^{4t}$.

3. Exponential growth with $k = -3$ so we can use Equation (1.4) to arrive at the general solution of $y = Ae^{-3t}$. The initial condition gives $5 = A$ so the solution is $y = 5e^{-3t}$.

5. Exponential growth with $k = 2$ so we can use Equation (1.4) to arrive at the general solution of $y = Ae^{2t}$. The initial condition gives $2 = Ae^2, A = \dfrac{2}{e^2}$ so the solution is $y = \dfrac{2}{e^2}e^{2t}$.

7. Integrating give the general solution $y = -3t + c$. The initial condition gives $3 = c$ and so the solution is $y = -3t + 3$.

9. The equation for population must be $y(t) = 100e^{kt}$

We know that in 4 hours, the population doubles, so
$200 = y(4) = 100e^{k4}$

Solving for k give $k = (\ln 2)/4$ and $y(t) = 100e^{t(\ln 2)/4}$

To determine when the population reaches $6,000$, we solve $y(t) = 6,000$ or
$6000 = 100e^{t(\ln 2)/4}$
Solving gives
$t = \dfrac{4\ln 60}{\ln 2} \approx 23.628$ hours.

11. The equation for population must be $y(t) = 400e^{kt}$

We know that in 1 hour, the population is 800, so $800 = y(1) = 400e^k$. Solving for k gives $k = \ln 2$.
$y(t) = 400e^{t\ln 2}$

After 10 hours, the population is $y(8) = 400e^{10\ln 2} = 409,600$ cells.

13. The equation for population is $y = Ae^{0.44t}$

The initial population is A, so we want to find when the population is $2A$. So we solve the equation
$2A = Ae^{0.44t}$
which gives solution
$t = \dfrac{\ln 2}{0.44} \approx 1.5753$ hours.

15. With t measured in minutes, and $y = Aekt = 10^8e^{kt}$ on the time interval $(0, T)$ (during which no treatment is given), the condition on T is that 10% of the population at time T (surviving after the treatment) will be the same as the initial population.

In other words, $10^8 = (.1)10^8e^{kT}$. This gives $e^{kT} = 10$ and $T = \ln(10)/k$.

To get k we use the given doubling time $t_d = 20$. Since we always have $t_d = \ln(2)/k$, this leads to $k = \ln(2)/20$ and
$T = \dfrac{\ln(10)}{\ln(2)/20} = \dfrac{20\ln(10)}{\ln(2)}$
≈ 66.44 minutes.

17. Given $y(t) = Ae^{rt}$, the doubling time t_d obeys $2A = Ae^{rt_d}, 2 = e^{rt_d}$
$rt_d = \ln 2, t_d = \dfrac{\ln 2}{r}$ as desired.

19. We apply the formula of Exercise 18 to get half-life equal to
$\dfrac{\ln 2}{1.3863} \approx 0.500$ days.

21. Using the formula in Exercise 18, we have $3 = -(\ln 2)/r$ and therefore $r = -(\ln 2)/3$.

Thus the formula for amount of substance is
$y(t) = Ae^{\ t(\ln 2)/3}$

The initial condition gives $A = 0.4$ and so
$y(t) = 0.4e^{-t(\ln 2)/3}$

Solve for t in the equation
$0.01 = 0.4e^{-t(\ln 2)/3}$ gives
$t = \dfrac{3\ln(40)}{\ln(2)} \approx 15.97$ hours.

Thus the amount will drop below 0.01 mg after 15.97 hours.

23. Using the formula in Exercise 18, we find the decay constant is
$r = -\dfrac{\ln 2}{28}$

Thus the formula for the amount of substance is
$y(t) = Ae^{rt}$

After 50 years,
$y(50) = Ae^{50r} \approx 0.29A$.

Thus, this is about 29% of the original amount of strontium-90.

25. The half-life is 5730 years, so
$r = -\dfrac{\ln 2}{5730}$
Solving for t in
$y(t) = 0.20A = Ae^{-rt}$ gives
$t = \dfrac{5730\ln(5)}{\ln(2)} \approx 13{,}305$ years.

27. Newton's Law of Cooling gives
$y(t) = Ae^{kt} + T_a$ with $T_a = 70$.
We have $y(0) = 200$ so
$200 = A + 70$ and $A = 130$
We have $y(1) = 180$ so
$180 = y(1) = 130e^k + 70$ and
$k = \ln\left(\dfrac{110}{130}\right)$.

The temperature will be 120 when
$120 = y(t) = 130e^{\ln(110/130)t} + 70$ and
$t = \dfrac{\ln(5/13)}{\ln(11/13)} \approx 5.720$ minutes.

29. Using Newton's Law of Cooling
$y = Ae^{kt} + T_a$
with $T_a = 70, y(0) = 50$, we get
$50 = Ae^0 + 70, A = -20$
so that $y(t) = -20e^{kt} + 70$.

If, after two minutes, the temperature is 56 degrees, $56 = -20e^{k2} + 70$
$e^{2k} = \dfrac{14}{20} = 0.7$
$2k = \ln 0.7, k = \dfrac{1}{2}\ln 0.7$

Therefore, $y(t) = -20e^{(\ln 0.7)t/2} + 70$.

31. Using Newton's Law of Cooling with ambient temperature 70 degrees, initial temperature 60 degrees, and with time t (in minutes) elapsed since 10:07, we have
$y(t) = Ae^{kt} + 70, 60 = Ae^{0k} + 70 = A + 70, A = -10$
and $y(t) = -10e^{kt} + 70$ (for the martini).

Two minutes later, its temperature is 61 degrees. Hence

$61 = -10e^{k2} + 70, e^{2k} = \frac{9}{10}$
$2k = \ln \frac{9}{10}, k = \frac{1}{2} \ln \frac{9}{10} = \frac{1}{2} \ln .9$

Therefore, $y(t) = -10e^{\left(\frac{1}{2}\ln .9\right)t} + 70$

The temperature is 40 degrees at elapsed time t only if
$40 = -10e^{\left(\frac{1}{2}\ln .9\right)t} + 70$
$t = \frac{2\ln 3}{\ln .9} \approx -20.854$
or about 21 minutes *before* 10:07 p.m. The time was 9:46 p.m.

33. With t the time elapsed since serving, with the ambient temperature 68 degrees and if the temperature is 160 degrees when $t = 20$, then
$y(t) = Ae^{kt} + T_a, 160 = Ae^{k\cdot 20} + 68$
$Ae^{20k} = 92$

After 22 minutes the temperature is 158 degrees,
$158 = Ae^{22k} + 68, Ae^{22k} = 90$
$e^{2k} = \frac{Ae^{22k}}{Ae^{20k}} = \frac{90}{92}, k = \frac{1}{2}\ln\frac{90}{92}$
Therefore,
$y(t) = Ae^{\frac{1}{2}\left(\ln \frac{90}{92}\right)t} + 68$

Using the first set of numbers,
$Ae^{20\cdot\frac{1}{2}\ln\frac{90}{92}} = 92$
$A = \frac{92}{e^{10\ln\frac{90}{92}}} \approx 114.615$

$y(t) = 114.615e^{\frac{1}{2}\left(\ln \frac{90}{92}\right)t} + 68$

The serving temperature is
$y(0) = 114.615e^0 + 68$
$= 182.615$ minutes.

35. Annual:
$A = 1000(1 + 0.08)^1 \approx \$1080.00$

Monthly:
$$A = 1000\left(1 + \frac{0.08}{12}\right)^{12} \approx \$1083.00$$

Daily:
$$A = 1000\left(1 + \frac{0.08}{365}\right)^{365} \approx \$1083.28$$

Continuous:
$A = 1000e^{(0.8)1} \approx \$1083.29$

37. Person A:
$A = 10{,}000e^{.12\cdot 20} = \$110{,}231.76$

Person B:
$B = 20,000e^{.12\cdot 10} = \$66{,}402.34$

ing gives $r = \ln 2/2 \approx 6.93\%$

39. Let t be the number of years after 1985. Then, assuming continuous compounding at rate r,
$9800 = 34e^{r\cdot 10}, e^{10r} = \frac{9800}{34}$
$r = \frac{1}{10}\ln\left(\frac{9800}{34}\right) \approx .566378$

Therefore,
$$A = 34e^{\frac{1}{10}\ln\left(\frac{9800}{34}\right)t} = 34\left(\frac{9800}{34}\right)^{t/10}$$

In 2005, $t = 20$ and
$$A = 34\left(\frac{9800}{34}\right)^2 = \$2{,}824{,}705.88$$

41. The problem with comparing tax rates for the income bracket $[16K, 20K]$ over a thirteen year time interval, is that due to inflation, the persons in this income bracket in 1988 have less purchasing power than those in the same bracket in 1975, and a lower tax rate may or may not compensate. To quantify and illustrate, assume a 5.5% annual inflation rate. This would translate into a loss of purchasing power amounting to $41/(1.055)^{13} = 1/(2.006) \approx 1/2$, which is essentially to say that in terms of comparable purchasing power, the income bracket $[16K, 20K]$ in 1988 corresponds to an income bracket of $[8K, 10K]$ in 1975. One should then go back and look at the tax rate for the latter bracket in 1975. Only if that tax rate exceeds the 1988 rate (15%) for the bracket $[16K, 20K]$

should one consider that taxes have genuinely gone down.

43. $T_1 = 30,000 \cdot 0.15 + (40{,}000 - 30{,}000) \cdot 0.28 = \$7300$

$T_2 = 30{,}000 \cdot 0.15 + (42{,}000 - 30{,}000) \cdot 0.28 = \$7860$

$T_1 + .05T_1 = \$7665$

The tax T_2 on the new salary is greater than the adjusted tax $(1.05T_1)$ on the old salary.

45. With a constant depreciation rate of 10%, the value of the \$40,000 item after ten years would be
$40{,}000(e^{-(0.1)10}) = 40{,}000e^{-1}$
$\approx \$14{,}715.18$

and after twenty years
$40{,}000(e^{-(0.1)20}) = 40{,}000e^{-2}$
$\approx \$5{,}413.41$

By the straight line method, assuming a value of zero after 20 years, the value would be \$20,000 after ten years.

47. Over an interval of elapsed time Δt, a drop from 1000 to 500 would be represented by a slope of $\dfrac{500}{\Delta t}$
while a drop from 10 to 5 would be represented by a slope of $\dfrac{5}{\Delta t}$.

A drop from $\ln 1000$ to $\ln 500$ would result in a slope of
$$\frac{\ln 1000 \ln 500}{\Delta t} = \frac{\ln \frac{1000}{500}}{\Delta t} = \frac{\ln 2}{\Delta t}$$

while a drop from $\ln 10$ to $\ln 5$ would result in a slope of
$$\frac{\ln 10 - \ln 5}{\Delta t} = \frac{\ln \frac{10}{5}}{\Delta t} = \frac{\ln 2}{\Delta t}.$$
So the slopes of the logarithmic drops are the same. If a population were changing at a constant percentage rate, the graph of population versus time would appear exponential while the graph of the logarithm of population versus time would appear linear.

49. Fitting a line to the first two data points on the plot of time vs. the natural log of the population ($y = \ln(P(x))$) produces the linear function
$y = 1.468x + 0.182$,
which is equivalent to fitting the original date with the exponential function
$P(x) = e^{1.468x+0.182}$ or
$P(x) = 1.200e^{1.468x}$

51. As in Exercise 49, we let x denote time and
$y = \ln P$.

We pick the second and fourth data point to fit a line to (any two data points are fine to use and will give slightly different answers). In this case, the points are
$(1, \ln 15)$ $(3, \ln 33)$.
The equation of the line connecting these two points is
$\ln P = y = 0.394x + 3.102$
Exponentiating this equation gives
$P = e^y = e^{0.394x+3.102} = 22.242\, e^{0.394x}$

53. As in Exercise 49, we let x denote time (with $x = 0$ corresponding to the year 1960) and let $y = \ln P$.

Looking at the graph of the modified data, we decide to use the first and last data points. In this case, the points are
$(0, \ln 7.5)$ $(30, \ln 1.6)$
The equation of the line connecting these two points is
$\ln P = y = -0.0515x + 2.0149$
Exponentiating this equation gives
$P = e^y = e^{-0.0515x+2.0149}$
$= 7.5\, e^{-0.0515x}$

55. With known conclusion
$y = Ae^{-rt}$, $A = 150$, $t = 24$, and $r = \ln(2)/t_h$ we find that with $t_h = 31$ we get
$y = 150(1/2)^{(24/31)} = 87.7$,
and with $t_h = 46$ we get
$y = 150(1/2)^{(24/46)} = 104.5$.

The difference is about 17 days, at 19% not a dramatically large percentage of the smaller base of 88(105/88 = 1.19). If one had expected the two numbers to be proportional to the half lives, one would have expected the difference to come in at 48% (46/31 = 1.48) and would definitely consider the 19% to be far less than anticipated.

57. In this case, with $t_h = 4$, $A = 1$, $y = Ae^{-rt}$, and $r = \ln(2)/4$, one finds $y = (1/2)^{(t/4)}$.

The curve is a typical exponential, declining from a value of 1 at $t = 0$ to $1/2^6 = 1/64 = .016$ at $t = 24$.

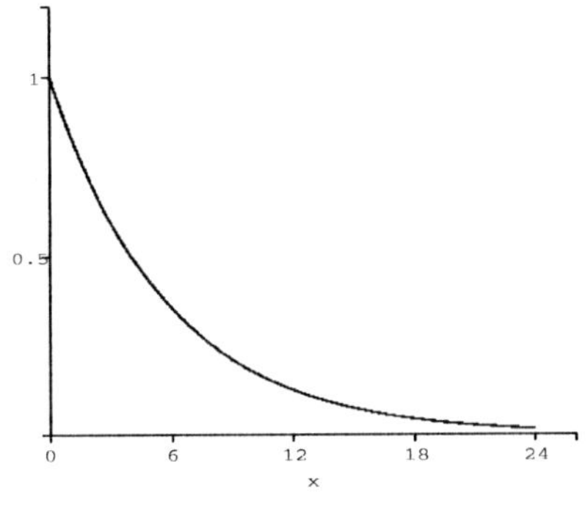

59. With r the rate of continuous compounding, the value of an initial amount X after t years is Xe^{rt}. If the goal is P, then the relation is $P = Xe^{rt}$ or $X = Pe^{-rt}$.

With $r = .08$, $t = 10$, $P = 10{,}000$, we find
$X = 10{,}000e^{-.8} = \$4493.29$.

61. $\displaystyle\int_0^T e^{-rt}dt = \frac{1-e^{-rT}}{r}$

$\displaystyle\int_0^T te^{-rt}dt = \frac{-Te^{-rT}}{r} + \frac{1-e^{-rT}}{r^2}$,

With $r = .05$ and $T = 3$, we find

for (A): $60{,}000(20)\ (1 - e^{-.15})$
$= \$167{,}150$

for (B): we get the above *plus*
$(3000) - 60e^{-.15} + 400(1 - e^{-.15})$
$= 12{,}223$ for a total of \$179,373

for (C), the exponentials cancel, and the answer is simply

$$\int_0^3 60000dt = \$180{,}000.$$

63. The comparison is to be made between three years of accumulation of \$1,000,000 versus the accumulation of four annual payments of \$280,000 at times 0, 1, 2, 3, then the respective figures are

$1{,}000{,}000(1.08)^3 = 1{,}259{,}712$

versus

$280{,}000(1.083 + 1.082 + 1.08 + 1)$
$= 1{,}261{,}711$.

One should take the annuity.

7.2 Separable Differential Equations

1. Separable.
$$\frac{y'}{\cos y} = 3x + 1$$

3. Not separable.

5. Separable.
$y' = y(x^2 + \cos x)$
$$\frac{y'}{y} = x^2 + \cos x$$

7. Not separable.

9. $\frac{1}{y}y' = x^2 + 1$

$$\int \frac{1}{y}\,dy = \int (x^2+1)\,dx$$

$$\ln|y| = \frac{x^3}{3} + x + c$$

$$y = e^{x^3/3+x+c}$$

$$y = Ae^{x^3/3+x}$$

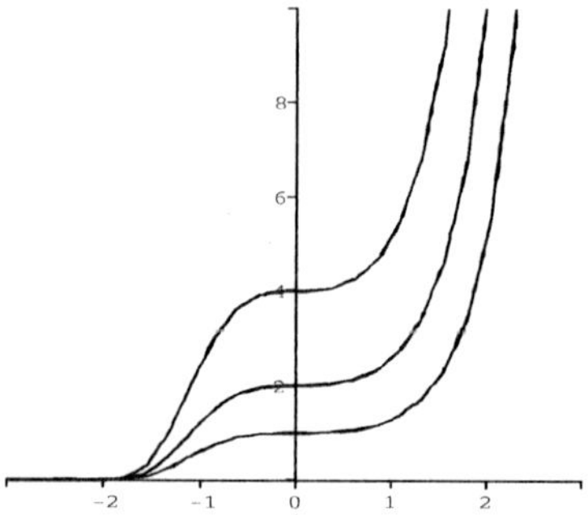

11. $\frac{1}{y^2}y' = 2x^2$

$$\int \frac{1}{y^2}\,dy = \int 2x^2\,dx$$

$$-\frac{1}{y} = \frac{2x^3}{3} + c$$

$$y = -\frac{1}{2x^3/3+c}$$

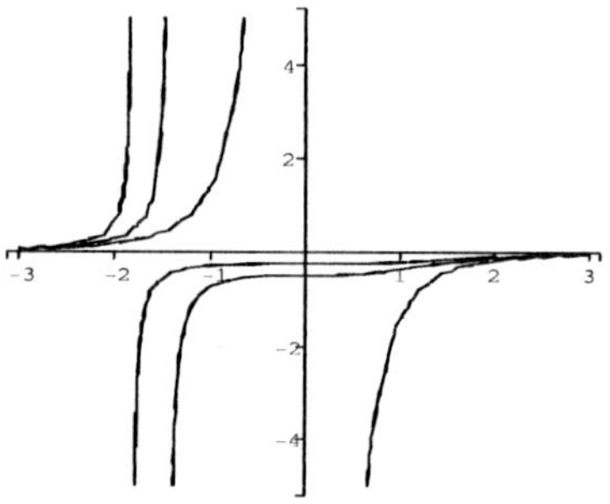

13. $yy' = \frac{6x^2}{1+x^3}$

$$\int y\,dy = \int \frac{6x^2}{1+x^3}\,dx$$

$$\frac{1}{2}y^2 = 2\ln|1+x^3| + c$$

$$y = \pm\sqrt{4\ln|1+x^3|+c}$$

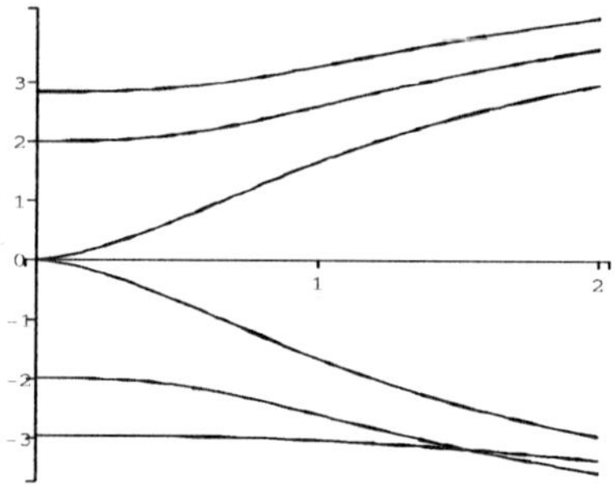

15. $ye^{-y}y' = 2xe^{-x}$

$$\int ye^{-y}\,dy = \int 2xe^{-x}dx$$

Integrating by parts,

$$\int xe^{-x}\,dx = -xe^{-x} - e^{-x} + c$$

Therefore,

$$-ye^{-y} - e^{-y} = -2xe^{-x} - 2e^{-x} + c$$

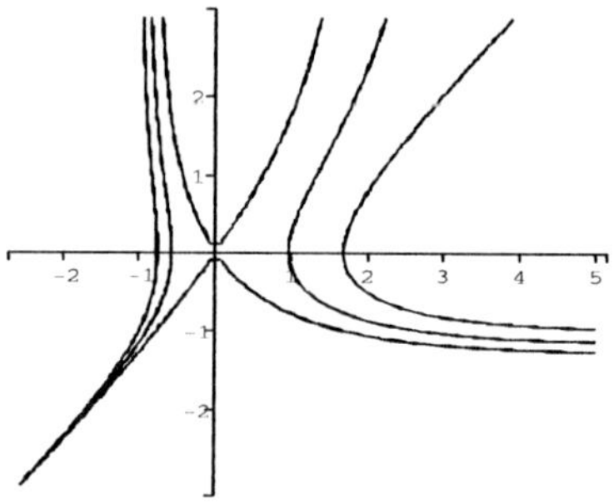

17. $\frac{1}{y^2-y}y' = 1$

$$\int \frac{1}{y^2-y}\,dy = \int 1\,dx$$

Integrating by partial fraction decomposition,

$$\int \frac{1}{y^2-y}\,dy = \int \left(\frac{1}{y-1} - \frac{1}{y}\right)\,dy$$

$$= \ln|y-1| - \ln|y| + c$$

Therefor, $\ln\left(\frac{y-1}{y}\right) = x + c$

$$\frac{y-1}{y} = ke^x$$

$$y = \frac{1}{1-ke^x}$$

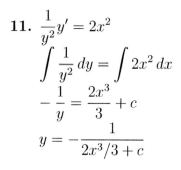

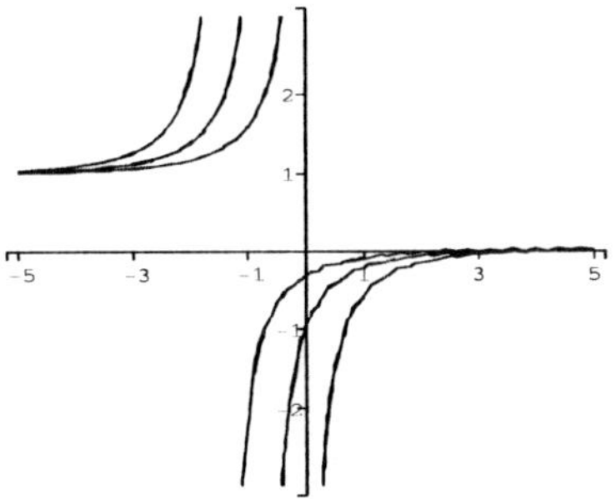

19. $\frac{1}{y}y' = \frac{x}{1+x^2}$

$\int \frac{1}{y}\,dy = \int \frac{x}{1+x^2}\,dx$

$\ln|y| = \frac{1}{2}\ln|1+x^2| + c$

$y = e^{\frac{1}{2}\ln|1+x^2|+c} = k\sqrt{1+x^2}$

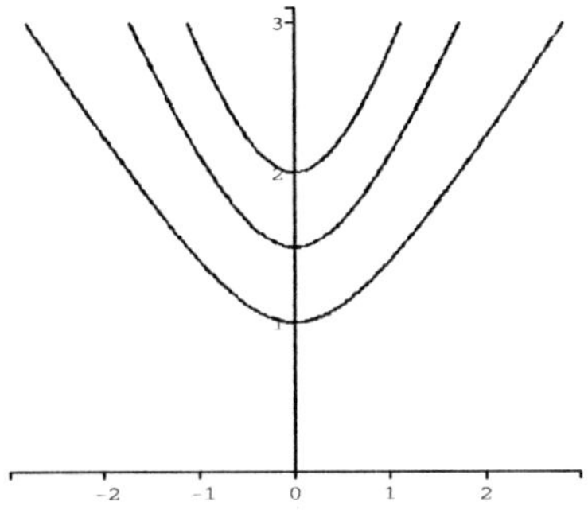

21. $\sec^2 y y' = \frac{1}{4x-3}$

$\int \sec^2 y dy = \int \frac{1}{4x-3}dx$

$\tan y = \frac{1}{4}\ln|4x-3| + c$

$y = \tan^{-1}\left[\frac{1}{4}\ln|4x-3| + c\right]$

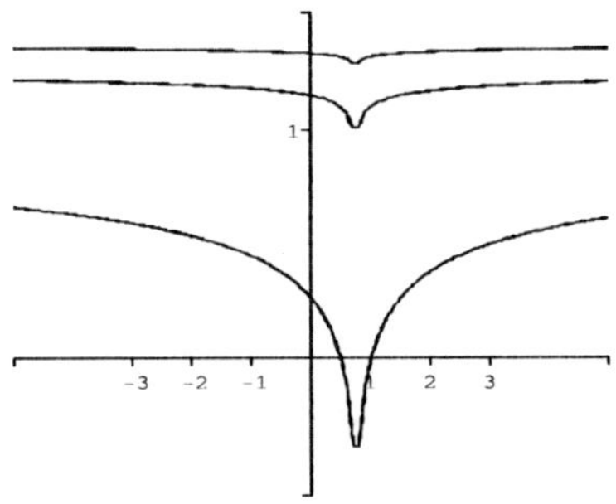

23. $\frac{y'}{y} = 3(x+1)^2$

$\ln y = (x+1)^3 + c$

$y = ke^{(x+1)^3}$

Using the initial condition,

$1 = ke, k = \frac{1}{e}$

$y = \frac{1}{e}\,e^{(x+1)^3}$

25. $yy' = 4x^2$

$\frac{y^2}{2} = \frac{4x^3}{3} + c$

Using the initial condition,

$\frac{2^2}{2} = c = 2$

$\frac{y^2}{2} = \frac{4x^3}{3} + 2$

27. $\frac{y'}{4y} = \frac{1}{x+3}$

$\frac{\ln|y|}{4} = \ln|x+3| + c$

$\ln|y| = 4\ln|x+3| + c$

$|y| = k(|x+3|)^4$

Using the initial condition,

$|1| = k(1)^4$

$k = 1$

$|y| = (|x+3|)^4$

29. $\cos y\, y' = 4x$

$\sin y = 2x^2 + c.$

Using the initial condition,

$0 = \sin(0) = \sin y(0) = 0 + c = c$

$\sin y = 2x^2$

$y = \arcsin(2x^2)$

$(-1/\sqrt{2} < x < 1/\sqrt{2})$

31. For this problem we have $M = 2$ and $k = 3$. Using these and the initial condition, we solve for A.

$1 = \frac{2Ae^{3(2)(0)}}{1+Ae^{3(2)(0)}} = \frac{2A}{1+A}$

$A = 1$

$y = \frac{2e^{6t}}{1+e^{6t}}$

33. For this problem we have $M = 5$ and $k = 2$. Using these and the initial condition, we solve for A.

$$4 = \frac{5Ae^{10(0)}}{1 + Ae^{10(0)}} = \frac{5A}{1+A}$$

$A = 4$

$$y = \frac{20e^{10t}}{1 + 4e^{10t}}$$

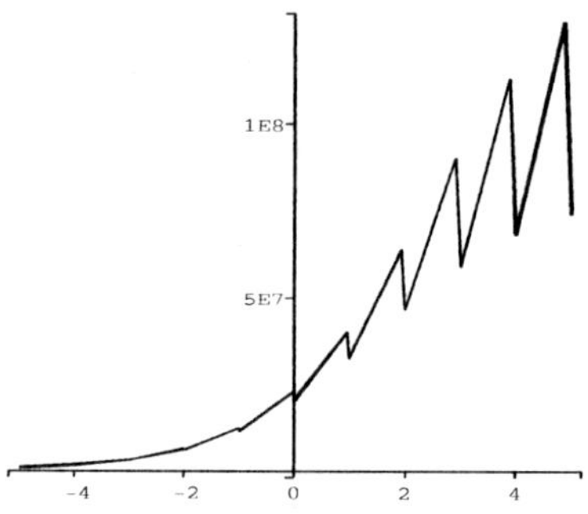

35. For this problem we have $M = 1$ and $k = 1$. Using these and the initial condition, we solve for A.

$$\frac{3}{4} = \frac{Ae^{(0)}}{1 + Ae^{(0)}} = \frac{A}{1+A}$$

$A = 3$

$$y = \frac{3e^t}{1 + 3e^t}$$

37. Substituting $r = Mk$ in

$$y' = ry\left(1 - \frac{y}{M}\right)$$

we get

$$y' = Mk\left(1 - \frac{y}{M}\right) = ky(M - y)$$

$$\frac{1}{y(M-y)}y' = k$$

Adapting the solution $y = \dfrac{MAe^{Mkt}}{1 + Ae^{Mkt}}$ in (2.7) with $r = Mk$, we find $y = \dfrac{MAe^{rt}}{1 + Ae^{rt}}$

In this case with $r = .71$, $M = 8 \times 10^7$ and $y(0) = 2 \times 10^7$, we find $2 \times 10^7 = y(0) = \dfrac{8 \times 10^7 A}{1 + A}$.

Therefore $\dfrac{A}{1+A} = \dfrac{2}{8} = \dfrac{1}{4}$, $A = 1/3$, and after routine simplification we find $y(t) = \dfrac{(8 \times 10^7)e^{.71t}}{3 + e^{.71t}}$

39. $\left|\dfrac{y}{M-y}\right| = Ae^{Mkt}$ with $A > 0$.

Under the circumstances $y > M$, the ratio is negative, and the resolution is $\dfrac{y}{M-y} = -Ae^{Mkt}$.
This further resolves as $y = -MAe^{Mkt} + yAe^{Mkt}$, which eventually becomes

$$y = \frac{MAe^{Mkt}}{Ae^{Mkt} - 1} = \frac{MAe^{rt}}{Ae^{rt} - 1}.$$

41. Starting from

$$y = \sqrt[3]{x^3 + \frac{21}{2}x^2 + 9x + 3c}$$

with $y(0) = 0$, we have $c = 0$.

Therefore,

$$y = \sqrt[3]{x^3 + \frac{21}{2}x^2 + 9x}$$

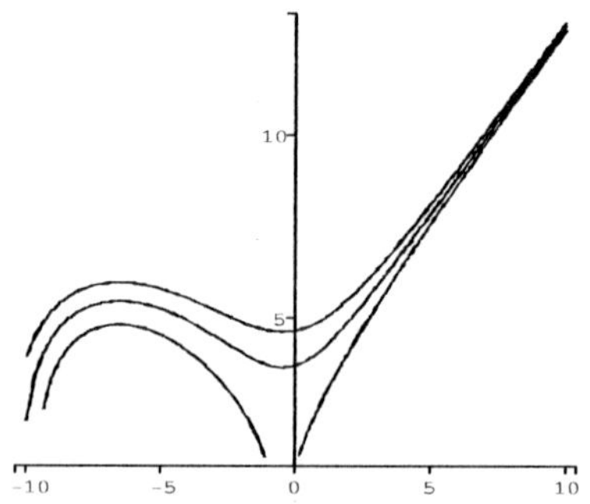

43. Given $y' = \dfrac{x^2 + 7x + 3}{y^2}$, that $y'(x)$ does not exist for a given x if $y(x) = 0$.
We see that $y(x) = 0$ if

$$-3c = x^3 + \left(\frac{21}{2}\right) x^2 + 9x$$

The cubic polynomial on the right, call it $h(x)$, has its derivative given by
$h'(x) = 3x^2+21x+9 = 3(x2+7x+3)$, and the roots of $h'(x)$ are

$$x_1 = \frac{-7-\sqrt{37}}{2} \approx -6.5414$$

$$x_2 = \frac{-7+\sqrt{37}}{2} \approx -.4586$$

The effect is that $h(x)$ has a relative maximum at x_1 and a relative minimum at x_2, and so the equation $-3c = h(x)$ has three solutions when $-3c$ lies between the relative minimum and the relative maximum, i.e. if $h(x_2) < -3c < h(x_1)$, or when

$$\frac{-h(x_1)}{3} < c < \frac{-h(x_2)}{3}$$

Therefor,

$$c_1 = -\left(\frac{217+37\sqrt{37}}{12}\right) \approx -36.84$$

$$c_2 = \frac{-217+37\sqrt{37}}{12} \approx .67185.$$

45. When $c = c_2, h(x) = -3c_2$
In effect, $h(x) + 3c_2$

$$= (x-x_2)^2\left(x + \frac{3c_2}{(x_2)^2}\right).$$

Now, in the solution $y(x)$ to the differential equation, we have

$$3y^2(x)y'(x) = \frac{d}{dx}(y^3)$$
$$= \frac{d}{dx}(h(x)+3c_2)$$
$$= h'(x) = (x-x_1)(x-x_2)$$

, while
$$y^2(x) = [y^3(x)]^{2/3} = [h(x)+3c_2]^{2/3}$$
$$= (x-x_2)^{4/3}(x-x_3)^{2/3}.$$

Now we can see that

$$y'(x) = \frac{h'(x)}{3y^2(x)}$$

$$= \frac{(x-x_1)}{3(x-x_2)^{1/3}(x-x_3)^{2/3}}$$

and this will become unbounded if x approaches either x_2 or x_3. These are the two points of vertical tangency.

47. Let A be the accumulated value at time t and d be the amount of the deposits made yearly, then A satisfies
$A' = 0.06A + d$
This differential equation separates to

$$\frac{A'}{0.06A+d} = 1$$

and integrates to

$$\frac{\ln(0.06A+d)}{0.06} = t + c$$

or

$$0.06A + d = ke^{0.06t}$$

At time $t = 0$, A is the unknown initial investment P, hence $k = .06P + 2000$, and so
$.06A + 2000 = (.06P + 2000)e^{.06t}$.
If we want $A = 1{,}000{,}000$ at $t = 20$, we must have
$62000 = (.06P + 2000)e^{1.2}$

$$P = \frac{62000e^{-1.2} - 2000}{.06} \approx \$277{,}901$$

49. We start with $A'(t) = 0.08A(t) - 12P$
$A(0) = 150,000$
where P is the payment made each month. Solving this differential equation:

$$\frac{A'}{0.08A - 12P} = 1$$

$$\frac{\ln(0.08A-12P)}{0.08} = t + c$$

$$0.08A - 12P = ke^{0.08t}$$

Using the initial condition gives
$k = 12000 - 12P$

We set $A(30) = 0$ and solve for P:
$-12P = (12000 - 12P)e^{2.4}$

$$P = \frac{12000e^{2.4}}{12(e^{2.4}-1)} \approx \$1099.77$$

Total amount paid:
$(30)(12)(1099.77) = \$395,917$

Total interest:
$395,917 - 150,000 = \$245,917$

51. Reworking Exercise 49:
$A'(t) = 0.08A(t) - 12P$
$A(0) = 150,000$
where P is the payment made each month. Solving this differential equation:
$$\frac{A'}{0.08A - 12P} = 1$$
$0.08A - 12P = ke^{0.08t}$
$k = 12000 - 12P$

We set $A(15) = 0$ and solve for P:
$-12P = (12000 - 12P)e^{1.2}$
$$P = \frac{12000e^{1.2}}{12(e^{1.2} - 1)} \approx \$1430.01$$

The monthly payments are increased by about \$330.

Total amount paid:
$(15)(12)(1430.01) = \$257,582$

The total amount is decreased by about \$138,335.

Total interest:
$257,582 - 150,000 = \$107,582$

53. Starting with
$A' = .08A + 10{,}000$ with the initial condition $A(0) = 0$.

Solving gives
$.08A + 10{,}000 = 10{,}000e^{.08t}$.

At time $t = 10$ we have
$$A = \frac{10{,}000(e^{.8} - 1)}{.08} = \$153{,}193$$
This would be the amount in his fund at age 40, and it would accumulate in the next 25 years to
$153{,}193e^{(.08)25} = \$1{,}131{,}949$.

55. Following the conditions of Exercise 53, replacing however the 8% by an unknown force r, we come after ten years of payment and twenty-five additional years of accumulation to
$$10{,}000\frac{(e^{10r} - 1)}{r}e^{25r}.$$
For contrast, if the payment rate 10,000 is replaced by 20,000, and the payment interval of ten years is replaced by twenty-five years, we come to an accumulation after the twenty-five years of
$$20{,}000\frac{(e^{25r} - 1)}{r}.$$
This number is to be compared to the previous. Equating the two expressions leads to
$2(e^{25r} - 1) = e^{35r} - e^{25r}$ or
$3e^{25r} - 2 = e^{35r}$.

The equation can only be solved with the help of some form of technology, but the answer of r about .105(10.5%) can at least be checked.

57. When the given numbers are substituted for the given symbols, the differential equation becomes
$$x' = (.4 - x)(.6 - x) - .625x^2$$
$$= \frac{3}{8}x^2 - x + \frac{6}{25}$$
$$= \frac{3}{8}\left(x - \frac{12}{5}\right)\left(x - \frac{4}{15}\right).$$
When separated it takes the form
$$\frac{x'}{(x - b)(x - a)} = r$$
in which $b = 12/5$, $a = 4/15 < b$, and $r = 3/8$.
By partial fractions we find
$$\frac{1}{(x - b)(x - a)}$$
$$= \frac{1}{(b - a)}\left\{\frac{1}{(x - b)}\frac{1}{(x - a)}\right\}$$
and after integration we find
$$\frac{1}{(b - a)}\ln\left|\frac{x - b}{x - a}\right| = rt + c_1$$
or in this case with
$b - a = (36/15) - (4/15) = 32/15$,

$$\ln\left|\frac{x-12/5}{x-4/15}\right| = \frac{32}{15}\left(\frac{3}{8}t+c_1\right)$$
$$= \frac{4}{5}t + c_2\left(c_2 = \frac{32}{15}c_1\right).$$

Using the given initial condition $x = .2 = 1/5$ when $t = 0$, we find $c_2 = \ln|(11/5)/(1/15)| = \ln(33)$,

$$\ln\left|\frac{x-12/5}{33\,(x-4/15)}\right|$$
$$= \frac{4}{5}t \text{ and}$$

$$\frac{x-12/5}{33\,(x-4/15)} = \pm e^{\frac{4}{5}t} = e^{\frac{4}{5}t}$$

(the choice of sign is + since the left side is 1 when $x = 1/5$).

Concluding the algebra we find

$$\frac{5x-12}{11(15x-4)} = e^{\frac{4}{5}t},$$

$$5x - 12 = 11(15x-4)e^{\frac{4}{5}t},$$

$$x = \frac{12 - 44e^{\frac{4}{5}t}}{5 - 11(15)e^{\frac{4}{5}t}} = \frac{4}{5}\left(\frac{3 - 11e^{\frac{4}{5}t}}{1 - 33e^{\frac{4}{5}t}}\right),$$

and it is apparent that $x \to \dfrac{4}{15}$ as $t \to \infty$.

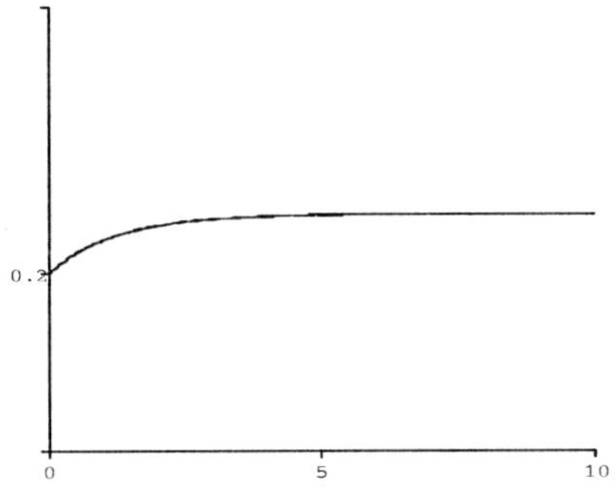

59. After beginning

$$x' = .6(.5-x)(.6-x) - .4x(0+x)$$
$$= .6(.3 - 1.1x + x^2) - .4x^2$$
$$= .2x^2 - .66x + .18$$
$$= \frac{1}{5}\left(x^2 - \frac{33}{10}x + \frac{9}{10}\right)$$
$$= \frac{1}{5}(x-3)\left(x - \frac{3}{10}\right).$$

The parameters b, a, r are respectively $3, 3/10, 1/5$. We jump ahead to

$$\ln\left|\frac{x-3}{x-3/10}\right| = \frac{27}{10}\left(\frac{t}{5}+c_1\right)$$
$$= \frac{27}{50}t + c_2.$$

In this case with $x = .2 = 1/5$ when $t = 0$, we find

$$c_2 = \ln\left|\frac{(1/5)-3}{(1/5)-(3/10)}\right| = \ln 28,$$

$$\frac{x-3}{28\,(x-3/10)} = \pm e^{\frac{27}{50}} = e^{\frac{27}{50}t},$$

and the conclusion is

$$5(x-3) = 14(10x-3)e^{\frac{27}{50}t},$$

$$x = \frac{15 - 42e^{\frac{27}{50}t}}{5 - 140e^{\frac{27}{50}t}} = \frac{42 - 15e^{-\frac{27}{50}t}}{140 - 5e^{-\frac{27}{50}t}}$$

$$= \frac{3}{5}\left(\frac{14 - 5e^{-\frac{27}{50}t}}{28 - e^{-\frac{27}{50}t}}\right).$$

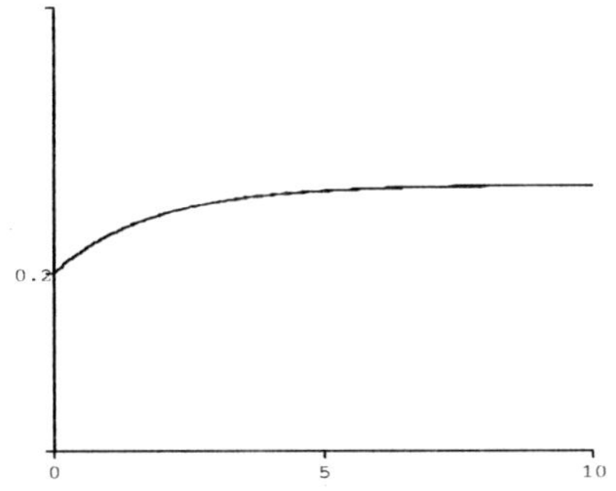

61. (a) This time we are given outright $x' = r(a-x)(b-x)$ with $r = .4$, $a = 6$, $b = 8$.

$$\ln\left|\frac{x-b}{x-a}\right| = (b-a)rt + c_2,$$

in this case

$$\ln\left|\frac{x-8}{x-6}\right| = \frac{4}{5}t + c_2,$$

and without using any initial condition, $\dfrac{x-8}{x-6} = ke^{\frac{4}{5}t}$, ($k = \pm e^{c_2}$ or zero),

$x - 8 = (x-6)ke^{\frac{4}{5}t}$,

$x = \dfrac{8 - 6ke^{\frac{4}{5}t}}{1 - ke^{\frac{4}{5}t}} = \dfrac{6k - 8e^{-\frac{4}{5}t}}{k - e^{-\frac{4}{5}t}}$.

Under the circumstances, we have $k = 4/3$ and

$x = 24\dfrac{1 - e^{-\frac{4}{5}t}}{4 - 3e^{-\frac{4}{5}t}}$.

It is now apparent that $x \to 6$ as $t \to \infty$, and this makes perfect sense again because of the molecular exchange: the reaction continues as long as there is any "free" A and B, but in this case there is less A initially ($a = 6$), and the process essentially stops when the A is essentially gone, i.e. when $x \to a(= 6)$.

(b) The solution to the differential equation is (with $a = 6$, $b = 8$ and $r = 0.6$):

$\ln\left|\dfrac{x-8}{x-6}\right| = \dfrac{6}{5}t + c_2$

or

$x = \dfrac{6ke^{6t/5} - 8}{ke^{6t/5} - 1}$

We use the initial condition $x(0) = 0$ which gives $k = \dfrac{8}{3}$ and therefore

$x = \dfrac{8e^{6t/5} - 8}{\frac{4}{3}e^{6t/5} - 1} = \dfrac{24\left(e^{6t/5} - 1\right)}{4e^{6t/5} - 3}$

63. We find

$y' = .025y(8-y) - .2 = -.025(y^2 - 8y + 8)$

$= -\dfrac{1}{40}(y-b)(y-a)$, in which $b = 4 + \sqrt{8}, a = 4 - \sqrt{8}$.

This leads to

$\ln\left|\dfrac{y-b}{y-a}\right| = -\dfrac{1}{40}\left(2\sqrt{8}\right)t + c_2$

and with $y(0) = 8$ we have

$\ln\left|\dfrac{8-b}{8-a}\right| = c_2$,

$\ln\left|\dfrac{(y-b)(8-a)}{(y-a)(8-b)}\right| = \dfrac{-t\sqrt{8}}{20}$,

$\dfrac{y-b}{y-a} = \dfrac{8-b}{8-a}e^{-\frac{t\sqrt{8}}{20}}$.

We can see that as $t \to \infty$ the right side goes to zero, hence also the left side, and hence

$y \to b = 4 + \sqrt{8} = 6.828427$

This represents an eventual fish population of about 682,800.

65. The equilibrium solutions are the algebraic solutions to the quadratic equation

$.025P(8-P) - R = 0$, or

$P^2 - 8P - 40R = 0$.

In the process of studying Exercise 63 ($R = .2$) we found it convenient to factor the left side (P was y at the time) and the roots were $b = 4 + \sqrt{8}$ and $a = 4 - \sqrt{8}$.

In Exercise 64, the corresponding equation ($R = .6$) would be

$0 = P^2 - 8P + 40R = P^2 - 8P + 24$.

But this equation has no real roots, hence no equilibrium populations.

67. $P' = .05P(8-P) - .6$

$= -\dfrac{1}{20}(P^2 - 8P + 12)$

$= -\dfrac{1}{20}(P-6)(P-2)$

Following well-established procedures, we come to

$\ln\left|\dfrac{P-6}{P-2}\right| = -\dfrac{1}{5}t + c_2,$

$\dfrac{P-6}{P-2} = Ae^{-\frac{t}{5}}, (A = \pm e^{c_2} \text{ or zero})$

We learn from this relation that the ratio $(P-6)/(P-2)$ never changes sign, always negative if the initial condition has $P(0)$ in the interval $(2, 6)$. Clearly in this case the exponential approaches zero as $t \to \infty$ and P approaches 6. This last conclusion is true even if $P(0) > 6$.

If on the other hand $0 \le P(0) < 2$, the ratio is forever positive, and we find eventually

$\dfrac{P-6}{P-2} = \dfrac{P(0)-6}{P(0)-2}e^{-t/5}.$

Here the right side is a positive decreasing function of t and so must be the left side. The effect is that P itself is decreasing (not obvious) and reaches the value zero when

$e^{-t/5} = 3\dfrac{P(0)-2}{P(0)-6}$ or when

$t = 5\ln\dfrac{6-P(0)}{3[2-P(0)]} = 5\ln\dfrac{6-P(0)}{6-3P(0)}$

In the ratio inside the (second) ln, the numerator is clearly more than the denominator, which is itself positive. This is some moment of positive time, after which the population is zero and no further activity occurs.

69. The differential equation is
$r'(t) = k[r(t) - S]$.

This separates as
$\dfrac{r'}{r-S} = k$, and solves as
$\ln(r - S) = kt + c.$

In this case $S = 1000$, $r(0) = 14{,}000$, and $r(4) = 8{,}000$.
Putting $t = 0$, we see that the constant c is $\ln 13{,}000$, we learn

$\ln\dfrac{r-1000}{13{,}000} = kt,$
and putting $t = 4$,
$\ln\dfrac{7}{13} = \ln\dfrac{7{,}000}{13{,}000} = 4k.$

Assembling the available information, we find

$\ln\dfrac{r-1000}{13{,}000} = kt$
$= \dfrac{t}{4}(4k) = \dfrac{t}{4}\ln\dfrac{7}{13}$, and

$r = 1{,}000 + 13{,}000\left(\dfrac{7}{13}\right)^{t/4},$

or equivalently
$r = 1 + 13e^{-.15476t}$ thousands.

71. If $P' = kP^{1.1}$, we separate as $P^{-1.1}P' = k$, and get
$\dfrac{-10}{P^{1/10}} = \dfrac{P^{-.1}}{(-.1)} = kt + c$
$= kt - \dfrac{10}{P(0)^{1/10}}$

$P = \left(\dfrac{kt}{10} - \dfrac{1}{P(0)^{1/10}}\right)^{-10}.$

We see that P approaches infinity as t approaches $\dfrac{10}{kP(0)^{1/10}}$.

73. From the differential equation, with $z = y'/y$, we find $z = k(M - y)$. This is a line in the (y, z)-plane. The z-intercept is M and the slope is $-k$.

75. $\dfrac{dv}{dt} = 32 - \dfrac{4v^2}{10}$
$= -\dfrac{4}{10}\left(v - \sqrt{80}\right)\left(v + \sqrt{80}\right)$

We see that here $k = -4/10$, is a negative number. The parameters b, a $(b > a)$ are
$b = \sqrt{80}$, $a = -\sqrt{80}$.

The solution is
$\ln\left|\dfrac{v-\sqrt{80}}{v+\sqrt{80}}\right| = 2kt\sqrt{80} + c_2.$
Given that $v(0) = 0$, we find $c2 = 0$

and
$$\frac{\sqrt{80}-v}{\sqrt{80}+v} = e^{2kt\sqrt{80}}.$$

Because $k < 0$, the right hand side goes to zero as t goes to infinity. Therefore $v \to \sqrt{80}$. This is the terminal velocity.

77. If $y' = ky(M-y)$, then by the product rule
$y'' = k[y'(M-y)-yy'] = ky'[M-2y]$.

This will be zero when $y = M/2$. In what follows, we make exception of the two equilibrium solutions $y \equiv 0$ and $y \equiv M$. With any other solution, $y \neq 0$, $y \neq M$, and $y' \neq 0$. Thus whatever time t_0 (if any) at which y becomes $M/2$ is sure to be an inflection time. Moreover, there can be *no circumstances of inflection other than* $y = M/2$. Such a time $t_0 > 0$ is bound to occur if and only if $0 < y(0) < M/2$, in which case the time t_0 is unique.

79. $y'' = -ay' \ln\left(\frac{y}{b}\right) - ay'$
$= ay'\left[\ln\left(\frac{y}{b}\right) - 1\right]$

Thus we have $y'' = 0$ if and only if
$y' = 0$ or if $y = \dfrac{b}{e}$.

Notice that $y' = 0$ can occur when $y = 0$ or when $y = b$. But, as in Exercise 77, these are equilibrium solutions — any solution with $y' = 0$ is a constant and can not have any inflection points.

7.3 Direction Fields and Euler's Method

1.

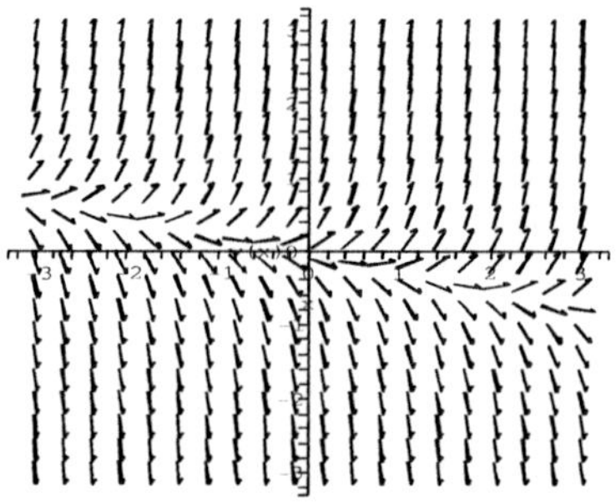

3.

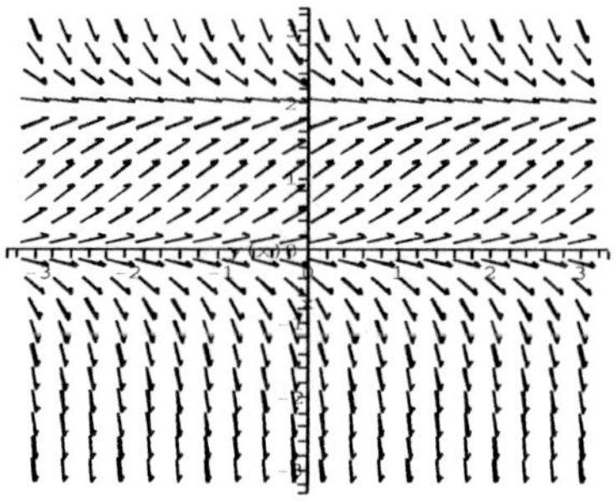

5.

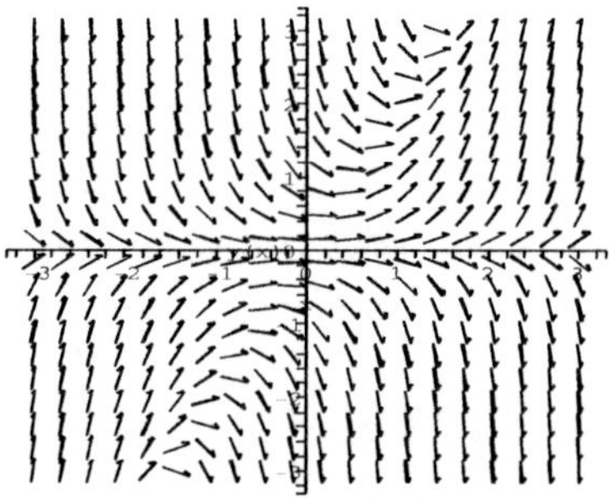

7. Field C.

9. Field D.

11. Field A.

13. For $h = 0.1$:

n	x_n	$y(x_n)$	$f(x_n, y_n)$
0	0.0	1	0
1	0.1	1	0.2
2	0.2	1.02	.408
3	0.3	1.0608	.63648
10	1.0	2.334633363	4.669266726
20	2.0	29.49864321	117.9945728

For $h = 0.05$:

n	x_n	$y(x_n)$	$f(x_n, y_n)$
0	0.00	1	0
1	0.05	1	.10
2	0.10	1.0050	.201000
3	0.15	1.01505000	.3045150000
20	1.00	2.510662314	5.021324628
40	2.00	39.09299942	156.3719977

15. First for $h = 0.1$:

n	x_n	$y(x_n)$	$f(x_n, y_n)$
0	0.0	1	3
1	0.1	1.3	3.51
2	0.2	1.651	3.878199
3	0.3	2.0388199	3.998493015
10	1.0	3.847783601	.58569576
20	2.0	3.999018724	0.00392415

For for $h = 0.05$:

n	x_n	$y(x_n)$	$f(x_n, y_n)$
0	0.00	1	3
1	0.05	1.15	3.2775
2	0.10	1.313875	3.529232484
3	0.15	1.490336624	3.740243243
20	1.00	3.818763110	.69210075
40	2.00	3.997787406	0.00884548

17. For $h = 0.1$:

n	x_n	$y(x_n)$	$f(x_n, y_n)$
0	0.0	3	−3
1	0.1	2.7	−2.604837418
2	0.2	2.439516258	−2.258247011
3	0.3	2.213691557	−1.954509778
10	1.0	1.300430235	−.6683096762
20	2.0	.9587323942	−0.0940676774

For $h = 0.05$:

n	x_n	$y(x_n)$	$f(x_n, y_n)$
0	0.0	3	−3
1	0.05	2.85	−2.801229424
2	0.10	2.709938529	−2.614775947
3	0.15	2.579199732	−2.439907708
20	1.00	1.334942742	−.7028221832
40	2.00	.9795316061	−.1148668893

19. For $h = 0.1$:

n	x_n	$y(x_n)$	$f(x_n, y_n)$
0	0.0	1	1.0
1	0.1	1.10	1.095445115
2	0.2	1.209544512	1.187242398
3	0.3	1.328268752	1.276036344
10	1.0	2.395982932	1.842819289
20	2.0	4.568765342	2.562960269

For $h = 0.05$:

n	x_n	$y(x_n)$	$f(x_n, y_n)$
0	0.00	1	1
1	0.05	1.05	1.048808848
2	0.10	1.102440442	1.096558454
3	0.15	1.157268365	1.143358371
20	1.00	2.420997836	1.849593965
40	2.00	4.620277218	2.572989937

21. (a) The exact solution to Exercise 13 is
$y(x) = e^{x^2}$
$y(1) \approx 2.718281828$
$y(2) \approx 54.59815003$

(b) The exact solution to Exercise 14 is
$y(x) = \sqrt{x^2 + 4}$
$y(1) \approx 2.236067977$
$y(2) \approx 2.828427124$

23.

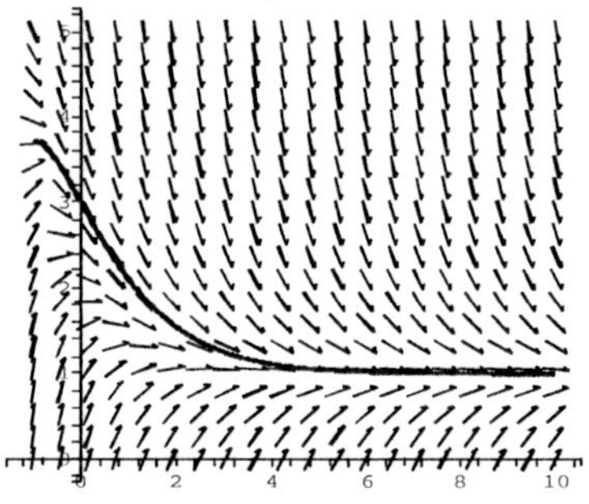

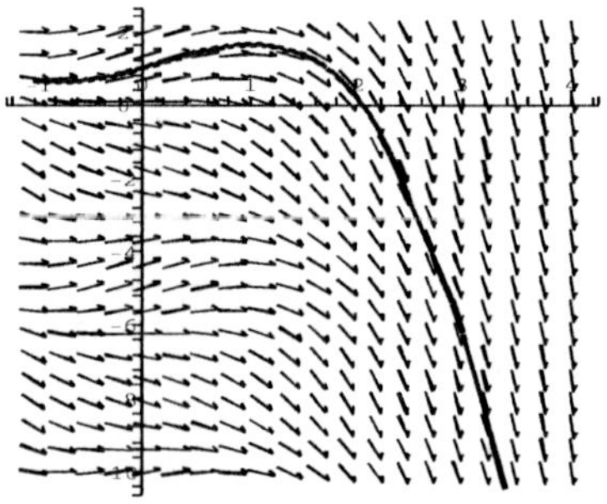

25. Equilibrium solutions come from $y' = 0$, which only occur when $y = 0$ or $y = 2$. From the direction field, $y = 0$ is seen to be an unstable equilibrium and $y = 2$ is seen to be a stable equilibrium.

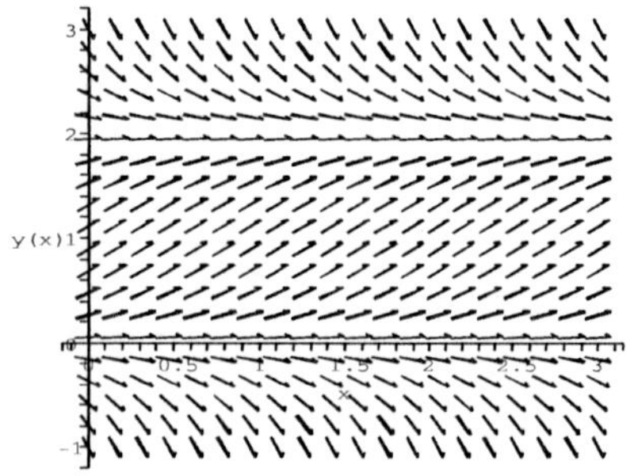

27. Equilibrium solutions come from $y' = 0$, which only occur when $y = 0$ or $y = \pm 1$. From the direction field, $y = 0$ and $y = -1$ are seen to be an unstable equilibrium and $y = 1$ is seen to be a stable equilibrium.

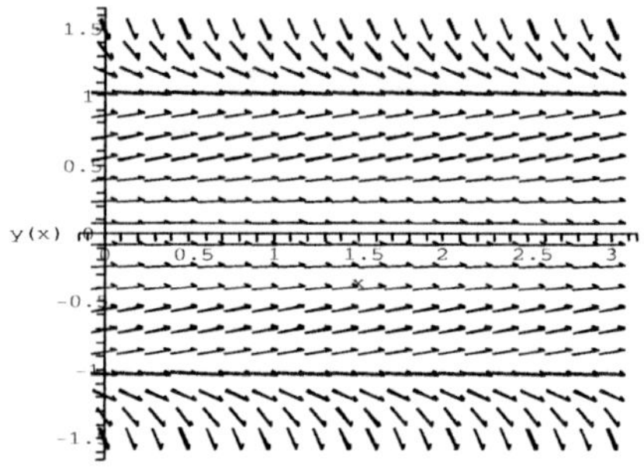

29. Equilibrium solutions come from $y' = 0$, which only occur when $y = 1$. From the direction field, $y = 1$ is seen to be a stable equilibrium.

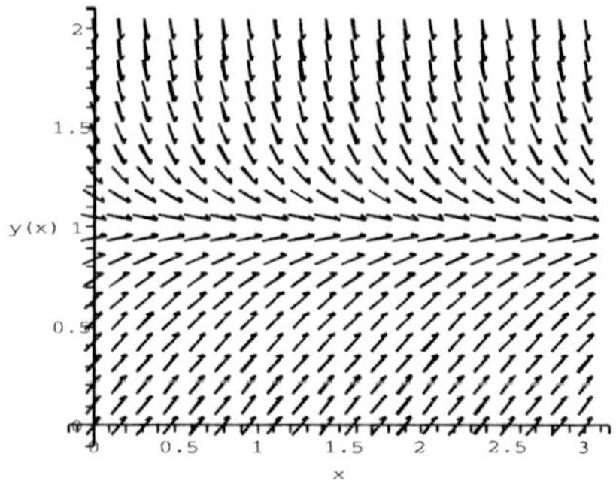

31. The equilibrium solutions are the constant solutions to the DE. If indeed g is a certain constant k, then
$0 = g' = -k + 3k^2/(1+k^2) = -k(k^2 - 3k+1)/(k^2+1)$. Thus $k = 0$ is clearly one solution, while the two roots of the quadratic in the numerator are also solutions. These are the numbers

$$a = \frac{3-\sqrt{5}}{2} \approx .3820 \quad \text{and}$$

$$b = \frac{3+\sqrt{5}}{2} \approx 2.6810.$$

Of the three, 0 and b are stable, while a is unstable. As a result of this stability feature,
$\lim_{t\to\infty} g(t) = 0$ if $0 \le g(0) < a$, while
$\lim_{t\to\infty} g(t) = b$ if $a < g(0)$.

As the problem evolves, g depends not only on time t, but on a certain real parameter x. We could write $g = g_x(t)$, and the dependence on x is through the initial condition: $g_x(0) = \frac{3}{2} + \frac{3\sin(x)}{2}$.

With x restricted to the interval $[0, 4\pi]$ (4π being about 12.5664), the first event ($g_x(0) < a$, equivalently $\lim_{t\to\infty} g_x(t) = 0$, equivalently eventual black - stripe zone) occurs when x lies in one of the two intervals $(3.9827, 5.4421)$ or $(10.2658, 11.7253)$. More precisely, these are the intervals with endpoints

$$\frac{3\pi}{2} \pm \cos^{-1}\left(\frac{\sqrt{5}}{3}\right) \quad \text{and}$$

$$\frac{7\pi}{2} \pm \cos^{-1}\left(\frac{\sqrt{5}}{3}\right).$$

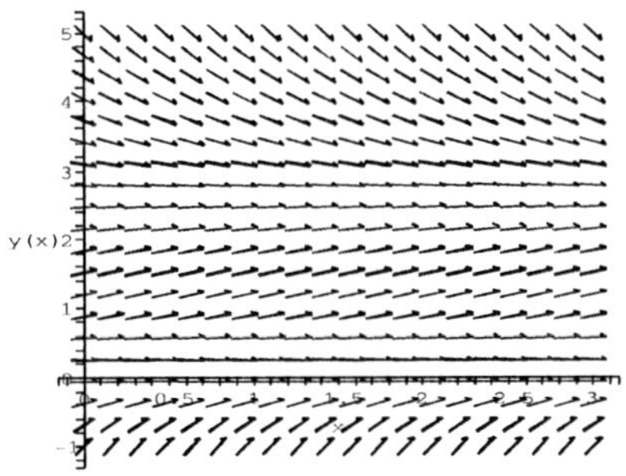

33. Using Euler's method:

For $h = 0.1$:

n	x_n	$y(x_n)$	$f(x_n, y_n)$
0	0.00	3.0000	8.0000
1	0.10	3.8000	13.4400
2	0.20	5.1440	25.4607
3	0.30	7.6901	58.1372
4	0.40	13.5038	181.3525
5	0.50	31.6390	1000.0295

For $h = 0.05$:

n	x_n	$y(x_n)$	$f(x_n, y_n)$
0	0.00	3.0000	8.0000
1	0.05	3.4000	10.5600
2	0.10	3.9280	14.4292
3	0.15	4.6495	20.6175
4	0.20	5.6803	31.2662
10	0.50	218.1215	47576.0009

For $h = 0.01$:

n	x_n	$y(x_n)$	$f(x_n, y_n)$
0	0.00	3.0000	8.0000
1	0.01	3.0800	8.4864
9	0.09	3.9396	14.5203
10	0.10	4.0848	15.6855
20	0.20	6.5184	41.4900
21	0.21	6.9333	47.0711
30	0.30	15.8434	250.0139

x	Exact
0.000	3.0000
0.100	4.1374
0.200	6.8713
0.300	21.4869
0.400	−18.7351
0.500	−6.5688

35. The vertical asymptote in the solution occurs when the denominator vanishes, which is to say when $e^{2x} = 1/k$, or $x = -\ln(k)/2$. In our case, with $y(0) = 3$, we have $k = 1/2$ and the vertical asymptote at $x = \ln(2)/2 = .3466$.

The field diagram cannot give any fore-warning of the vertical asymptote. Dependent as the field equations are only on y, they can only hint at things which likewise depend on y. The location of the vertical asymptote, by its very nature an x-measurement, is instead dependent directly on the solution-parameter k and indirectly on the initial condition.

In this case where the actual x-value does not enter the calculations, the

Euler process merely generates the numbers in the recursive sequence $y_n = hy_{n-1}^2 + y_{n-1} - h$ subject to an initial condition of $y_o = 3$. The numbers in such a sequence will increase to infinity, with growth rate depending on h. The simultaneous determination of x_n through the law $x_n = hn$ has nothing to do with the geometry of the solution to the differential equation. "Jumping over the asymptote" is the pseudo-event which happens when n passes from below $\frac{.3466}{h}$ to above, has no special relation to the Euler y-numbers, and no relation whatever to the solution of the differential equation.

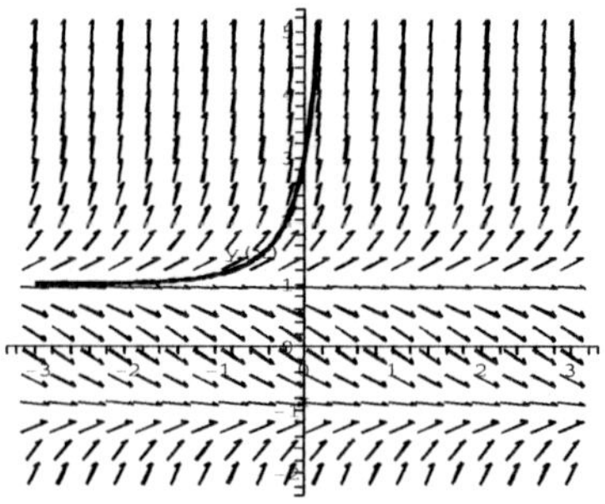

37. The general solution is

$$y = \frac{x^3}{3} - 2x^2 + 2x + c.$$

Using the initial condition $y(3) = 0$ gives $y(0) = c = 3$ and therefore

$$y = \frac{x^3}{3} - 2x^2 + 2x + 3$$

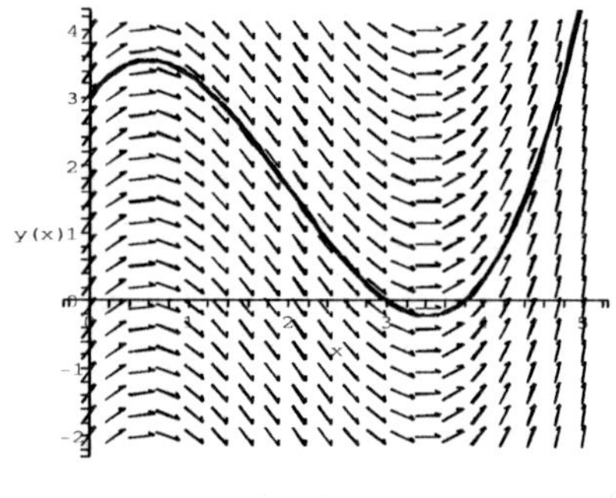

39. Using a CAS gives $y(0) \approx -5.55$.

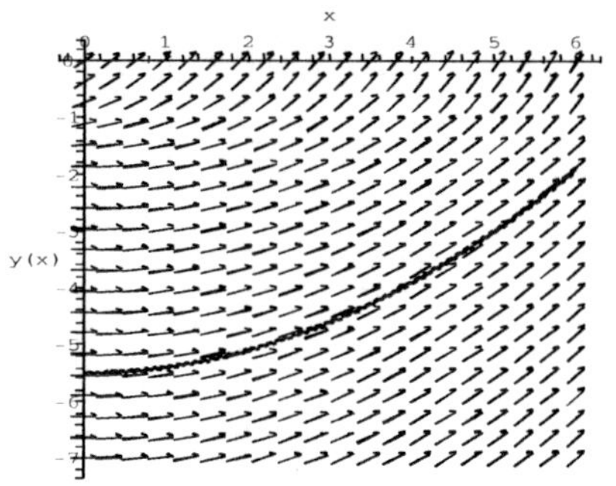

7.4 Systems of First-Order Differential Equations

1. Equilibrium points are those that satisfy $x'(t) = 0$ and $y'(t) = 0$. Substituting into the equations, we have

$$0 = 0.2x - 0.2x^2 - 0.4xy$$
$$0 = -0.1y + 0.2xy$$

$$0 = x(0.2 - 0.2x - 0.4y)$$
$$0 = y(-0.1 + 0.2x)$$

$$x = 0 \text{ or } 0.2 - 0.2x - 0.4y = 0$$
$$y = 0 \text{ or } x = 0.5$$

The equilibrium points are
$(0, 0)$, corresponding to the case where there are no predators or prey
$(1, 0)$, corresponding to the case where there are 200 prey but no predators
$(0.5, 0.25)$, corresponding to the having both populations constant, with two times as many prey as predators.

3. Equilibrium points are those that satisfy $x'(t) = 0$ and $y'(t) = 0$. Substituting into the equations, we have

$$0 = 0.3x - 0.1x^2 - 0.2xy$$
$$0 = -0.2y + 0.1xy$$

$$0 = x(0.3 - 0.1x - 0.2y)$$
$$0 = y(-0.2 + 0.1x)$$

$x = 0$ or $0.3 - 0.1x - 0.2y = 0$
$y = 0$ or $x = 2$

The equilibrium points are
$(0, 0)$, corresponding to the case where there are no predators or prey
$(3, 0)$, corresponding to the case where there are 300 prey but no predators
$(2, 0.5)$, corresponding to the having both populations constant, with four times as many prey as predators.

nstant, with four times as many prey as predators.

5. Equilibrium points are those that satisfy $x'(t) = 0$ and $y'(t) = 0$. Substituting into the equations, we have

$0 = 0.2x - 0.1x^2 - 0.4xy$
$0 = -0.3y + 0.1xy$

$0 = x(0.2 - 0.1x - 0.4y)$
$0 = y(-0.3 + 0.2x)$

$x = 0$ or $0.2 - 0.1x - 0.4y = 0$
$y = 0$ or $x = 1.5$

The equilibrium points are
$(0, 0)$, corresponding to the case where there are no predators or prey
$(2, 0)$, corresponding to the case where there are 200 prey but no predators
$(1.5, 0.125)$, corresponding to the having both populations constant, with twelve times as many prey as predators.

7. In Exercise 1,
$$\frac{dy}{dx} = \frac{0.2x - 0.2x^2 - 0.4xy}{-0.1y + 0.2xy}$$
From the following phase portrait, we observe that
$(0, 0)$ is an unstable equilibrium,
$(1, 0)$ is a stable equilibrium,
$(0.5, 0.25)$ is an unstable equilibrium.

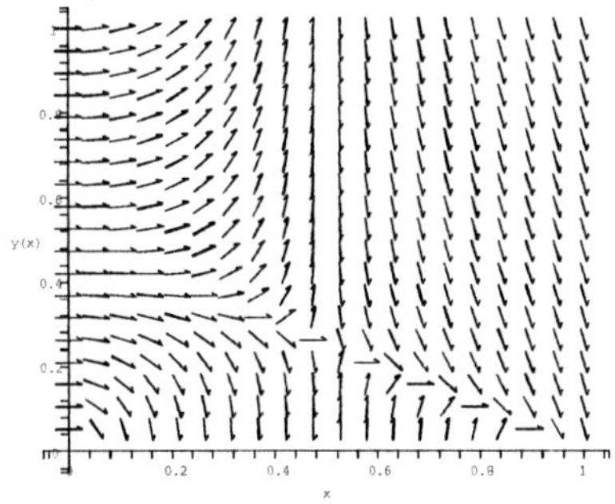

9. In Exercise 5,
$$\frac{dy}{dx} = \frac{0.2x - 0.1x^2 - 0.4xy}{-0.3y + 0.1xy}$$
From the following phase portrait, we observe that
$(0, 0)$ is an unstable equilibrium,
$(2, 0)$ is an unstable equilibrium,
$(1.5, 0.125)$ is a stable equilibrium.

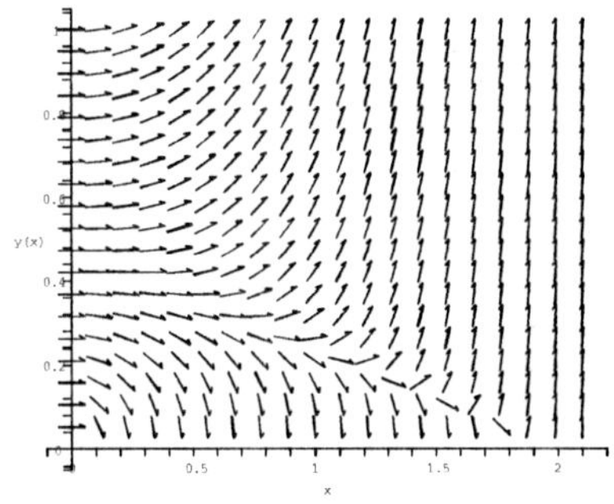

11. The point $(0, 0)$ is an unstable equilibrium.

13. The point $(0.5, 0.5)$ is a stable equilibrium.

15. The point $(1, 0)$ is a stable equilibrium.

17. Equilibrium points are those that satisfy $x'(t) = 0$ and $y'(t) = 0$. Substituting into the equations, we have

$0 = 0.3x - 0.2x^2 - 0.1xy$
$0 = 0.2y - 0.1y^2 - 0.1xy$

$0 = x(0.3 - 0.2x - 0.1y)$
$0 = y(0.2 - 0.1y - 0.1x)$

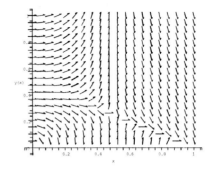

$x = 0$ or $0.3 - 0.2x - 0.1y = 0$
$y = 0$ or $0.2 - 0.1y - 0.1x = 0$

$x = 0$ or $2x + y = 3$
$y = 0$ or $x + y = 2$

The equilibrium points are
$(0, 0)$, corresponding to the case where neither species exists,
$(0, 2)$, corresponding to the case where species Y exists but species X does not,
$(1.5, 0)$, corresponding to the case where species X exists but species Y does not,
$(1, 1)$, corresponding to the have both species exist, with species Y as many as species X.

19. Equilibrium points are those that satisfy $x'(t) = 0$ and $y'(t) = 0$. Substituting into the equations, we have

$0 = 0.3x - 0.2x^2 - 0.2xy$
$0 = 0.2y - 0.1y^2 - 0.2xy$

$0 = x(0.3 - 0.2x - 0.2y)$
$0 = y(0.2 - 0.1y - 0.2x)$

$x = 0$ or $0.3 - 0.2x - 0.2y = 0$
$y = 0$ or $0.2 - 0.1y - 0.2x = 0$

$x = 0$ or $x + y = 1.5$
$y = 0$ or $2x + y = 2$

The equilibrium points are
$(0, 0)$, corresponding to the case where neither species exists,
$(0, 2)$, corresponding to the case where species Y exists but species X does not,
$(1.5, 0)$, corresponding to the case where species X exists but species Y does not,
$(0.5, 1)$, corresponding to the have both species exist, with species Y twice as many as species X.

21. Equilibrium points are those that satisfy $x'(t) = 0$ and $y'(t) = 0$. Substituting into the equations, we have

$0 = 0.2x - 0.2x^2 - 0.1xy$
$0 = 0.1y - 0.1y^2 - 0.2xy$

$0 = x(0.2 - 0.2x - 0.1y)$
$0 = y(0.1 - 0.1y - 0.2x)$

$x = 0$ or $0.2 - 0.2x - 0.1y = 0$
$y = 0$ or $0.1 - 0.1y - 0.2x = 0$

$x = 0$ or $2x + y = 2$
$y = 0$ or $2x + y = 1$

The equilibrium points are
$(0, 0)$, corresponding to the case where neither species exists,
$(0, 1)$, corresponding to the case where species Y exists but species X does not,
$(1, 0)$, corresponding to the case where species X exists but species Y does not.

23. This is because both species are hurt by the presence of the other.

25. In Exercise 17,
$$\frac{dy}{dx} = \frac{0.3x - 0.2x^2 - 0.1xy}{0.2y - 0.1y^2 - 0.1xy}$$

From the following phase portrait, we observe that
$(0, 0)$ is an unstable equilibrium,
$(0, 2)$ is an unstable equilibrium,
$(1.5, 0)$ is an unstable equilibrium,
$(1, 1)$ is a stable equilibrium.

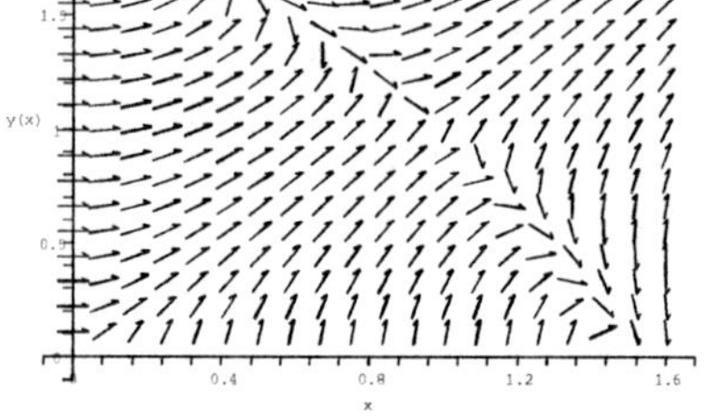

27. (a)

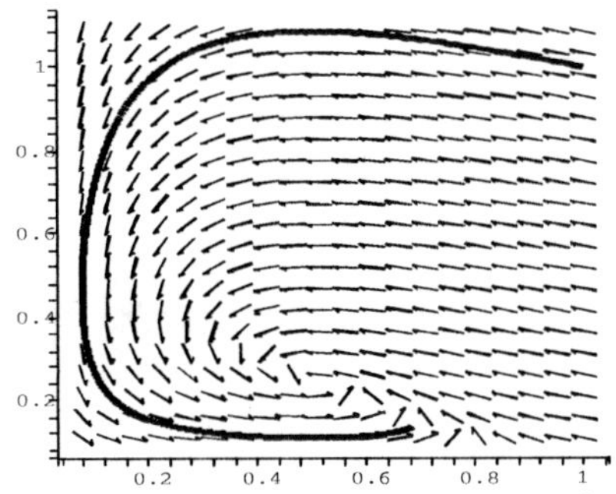

(b)

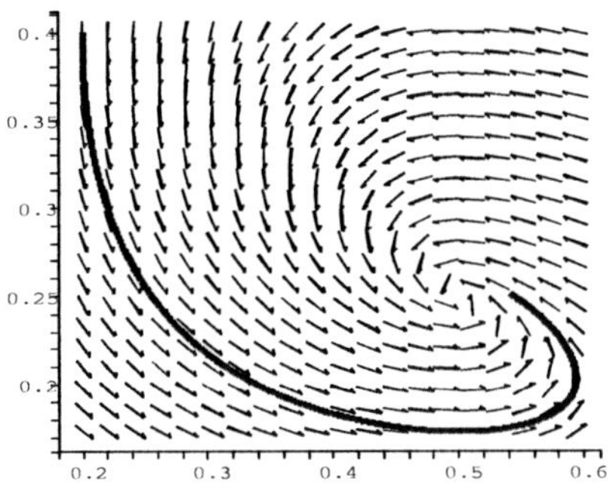

(c)

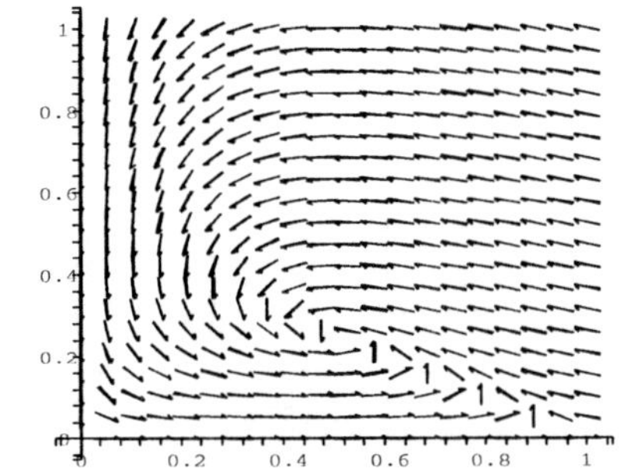

29. Write $u = y, v = y'$. We then have
$u' = v$
$v' = -2xv - 4u + 4x^2$

31. Write $u = y, v = y'$. We then have
$u' = v$
$v' = \cos x\, v - xu^2 + 2x$

33. Write $u_1 = y$, $u_2 = y'$, and $u_3 = y''$,
$u_1' = u_2$
$u_2' = u_3$
$u_3' = -2xu_3 + 4u_2 - 2u_1 + x^2$

35. Write $u_1 = y$, $u_2 = y', u_3 = y'', u_4 = y'''$,
$u_1' = u_2$
$u_2' = u_3$
$u_3' = u_4$
$u_4' = 2u_4 - xu_2 + 2 - e^x$

37. An approximate solution is $x(1) \approx 0.253718$, $y(1) \approx 0.167173$.

n	x_n	y_n
0	0.2	0.2
1	0.2048	0.1964
2	0.2097201152	0.19287422728
3	0.2147629013	0.1894212388
5	0.2252268589	0.1827279868
10	0.2537179001	0.1671729953

39. Equilibrium points are those that satisfy $x'(t) = 0$ and $y'(t) = 0$. Substituting into the equations, we have

$0 = (x^2 - 4)(y^2 - 9)$
$0 = x^2 - 2xy$

$0 = (x + 2)(x - 2)(y + 3)(y - 3)$
$0 = x(x - 2y)$

$x = 2, x = -2, y = 3, y = -3$
$x = 0, x = 2y$

The equilibrium points are $(2, 1), (-2, -1), (6, 3), (-6, -3), (0, 3), (0, -3)$.

41. Equilibrium points are those that satisfy $x'(t) = 0$ and $y'(t) = 0$. Substituting into the equations, we have

$0 = (2 + x)(y - x)$
$0 = (4 - x)(x + y)$

$x = -2$ or $x = y$
$x = 4$ or $x = -y$

The equilibrium points are $(0, 0), (-2, 2), (4, 4)$.

43. For equilibrium solutions, set $x' = y' = 0$ to get
$0 = 0.4x - 0.1x^2 - 0.2xy$
$0 = -0.5y + 0.1xy$

$0 = 0.1x(4 - x - 2y)$
$0 = 0.1y(-5 + x)$

Equilibrium points are $(0, 0)$, $(4, 0)$. Neither of these solutions has non-zero values for both populations, so the species cannot coexist.

Now suppose that the death rate of species Y is D instead of 0.5, and let us search for equilibrium solutions where both population values are non-zero. The equations are now
$x' = 0.4x - 0.1x^2 - 0.2xy$
$y' = -Dy + 0.1xy$
where $D > 0$.

$0 = 0.1x(4 - x - 2y)$
$0 = 0.1y(x - 10D)$

Since we are searching for non-zero solutions,
$0 = 4 - x - 2y$
$0 = x - 10D$

Solving the second equation gives $x = 10D$, and substituting this expression into the first equation gives
$0 = 4 - 10D - 2y = 2 - 5D - y$
$y = 2 - 5D$

The equilibrium solution for y will be positive provided that $2 - 5D > 0$, which means that $D < 0.4$.

45. Assume that all coefficients are positive. The equations that define equilibrium are
$0 = x(b - cx - k_1y)$
$0 = y(-d + k_2y)$

For the species to coexist, both x and y must be nonzero, and so the equations reduce to
$0 = b - cx - k_1y$
$0 = -d + k_2y$

Solving the second equation, we get $y = \frac{d}{k_2}$. Substituting the result into the first equation,

$0 = b - cx - k_1\frac{d}{k_2}$
$cx = b - \frac{dk_1}{k_2} = \frac{bk_2 - dk_1}{k_2}$
$x = \frac{bk_2 - dk_1}{ck_2}$

Thus, $x > 0$ if and only if $bk_2 - dk_1 > 0$, which is equivalent to $bk_2 > dk_1$.

Ch. 7 Review Exercises

1. We separate variables and integrate.
$\frac{1}{y}y' = 2$
$\int \frac{1}{y}\,dy = \int 2\,dx$
$\ln|y| = 2x + c$
$y = ke^{2x}$

The initial condition gives $3 = k$ so the solution is
$y = 3e^{2x}$

3. We separate variables and integrate.
$yy' = 2x$
$\int y\,dy = \int 2x\,dx$
$\frac{y^2}{2} = x^2 + c$
$y = \sqrt{2x^2 + c}$

The initial condition gives us $2 = \sqrt{c}, c = 4$ so the solution is
$y = \sqrt{2x^2 + 4}$

5. We separate variables and integrate.
$\frac{1}{\sqrt{y}}y' = \sqrt{x}$
$\int y^{-1/2}\,dy = \int x^{1/2}\,dx$
$2y^{1/2} = \frac{2}{3}x^{3/2} + c$
$y = \left(\frac{x^{3/2}}{3} + c\right)^2$

The initial condition gives us
$4 = \left(\frac{1}{3} + c\right)^2, c = \frac{5}{3}$ so the solution

is
$$y = \left(\frac{x^{3/2}}{3} + \frac{5}{3}\right)^2$$

7. With t measured in hours, we have
$y = Ae^{kt}$, $A = y(0) = 10^4$.

If the doubling time is 2, then
$2 = e^{2k}$, $k = \ln(2)/2$, and
$y = 10^4 e^{t\ln(2)/2} = 10^4 2^{t/2}$.

To reach $y = 10^6$ at a certain unknown time t, we need
$2^{t/2} = 100$,
$$t = \frac{2\ln(100)}{\ln(2)} \approx 13.3 \text{ hours.}$$

9. With t measured in hours, x in milligrams, we get
$$x = 2\left(\frac{1}{2}\right)^{t/2} = \frac{2}{2^{t/2}}.$$
To get to $x = .1$ at a certain unknown time t, we need
$$2^{t/2} = \frac{2}{.1} = 20,$$
$$t = \frac{2\ln(20)}{\ln(2)} \approx 8.64 \text{ hours.}$$

11. The equation for the doubling time t_d in this case is
$2 = e^{.08t_d}$, hence
$$t_d = \frac{\ln(2)}{.08} \approx 8.66 \text{ years.}$$

13. For temperature T at time t, and ambient temperature T_a, we have
$$\frac{T - T_a}{T(0) - T_a} = e^{kt}.$$
In this case with $T_a = 68, T(0) = 180$ and $T(1) = 176$, we have
$$\frac{108}{112} = \frac{176 - 68}{180 - 68} = e^k,$$
$$k = \ln\left(\frac{108}{112}\right) = \ln\left(\frac{27}{28}\right),$$
$$\frac{T - 68}{112} = e^{tk} = e^{t\ln(27/28)} = \left(\frac{27}{28}\right)^t,$$
$$T = 68 + 112\left(\frac{27}{28}\right)^t.$$
To reach $T = 120$ at unknown time t, we need
$$t = \frac{\ln(52/112)}{\ln(27/28)} \approx 21.1 \text{ minutes.}$$

15. $\dfrac{y'}{y} = 2x^3$
$$\ln|y| = \frac{x^4}{2} + c$$
$$y = Ae^{x^4/2}$$

17. $(y^2 + y)y' = \dfrac{4}{1 + x^2}$
$$\int (y^2 + y)\, dy = \int 41 + x^2\, dx$$
$$\frac{y^3}{3} + \frac{y^2}{2} = 4\tan^{-1} x + c$$

It is impossible (without using a CAS) to write out the explicit formula of y in terms of x.

19. Equilibrium solutions occur where $y' = 0$ which occurs when $y = 0$ and $y = 2$. $y = 0$ is unstable and $y = 2$ is stable which can be seen by drawing the direction field.

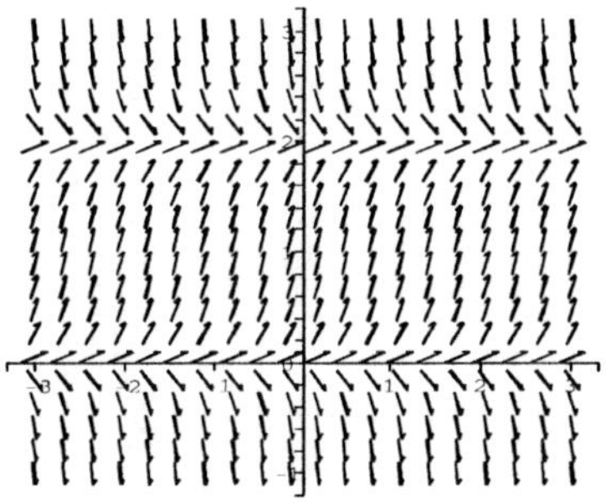

21. Equilibrium solutions occur where $y' = 0$ which occurs when $y = 0$, and it is stable.

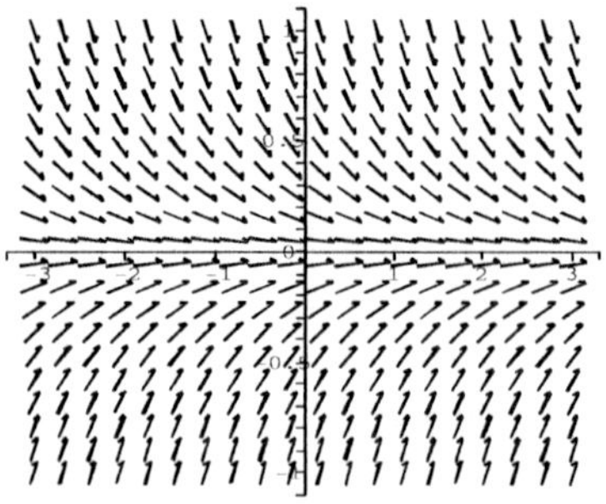

23.

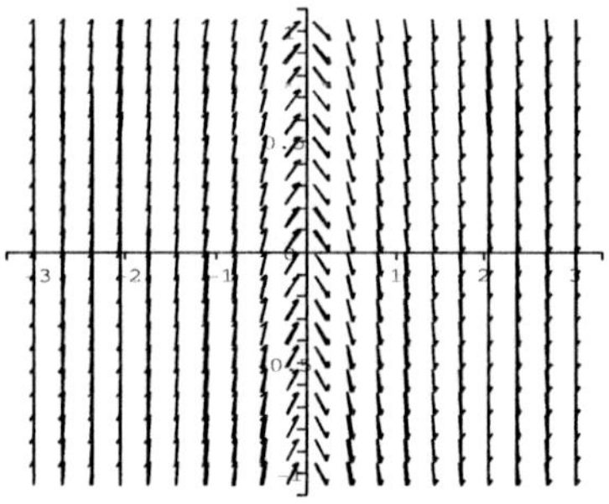

25.

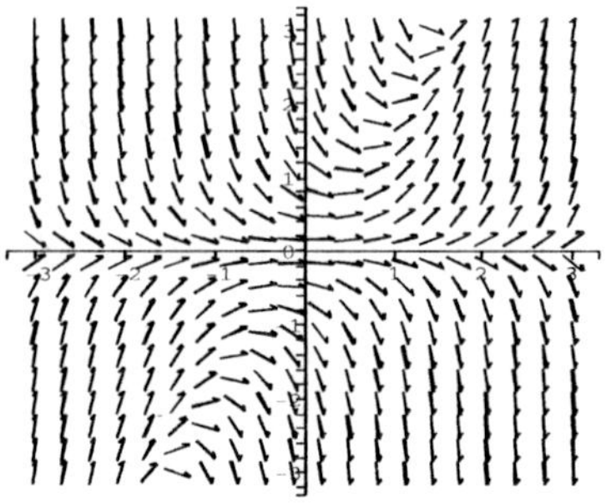

27. The differential equation is

$x' = (.3 - x)(.4 - x) - .25x^2$

$= .12 - .7x + .75x^2$

$= \frac{3}{4}(x^2 - \frac{14}{15}x + \frac{4}{25})$

$= \frac{3}{4}(x - r)(x - s)$. in which

$r = \frac{7 + \sqrt{13}}{15}, s = \frac{7 - \sqrt{13}}{15},$

$r - s = \frac{2\sqrt{13}}{15}.$

When separated it takes the form

$$\frac{x'}{(x-r)(x-s)} = k$$

in which $k = 3/4$.

By partial fractions we find

$$\frac{1}{(x-r)(x-s)} = \frac{1}{(r-s)}\left\{\frac{1}{(x-r)}\frac{1}{(x-s)}\right\}$$

and after integration we find

$$\frac{1}{(r-s)}\ln\left|\frac{x-r}{x-s}\right| = kt + c_1$$

or in this case

$$\ln\left|\frac{x-r}{x-s}\right| = \frac{2\sqrt{13}}{15}\left(\frac{3}{4}t + c_1\right) = wt + c_2$$

$$\left(w = \frac{\sqrt{13}}{10} \approx .36056, c_2 = \frac{2\sqrt{13}}{15}c_1\right).$$

Using the initial condition $x(0) = c$, we find $c2 = \ln|(c-r)/(c-s)|$,

$$\ln\left|\frac{(c-s)(x-r)}{(c-r)(x-s)}\right| = wt \text{ and}$$

$$\frac{x-r}{x-s} = \pm\frac{c-r}{c-s}e^{wt} = \frac{c-r}{c-s}e^{wt},$$

$$x = \frac{s(r-c)e^{wt} + r(c-s)}{(r-c)e^{wt} + (c-s)} = \frac{r\left(\frac{c-s}{r-c}\right)e^{-wt} + s}{\left(\frac{c-s}{r-c}\right)e^{-wt} + 1}$$

The choice of sign is + since the left side of the middle equation is $(c - r)/(c - s)$ when $t = 0$ and $x = c$. The last expression is one of many possible ways to normalize. It is apparent that $x \to s \approx .22630$ as $t \to \infty$

Numerically, when $c = 0.1$, this comes to

$x = \frac{.22630 - .14710e^{-.36056t}}{1 - .20806e^{-.36056t}}$. and the graph looks like

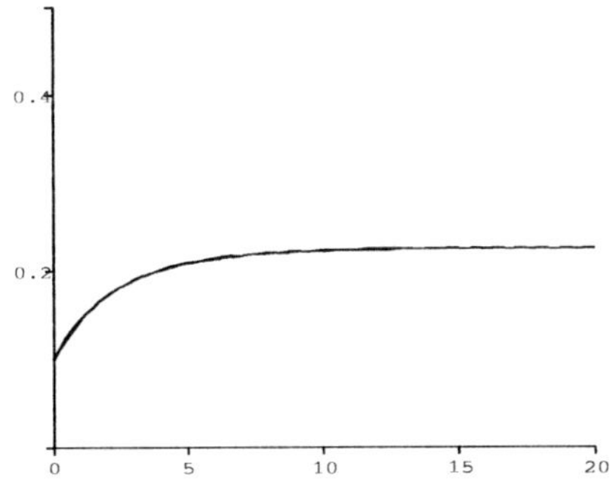

When $c = 0.4$, this comes to
$x = \dfrac{.22630 + .39999e^{-.36056t}}{1 + .56574e^{-.36056t}}$. and the graph looks like

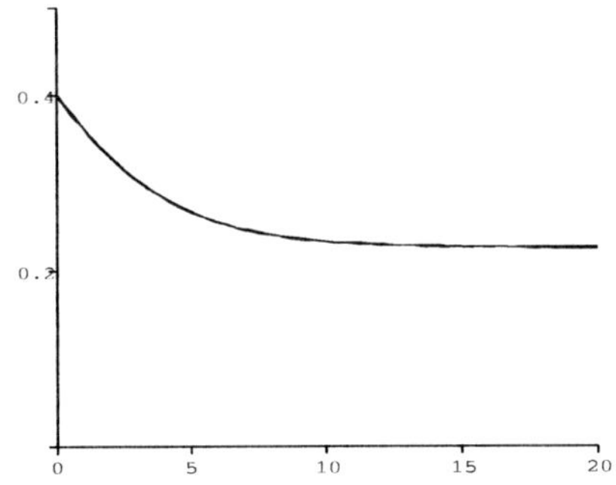

29. The DE $\dfrac{x'}{(x-a)^2} = r$ integrates to
$\dfrac{-1}{x-a} = rt + c$. and then
$c = \dfrac{1}{a}$,
$x - a = \dfrac{-1}{c+rt} = \dfrac{-a}{1+art}$,
$x = a\left(1 - \dfrac{1}{1+art}\right) = \dfrac{a^2rt}{1+art}$.

One can see that all values of x lie between 0 and a, and that

$$\lim_{t\to\infty} x(t) = a$$

All the initial amounts of the A,B substances (both a in this case) will eventually be converted to the X substance which ultimately will have the same concentration as the original concentrations of the other two substances.

31. With A the amount in the account at time t the DE is
$A'(t) = .10A + 20{,}000$ with an IC of $A(0) = 100{,}000$. The DE separates and integrates easily, yielding
$10\ln|.10A + 20{,}000| = t + c$
$c = 10\ln(30{,}000)$,
$.10A + 20{,}000 = 30{,}000e^{t/10}$.

If the fortune is to reach 1,000,000 at unknown time t, we must have
$120{,}000 = 30{,}000e^{t/10}$
$\dfrac{t}{10} = \ln\dfrac{12}{3} = \ln(4)$,
$t = 10\ln(4) \approx 13.86$ years.

33. It is a predator-prey model.

For equilibrium solutions, set $x' = y' = 0$ to get
$0 = 0.1x - 0.1x^2 - 0.2xy$
$0 = -0.1y + 0.1xy$
which are equivalent to
$0 = 0.1x(1 - x - 2y)$
$0 = 0.1y(-1 + x)$

$x = 0$ or $x = 1 - 2y$
$y = 0$ or $x = 1$.

The equilibrium solutions are $(0, 0)$ (no prey or predators) and $(1, 0)$ (prey but no predators).

35. It is a competing species model.

For equilibrium solutions, set $x' = y' = 0$ to get
$0 = 0.5x - 0.1x^2 - 0.2xy$
$0 = 0.4y - 0.1y^2 - 0.2xy$
which are equivalent to
$0 = 0.1x(5 - x - 2y)$
$0 = 0.1y(4 - y - 2x)$

$x = 0$ or $x + 2y = 5$
$y = 0$ or $2x + y = 4$.

The equilibrium solutions are $(0, 0)$ (none of either species), $(0, 4)$ (none of first species, some of second), $(5, 0)$ (some of first species, none of second),

$(1, 2)$ (twice as many of second species as first species)

37. In Exercise 33,

$$\frac{dy}{dx} = \frac{0.1x - 0.1x^2 - 0.2xy}{-0.1y + 0.1xy}$$

From the direction field, we see that $(0, 0)$ is unstable and $(1, 0)$ is stable.

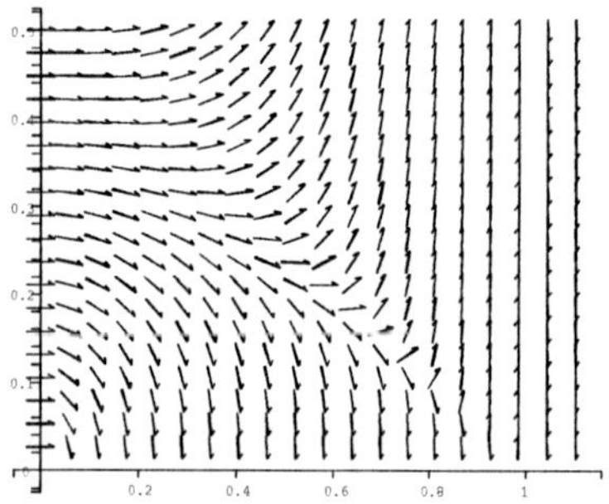

39. Write $u = y, v = y'$, then

$u' = v$

$v' = 4x^2v - 2u + 4xu - 1$

Chapter 8

Infinite Series

8.1 Sequences of Real Numbers

1. $1, \frac{3}{4}, \frac{5}{9}, \frac{7}{16}, \frac{9}{25}, \frac{11}{36}$

3. $4, 2, \frac{2}{3}, \frac{1}{6}, \frac{1}{30}, \frac{1}{180}$

5. (a) $\lim_{n\to\infty} \frac{1}{n^3} = 0$

(b) As n gets large, n^3 gets large, so $1/n^3$ goes to 0.

(c)

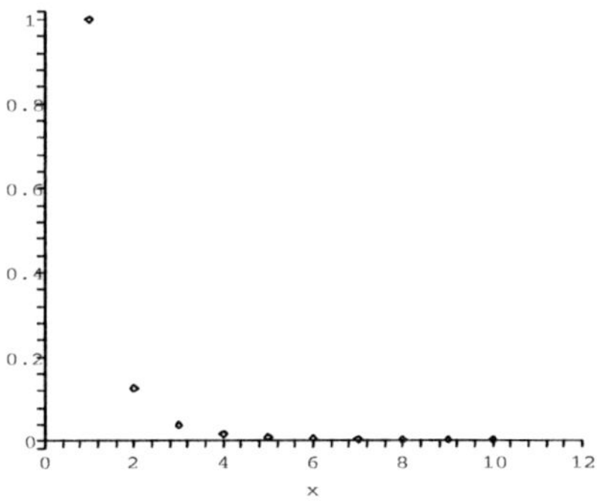

7. (a) $\lim_{n\to\infty} \frac{n}{n+1} = \lim_{n\to\infty} \frac{1}{1+\frac{1}{n}} = 1$

(b) As n gets large, $n/(n+1)$ gets close to 1.

(c)

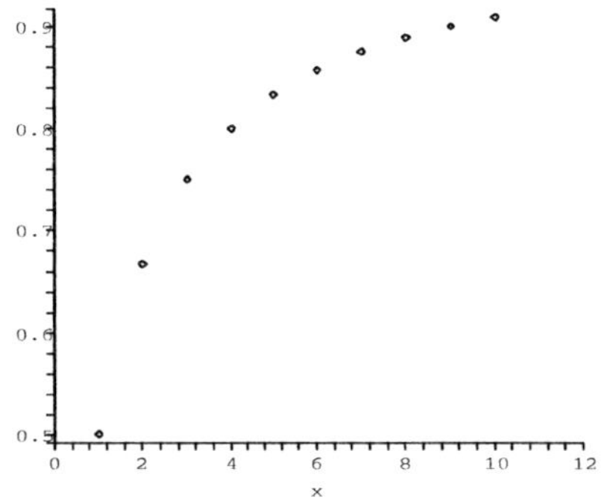

9. (a) $\lim_{n\to\infty} \frac{2}{\sqrt{n}} = 0$

(b) As n gets large, $\sqrt{n}$ gets large, so $\frac{2}{\sqrt{n}}$ goes to 0.

(c)

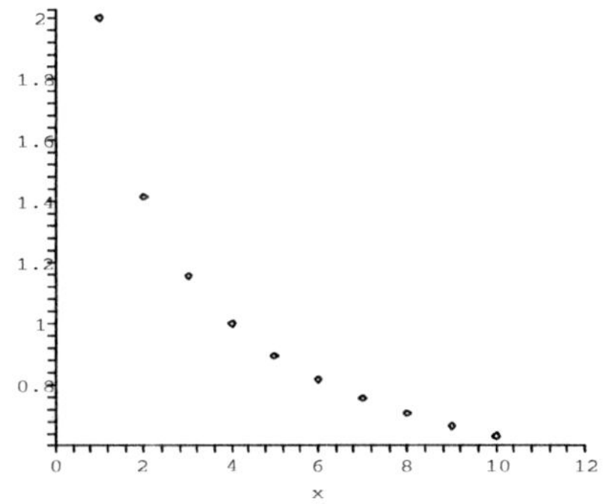

11. $\lim_{n\to\infty} \frac{3n^2+1}{2n^2-1} = \lim_{n\to\infty} \frac{3+\frac{1}{n^2}}{2-\frac{1}{n^2}} = \frac{3}{2}$

13. $\lim_{n\to\infty} \frac{n^2+1}{n+1} = \lim_{n\to\infty} \frac{n+\frac{1}{n}}{1+\frac{1}{n}} = \infty$

15. $\lim_{n\to\infty} (-1)^n \frac{n+2}{3n-1} = \lim_{n\to\infty} (-1)^n \frac{1+\frac{2}{n}}{3-\frac{1}{n}}$
$= \pm\frac{1}{3}$, the limit does not exist; diverges

17. $\lim_{n\to\infty} \left|(-1)^n \frac{n+2}{n^2+4}\right| = \lim_{n\to\infty} \frac{n+2}{n^2+4}$
$= \lim_{n\to\infty} \frac{\frac{1}{n}+\frac{2}{n^2}}{1+\frac{4}{n^2}} = 0$

Hence $\lim_{n\to\infty} \frac{(-1)^2(n+2)}{n^2+4} = 0$

19. $\lim_{x\to\infty} \frac{x}{e^x} = \lim_{x\to\infty} \frac{1}{e^x}$ by l'Hopital's Rule and $\lim_{x\to\infty} \frac{1}{e^x} = 0$, so by Theorem 1.2 $\lim_{n\to\infty} \frac{n}{e^n} = 0$

21. $\lim_{n\to\infty} \frac{e^n+2}{e^{2n}-1} = \lim_{n\to\infty} \frac{\frac{1}{e^n}+\frac{2}{e^{2n}}}{1-\frac{1}{e^{2n}}} = 0$

23. $\lim_{x\to\infty} \frac{x2^x}{3^x} = \lim_{x\to\infty} \frac{x}{\frac{3^x}{2^x}} = \lim_{x\to\infty} \frac{x}{\left(\frac{3}{2}\right)^x}$

$= \lim_{x\to\infty} \frac{1}{\left(\frac{3}{2}\right)^x \ln\frac{3}{2}} = 0$, by l'Hopital's Rule, since $\lim_{x\to\infty} \left(\frac{3}{2}\right)^x = \infty$.

Hence $\lim_{n\to\infty} \frac{n2^n}{3^n} = 0$, by Theorem 1.2.

25. Since $\frac{n!}{2^n} = \frac{1}{2}\cdot\frac{2}{2}\cdots\frac{n}{2} \geq \frac{1}{2}\cdot\frac{n}{2} = \frac{n}{4}$ and $\lim_{n\to\infty} \frac{n}{4} = \infty$, we have $\lim_{n\to\infty} \frac{n!}{2^n} = \infty$.

27. Since $\lim_{n\to\infty} \frac{2n+1}{n} = 2$ so $\lim_{n\to\infty} [\ln(2n+1) - \ln n]$

$= \lim_{n\to\infty} \ln \frac{2n+1}{n} = \ln 2$

29. $-1 \leq \cos n \leq 1 \Rightarrow \frac{-1}{n^2} \leq \frac{\cos n}{n^2} \leq \frac{1}{n^2}$ for all n, and $\lim_{n\to\infty} \frac{-1}{n^2} = \lim_{n\to\infty} \frac{1}{n^2} = 0$ so by the Squeeze Theorem, $\lim_{n\to\infty} \frac{\cos n}{n^2} = 0$

31. $0 \leq |a_n| = \frac{1}{ne^n} \leq \frac{1}{n}$ and $\lim_{n\to\infty} \frac{1}{n} = 0$ so by the Squeeze Theorem and Corollary 1.1, $\lim_{n\to\infty} \frac{(-1)^n}{ne^n} = 0$

33. $\frac{a_{n+1}}{a_n} = \left(\frac{n+4}{n+3}\right)\cdot\left(\frac{n+2}{n+3}\right)$

$= \frac{n^2+6n+8}{n^2+6n+9} < 1$ for all n, so $a_{n+1} < a_n$ for all n, so the sequence is decreasing.

35. $\frac{a_{n+1}}{a_n} = \left(\frac{e^{n+1}}{n+1}\right)\cdot\left(\frac{n}{e^n}\right) = \frac{e\cdot n}{n+1} > 1$ for all n, so $a_{n+1} > a_n$ for all n, so $\{a_n\}_{n=1}^{\infty}$ is increasing.

37. $\frac{a_{n+1}}{a_n} = \left(\frac{2^{n+1}}{(n+2)!}\right)\cdot\left(\frac{(n+1)!}{2^n}\right)$

$= \frac{2}{n+2} < 1$ for all n, so $a_{n+1} < a_n$ for all n, so the sequence is decreasing.

39. $|a_n| = \left|\frac{3n^2-2}{n^2+1}\right| = \frac{3n^2-2}{n^2+1}$

$< \frac{3n^2}{n^2+1} < \frac{3n^2}{n^2} = 3$

41. $|a_n| = \left|\frac{\sin(n^2)}{n+1}\right| \leq \frac{1}{n+1} \leq \frac{1}{2}$ for $n > 1$.

43. $a_{1000} \approx 7.374312390$,
$e^2 \approx 7.389056099$

$b_{1000} \approx .135064522$,
$e^{-2} \approx .135335283$.

45. If side $s = 12''$ and the diameter $D = \frac{12}{n}$ then the number of disks that fit along one side is $\frac{12}{\left(\frac{12}{n}\right)} = n$. Thus, the total number of disks is $\frac{12}{\left(\frac{12}{n}\right)}\cdot\frac{12}{\left(\frac{12}{n}\right)} = n\cdot n = n^2$

a_n = wasted area in box with n^2 disks

$= 12\cdot 12 - n^2\left(\frac{6}{n}\right)^2\pi$

$= 144 - 36\pi \approx 30.9$

47. $a_0 = 3.049$,
$a_1 = 3.049 + .005(3.049)^{2.01} = 3.096$
$a_2 = 3.096 + .005(3.096)^{2.01} = 3.144$
$a_3 = 3.144 + .005(3.144)^{2.01} = 3.194$
$a_{10} = 3.594 <$ population in 1970
$a_{20} = 4.376 <$ population in 1980

$a_{30} = 5.589 >$ ...
= population ...
Estimated po...
$a_{75} = 4.131 >$ ...

49. In the 3rd m...
bits have ne...
In the 4th r...
adult rabbi...
borns, so a...
In general,

51. For the fir...
$a_1 = 1,$
$a_2 = 2.5,$
$a_3 = 2.05$...
$a_4 = 2.00$...
$a_5 = 2.00$...
$a_6 = 2.00$...
$a_7 = 2.0$...
$a_8 = 2.0$...

The eq...
two so...
Since ...
positiv...
lution ...

53. Use in...
First ...
$a_2^2 = $...
so a...
true ...
Nov ...
true ...
and ...
for ...
a_k...
a_k^2...
a_k^2...
(...
...

$a_{k+1} > a_k$, it follows that
$(a_{k+1}^2 - 2)^2$ and therefore
$(\ldots - 2)^2 > (a_{k+1}^2 - 2)^2$
$\ldots > a_{k+1}^2 - 2$
$a_{k+1}^2, a_{k+2} > a_{k+1}$

... y induction, a_n is increasing.

... e'll prove that $a_n < 2$ by in-
... First note that $a_1 < 2$, and
... e that $a_k < 2$. Then $a_{k+1} =$
$\sqrt{a_k}$
$\ldots - 2 = \sqrt{a_k}, (a_{k+1}^2 - 2)^2 = a_k$
$\ldots - 2)^2 < 2, a_{k+1}^2 - 2 < \sqrt{2}$
$\ldots < 2 + \sqrt{2}, a_{k+1}^2 < 4$
$\ldots < 2$

... s, by induction, $a_n < 2$.

... e a_n is increasing and bounded
... ve by 2, a_n converges. To estimate
... limit, we'll approximate the solu-
... n of $x = \sqrt{2 + \sqrt{x}}$:
$\ldots = 2 + \sqrt{x}, (x^2 - 2)^2 = x$
$\ldots - 4x^2 + 4 = x, 0 = x^4 - 4x^2 - x + 4$
$\ldots = (x - 1)(x^3 + x^2 - 3x - 4)$

... ince $a_n > \sqrt{2}$, it follows that $x \neq 1$.
... herefore, $0 = x^3 + x^2 - 3x - 4$. Using
... CAS, the solution is $x \approx 1.8312$.

$p > 1 \Rightarrow \lim_{n \to \infty} \dfrac{1}{p^n} = 0$

$p = 1 \Rightarrow \lim_{n \to \infty} \dfrac{1}{p^n} = 1$

$0 < p < 1 \Rightarrow \lim_{n \to \infty} \dfrac{1}{p^n}$ does not exist

$-1 < p < 0 \Rightarrow \lim_{n \to \infty} \dfrac{1}{p^n}$ does not exist

$p = -1 \Rightarrow \lim_{n \to \infty} \dfrac{1}{p^n}$ does not exist

$p < -1 \Rightarrow \lim_{n \to \infty} \dfrac{1}{p^n} = 0$

Therefore, the sequence $a_n = 1/p^n$ converges for $p < -1$ and $p \geq 1$.

57. $a_n = \frac{1}{n^2}(1+2+3+\cdots+n)$

$= \frac{1}{n^2}\left(\frac{n(n+1)}{2}\right) = \frac{n+1}{2n}$

$= \frac{1}{2}\left(1+\frac{1}{n}\right)$

$\lim_{n\to\infty} a_n = \frac{1}{2}\lim_{n\to\infty}\left(1+\frac{1}{n}\right) = \frac{1}{2}$

Thus, the sequence a_n converges to 1/2.

Note that

$\int_0^1 x\,dx = \lim_{n\to\infty}\left(\sum_{k=1}^{n}\frac{1}{n}\frac{k}{n}\right)$

Therefore the sequence a_n converges to $\int_0^1 x\,dx$.

59. Begin by joining the centers of circles C_1 and C_2 with a line segment. The length of this line segment is the sum of the radii of the two circles, which is $r_1 + r_2$. Thus, the squared of the length of the line segment is $(r_1+r_2)^2$. Now, the coordinates of the centers of the circles are (c_1, r_1) and (c_2, r_2). Using the formula for the distance between two points, the square of the length of the line segment joining the two centers is $(c_2-c_1)^2+(r_2-r_1)^2$. Equating the two expressions, we get

$(c_2-c_1)^2+(r_2-r_1)^2 = (r_1+r_2)^2$

Expanding and simplifying this relationship, we get

$(c_2-c_1)^2 = (r_1+r_2)^2 - (r_2-r_1)^2$

$= r_1^2 + 2r_1r_2 + r_2^2 - (r_2^2 - 2r_1r_2 + r_1^2)$

$= 4r_1r_2$

$|c_2 - c_1| = 2\sqrt{r_1r_2}$

The same reasoning applied to the other two pairs of centers yields analogous results. Without going through the motions again, you can simply take the results above and first replace all subscripts "2" by "3" to get the results for circles C_1 and C_3. Then take these new results and replace all subscripts "1" by "2" to get the results for circles C_2 and C_3. The results are

$(c_3-c_1)^2+(r_3-r_1)^2 = (r_1+r_3)^2$

$|c_3-c_1| = 2\sqrt{r_1r_3}$

$(c_3-c_2)^2+(r_3-r_2)^2 = (r_2+r_3)^2$

$|c_3-c_2| = 2\sqrt{r_2r_3}$

Finally, $|c_1-c_2| = |c_1-c_3| + |c_3-c_2|$

$2\sqrt{r_1r_2} = 2\sqrt{r_1r_3} + 2\sqrt{r_2r_3}$

$\sqrt{r_1r_2} = \sqrt{r_3}\left(\sqrt{r_1}+\sqrt{r_2}\right)$

$\sqrt{r_3} = \frac{\sqrt{r_1r_2}}{\sqrt{r_1}+\sqrt{r_2}}$

61. The distance between the two points $(0, c)$ and (x_0, y_0), where $y_0 = x_0^2$, is r,

$\sqrt{x_0^2 + (x_0^2 - c)^2} = r$

$x_0^2 + x_0^4 - 2cx_0^2 + c^2 = r^2$

$y_0^2 + (1-2c)y_0 + (c^2 - r^2) = 0$

We want the solution y_0 to the above equation to be unique, so that

$(1-2c)^2 - 4(c^2 - r^2) = 0$

$1 - 4c + 4c^2 - 4c^2 + 4r^2 = 0$

$1 - 4c + 4r^2 = 0, c = \frac{1}{4} + r^2$

8.2 Infinite Series

1. $\sum_{k=0}^{\infty} 3\left(\frac{1}{5}\right)^k$ is a geometric series with $a = 3$ and $|r| = \frac{1}{5} < 1$, so it converges to $\frac{3}{1-1/5} = \frac{15}{4}$

3. $\sum_{k=0}^{\infty}\frac{1}{2}\left(-\frac{1}{3}\right)^k$ is a geometric series with $a = \frac{1}{2}$ and $|r| = \frac{1}{3} < 1$, so it converges to $\frac{1/2}{1-(-1/3)} = \frac{3}{8}$.

5. $\sum_{k=0}^{\infty}\frac{1}{2}(3)^k$ is a geometric series with

$|r| = 3 > 1$, so it diverges.

7. Using partial fractions,

$$S_n = \sum_{k=1}^{n} \frac{4}{k(k+2)} = \sum_{k=1}^{n} \left(\frac{2}{k} - \frac{2}{k+2}\right)$$
$$= \left(2 - \frac{2}{3}\right) + \left(1 - \frac{2}{4}\right) + \left(\frac{2}{3} - \frac{2}{5}\right)$$
$$+ \cdots + \left(\frac{2}{n} - \frac{2}{n+2}\right)$$
$$= 2 + 1 - \frac{2}{n+1} - \frac{2}{n+2}$$
$$= 3 - \frac{4n+6}{n^2+3n+2}$$

and

$$\lim_{n\to\infty} S_n$$
$$= \lim_{n\to\infty} \left(3 - \frac{4n+6}{n^2+3n+2}\right) = 3.$$

Thus, the series converges to 3.

9. $\lim\limits_{k\to\infty} \dfrac{3k}{k+4} = 3 \neq 0$, so by the kth-Term Test for Divergence, the series diverges.

11. $\sum\limits_{k=1}^{\infty} \dfrac{2}{k} = 2\sum\limits_{k=1}^{\infty} \dfrac{1}{k}$ and from Example 2.7, $\sum\limits_{k=1}^{\infty} \dfrac{1}{k}$ diverges, so $2\sum\limits_{k=1}^{\infty} \dfrac{1}{k}$ diverges.

13. Using partial fractions

$$S_n = \sum_{k=1}^{n} \frac{2k+1}{k^2(k+1)^2}$$
$$= \sum_{k=1}^{n} \left[\frac{1}{k^2} - \frac{1}{(k+1)^2}\right]$$
$$= \left(1 - \frac{1}{4}\right) + \left(\frac{1}{4} - \frac{1}{9}\right) + \left(\frac{1}{9} - \frac{1}{16}\right)$$
$$+ \cdots + \left(\frac{1}{(n-1)^2} - \frac{1}{n^2}\right)$$
$$+ \left(\frac{1}{n^2} - \frac{1}{(n+1)^2}\right)$$
$$= 1 - \frac{1}{(n+1)^2}$$

and

$$\lim_{n\to\infty} S_n = \lim_{n\to\infty} \left(1 - \frac{1}{(n+1)^2}\right) = 1$$

Thus the series converges to 1.

15. $\sum\limits_{k=2}^{\infty} \dfrac{1}{3^k}$ is a geometric series with $a = \dfrac{1}{9}$, and $|r| = \dfrac{1}{3} < 1$ so it converges to $\dfrac{1/9}{1-1/3} = \dfrac{1}{6}$.

17. $\sum\limits_{k=0}^{\infty} \left(\dfrac{1}{2^k} - \dfrac{1}{k+1}\right)$

$$= \sum_{k=0}^{\infty} \frac{1}{2^k} - \sum_{k=0}^{\infty} \frac{1}{k+1}.$$

The first series is a convergent geometric series but the second series is the divergent harmonic series, so the original series diverges.

19. $\sum\limits_{k=0}^{\infty} \left(\dfrac{2}{3^k} + \dfrac{1}{2^k}\right) = \sum\limits_{k=0}^{\infty} \dfrac{2}{3^k} + \sum\limits_{k=0}^{\infty} \dfrac{1}{2^k}.$

The first series is a geometric series with

$a = 2$ and $|r| = \dfrac{1}{3} < 1$ so it converges to $\dfrac{2}{1-1/3} = 3.$

The second series is a geometric series with

$a = 1$ and $|r| = \dfrac{1}{2} < 1$, so it converges to $\dfrac{1}{1-1/2} = 2.$

Thus

$$\sum_{k=0}^{\infty} \left(\frac{2}{3^k} + \frac{1}{2^k}\right) = 3 + 2 = 5.$$

21. $\lim\limits_{k\to\infty} |a_k| = \lim\limits_{k\to\infty} \dfrac{3k}{k+1} = 3 \neq 0$

So by the kth-Term Test for Divergence, the series diverges.

23. The series appears to converge.

n	$\sum_{k=1}^{n} \frac{1}{\sqrt{k}}$
2	1.250000000
4	1.423611111
8	1.527422052
16	1.584346533
32	1.614167263
64	1.629430501
128	1.637152005
256	1.641035436
512	1.642982848
1024	1.643957981

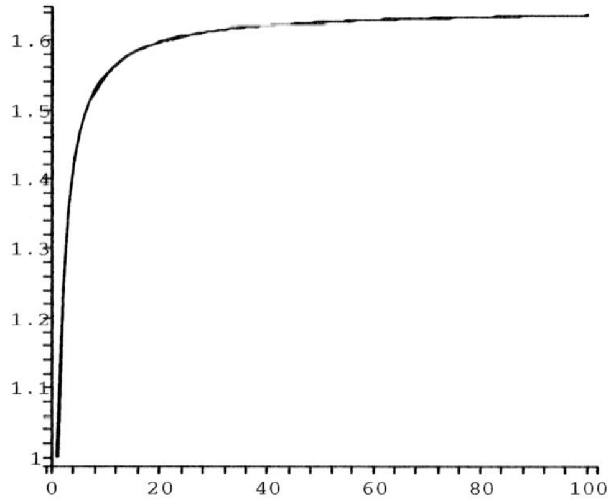

25. The series appears to converge.

n	$\sum_{k=1}^{n} \frac{1}{\sqrt{k}}$
2	4.500000000
4	5.125000000
8	5.154836310
16	5.154845485
32	5.154845485
1024	5.154845485

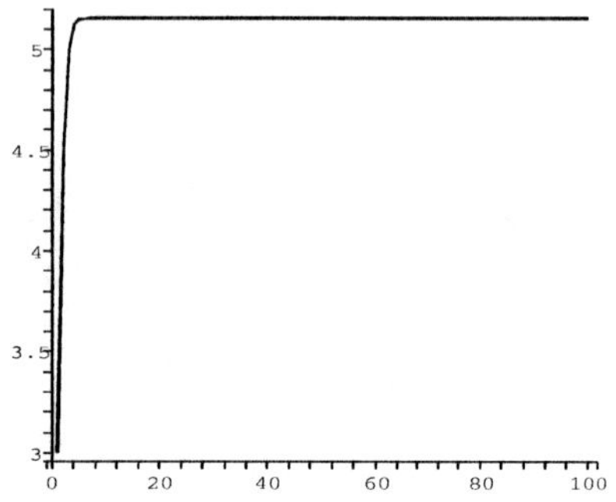

27. Assume $\sum_{k=1}^{\infty} a_k$ converges to L. Then for any m,

$$L = \sum_{k=1}^{\infty} a_k = \sum_{k=1}^{m-1} a_k + \sum_{k=m}^{\infty} a_k$$

$$= S_{m-1} + \sum_{k=m}^{\infty} a_k.$$

So $\sum_{k=m}^{\infty} a_k = L - S_{m-1}$, and thus converges.

29. Assume $\sum_{k=1}^{\infty} a_k$ converges to A and $\sum_{k=1}^{\infty} b_k$ converges to B. Then the sequences of partial sums converge, and letting

$$S_n = \sum_{k=1}^{n} a_k \text{ and } T_n = \sum_{k=1}^{n} b_k,$$

we have $\lim_n S_n = A$ and $\lim_n T_n = B$.

Let $Q_n = \sum_{k=1}^{n} (a_k + b_k)$, the sequence of partial sums for $\sum_{k=1}^{\infty} (a_k + b_k)$. Since S, T, and Q are all finite sums,

$$Q_n = S_n + T_n.$$

Then by Theorem 1.1(i),
$A+B = \lim_n S_n + \lim_n T_n = \lim_n (S_n+T_n)$

$$= \lim_n Q_n = \sum_{k=1}^{\infty} (a_k + b_k)$$

The proofs for $\sum_{k=1}^{\infty} (a_k - b_k)$ and $\sum_{k=1}^{\infty} ca_k$ are similar.

31. Let $S_n = \sum_{k=1}^{\infty} \frac{1}{k}$. Then

$S_1 = 1$ and $S_2 = 1 + \frac{1}{2} = \frac{3}{2}$. Since

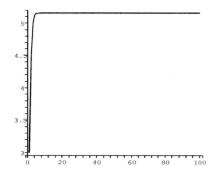

$S_8 > \frac{5}{2}$, we have

$$S_{16} = S_8 + \frac{1}{9} + \frac{1}{10} + \cdots + \frac{1}{16}$$
$$> S_8 + 8\left(\frac{1}{16}\right)$$
$$= S_8 + \frac{1}{2} > \frac{5}{2} + \frac{1}{2} = 3.$$

So $S_{16} > 3$.

$$S_{32} = S_{16} + \frac{1}{17} + \frac{1}{18} + \cdots + \frac{1}{32}$$
$$> S_{16} + 16\left(\frac{1}{32}\right)$$
$$= S_{16} + \frac{1}{2} > 3 + \frac{1}{2} = \frac{7}{2}.$$ So $S_{32} > \frac{7}{2}$

If $n = 64$, then $S_{64} > 4$. If $n = 256$, then $S_{256} > 5$. If $n = 4^{m-1}$, then $S_n > m$.

33.

$$.9 + .09 + .009 + \cdots = \sum_{k=0}^{\infty} .9(.1)^k$$

which is a geometric series with $a = .9$ and $|r| = .1 < 1$ so it converges to $\frac{.9}{1 - .1} = 1$.

35. $$0.1818\overline{18} = \frac{18}{100} + \frac{18}{10000} + \ldots = 18\sum_{k=1}^{\infty} \frac{1}{100^k}.$$

This is a geometric series with $a = \frac{18}{100}$ and $|r| = \frac{1}{100}$, so the sum is $\frac{18/100}{1 - 1/100} = \frac{2}{11}$.

37. The amount of overhang is $\sum_{k=0}^{n-1} \frac{L}{2(n-k)}$. So if n = 8, then

$$\sum_{k=0}^{7} \frac{L}{2(8-k)} = 1.3589L.$$

When $n = 4$,

$$L\sum_{k=0}^{n-1} \frac{1}{2(n-k)} = L\sum_{k=0}^{3} \frac{1}{2(4-k)}$$
$$= 1.0417L > L$$
$$\lim_{n\to\infty} \sum_{k=0}^{n-1} \frac{L}{2(n-k)} = \lim_{n\to\infty} \sum_{k=1}^{n} \frac{L}{2k}$$
$$= \frac{L}{2} \lim_{n\to\infty} \sum_{k=1}^{n} \frac{1}{k} = \infty$$

39. If $0 < 2r < 1$ then

$$1 + 2r + (2r)^2 + \cdots = \sum_{k=0}^{\infty} (2r)^k$$

which is a geometric series with $a = 1$ and $|2r| = 2r < 1$, so it converges to $\frac{1}{1-2r}$.

If $r = \frac{1}{1000}$, then

$$1 + .002 + .000004 + \cdots$$
$$= \frac{1}{1 - 2\left(\frac{1}{1000}\right)}$$
$$= \frac{500}{499} = 1.002004008\ldots$$

41. $$p^2 + 2p(1-p)p^2 + [2p(1-p)]^2 p^2 + \cdots = \sum_{k=0}^{\infty} p^2 [2p(1-p)]^k$$

is a geometric series with $a = p^2$ and $|r| = 2p(1-p) < 1$ because $2p(1-p)$ is a probability and therefore must be between 0 and 1. So the series converges to

$$\frac{p^2}{1 - [2p(1-p)|} = \frac{p^2}{1 - 2p + 2p^2}.$$

If $p = .6$,

$$\frac{.6^2}{1 - 2(.6) + 2(.6)^2} = .692 > .6.$$

If $p > \frac{1}{2}$, $\frac{p^2}{1 - 2p(1-p)} > p$.

43.

$$d + de^{-r} + de^{-2r} + \cdots = \sum_{k=0}^{\infty} d(e^{-r})^k$$

which is a geometric series with $a = d$ and $|e^{-r}| = e^{-r} < 1$ if $r > 0$.

So $\sum_{k=0}^{\infty} d(e^{-r})^k = \dfrac{d}{1-e^{-r}}$.

If $r = .1$,

$$\sum_{k=0}^{\infty} d(e^{-.1})^k = \frac{d}{1-e^{-.1}} = 2$$

so $d = 2(1 - .905) \approx .19$

45. $\sum_{k=0}^{\infty} 100{,}000\left(\dfrac{3}{4}\right)^k$
$= \dfrac{100{,}000}{1-3/4} = \$\,400{,}000.$

47. $\sum_{k=1}^{\infty} \dfrac{1}{k}$ and $\sum_{k=1}^{\infty} \dfrac{-1}{k}$

49. The geometric series

$$1 + r + r^2 + r^3 + \cdots$$

converges to $S = \dfrac{1}{1-r}$ provided that $-1 < r < 1$.

But $-1 < r$ implies that $-r < 1$, and so $1 - r < 2$, and therefore $\dfrac{1}{1-r} > \dfrac{1}{2}$, which means that $S > \dfrac{1}{2}$.

51. Since $\sum a_k$ converges, $\lim_{k\to\infty} a_k = 0$, by Theorem 2.2. Therefore, $\lim_{k\to\infty} \dfrac{1}{a_k}$ does not exist. In particular,
$\lim_{k\to\infty} \dfrac{1}{a_k} \neq 0$.

Therefore, $\sum \dfrac{1}{a_k}$ diverges, by the converse of Theorem 2.2.

53. Since $0 < p < 1$, therefore $-1 < p < 0$, and $0 < 1 - p < 1$. Thus, the given series is geometric with common ratio $r = 1 - p$, and so converges:

$$\sum_{n=1}^{\infty} p(1-p)^{n-1} = p\sum_{n=1}^{\infty} r^{n-1}$$
$$= p\frac{1}{1-r} = \frac{p}{1-(1-p)} = \frac{p}{p} = 1.$$

The sum represents the probability that you eventually win a game.

8.3 The Integral Test and Comparison Tests

1. $\sum_{k=1}^{\infty} \dfrac{4}{\sqrt[3]{k}}$ is a divergent p-series, because $p = \dfrac{1}{3} < 1$.

3. $\sum_{k=4}^{\infty} \dfrac{1}{k^{11/10}}$ is a convergent p-series, since $p = \dfrac{11}{10} > 1$.

5. Using the Limit Comparison Test, let $a_k = \dfrac{k+1}{k^2+2k+3}$ and $b_k = \dfrac{1}{k}$, so
$\lim_{k\to\infty} \dfrac{a_k}{b_k} = \lim_{k\to\infty} \left(\dfrac{k+1}{k^2+2k+3}\right)\left(\dfrac{k}{1}\right)$
$= \lim_{k\to\infty} \dfrac{k^2+k}{k^2+2k+3} = 1 > 0$, and since $\sum_{k=3}^{\infty} \dfrac{1}{k}$ is the divergent harmonic series, $\sum_{k=3}^{\infty} \dfrac{k+1}{k^2+2k+3}$ diverges.

7. Using the Limit Comparison Test, let $a_k = \dfrac{4}{2+4k}$ and $b_k = \dfrac{1}{k}$, so
$\lim_{k\to\infty} \dfrac{a_k}{b_k} = \lim_{k\to\infty} \left(\dfrac{4}{2+4k}\right)\left(\dfrac{k}{1}\right)$
$= \lim_{k\to\infty} \dfrac{4k}{2+4k} = 1 > 0$,

and since $\sum_{k=1}^{\infty} \dfrac{1}{k}$ is the divergent harmonic series, $\sum_{k=1}^{\infty} \dfrac{4}{2+4k}$ diverges.

9. Let $f(x) = \dfrac{2}{x \ln x}$. Then f is continuous and positive on $[2, \infty]$ and

$$f'(x) = \frac{-2(1 + \ln x)}{x^2(\ln x)^2} < 0$$

for $x \in [2, \infty)$, so f is decreasing. So we can use the Integral Test,

$$\int_2^\infty \frac{2}{x \ln x}\,dx = \lim_{R\to\infty} \int_2^R \frac{2}{x \ln x}\,dx$$
$$= 2 \lim_{R\to\infty} [\ln(\ln x)]_2^R$$
$$= 2 \lim_{R\to\infty} [\ln(\ln R) - \ln(\ln 2)]$$
$$= \infty,$$

so $\displaystyle\sum_{k=2}^{\infty} \frac{2}{k \ln k}$ diverges.

11. Using the Limit Comparison Test, let $a_k = \dfrac{2k}{k^3 + 1}$ and $b_k = \dfrac{1}{k^2}$.

Then

$$\lim_{k\to\infty} \frac{a_k}{b_k} = \lim_{k\to\infty} \left(\frac{2k}{k^3+1}\right) \cdot \left(\frac{k^2}{1}\right)$$
$$= \lim_{k\to\infty} \frac{2k^3}{k^3+1} = 2 > 0$$

and $\displaystyle\sum_{k=1}^{\infty} \frac{1}{k^2}$ is a convergent p-series $(p = 2 > 1)$, so $\displaystyle\sum_{k=1}^{\infty} \frac{2k}{k^3+1}$ converges.

13. Let $f(x) = \dfrac{e^{1/x}}{x^2}$.

Then f is continuous and positive on $[1, \infty)$ and

$$f'(x) = \frac{-e^{1/x}(1 + 2x)}{x^4} < 0$$

for all $x \in [1, \infty)$, so f is decreasing. Therefore, we can use the Integral Test.

$$\int_1^\infty \frac{e^{1/x}}{x^2}\,dx = \lim_{R\to\infty} \int_1^R \frac{e^{1/x}}{x^2}\,dx$$
$$= \lim_{R\to\infty} -e^{1/x}\Big|_1^R = \lim_{R\to\infty} \left(e - e^{1/R}\right)$$
$$= e - 1$$

So the series $\displaystyle\sum_{k=1}^{\infty} \frac{e^{1/k}}{k^2}$ converges, and so does $\displaystyle\sum_{k=3}^{\infty} \frac{e^{1/k}}{k^2}$.

15. Let $f(x) = \dfrac{e^{-\sqrt{x}}}{\sqrt{x}}$. Then f is continuous and positive on $[1, \infty)$ and

$$f'(x) = \frac{-(\sqrt{x} - 1)}{2x^{3/2}e^{\sqrt{x}}} < 0$$

for $x \in [1, \infty)$. So f is decreasing. Therefore, we can use the Integral Test.

$$\int_1^\infty \frac{e^{-\sqrt{x}}}{\sqrt{x}}\,dx = \lim_{R\to\infty} \int_1^R \frac{e^{-\sqrt{x}}}{\sqrt{x}}\,dx$$
$$= \lim_{R\to\infty} \left[-2e^{-\sqrt{x}}\right]_1^R = \lim_{R\to\infty} \left[\frac{2}{e} - \frac{2}{e^{\sqrt{R}}}\right]$$
$$= \frac{2}{e}.$$

So $\displaystyle\sum_{k=1}^{\infty} \frac{e^{\sqrt{k}}}{\sqrt{k}}$ converges.

17. Using the Limit Comparison Test, let $a_k = \dfrac{2k^2}{k^{5/2} + 2}$ and $b_k = \dfrac{1}{\sqrt{k}}$. Since

$$\lim_{k\to\infty} \frac{a_k}{b_k} = \lim_{k\to\infty} \frac{2k^2}{k^{5/2}+2} \frac{\sqrt{k}}{1}$$
$$= \lim_{k\to\infty} \frac{2k^{5/2}}{k^{5/2}+2} = 2 > 0$$

and $\displaystyle\sum_{k=1}^{\infty} \frac{1}{\sqrt{k}}$ is a divergent p-series, so $\displaystyle\sum_{k=1}^{\infty} \frac{2k^2}{k^{5/2}+2}$ diverges.

19. Let $a_k = \dfrac{4}{\sqrt{k^3+1}}$ and $b_k = \dfrac{1}{k^{3/2}}$.

Since $\displaystyle\lim_{k\to\infty} \frac{a_k}{b_k} = \lim_{k\to\infty} \frac{4}{\sqrt{k^3+1}} \frac{k^{3/2}}{1}$

$$= \lim_{k\to\infty} \frac{4}{\sqrt{1 + k^{-3}}} = 4 > 0$$

and $\displaystyle\sum_{k=1}^{\infty} \frac{1}{k^{3/2}}$ is a convergent p-series,

$\sum_{k=0}^{\infty} \frac{4}{\sqrt{k^3+1}}$ converges, by the Limit Comparison Test.

21. Let $f(x) = \frac{\tan^{-1} x}{1+x^2}$ which is continuous and positive on $[1, \infty)$ and $f'(x) = \frac{1 - 2x\tan^{-1} x}{(1+x^2)^2} < 0$ for $x \in [1, \infty)$, so f is decreasing. So we can use the Integral Test.

$$\int_1^{\infty} \frac{\tan^{-1} x}{1+x^2}\, dx = \lim_{R\to\infty} \int_1^R \frac{\tan^{-1} x}{1+x^2}\, dx$$
$$= \lim_{R\to\infty} \frac{1}{2}(\tan^{-1} x)^2 \Big|_1^R$$
$$= \lim_{R\to\infty} \left[\frac{1}{2}(\tan^{-1} R)^2 - \frac{1}{2}(\tan^{-1} 1)^2\right]$$
$$= \frac{1}{2}\left(\frac{\pi}{2}\right)^2 - \frac{1}{2}\left(\frac{\pi}{4}\right)^2 = \frac{3\pi^2}{32},$$

so $\sum_{k=1}^{\infty} \frac{\tan^{-1} k}{1+k^2}$ converges.

23. Since $\left|\cos^2 k\right| \le 1$, $\frac{1}{\cos^2 k} > 1$, so by the $k-$th Term Test for Divergence, $\sum_{k=1}^{\infty} \frac{1}{\cos^2 k}$ diverges.

25. Let $f(x) = \frac{\ln x}{x}$ which is continuous and positive on $[2, \infty)$ and

$$f'(x) = \frac{1 - \ln x}{x^2} < 0$$

for $x \in [2, \infty)$, so f is decreasing. Therefore, we can use the Integral Test.

$$\int_2^{\infty} \frac{\ln x}{x}\, dx = \lim_{R\to\infty} \int_2^R \frac{\ln x}{x}\, dx$$
$$= \lim_{R\to\infty} \frac{(\ln x)^2}{2}\Big|_2^R$$
$$= \lim_{R\to\infty} \left[\frac{(\ln R)^2}{2} - \frac{(\ln 2)^2}{2}\right]$$
$$= \infty,$$

So $\sum_{k=2}^{\infty} \frac{\ln k}{k}$ diverges.

27. Using the Limit Comparison Test, let $a_k = \frac{k^4 + 2k - 1}{k^5 + 3k^2 + 1}$ and $b_k = \frac{1}{k}$.

Then $\lim_{k\to\infty} \frac{a_k}{b_k} = \lim_{k\to\infty} \left(\frac{k^4 + 2k - 1}{k^5 + 3k^2 + 1}\right)\left(\frac{k}{1}\right)$
$= \lim_{k\to\infty} \frac{k^5 + 2k^2 - k}{k^5 + 3k^2 + 1} = 1 > 0.$

Since $\sum_{k=1}^{\infty} \frac{1}{k}$ is the divergent harmonic series, $\sum_{k=1}^{\infty} \frac{k^4 + 2k1}{k^5 + 3k^2 + 1}$ diverges, and so does $\sum_{k=4}^{\infty} \frac{k^4 + 2k1}{k^5 + 3k^2 + 1}$.

29. $\lim_{k\to\infty} \frac{k+1}{k+2} = 1 \ne 0$, so by the k-th Term Test for Divergence, $\sum_{k=3}^{\infty} \frac{k+1}{k+2}$ diverges.

31. Using the Limit Comparison Test, let $a_k = \frac{k+1}{k^3+2}$ and $b_k = \frac{1}{k^2}$. Then
$\lim_{k\to\infty} \frac{a_k}{b_k} = \lim_{k\to\infty} \left(\frac{k+1}{k^3+2}\right)\left(\frac{k^2}{1}\right)$
$= \lim_{k\to\infty} \frac{k^3 + k^2}{k^3 + 2} = 1 > 0$
and $\sum_{k=1}^{\infty} \frac{1}{k^2}$ is a convergent p-series $(p = 2 > 1)$, so $\sum_{k=8}^{\infty} \frac{k+1}{k^3+2}$ converges.

33. $\frac{1}{(k+1)\sqrt{k} + k\sqrt{k+1}} < \frac{1}{k\sqrt{k} + k\sqrt{k}}$
$\frac{1}{(k+1)\sqrt{k} + k\sqrt{k+1}} < \frac{1}{2k^{3/2}}$
The series $\sum_{k=1}^{\infty} \frac{1}{2k^{3/2}}$ converges, as it is a p-series with $p > 1$. Thus, by the Comparison Test, the original series converges.

35. We are only concerned with what happens to the terms as $k \to \infty$, so the first few terms don't matter.

37. Since $\lim_{k\to\infty} \frac{a_k}{b_k} = 0$, there exists N, so that for all $k > N$, $\left|\frac{a_k}{b_k} - 0\right| < 1$. Since a_k and b_k are positive, $\frac{a_k}{b_k} < 1$, so $a_k < b_k$. Thus, since $\sum_{k=1}^{\infty} b_k$ converges, $\sum_{k=1}^{\infty} a_k$ converges by the Comparison Test.

39. Case 1: Suppose that

$$\lim_{k\to\infty} \frac{|a_k|}{|b_k|} = 0.$$

Then there exists a natural number N such that, for all $k \geq N$, $|a_k| < |b_k|$. Thus, for all $k \geq N$, $|a_k b_k| < b_k^2$. Thus, by the Comparison Test, $\sum_{k=1}^{\infty} |a_k b_k|$ converges.

Case 2: Suppose that

$$\lim_{k\to\infty} \frac{|b_k|}{|a_k|} = 0.$$

Using the same reasoning as in Case 1, there exists a natural number N such that, for all $k \geq N$, $|b_k| < |a_k|$. Thus, for all $k \geq N$, $|a_k b_k| < a_k^2$. Thus, by the Comparison Test, $\sum_{k=1}^{\infty} |a_k b_k|$ converges.

The only other possibility is that $\lim_{k\to\infty} \frac{|a_k|}{|b_k|} = L$, where $L > 0$. Then it follows that $\lim_{k\to\infty} \frac{a_k^2}{|a_k b_k|} = L$. Thus, by the Limit Comparison Test, $\sum_{k=1}^{\infty} |a_k b_k|$ converges.

41. If $p \leq 1$, then the series diverges because $\frac{1}{k(\ln k)^p} \geq \frac{1}{k \ln k}$ (at least for $k > 2$) and $\sum_{k=2}^{\infty} \frac{1}{k \ln k}$ diverges (from Exercise 9). If $p > 1$, then let $f(x) = \frac{1}{x(\ln x)^p}$. f is continuous and positive on $[2, \infty)$, and

$$f'(x) = \frac{-(\ln(x))^{-p-1}(p + \ln x)}{x^2} < 0$$

so f is decreasing. Thus, we can use the Integral Test.

$$\int_2 \frac{1}{x(\ln x)^p}\,dx$$
$$= \lim_{R\to\infty} \int_2^R \frac{1}{x(\ln x)^p}\,dx$$
$$= \lim_{R\to\infty} \left.\frac{1}{(1-p)(\ln x)^{p-1}}\right|_2^R$$
$$= \lim_{R\to\infty} \left[\frac{1}{(1-p)(\ln R)^{p-1}} \frac{1}{(1-p)(\ln 2)^{p-1}}\right]$$
$$= -\frac{1}{(1-p)(\ln 2)^{p-1}},$$

so $\sum_{k=2}^{\infty} \frac{1}{k(\ln k)^p}$ converges when $p > 1$.

43. If $p \leq 1$, then $k^p \leq k$ for all $k \geq 1$, so $\frac{1}{k^p} \geq \frac{1}{k}$ for all $k \geq 1$, and $\frac{\ln k}{k^p} \geq \frac{\ln k}{k} \geq \frac{1}{k}$ for all $k > 2$, and $\sum_{k=2}^{\infty} \frac{1}{k}$ diverges.

So by the Comparison Test, $\sum_{k=2}^{\infty} \frac{\ln k}{k^p}$ diverges.

If $p > 1$, then let $f(x) = \frac{\ln x}{x^p}$. Then f is continuous and positive on $[2, \infty)$ and

$$f'(x) = \frac{k^{p-1}(1 - p \ln k)}{k^2 p} < 0$$

for $k > 2$, so f is decreasing. Thus, we can use the Integral Test.

$$\int_2^{\infty} \frac{\ln x}{x^p}\, dx = \lim_{R\to\infty} \int_2^R \frac{\ln x}{x^p}\, dx$$
$$= \lim_{R\to\infty} \frac{\ln x}{(1-p)x^{p-1}} - \frac{1}{(1-p)^2 x^{p-1}}\Big|_2^R$$
$$= \lim_{R\to\infty} \left[\frac{\ln R}{(1-p)R^{p-1}} - \frac{1}{(1-p)^2 R^{p-1}}\right] - \frac{\ln 2}{(1-p)2^{p-1}} + \frac{1}{(1-p)^2 2^{p-1}}$$
$$= \frac{1}{(1-p)^2 2^{p-1}} - \frac{\ln 2}{(1-p)2^{p-1}}$$

because

$$\lim_{R\to\infty} \frac{\ln R}{(1-p)R^{p-1}} = \lim_{R\to\infty} \frac{\frac{1}{R}}{-(1-p)^2 R^{p-2}}$$
$$= \lim_{R\to\infty} \frac{-1}{(1-p)^2 R^{p-1}} = 0$$

by l'Hopital's Rule and

$$\lim_{R\to\infty} \frac{1}{(1-p)^2 R^{p-1}} = 0$$

because $p > 1$.

Thus by the Integral Test, $\sum_{k=R}^{\infty} \frac{\ln k}{k^p}$ converges when $p > 1$.

45. $R_{100} \leq \frac{1}{3 \cdot 100^3} = 3.33 \times 10^{-7}$

47. $R_{100} \leq \frac{6}{7 \cdot 50^7} = 1.097 \times 10^{-12}$

49. To estimate $\left(\frac{1}{2}e^{-40^2}\right)$, take the logarithm: $\log\left(\frac{1}{2}e^{-40^2}\right) \approx -695.2$, so the error is less than 10^{-695}.

51. $R_n \leq \frac{3}{3n^2} < 10^{-6}$, so $n > 100$ will do.

53. $R_n \leq \frac{1}{2}e^{-n^2} < 10^{-6}$. Taking the natural logarithm, we need $n > \sqrt{\ln 500000}$ or $n \geq 4$.

55. (a) can't tell

(b) converges

(c) converges

(d) can't tell

57.

$$1 + \frac{1}{3} + \frac{1}{5} + \frac{1}{7} + \cdots = \sum_{k=0}^{\infty} \frac{1}{2k+1}.$$

Using the Limit Comparison Test, let $a_k = \frac{1}{2k+1}$ and $b_k = \frac{1}{k}$. Then
$$\lim_{k\to\infty} \frac{a_k}{b_k} = \lim_{k\to\infty} \left(\frac{1}{2k+1}\right)\left(\frac{k}{1}\right)$$
$$= \lim_{k\to\infty} \frac{k}{2k+1} = \frac{1}{2} > 0$$
and since $\sum_{k=1}^{\infty} \frac{1}{k}$ is the divergent harmonic series, then $\sum_{k=1}^{\infty} \frac{1}{2k+1}$ diverges.

59. If $x > 1$, $\zeta(x) = \sum_{k=1}^{\infty} \frac{1}{k^x}$ is a convergent p-series. If $x \leq 1$, $\zeta(x)$ is a divergent p-series.

61. $\zeta(4) = \sum_{k=1}^{\infty} \frac{1}{k^4} \approx \sum_{k=1}^{100} \frac{1}{k^4} \approx 1.0823$

63. $\zeta(8) = \sum_{k=1}^{\infty} \frac{1}{k^8} \approx \sum_{k=1}^{100} \frac{1}{k^8} \approx 1.0004$

65. First we'll compare $(\ln k)^{\ln k}$ to k^2. Let $L = \lim_{k\to\infty} \frac{(\ln k)^{\ln k}}{k^2}$ and calculate as follows:

$$L = \lim_{k\to\infty} \frac{(\ln k)^{\ln k}}{k^2}$$

$$\ln L = \lim_{k\to\infty} \ln\left[\frac{(\ln k)^{\ln k}}{k^2}\right]$$
$$= \lim_{k\to\infty} \left[\ln\left[(\ln k)^{\ln k}\right] - \ln(k^2)\right]$$
$$= \lim_{k\to\infty} \left[(\ln k)\ln(\ln k) - 2\ln k\right]$$
$$= \lim_{k\to\infty} (\ln k)\left[\ln(\ln k) - 2\right]$$
$$= \infty$$

Thus,

$$\lim_{k\to\infty}\frac{(\ln k)^{\ln k}}{k^2}=\infty$$

This means that eventually $(\ln k)^{\ln k} > k^2$. In other words, there exists a natural number N such that for all $k \geq N$,

$$\frac{1}{(\ln k)^{\ln k}}<\frac{1}{k^2}$$

Thus, by the Comparison test, $\sum_{k=2}^{\infty}\frac{1}{(\ln k)^{\ln k}}$ converges.

Now we'll use the Comparison test to show that $\sum_{k=2}^{\infty}\frac{1}{(\ln k)^k}$ converges. We begin by noting that

$\ln k < k$
$\Rightarrow (\ln k)^{\ln k} < (\ln k)^k$
$\Rightarrow \frac{1}{(\ln k)^k} < \frac{1}{(\ln k)^{\ln k}}$

Therefore, by the Comparison test, $\sum_{k=2}^{\infty}\frac{1}{(\ln k)^k}$ converges.

67. $\int_1^{\infty}\frac{x}{2^x}\,dx=\lim_{R\to\infty}\int_1^R\frac{x}{2^x}\,dx$
$=\lim_{R\to\infty}\frac{1}{2^x}\left(-\frac{1}{\ln^2 2}-\frac{x}{\ln 2}\right)\Big|_2^R$

This integral converges since $\lim_{R\to\infty}\frac{R}{2^R}=0$ by L'Hopital's rule.

$$\sum_{k=1}^{20}k\left(\frac{1}{2}\right)^k\approx 2.0$$

69. $\sum_{k=1}^{\infty}\frac{9k}{10^k}=9\sum_{k=1}^{\infty}k\frac{1}{10^k}=$

$$9\left(\frac{1}{10}\right.$$
$$+\frac{1}{10^2}+\frac{1}{10^2}$$
$$+\frac{1}{10^3}+\frac{1}{10^3}+\frac{1}{10^3}$$
$$+\frac{1}{10^4}+\frac{1}{10^4}+\frac{1}{10^4}+\frac{1}{10^4}$$
$$\left.+\frac{1}{10^5}+\frac{1}{10^5}+\frac{1}{10^5}+\frac{1}{10^5}+\frac{1}{10^5}\cdots\right)$$
$$=9\left(\frac{1}{10}+\frac{1}{10^2}+\frac{1}{10^3}+\frac{1}{10^4}+\frac{1}{10^5}+\cdots\right.$$
$$+\frac{1}{10^2}+\frac{1}{10^3}+\frac{1}{10^4}+\frac{1}{10^5}+\cdots$$
$$\left.+\frac{1}{10^3}+\frac{1}{10^4}+\frac{1}{10^5}\cdots\right)$$
$$=9\left(\sum_{k=1}^{\infty}\frac{1}{10^k}+\sum_{k=2}^{\infty}\frac{1}{10^k}+\sum_{k=3}^{\frac{1}{10^k}}+\cdots\right)$$

Each of these sums is a geometric series with $r = 1/10$, so we get

$$9\left(\frac{1/10}{1-\frac{1}{10}}+\frac{1/10^2}{1-\frac{1}{10}}+\frac{1/10^3}{1-\frac{1}{10}}+\cdots\right)$$
$$=\frac{9}{9/10}\sum_{k=1}^{\infty}\frac{1}{10^k}=\left(\frac{9}{9/10}\right)\left(\frac{1/10}{1-\frac{1}{10}}\right)$$
$$=\left(\frac{9}{9/10}\right)\left(\frac{1/10}{9/10}\right)=\frac{10}{9}$$

71. $1+\frac{10}{9}+\frac{10}{8}+\cdots+\frac{10}{1}\equiv 29.29$

73. For n cards in the set, you will need an average of

$$\frac{n}{n-1}+\frac{n}{n-2}+\cdots\frac{n}{1}$$

attempts. The ratio is $\sum_{k=1}^{n-1}\frac{1}{k}$.

8.4 Alternating Series

1. $\lim_{k\to\infty}a_k=\lim_{k\to\infty}\frac{3}{k}=0$, and

$$0<a_{k+1}=\frac{3}{k+1}\leq\frac{3}{k}=a_k$$

for all $k \geq 1$, so by the Alternating Series Test, the original series converges.

3. $\lim_{k\to\infty} a_k = \lim_{k\to\infty} \frac{4}{\sqrt{k}} = 0$ and

$$0 < a_{k+1} = \frac{4}{\sqrt{k+1}} < \frac{4}{\sqrt{k}} = a_k$$

for all $k \geq 1$. Thus by the Alternating Series Test, the original series converges.

5.

$$\lim_{k\to\infty} a_k = \lim_{k\to\infty} \frac{k}{k^2+2} = 0$$

and

$$\frac{a_{k+1}}{a_k} = \frac{k+1}{(k+1)^2+2} \cdot \frac{k^2+2}{k}$$
$$= \frac{k^3+k^2+2k+2}{k^3+2k^2+3k} \leq 1$$

for all $k \geq 2$, so $a_{k+1} < a_k$ for all $k \geq 2$. Thus by the Alternating Series Test, the original series converges.

7. By l'Hospital's Rule,

$$\lim_{k\to\infty} a_k = \lim_{k\to\infty} \frac{k}{2^k} = \lim_{k\to\infty} \frac{1}{2^k \ln 2} = 0$$

and

$$\frac{a_{k+1}}{a_k} = \frac{k+1}{2^{k+1}} \cdot \frac{2^k}{k} = \frac{k+1}{2k}$$

for all $k \geq 5$, so $a_{k+1} \leq a_k$ for all $k \geq 5$.

Thus by the Alternating Series Test, the original series converges.

9. $\lim_{k\to\infty} a_k = \lim_{k\to\infty} \frac{4^k}{k^2} = \lim_{k\to\infty} \frac{4^k \ln 4}{2k}$
$= \lim_{k\to\infty} \frac{4^k (\ln 4)^2}{2} = \infty$ (by l'Hopital's Rule)

So by the k-th Term Test for Divergence, $\sum_{k=1}^{\infty} (-1)^k \frac{4^k}{k^2}$ diverges.

11.

$$\lim_{k\to\infty} a_k = \lim_{k\to\infty} \frac{2k}{k+1} = 2,$$

so by the k-th Term Test for Divergence, $\sum_{k=1}^{\infty} \frac{2k}{k+1}$ diverges.

13.

$$\lim_{k\to\infty} a_k = \lim_{k\to\infty} \frac{2k}{\sqrt{k+1}} = 0$$

and

$$\frac{a_{k+1}}{a_k} = \frac{3}{\sqrt{k+2}} \cdot \frac{\sqrt{k+1}}{3} = \frac{\sqrt{k+1}}{\sqrt{k+2}} < 1$$

for all $k \geq 3$, so $a_{k+1} < a_k$ for all $k \geq 3$. Thus by the Alternating Series Test, $\sum_{k=3}^{\infty} (-1)^k \frac{3}{\sqrt{k+1}}$ converges.

15.

$$\lim_{k\to\infty} a_k = \lim_{k\to\infty} \frac{2}{\sqrt{k!}} - 0$$

and

$$\frac{a_{k+1}}{a_k} = \frac{2}{\sqrt{k+1!}} \cdot \frac{k!}{2} = \frac{1}{k+1} < 1$$

for all $k \geq 1$, so $a_{k+1} < a_k$ for all $k \geq 1$.

Thus by the Alternating Series Test, $\sum_{k=1}^{\infty} (-1)^{k+1} \frac{2}{k!}$ converges.

17.

$$a_k = \frac{k!}{2^k} = \frac{1 \cdot 2 \cdot 3 \ldots k}{2 \cdot 2 \cdot 2 \ldots 2} \geq \frac{1}{2} \cdot \frac{k}{2} = \frac{k}{4}$$

and

$$\lim_{k\to\infty} = k = \infty$$

so by the k-th Term Test for Divergence $\sum_{k=2}^{\infty} (-1)^k \frac{k!}{2^k}$ diverges.

19.

$$\lim_{k\to\infty} a_k = \lim_{k\to\infty} 2e^{-k} = 0$$

and

$$\frac{a_{k+1}}{a_k} = \frac{2e^{-(k+1)}}{2e^{-k}} = \frac{1}{e} < 1$$

for all $k \geq 0$,, so $a_{k+1} < a_k$ for all $k \geq 0$. Thus by the Alternating Series Test, $\sum_{k=5}^{\infty} (-1)^{k+1} 2e^{-k}$ converges.

21. Note that

$$\lim_{k\to\infty} a_k = \lim_{k\to\infty} \ln k = \infty,$$

so by the k-th Term Test for Divergence, $\sum_{k=2}^{\infty} (-1)^k \ln k$ diverges.

23.

$$\lim_{k\to\infty} a_k = \lim_{k\to\infty} \frac{1}{2^k} = 0$$

and

$$\frac{a_{k+1}}{a_k} = \frac{1}{2^{k+1}} \cdot \frac{2^k}{1} = \frac{1}{2} < 1$$

for all $k \geq 0$,
so $a_{k+1} < a_k$ for all $k \geq 0$.
Thus by the Alternating Series Test, $\sum_{k=0}^{\infty} (-1)^{k+1} \frac{1}{2^k}$ converges.

25. $|S - S_k| \leq a_{k+1} = \dfrac{4}{(k+1)^3} \leq .01$
and $a_7 = 4/8^3 < 0.01$, so
$S \approx S_7 \approx 3.61$.

27. $|S - S_k| \leq a_{k+1} = \dfrac{k+1}{2^{k+1}} \leq .01$
If $k = 9$, $a_{10} = \dfrac{10}{2^{10}} \approx .00977 < .01$
So $S \approx S_9 \approx -.22$.

29. $|S - S_k| \leq a_{k+1} = \dfrac{3}{(k+1)!} \leq .01$
If $k = 5$, $a_6 = \dfrac{3}{6!} \approx .0042 < .01$
So $S \approx S_5 \approx 1.10$.

31. $|S - S_k| \leq a_{k+1} = \dfrac{4}{(k+1)^4} \leq .01$,
so $400 \leq (k+1)^4$, then $\sqrt[3]{400} \leq k+1$,
so $k \geq \sqrt[3]{400} - 1 \approx 3.47$ so $k \geq 4$.
Thus $S \approx S_4 \approx -0.21$.

33. $|S - S_k| \leq a_{k+1} = \dfrac{2}{k+1} < .0001$,
so $k + 1 \geq 20{,}000$ and then $k \geq 19{,}999$. Thus $k = 20{,}000$.

35. $|S - S_k| \leq a_{k+1} = \dfrac{10^{k+1}}{(k+1)!}$ Since $a_{34} = \dfrac{10^{34}}{34!} \approx 0.00003387$ is the first term such that $a_k < 0.0001$, 34 terms are needed.

37. If the derivative of a function $f(k) = a_k$ is negative, it means that the function is decreasing, i.e., each successive term is smaller than the one before. If $f(k) = a_k = \dfrac{k}{k^2+2}$, then
$f'(k) = \dfrac{-k^2+2}{(k^2+2)^2} < 0$ for all $k \geq 2$
so a_k is decreasing.

39. The sum of the odd terms is

$$1 + \frac{1}{3} + \frac{1}{5} + \frac{1}{7} + \cdots,$$

which diverges to ∞. The sum of the even terms is
$-\sum_{k=1}^{\infty} \dfrac{1}{(2k)^2}$, which is a convergent p-series. So the series diverges to ∞.

41.

$$\frac{3}{4} - \frac{3}{4}\left(\frac{3}{4}\right) - \cdots = \sum_{k=0}^{\infty} \left(\frac{3}{4}\right)\left(-\frac{3}{4}\right)^k$$

which is a geometric series with $a = \dfrac{3}{4}$ and $|r| = \dfrac{3}{4} < 1$. So

$$\sum_{k=0}^{\infty} \left(\frac{3}{4}\right)\left(-\frac{3}{4}\right)^k = \frac{3/4}{1+3/4} = \frac{3}{7}.$$

The person ends up $\frac{3}{7}$ of the distance from home.

43.

$$S_{2n} = \sum_{k=1}^{2n} \frac{1}{k} - \sum_{k=1}^{n} \frac{1}{n} = \sum_{k=n+1}^{2n} \frac{1}{k}$$

Let $k = r + n$, so that $k = n + 1 \Leftrightarrow r = 1$ and $k = 2n \Leftrightarrow r = n$. Thus,

$$S_{2n} = \sum_{r=1}^{n} \frac{1}{r+n}$$

Now relabel r as k to get

$$S_{2n} = \sum_{k=1}^{n} \frac{1}{k+n} = \frac{1}{n} \sum_{k=1}^{n} \frac{1}{1 + \frac{k}{n}}$$

The previous line is a Riemann sum for $\int_1^2 \frac{1}{x}\, dx$. Thus,

$$\lim_{n\to\infty} S_{2n} = \int_1^2 \frac{1}{x}\, dx = [\ln x]_1^2$$
$$= \ln 2 - \ln 1 = \ln 2$$

45. An example is

$$a_k = b_k = \frac{(-1)^k}{\sqrt{k}}.$$

Then $\sum_{k=1}^{\infty} a_k$ and $\sum_{k=1}^{\infty} b_k$ converge by the Alternating Series Test. However,

$$\sum_{k=1}^{\infty} a_k b_k = \sum_{k=1}^{\infty} \frac{(-1)^{2k}}{k} = \sum_{k=1}^{\infty} \frac{1}{k}$$

which diverges.

8.5 Absolute Convergence and the Ratio Test

1. By the Ratio Test,

$$\lim_{k\to\infty} \left| \frac{a_{k+1}}{a_k} \right| = \lim_{k\to\infty} \frac{3}{(k+1)!} \cdot \frac{k!}{3}$$
$$= \lim_{k\to\infty} \frac{1}{k+1} = 0 < 1$$

so $\sum_{k=0}^{\infty} (-1)^k \frac{3}{k!}$ converges absolutely.

3. By the Ratio Test,

$$\lim_{k\to\infty} \left| \frac{a_{k+1}}{a_k} \right| = \lim_{k\to\infty} \frac{2^{k+1}}{2^k}$$
$$= \lim_{k\to\infty} 2 = 2 > 1,$$

so $\sum_{k=0}^{\infty} (-1)^k 2^k$ diverges. (Or use k-th Term Test.)

5. By the Alternating Series Test,

$$\lim_{k\to\infty} a_k = \lim_{k\to\infty} \frac{k}{k^2+1} = 0$$

and
$$\frac{a_{k+1}}{a_k} = \frac{k+1}{(k+1)^2+1} \cdot \frac{k^2+1}{k}$$
$$= \frac{k^3+k^2+k+1}{k^3+2k^2+2k} < 1$$

for all $k \geq 1$, so $a_{k+1} < a_k$ for all $k \geq 1$, so the series converges.

But by the Limit Comparison Test, letting $a_k = \frac{k}{k^2+1}$ and $b_k = \frac{1}{k}$,

$$\lim_{k\to\infty} \frac{a_k}{b_k} = \lim_{k\to\infty} \frac{k^2}{k^2+1} = 1 > 0$$

and $\sum_{k=1}^{\infty} \frac{1}{k}$ is the divergent harmonic series.

Therefore $\sum_{k=1}^{\infty} |a_k| = \sum_{k=1}^{\infty} \frac{k}{k^2+1}$ diverges. Thus $\sum_{k=1}^{\infty} (-1)^{k+1} \frac{k}{k^2+1}$ converges conditionally.

7. By the Ratio Test,
$$\lim_{k\to\infty} \left|\frac{a_{k+1}}{a_k}\right| = \lim_{k\to\infty} \frac{3^{k+1}}{(k+1)!} \cdot \frac{k!}{3^k}$$
$$= \lim_{k\to\infty} \frac{3}{k+1} = 0 < 1,$$
so $\sum_{k=3}^{\infty} (-1)^k \frac{3^k}{k!}$ converges absolutely.

9. $\lim_{k\to\infty} a_k = \lim_{k\to\infty} \frac{k}{2k+1} = \frac{1}{2}$

so by the k-th Term Test for Divergence, $\sum_{k=2}^{\infty} (-1)^{k+1} \frac{k}{2k+1}$ diverges.

11. Using the Ratio Test,
$$\lim_{k\to\infty} \left|\frac{a_{k+1}}{a_k}\right| = \lim_{k\to\infty} \frac{(k+1)2^{k+1}}{3^{k+1}} \cdot \frac{3^k}{k2^k}$$
$$= \lim_{k\to\infty} \frac{2(k+1)}{3k} = \frac{2}{3} < 1,$$
so $\sum_{k=6}^{\infty} (-1)^k \frac{k2^k}{3^k}$ converges absolutely.

13. Using the Root Test,
$$\lim_{k\to\infty} \sqrt[k]{|a_k|} = \lim_{k\to\infty} \sqrt[k]{\left(\frac{4k}{5k+1}\right)^k}$$
$$= \lim_{k\to\infty} \frac{4k}{5k+1} = \frac{4}{5} < 1$$
so $\sum_{k=1}^{\infty} \left(\frac{4k}{5k+1}\right)^k$ converges absolutely.

15. $\sum_{k=1}^{\infty} \frac{1}{k}$ is the divergent harmonic series; so $\sum_{k=1}^{\infty} \frac{-2}{k}$ diverges.

17. Using the Alternating Series Test,
$$\lim_{k\to\infty} a_k = \lim_{k\to\infty} \frac{\sqrt{k}}{k+1} = 0 \text{ and}$$
$$\frac{a_{k+1}}{a_k} = \frac{\sqrt{k+1}}{k+2} \cdot \frac{k+1}{\sqrt{k}}$$
$$= \frac{(k+1)^{3/2}}{k^{3/2}+2k^{1/2}} < 1 \text{ for all } k \geq 1,$$
so $a_{k+1} < a_k$ for all $k \geq 1$ so the series converges.

But by the Limit Comparison Test, letting
$$a_k = \frac{\sqrt{k}}{k+1} \text{ and } b_k = \frac{1}{k^{1/2}},$$
$$\lim_{k\to\infty} \frac{a_k}{b_k} = \lim_{k\to\infty} \frac{\sqrt{k}}{k+1} \cdot \frac{k^{1/2}}{1}$$
$$= \lim_{k\to\infty} \frac{k}{k+1} = 1 > 0,$$
and $\sum_{k=0}^{\infty} \frac{1}{k^{1/2}}$ is a divergent p-series $\left(p = \frac{1}{2} < 1\right)$.

Therefore,

$\sum_{k=0}^{\infty} |a_k| = \sum_{k=0}^{\infty} \frac{\sqrt{k}}{k+1}$ diverges. So $\sum_{k=0}^{\infty} (-1)^{k+1} \frac{\sqrt{k}}{k+1}$ converges conditionally.

19. Using the Ratio Test,
$$\lim_{k\to\infty} \left|\frac{a_{k+1}}{a_k}\right| = \lim_{k\to\infty} \frac{(k+1)^2}{e^{k+1}} \cdot \frac{e^k}{k^2}$$
$$= \lim_{k\to\infty} \frac{(k+1)^2}{ek^2} = \frac{1}{e} < 1,$$
so $\sum_{k=7}^{\infty} \frac{k^2}{e^k}$ converges absolutely.

21. Using the Root Test,
$$\lim_{k\to\infty} \sqrt[k]{|a_k|} = \lim_{k\to\infty} \sqrt[k]{\left(\frac{e^3}{k^3}\right)^k}$$
$$= \lim_{k\to\infty} \frac{e^3}{k^3} = 0 < 1,$$

so $\sum_{k=2}^{\infty} \frac{e^{3k}}{k^{3k}}$ converges absolutely.

23. Since $|\sin k| \le 1$ for all k,

$$\left|\frac{\sin k}{k^2}\right| = \frac{|\sin k|}{k^2} \le \frac{1}{k^2},$$

and $\sum_{k=1}^{\infty} \frac{1}{k^2}$ is a convergent p-series ($p = 2 > 1$) so by the Comparison Test, $\sum_{k=1}^{\infty} \left|\frac{\sin k}{k^2}\right|$ converges, and by Theorem 5.1, $\sum_{k=1}^{\infty} \frac{\sin k}{k^2}$ converges absolutely.

25. Since $\cos k\pi = (-1)^k$ for all k,

$$\left|\frac{\cos k\pi}{k}\right| = \left|\frac{(-1)^k}{k}\right| = \frac{1}{k}$$

and $\sum_{k=1}^{\infty} \frac{1}{k}$ is the divergent harmonic series, so $\sum_{k=1}^{\infty} \left|\frac{\cos k\pi}{k}\right|$ diverges by the Comparison Test.

But, using the Alternating Series Test,

$$\lim_{k\to\infty} a_k = \lim_{k\to\infty} \frac{1}{k} = 0$$

and

$$\frac{a_{k+1}}{a_k} = \frac{1}{k+1} \cdot \frac{k}{1} = \frac{k}{k+1} < 1$$

for all $k \ge 1$, so $a_{k+1} < a_k$ for all $k \ge 1$, so $\sum_{k=1}^{\infty} \frac{(-1)^k}{k}$ converges.

Thus $\sum_{k=1}^{\infty} \frac{(-1)^k}{k} = \sum_{k=1}^{\infty} \frac{\cos k\pi}{k}$ converges conditionally.

27. Using the Alternating Series Test,

$$\lim_{k\to\infty} a_k = \lim_{k\to\infty} \frac{1}{\ln k} = 0$$

and

$$\frac{a_{k+1}}{a_k} = \frac{1}{\ln(k+1)} \cdot \frac{\ln k}{1} = \frac{\ln k}{\ln(k+1)} < 1$$

for all $k \ge 2$, so $a_{k+1} < a_k$ for all $k \ge 2$. So $\sum_{k=2}^{\infty} \frac{(-1)^k}{\ln k}$ converges.

But by the Comparison Test, because $\ln k < k$ for all $k \ge 2$, $\frac{1}{\ln k} > \frac{1}{k}$ for all $k \ge 2$ and $\sum_{k=2}^{\infty} \frac{1}{k}$ is the divergent harmonic series. Therefore, $\sum_{k=2}^{\infty} |a_k| = \sum_{k=2}^{\infty} \frac{1}{\ln k}$ diverges.

Thus $\sum_{k=2}^{\infty} \frac{(-1)^k}{\ln k}$ converges conditionally.

29. $\sum_{k=1}^{\infty} |a_k| = \sum_{k=1}^{\infty} \frac{1}{k^{3/2}}$ which is a convergent p-series $\left(p = \frac{3}{2} > 1\right)$, so by Theorem 5.1, $\sum_{k=1}^{\infty} \frac{(-1)^k}{k\sqrt{k}}$ converges absolutely.

31. Consider $\sum_{k=1}^{\infty} \frac{1}{k^k}$. Using the Root Test,

$$\lim_{k\to\infty} \sqrt[k]{|a_k|} = \lim_{k\to\infty} \sqrt[k]{\left(\frac{1}{k}\right)^k}$$

$$= \lim_{k\to\infty} \frac{1}{k} = 0 < 1,$$

so $\sum_{k=1}^{\infty} \frac{1}{k^k}$ converges absolutely, thus

$3\sum_{k=3}^{\infty}\frac{1}{k^k} = \sum_{k=3}^{\infty}\frac{3}{k^k}$ converges absolutely.

33. Using the Ratio Test,
$$\lim_{k\to\infty}\left|\frac{a_{k+1}}{a_k}\right| = \lim_{k\to\infty}\frac{(k+1)!}{4^{k+1}}\cdot\frac{4^k}{k!}$$
$$= \lim_{k\to\infty}\frac{k+1}{4} = \infty > 1,$$
so $\sum_{k=6}^{\infty}(-1)^{k+1}\frac{k!}{4^k}$ diverges.

35. Using the Ratio Test,
$$\lim_{k\to\infty}\left|\frac{a_{k+1}}{a_k}\right| = \lim_{k\to\infty}\frac{(k+1)^{10}}{(2k+2)!}\cdot\frac{2k!}{k^{10}}$$
$$= \lim_{k\to\infty}\frac{(k+1)^{10}}{k^{10}(2k+1)(2k+2)}$$
$$= \lim_{k\to\infty}\frac{(k+1)^{10}}{4k^{12}+6k^{11}+2k^{10}}$$
$$= 0 < 1$$
so $\sum_{k=1}^{\infty}(-1)^{k+1}\frac{k^{10}}{(2k)!}$ converges absolutely.

37. Using the Ratio Test,
$$\lim_{k\to\infty}\left|\frac{a_{k+1}}{a_k}\right|$$
$$= \lim_{k\to\infty}\left|\frac{(-2)^{k+1}(k+2)}{5^{k+1}}\cdot\frac{5^k}{(2)^k(k+1)}\right|$$
$$= \lim_{k\to\infty}\frac{2(k+2)}{5(k+1)} = \frac{2}{5} < 1$$
so $\sum_{k=0}^{\infty}\frac{(-2)^k(k+1)}{5^k}$ converges absolutely.

39. $$\frac{\sqrt{8}}{9801}\sum_{k=0}^{0}\frac{(4k)!(1103+26390k)}{(k!)^4 396^{4k}}$$
$$= \frac{\sqrt{8}}{9801}(1103) \approx .318309878 \approx \frac{1}{\pi}$$

so $\pi \approx 3.14159273$

$$\frac{\sqrt{8}}{9801}\sum_{k=0}^{1}\frac{(4k)!(1103+26390k)}{(k!)^4 396^{4k}}$$
$$= \frac{\sqrt{8}}{9801}(1103) + \frac{\sqrt{8}}{9801}\cdot\frac{4!(27,493)}{396^4}$$
$$\approx .318309886183791 \approx \frac{1}{\pi}$$
For comparison, the value of $1/\pi$ to 15 places is

$$0.318309886183791,$$

so two terms of the series give this value correct to 15 places.

41. Using the Ratio Test
$$\lim_{k\to\infty}\left|\frac{a_{k+1}}{a^k}\right| = \lim_{k\to\infty}\frac{(k+1)!}{(k+1)^{k+1}}\cdot\frac{k^k}{k!}$$
$$= \lim_{k\to\infty}\frac{k^k}{(k+1)^k} = \lim_{k\to\infty}\left(\frac{k}{1+k}\right)^k$$
$$= \frac{1}{e} < 1$$
so $\sum_{k=1}^{\infty}\frac{k!}{k^k}$ converges absolutely.

43. Use the Ratio Test,
$$\lim_{k\to\infty}\left|\frac{a_{k+1}}{a_k}\right| = \lim_{k\to\infty}\left|\frac{p^{k+1}/(k+1)}{p^k/k}\right|$$
$$= \lim_{k\to\infty}\left|p\frac{k}{k+1}\right|$$
$$= |p|\lim_{k\to\infty}\left|\frac{k}{k+1}\right| = |p|$$
Thus, the series converges if $|p| < 1$; that is, if $-1 < p < 1$. If $p = 1$, we have the harmonic series, which diverges. If $p = -1$, we have the alternating harmonic series, which converges conditionally by the Alternating Series Test.

Thus, the series converges if $-1 \le p < 1$.

8.6 Power Series

1. $f(x) = \frac{2}{1-x} = 2\sum_{k=0}^{\infty}x^k = \sum_{k=0}^{\infty}(2)x^k.$

This is a geometric series that converges when $|x| < 1$, so the interval of convergence is $(-1, 1)$ and $r = 1$.

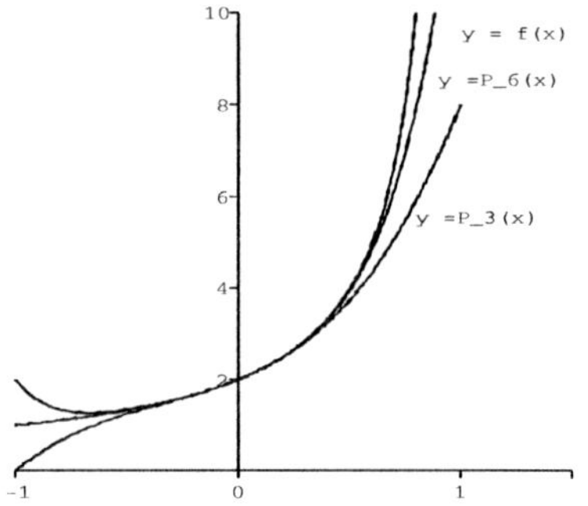

3. $f(x) = \dfrac{3}{1+x^2} = 3\left[\dfrac{1}{1-(-x^2)}\right]$

$= 3\displaystyle\sum_{k=0}^{\infty}(-x^2)^k = \sum_{k=0}^{\infty} 3(-1)^k x^{2k}$.

This is a geometric series that converges when $|x^2| < 1$, so the interval of convergence is $(-1, 1)$ and $r = 1$.

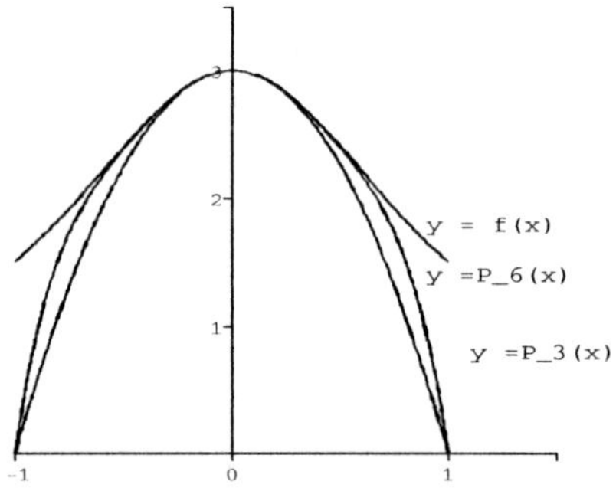

5. $f(x) = \dfrac{2x}{1-x^3} = 2x\displaystyle\sum_{k=0}^{\infty}(x^3)^k$

$= \displaystyle\sum_{k=0}^{\infty} 2x^{3k+1}$.

This is a geometric series that converges when $|x^3| < 1$, so the interval of convergence is $(-1, 1)$ and $r = 1$.

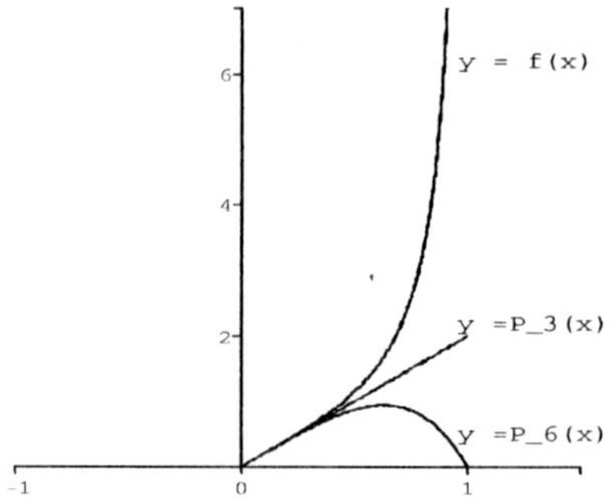

7. $f(x) = \dfrac{2}{4+x} = \dfrac{1}{2}\left[\dfrac{1}{1-(-x/4)}\right]$

$= \dfrac{1}{2}\displaystyle\sum_{k=0}^{\infty}\left(-\dfrac{x}{4}\right)^k$.

This is a geometric series that converges when $|x/4| < 1$, so the interval of convergence is $(-4, 4)$ and $r = 4$.

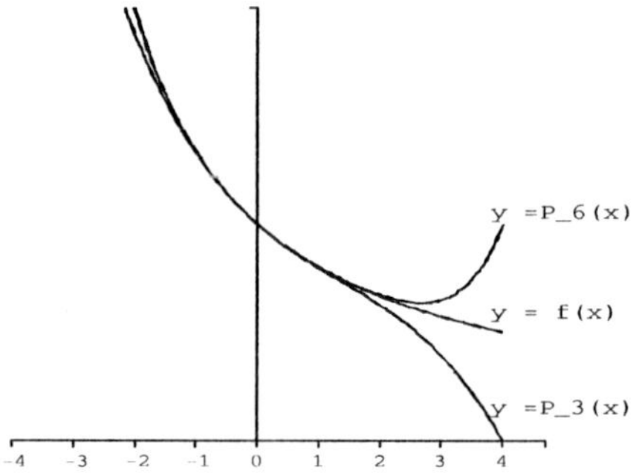

9. This is a geometric series
$\displaystyle\sum_{k=0}^{\infty}(x+2)^k = \dfrac{1}{1-(x+2)} = \dfrac{-1}{1+x}$
which converges for $|x+2| < 1$ so $-1 < x+2 < 1$ or $-3 < x < -1$, thus the interval of convergence is $(-3, -1)$ and $r = 1$.

11. This is a geometric series
$\displaystyle\sum_{k=0}^{\infty}(2x-1)^k = \dfrac{1}{1-(2x-1)}$
$= \dfrac{1}{2-2x}$
which converges for $|2x-1| < 1$ so $-1 < 2x-1 < 1$ or $0 < x < 1$, thus the interval of convergence is $(0, 1)$

and $r = \dfrac{1}{2}$.

13. This is a geometric series
$$\sum_{k=0}^{\infty}(-1)^k\left(\frac{x}{2}\right)^k = \sum_{k=0}^{\infty}\left(\frac{-x}{2}\right)^k$$
$$= \frac{1}{1+x/2} = \frac{2}{2+x}$$
which converges for $\left|\dfrac{-x}{2}\right| < 1$ so $1 < \dfrac{-x}{2} < 1$ or $2 > x > -2$, thus the interval of convergence is $(-2, 2)$ and $r = 2$.

15. Using the Ratio Test,
$$\lim_{k\to\infty}\left|\frac{a_{k+1}}{a_k}\right| = \lim_{k\to\infty}\left|\frac{2^{k+1}k!(x-2)^{k+1}}{2^k(k+1)!(x-2)^k}\right|$$
$$= 2|x-2|\lim_{k\to\infty}\frac{1}{k+1}$$
$= 0$ and $0 < 1$ for all x,
so the series converges absolutely for $x \in (-\infty, \infty)$. The interval of convergence is $(-\infty, \infty)$ and $r = \infty$.

17. Using the Ratio Test,
$$\lim_{k\to\infty}\left|\frac{a_{k+1}}{a_k}\right| = \lim_{k\to\infty}\left|\frac{(k+1)4^k x^{k+1}}{k4^{k+1}x^k}\right|$$
$$= \frac{|x|}{4}\lim_{k\to\infty}\frac{k+1}{k} = \frac{|x|}{4}$$
and $\dfrac{|x|}{4} < 1$ when $|x| < 4$ or $-4 < x < 4$. So the series converges absolutely for $x \in (-4, 4)$.

When $x = 4$,
$$\sum_{k=0}^{\infty}\frac{k}{4^k}4^k = \sum_{k=0}^{\infty}k \text{ and } \lim_{k\to\infty}k = \infty,$$
so the series diverges by the k-th Term Test for Divergence.

When $x = -4$,
$$\sum_{k=0}^{\infty}\frac{k}{4^k}(-4)^k = \sum_{k=0}^{\infty}(-1)^k k,$$
which diverges by the k-th Term Test for Divergence.

Thus the interval of convergence is $(-4,\ 4)$ and $r = 4$.

19. Using the Ratio Test,
$$\lim_{k\to\infty}\left|\frac{a_{k+1}}{a_k}\right|$$
$$= \lim_{k\to\infty}\left|\frac{(-1)^{k+1}k3^k(x-1)^{k+1}}{(-1)^k(k+1)3^{k+1}(x-1)^k}\right|$$
$$= \frac{|x-1|}{3}\lim_{k\to\infty}\frac{k}{k+1}$$
$$= \frac{|x-1|}{3}$$
and $\dfrac{|x-1|}{3} < 1$ when $-3 < x-1 < 3$ or $-2 < x < 4$ so the series converges absolutely for $x \in (-2, 4)$.

When $x = -2$,
$$\sum_{k=0}^{\infty}\frac{(-1)^k}{k3^k}(-3)^k = \sum_{k=0}^{\infty}\frac{1}{k}$$
which is the divergent harmonic series.

When $x = 4$,
$$\sum_{k=0}^{\infty}\frac{(-1)^k}{k3^k}3^k = \sum_{k=0}^{\infty}\frac{(-1)^k}{k}$$
which converges by the Alternating Series Test.

So the interval of convergence is $(-2, 4]$ and $r = 3$.

21. Using the Ratio Test,
$$\lim_{k\to\infty}\left|\frac{a_{k+1}}{a_k}\right| = \lim_{k\to\infty}\left|\frac{(k+1)!(x+1)^{k+1}}{k!(x+1)^k}\right|$$
$$= \lim_{k\to\infty}(k+1)\,|x+1||$$
$$= \begin{cases} 0 & \text{if } x = -1 \\ \infty & \text{if } x \neq -1 \end{cases}$$
so this series converges absolutely for $x = -1$ and $r = 0$.

23. Using the Ratio Test,
$$\lim_{k\to\infty}\left|\frac{a_{k+1}}{a_k}\right| = \lim_{k\to\infty}\left|\frac{(k+1)^2(x-3)^{k+1}}{k^2(x-3)^k}\right|$$
$$= |x-3|\lim_{k\to\infty}\frac{(k+1)^2}{k^2}$$
$= |x-3| < 1$
when $-1 < x-3 < 1$ or $2 < x < 4$,

so the series converges absolutely for $x \in (2,4)$.

If $x = 2$, $\sum_{k=2}^{\infty} k^2(-1)^k$ and $\lim_{k\to\infty} k^2 = \infty$, so the series diverges by the k-th Term Test for Divergence.

If $x = 4$, $\sum_{k=2}^{\infty} k^2$ and $\lim_{k\to\infty} k^2 = \infty$, so the series diverges by the k-th Term Test for Divergence.

Thus, the interval of convergence is $(2,4)$ and $r = 1$.

25. Using the Ratio Test,

$$\lim_{k\to\infty}\left|\frac{a_{k+1}}{a_k}\right| = \lim_{k\to\infty}\left|\frac{(k+1)!(2k)!x^{k+1}}{k!(2k+2)!x^k}\right|$$

$$= |x| \lim_{k\to\infty}\frac{k+1}{(2k+2)(2k+1)} = 0 < 1$$

for all x, so the series converges for all $x \in (-\infty,\infty)$.
Thus the interval of convergence is $(-\infty,\infty)$ and $r = \infty$.

27. Using the Ratio Test,

$$\lim_{k\to\infty}\left|\frac{a_{k+1}}{a_k}\right| = \lim_{k\to\infty}\left|\frac{2^{k+1}k^2(x+2)^{k+1}}{2^k(k+1)^2(x+2)^k}\right|$$

$$= 2|x+2| \lim_{k\to\infty}\frac{k^2}{(k+1)^2}$$

$= 2|x+2|$ and $2|x+2| < 1$

when $-\frac{1}{2} < x+2 < \frac{1}{2}$

or $-\frac{5}{2} < x < -\frac{3}{2}$,

so the series converges absolutely for $x \in \left(-\frac{5}{2}, -\frac{3}{2}\right)$.

If $x = -\frac{5}{2}$, then

$$\sum_{k=0}^{\infty}\frac{2^k}{k^2}\left(-\frac{1}{2}\right)^k = \sum_{k=0}^{\infty}\frac{(-1)^k}{k^2}$$

and

$$\lim_{k\to\infty}\frac{1}{k^2} = 0$$

and

$$\frac{a_{k+1}}{a_k} = \frac{k^2}{(k+1)^2} < 1$$

for all $k \ge 0$,
so $a_{k+1} < a_k$ for all $k \ge 0$, so the series converges by the Alternating Series Test.

If $x = -\frac{3}{2}$, then

$$\sum_{k=0}^{\infty}\frac{2^k}{k^2}\left(\frac{1}{2}\right)^k = \sum_{k=0}^{\infty}\frac{1}{k^2}$$

which is a convergent p-series ($p = 2 > 1$).

So the interval of convergence is $\left[-\frac{5}{2}, -\frac{3}{2}\right]$ and $r = \frac{1}{2}$.

29. Using the Ratio Test,

$$\lim_{k\to\infty}\left|\frac{a_{k+1}}{a_k}\right| = \lim_{k\to\infty}\left|\frac{4^{k+1}k^{1/2}x^{k+1}}{4^k(k+1)^{1/2}x^k}\right|$$

$$= 4|x| \lim_{k\to\infty}\frac{k^{1/2}}{(k+1)^{1/2}} = 4|x|$$

and $4|x| \le 1$ when

$-\frac{1}{4} < x < \frac{1}{4}$, so the series converges absolutely for $x \in \left(-\frac{1}{4}, \frac{1}{4}\right)$.

If $x = -\frac{1}{4}$, then

$$\sum_{k=0}^{\infty}\frac{4^k}{\sqrt{k}}\left(-\frac{1}{4}\right)^k = \sum_{k=0}^{\infty}\frac{(-1)^k}{\sqrt{k}}$$

and

$$\lim_{k\to\infty}\frac{1}{\sqrt{k}} = 0$$

and

$$\frac{a_{k+1}}{a_k} = \frac{\sqrt{k}}{\sqrt{k+1}} < 1$$

for all $k \ge 0$, so $a_{k+1} < a_k$ for all $k \ge 0$, therefore the series converges by the Alternating Series test.

If $x = \frac{1}{4}$,

$$\sum_{k=0}^{\infty} \frac{4^k}{\sqrt{k}} \left(\frac{1}{4}\right)^k = \sum_{k=0}^{\infty} \frac{1}{\sqrt{k}}$$

which is a divergent p-series $\left(p = \frac{1}{2} < 1\right)$.

Thus the interval of convergence is $\left[-\frac{1}{4}, \frac{1}{4}\right)$ and $r = \frac{1}{4}$.

31. We have seen that

$$\frac{3}{1+x^2} = \sum_{k=0}^{\infty} (-1)^k 3x^{2k}$$

with $r = 1$

Integrating both sides gives

$$\int \frac{3}{1+x^2}\,dx = \sum_{k=0}^{\infty} 3(-1)^k \int x^{2k}\,dx$$

$3\tan^{-1} x = \sum_{k=0}^{\infty} \frac{3(-1)^k x^{2k+1}}{2k+1} + c$ Taking $x = 0$

$$3\tan^{-1} 0 = \sum_{k=0}^{\infty} \frac{3(-1)^k (0)^{2k+1}}{2k+1} + c = c$$

so that

$$c = 3\tan^{-1}(0) = 0.$$

Thus

$$3\tan^{-1} x = \sum_{k=0}^{\infty} \frac{3(-1)^k x^{2k+1}}{2k+1}$$

with $r = 1$.

33. We have seen that

$$\frac{2}{1-x^2} = \sum_{k=0}^{\infty} 2x^{2k}$$

with $r = 1$. Taking the derivative of both sides gives

$$\frac{d}{dx}\left(\frac{2}{1-x^2}\right) = \sum_{k=0}^{\infty} 2\frac{d}{dx}\,x^{2k} \Big| \frac{4x}{(1-x^2)^2}$$

$$= \sum_{k=0}^{\infty} 2 \cdot 2kx^{2k-1} \frac{2x}{(1-x^2)^2}$$

$$= \sum_{k=0}^{\infty} 2kx^{2k-1} \text{ with } r = 1.$$

35. We have seen that

$$\frac{3x}{1+x^2} = \sum_{k=0}^{\infty} (-1)^k 3x^{2k+1}$$

with $r = 1$. Integrating both sides gives

$$\int \frac{3x}{1+x^2}\,dx = \sum_{k=0}^{\infty} (-1)^k 3 \int x^{2k+1}\,dx$$

$$\frac{3}{2}\ln(1+x^2) = \sum_{k=0}^{\infty} \frac{(-1)^k 3x^{2k+2}}{2k+2} + c$$

$$\ln(1+x^2) = \sum_{k=0}^{\infty} \frac{(-1)^k x^{2k+2}}{k+1} + c.$$

Taking $x = 0$,

$$\ln(1) = \sum_{k=0}^{\infty} \frac{(-1)^k (0)^{2k+2}}{k+1} + c = c$$

so that $c = \ln(1) = 0$.

Thus,

$$\ln(1+x^2) = \sum_{k=0}^{\infty} \frac{(-1)^k x^{2k+2}}{k+1}$$

with $r = 1$.

37. Since $|\cos(k^3 x)| \le 1$ for all x,

$$\left|\frac{\cos(k^3 x)}{k^2}\right| \le \frac{1}{k^2}$$

for all x, and $\sum_{k=0}^{\infty} \frac{1}{k^2}$ is a convergent p-series, so by the Comparison Test, $\sum_{k=0}^{\infty} \frac{\cos(k^3 x)}{k^2}$ converges absolutely for all x. So the interval of convergence is $(-\infty, \infty)$ and $r = \infty$.

The series of derivatives is

$$\sum_{k=0}^{\infty} \frac{d}{dx}\left[\frac{\cos(k^3x)}{k^2}\right] = \sum_{k=0}^{\infty}(-k)\sin(k^3x)$$

and $\lim_{k\to\infty}(-k)\sin(k^3x)$ diverges if $x \neq 0$, while

$$\sum_{k=0}^{\infty}(-k)\sin(k^3(0)) = \sum_{k=0}^{\infty} 0 = 0,$$

thus the series converges absolutely if $x = 0$.

39. Using the Ratio Test,

$$\lim_{k\to\infty}\left|\frac{a_{k+1}}{a_k}\right| = \lim_{k\to\infty}\left|\frac{e^{(k+1)x}}{e^{kx}}\right| = e^x$$

and $e^x < 1$ when $x < 0$, so the series converges absolutely for all $x \in (-\infty, 0)$.

When $x = 0$,

$$\sum_{x=0}^{\infty} e^0 = \sum_{x=0}^{\infty} 1$$

which diverges by the k-th Term Test for Divergence because $\lim_{k\to\infty} 1 = 1$. So the interval of convergence is $(-\infty, 0)$.

The series of derivatives is

$$\sum_{x=0}^{\infty} \frac{d}{dx} e^{kx}| = \sum_{x=0}^{\infty} ke^{kx}.$$

Using the Ratio Test,

$$\lim_{k\to\infty}\left|\frac{a_{k+1}}{a_k}\right| = \lim_{k\to\infty}\left|\frac{(k+1)e^{(k+1)x}}{ke^{kx}}\right|$$
$$= e^x \lim_{k\to\infty}\frac{k+1}{k}$$
$$= e^x$$

and $e^x < 1$ when $x < 0$, so the series converges absolutely for all $x \in (-\infty, 0)$.

When $x = 0$,

$$\sum_{k=0}^{\infty} ke^0 = \sum_{k=0}^{\infty} k$$

which diverges by the k-th Term Test for Divergence because $\lim_{k\to\infty} k = \infty$. So the interval of convergence is $(-\infty, 0)$.

41. Using the Ratio Test,

$$\lim_{k\to\infty}\left|\frac{a_{k+1}}{a_k}\right| = \lim_{k\to\infty}\left|\frac{(x-a)^{(k+1)}b^k}{b^{k+1}(x-a)^k}\right|$$
$$= \frac{|x-a|}{b}\lim_{k\to\infty} 1 = \frac{|x-a|}{b}$$

and $\frac{|x-a|}{b} < 1$ when $-b < x-a < b$ or $a-b < x < a+b$. So the series converges absolutely for $x \in (a-b, a+b)$.

If $x = a - b$,

$$\sum_{k=0}^{\infty}\frac{(a-b-a)^k}{b^k}$$
$$= \sum_{k=0}^{\infty}\frac{(-1)^k b^k}{b^k} = \sum_{k=0}^{\infty}(-1)^k$$

which diverges by the k-th Term Test for Divergence because $\lim_{k\to\infty} 1 = 1$.

If $x = a + b$,

$$\sum_{k=0}^{\infty}\frac{(a+b-a)^k}{b^k} = \sum_{k=0}^{\infty}\frac{b^k}{b^k} = \sum_{k=0}^{\infty} 1$$

which diverges by the k-th Term Test for Divergence because $\lim_{k\to\infty} 1 = 1$.

So the interval of convergence is $(a-b, a+b)$ and $r = b$.

43. If the radius of convergence of $\sum_{k=0}^{\infty} a_k x^k$ is r, then $-r < x < r$. For any constant c, $-r-c < x-c < r-c$. Thus, the radius of convergence of $\sum_{k=0}^{\infty} a_k(x-c)^k$ is

$$\frac{(r-c)-(-r-c)}{2} = r.$$

45. $\dfrac{1}{1-x} = \sum_{k=0}^{\infty} x^k$

$\dfrac{d}{dx}\left(\dfrac{1}{1-x}\right) = \dfrac{1}{(1-x)^2}$

$= \sum_{k=0}^{\infty} \dfrac{d}{dx} x^k = \sum_{k=0}^{\infty} k x^{k-1}$

$\dfrac{x+1}{(1-x)^2} = \dfrac{x}{(1-x)^2} + \dfrac{1}{(1-x)^2}$

$= \sum_{k=0}^{\infty} k x^k + \sum_{k=0}^{\infty} k x^{k-1}$

$= \sum_{k=0}^{\infty} (2k+1) x^k$

$= 1 + 3x + 5x^2 + 7x^3 + \cdots$

Since $\sum_{k=0}^{\infty} x^k$ converges for $|x| < 1$, $\dfrac{d}{dx}\left(\sum_{k=0}^{\infty} x^k\right)$ and $\dfrac{d}{dx}\left(\sum_{k=0}^{\infty} k^x\right)$ converge for $|x| < 1$. Hence $\dfrac{x+1}{(1-x)^2}$ converges for $|x| < 1$, so $r = 1$.

If $x = \dfrac{1}{1000}$, then

$\dfrac{1{,}001{,}000}{998{,}000}$

$= 1 + \dfrac{3}{1000} + \dfrac{5}{(1000)^2} + \dfrac{7}{(1000)^3} + \cdots$

$= 1.003005007\ldots$

47. If $x = 1$,

$\ldots + 1 + 1 + 1 + 1 + 1 + \cdots \neq 0$.

$\dfrac{1}{1 - \frac{1}{x}} = \sum_{k=0}^{\infty}\left(\dfrac{1}{x}\right)^k$ is a geometric series which converges for $\dfrac{1}{|x|} < 1$ or $|x| > 1$.

$\dfrac{x}{1-x} = \sum_{k=0}^{\infty} x^{k+1}$ is a geometric series which converges for $|x| < 1$. Euler's mistake was that there are no x's for which both series converge.

49. Note that for $-1 < x < 1$,

$\dfrac{1}{1+x} = \sum_{k=0}^{\infty} (-1)^k x^k$.

We can differentiate both sides and get

$\dfrac{-1}{(1+x)^2} = \sum_{k=0}^{\infty} (-1)^k k x^{k-1}$.

Similarly, for $-1 < x < 1$, we have

$\dfrac{1}{1-x} = \sum_{k=0}^{\infty} x^k$.

Therefore,

$\dfrac{1}{(1-x)^2} = \sum_{k=0}^{\infty} k x^{k-1}$.

Putting these together, we get

$E(x) = \dfrac{kq}{(x-1)^2} - \dfrac{kq}{(x+1)^2}$

$= kq \sum_{k=0}^{\infty} k x^{k-1} + kq \sum_{k=0}^{\infty} (-1)^k k x^{k-1}$

$= \sum_{k=0}^{\infty} [1 + (-1)^k] k^2 q x^{k-1}$

$= \sum_{k \text{ even}} k^2 q x^{k-1}$

for $-1 < x < 1$.

8.7 Taylor Series

1. $f(x) = \cos x,\ f'(x) = -\sin x$
$f''(x) = -\cos x,\ f'''(x) = \sin x$
$f(0) = 1,\ f'(0) = 0$
$f''(0) = -1,\ f'''(0) = 0$

Therefore,

$\cos x = 1 - \dfrac{1}{2} x^2 + \cdots = \sum_{k=0}^{\infty} \dfrac{(-1)^k}{(2k)!} x^{2k}$

From the Ratio Test,

$\lim_{n \to \infty} \left|\dfrac{a_{n+1}}{a_n}\right| = \lim_{n \to \infty} \dfrac{(2n)! x^{2n+2}}{(2n+2)! x^{2n}}$

$= x^2 \lim_{n\to\infty} \frac{1}{(2n+1)(2n+2)} = 0$
we see that the interval of convergence is $(-\infty, \infty)$.

3. $f(x) = e^{2x}, f'(x) = 2e^{2x}$,
$f''(x) = 4e^{2x}, f'''(x) = 8e^{2x}$
$f(0) = 1, f'(0) = 2$
$f''(0) = 4, f'''(0) = 8$.

Therefore,

$$e^{2x} = 1 + x + \cdots = \sum_{k=0}^{\infty} \frac{2^k}{k!} x^k$$

From the Ratio Test,
$\lim_{n\to\infty} \left| \frac{a_{n+1}}{a_n} \right| = x \lim_{n\to\infty} \frac{2}{n+1} = 0$
we see that the interval of convergence is $(-\infty, \infty)$.

5. $f(x) = \ln(1+x), f'(x) = \frac{1}{1+x}$,
$f''(x) = \frac{-1}{(1+x)^2}, f'''(x) = \frac{2}{(1+x)^3}$,
$f(0) = 0, f'(0) = 1$
$f''(0) = -1, f'''(0) = 2$

Therefore,

$$\ln(1+x) = x - \frac{1}{2}x^2 + \cdots = \sum_{k=0}^{\infty} \frac{(-1)^k x^{k+1}}{k+1}$$

From the Ratio Test,
$\lim_{n\to\infty} \left| \frac{a_{n+1}}{a_n} \right|$
$= \lim_{n\to\infty} \left| \frac{(-1)^{n+1}(n+1)x^{n+2}}{(-1)^n(n+2)x^{n+1}} \right|$
and $|x| < 1$ when $-1 < x < 1$. So the series converges absolutely for all $x \in (-1, 1)$.

When $x = 1$,

$$\sum_{k=0}^{\infty} \frac{(-1)^k (1)^{k+1}}{k+1} = \sum_{k=0}^{\infty} \frac{(-1)^k}{k+1}$$

and $\lim_{k=0}^{\infty} \frac{1}{k+1} = 0$ and

$$\frac{a_{k+1}}{a_k} = \frac{k}{k+1} < 1$$

for all $k \geq 0$ so $a_{k+1} > a_k$, for all $k \geq 0$, so the series converges by the Alternating Series Test.

When $x = -1$,

$$\sum_{k=0}^{\infty} \frac{(-1)^k(-1)^{k+1}}{k+1} = \sum_{k=0}^{\infty} \frac{-1}{k+1}$$

which is the negative of the harmonic series, so diverges.

Hence the interval of convergence is $(-1, 1]$.

7. $f^{(k)}(x) = (-1)^k (k+1)!(1+x)^{-2-k}$. Therefore,

$$\frac{1}{(1+x)^2} = \sum_{k=0}^{\infty} (-1)^k (k+1) x^k$$

From the Ratio Test,

$$\lim_{k\to\infty} \left| \frac{a_{k+1}}{a_k} \right| = x \lim_{k\to\infty} \frac{k+2}{k+1} = |x|$$

and $|x| < 1$ when $-1 < x < 1$. So the series converges absolutely for $x \in (-1, 1)$.

When $x = -1$,
$\sum_{k=0}^{\infty} (k+1)$ and $\lim_{k\to\infty} (k+1) = \infty$, so the series diverges by the k-th Term Test for Divergence.

When $x = 1$,
$\sum_{k=0}^{\infty} (-1)(k+1)^k$ and $\lim_{k\to\infty} k+1 = \infty$, so the series diverges by the k-th Term Test for Divergence.

So the interval of convergence is $(-1, 1)$.

9. $f^{(k)}(x) = e^{x-1}, f^{(k)}(1) = 1$

$$e^{x-1} = 1+(x-1)+\cdots = \sum_{k=0}^{\infty} \frac{(x-1)^k}{k!}.$$

Using the Ratio Test,

$$\lim_{k\to\infty} \left|\frac{a_{k+1}}{a_k}\right| = |x-1| \lim_{k\to\infty} \frac{1}{k+1} = 0$$

So the interval of convergence is $(-\infty, \infty)$.

11. $f(x) = \ln x$ and for $k > 1$,
$f^{(k)}(x) = (-1)^{k+1}\frac{1}{x^k}$
$f(e) = 1, f^{(k)}(e) = (-1)^{k+1}\frac{1}{e^k}$

$$\ln x = 1 + \sum_{k=1}^{\infty} \frac{(-1)^{k+1}}{e^k k!}(x-e)^k$$

By the ratio test,

$$\lim_{k\to\infty} \left|\frac{a_{k+1}}{a_k}\right| = |x-e| \lim_{k\to\infty} \frac{1}{e(k+1)} = 0$$

the interval of convergence is $(-\infty, \infty)$.

13. $f^{(k)}(x) = (-1)^k (k+1)! \frac{1}{x^{k+1}}$,
$f^{(k)}(1) = (-1)^k (k+1)!$. Therefore,

$$\frac{1}{x} = \sum_{k=0}^{\infty} \frac{(-1)^k (x-1)^k}{k!(k+1)!}$$

From the ratio test,
$\lim_{k\to\infty} \left|\frac{a_{k+1}}{a_k}\right|$
$= |x-1| \lim_{k\to\infty} \frac{1}{(k+1)(k+2)} = 0,$
we see that the interval of convergence is $(-\infty, \infty)$.

15.

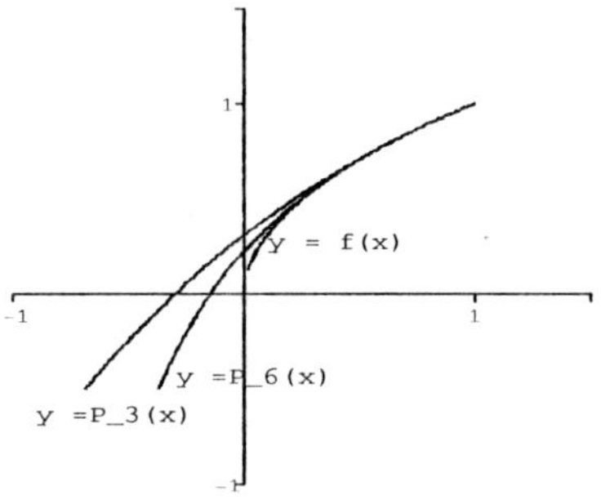

17.

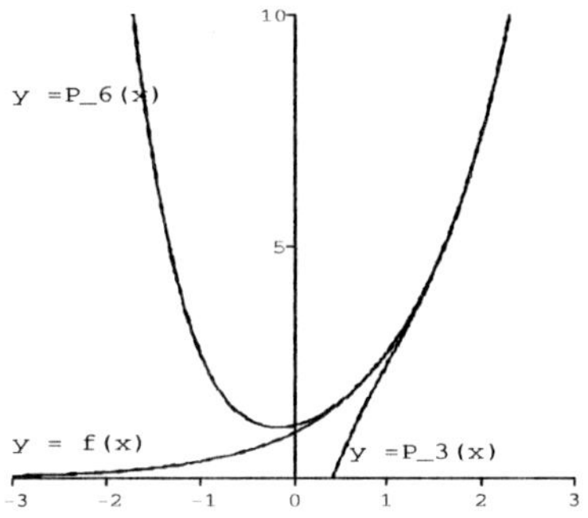

19.

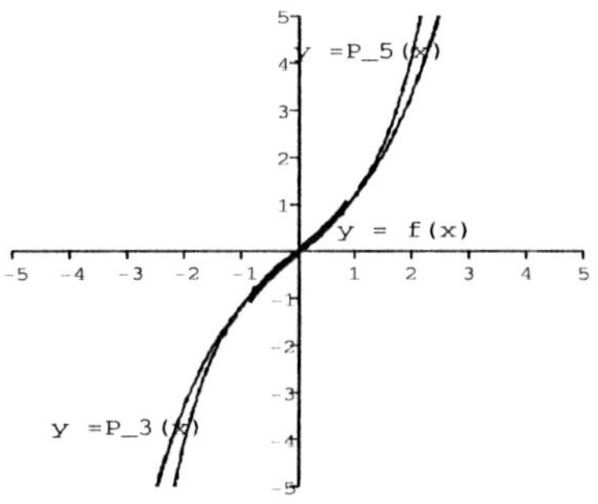

21. For any fixed x there exists a z between 0 and x such that

$$|R_n(x)| = \left|\frac{f^{(n+1)}(z)x^{n+1}}{(n+1)!}\right| \leq \left|\frac{x^{n+1}}{(n+1)!}\right| \to 0 \text{ as } n \to \infty,$$

since $\left|f^{(n+1)}(z)\right| \leq 1$ for all n.

23. For any fixed x there exists a z between x and 1 such that

$$|R_n(x)| = \left|\frac{f^{(n+1)}(z)(x-1)^{n+1}}{(n+1)!}\right|$$

.

Observe that for all n,

$\left|f^{(n+1)}(z)\right| = \dfrac{n!}{|z|^{n+1}}$, so

$$|R_n(x)| = \left|\frac{(x-1)^{n+1}}{|z|^{n+2}}\right| = \left|\frac{x-1}{z}\right|^{n+1}\frac{1}{z} \to 0$$

as $n \to \infty$.

25. (a) Expand $f(x) = \ln x$ into a Taylor series about $c = 1$. Recall

$$\ln(x) = \sum_{k=0}^{\infty} \frac{(-1)^{k+1}}{k}(x-1)^k$$

So $P_4(x) = \displaystyle\sum_{k=0}^{4} \frac{f^{(k)}(1)}{k!}(x-1)^k$

$= (x-1) - \dfrac{1}{2}(x-1)^2 + \dfrac{1}{3}(x-1)^3 - \dfrac{1}{4}(x-1)^4.$

Letting $x = 1.05$ gives

$$\ln(1.05) \approx P_4(1.05) = .04879$$

(b) $|\text{Error}| = |\ln(1.05) - P_4(1.05)|$

$= |R_n(1.05)|$

$= \left|\dfrac{f^{4+1}(x)}{(4+1)!}(1.05-1)^{4+1}\right|$

$= \dfrac{4!|z|^{-5}(0.05)^5}{5!}$,

so because $1 < z < 1.05$, $\dfrac{1}{z} < \dfrac{1}{1} = 1$. Thus we have

$$|\text{Error}| = \frac{(.05)^5}{5z^5} < \frac{(.05)^5}{5(1)^5} = \frac{(.05)^5}{5}.$$

(c) From part (b) we have for $1 < z < 1.05$,

$|R_n(1.05)|$

$= \left|\dfrac{f^{n+1}(z)}{(n+1)!}(1.05-1)^{n+1}\right|$

$= \dfrac{n!(.05)^{n+1}}{(n+1)!z^{n+1}} = \dfrac{(.05)^{n+1}}{(n+1)z^{n+1}}$

$< \dfrac{(.05)^{n+1}}{n+1}$

$< 10^{-10}$ if $n = 7$.

27. (a) Expand $f(x) = \sqrt{x}$ into a Taylor series about $c = 1$.

$P_4(x) = \dfrac{x}{2} - \dfrac{(x-1)^2}{8} + \dfrac{(x-1)^3}{16} - \dfrac{5(x-1)^4}{128}.$

Letting $x = 1.1$ gives

$$\sqrt{1.1} \approx P_4(1.1) = 1.0488$$

(b) $|\text{Error}| = |\sqrt{1.1} - P_4(1.1)|$

$= |R_n(1.1)|$

$= \left|\dfrac{f^{4+1}(x)}{(4+1)!}(1.1-1)^{4+1}\right|$

$= \dfrac{4!|z|^{-5}(0.1)^5}{5!}$,

so because $1 < z < 1.1$,

$$|\text{Error}| = \frac{(.1)^5}{5z^5} < \frac{(.1)^5}{5}.$$

(c) From part (b) we have for $1 < z < 1.1$,

$|R_n(1.1)|$

$= \left|\dfrac{f^{n+1}(z)}{(n+1)!}(1.1-1)^{n+1}\right|$

$= \dfrac{n!(.1)^{n+1}}{(n+1)!z^{n+1}} = \dfrac{(.1)^{n+1}}{(n+1)z^{n+1}}$

$< \dfrac{(.1)^{n+1}}{(n+1)} < 10^{-10}$ if $n = 9$.

29.

$$\sum_{k=0}^{\infty} \frac{1}{k!}x^k = e^x$$

Replacing x with 2, we have

$$\sum_{k=0}^{\infty} \frac{2^k}{k!} = e^2$$

31.

$$\sum_{k=0}^{\infty} \frac{(-1)^k}{2k+1}x^{2k+1} = \tan^{-1} x,$$

Replacing x with 1, we have

$$\sum_{k=0}^{\infty} \frac{(-1)^k}{2k+1} = \tan^{-1}(1) = \frac{\pi}{4}$$

33.

$$e^x = \sum_{k=0}^{\infty} \frac{x^k}{k!}$$

with $r = \infty$. Replacing x with $-3x$, we have

$$e^{-3x} = \sum_{k=0}^{\infty} \frac{(-1)^k (3x)^k}{k!}$$

with $r = \infty$.

35.

$$e^x = \sum_{k=0}^{\infty} \frac{x^k}{k!}$$

with $r = \infty$. We see that

$$\frac{e^x - 1}{x} = \frac{\sum_{k=0}^{\infty} x^k/k! - 1}{x} = \sum_{k=1}^{\infty} \frac{x^{k-1}}{k!} = \sum_{k=0}^{\infty} \frac{x^k}{(k+1)!}$$

with $r = \infty$.

37.

$$\sin x = \sum_{k=0}^{\infty} \frac{(-1)^k x^{2k+1}}{(2k+1)!}$$

with $r = \infty$. Replacing x with $2x$ and multiplying by x, we have

$$x \sin 2x = x \left(\sum_{k=0}^{\infty} \frac{(-1)^k (2x)^{2k+1}}{(2k+1)!} \right) = \sum_{k=0}^{\infty} \frac{(-1)^k 2^{2k+1} x^{2k+2}}{(2k+1)!}$$

with $r = \infty$.

39. Let us calculate $f'(0)$ and $f''(0)$ using the definition.

$$f'(0) = \lim_{h\to 0} \frac{f(h) - f(0)}{h} = \lim_{h\to 0} \frac{e^{-1/h^2} - 0}{h} = 0$$

When $x \neq 0$, $f'(x) = \frac{2}{x^3} e^{-1/x^2}$ and then,

$$f''(0) = \lim_{h\to 0} \frac{f'(h) - f'(0)}{h} = \lim_{h\to 0} \frac{2/h^3\, e^{-1/h^2} - 0}{h} = 2 \lim_{h\to 0} \frac{e^{-1/h^2} - 0}{h^4} = 0.$$

41. $f(x) = |x|$

$$f'(x) = \begin{cases} 1 & \text{if} \quad x > 0 \\ -1 & \text{if} \quad x < 0 \end{cases}$$

$f(1) = 1, f'(1) = 1,$
$f^{(n)}(1) = 0$ for all $n \geq 2$

Thus, the Taylor series for f about $c = 1$ is $1 + (x - 1) = x$. The Taylor series converges to f for all $x \geq 0$.

43. Since $\lim_{x\to a} \frac{f(x) - g(x)}{(x-a)^2} = 0$
it follows that $\lim_{x\to a} [f(x) - g(x)] = 0$
$\Rightarrow f(a) - g(a) = 0$
$\Rightarrow f(a) = g(a)$
Using L'Hospital's rule on the given limit, we get $\lim_{x\to a} \frac{f'(x) - g'(x)}{2(x-a)} = 0$
$\Rightarrow \lim_{x\to a} [f'(x) - g'(x)] = 0$
$\Rightarrow f'(a) - g'(a) = 0$
$\Rightarrow f'(a) = g'(a)$
Similarly, $\lim_{x\to a} \frac{f''(x) - g''(x)}{2} = 0$
$\Rightarrow \lim_{x\to a} [f''(x) - g''(x)] = 0$
$\Rightarrow f''(a) - g''(a) = 0$
$\Rightarrow f''(a) = g''(a)$

Thus, the first three terms of the Taylor series for f and g are identical.

45. The 5th term of the series is $\frac{1}{9!}$, which is less than 10^{-5}. So 4 terms will give the desired accuracy. In Simpson's Rule, the error is bounded by $\frac{1}{180n^4}$. Solving $\frac{1}{180n^4} \leq 10^{-5}$, we get

$n > 4.9$, which means n must be 6, since n is an even integer. The conclusion is that the Taylor method is more efficient.

47. $f(x) = e^x \sin x, f(0) = 0$
$f'(x) = e^x(\sin x + \cos x), f'(0) = 1$
$f''(x) = 2e^x \cos x, f''(0) = 2$
$f'''(x) = -2e^x(\sin x - \cos x), f'''(0) = 2$
$f^{(4)}(x) = -4e^x \sin x, f^{(4)}(0) = 0$
$f^{(5)}(x) = -4e^x(\sin x + \cos x)$,
$f^{(5)}(0) = -4$
$f^{(6)}(x) = -8e^x \cos x, f^{(6)}(0) = -8$
Thus, $e^x \sin x$

$$= x + \frac{2}{2!}x^2 + \frac{2}{3!}x^3 - \frac{4}{5!}x^5 - \frac{8}{6!}x^6 + \cdots$$

$$= x + x^2 + \frac{1}{3}x^3 - \frac{1}{30}x^5 - \frac{1}{90}x^6 + \cdots$$

Now, the Taylor series for e^x and $\sin x$ are

$$e^x = 1 + x + \frac{x^2}{2} + \frac{x^3}{6} + \frac{x^4}{24} + \frac{x^5}{120} + \cdots$$

$$\sin x = x - \frac{x^3}{3!} + \frac{x^5}{5!} - \cdots$$

Multiplying together the series for e^x and $\sin x$ and collecting like terms gives $x - \frac{x^3}{3!} + \frac{x^5}{5!} + x^2 - \frac{x^4}{3!} + \frac{x^6}{5!} + \frac{x^3}{2}$

$$- \frac{x^5}{12} + \frac{x^4}{6} - \frac{x^6}{36} + \frac{x^5}{4!} + \frac{x^6}{5!} + \cdots$$

$$= x + x^2 + x^3\left[-\frac{1}{3!} + \frac{1}{2}\right]$$

$$+ x^5\left[\frac{1}{120} - \frac{1}{12} + \frac{1}{24}\right]$$

$$+ x^6\left[\frac{1}{5!} + \frac{1}{5!} - \frac{1}{36}\right]$$

$$= x + x^2 + \frac{x^3}{3} - \frac{x^5}{30} - \frac{x^6}{90}$$

The results are the same in each case.

49. $f(x) = \begin{cases} \dfrac{\sin x}{x} & \text{if } x \neq 0 \\ 1 & \text{if } x = 0 \end{cases}$

First note that for $x \neq 0$,

$$f'(x) = \frac{x \cos x - \sin x}{x^2}$$

$$f''(x) = \frac{(2 - x^2)\sin x - 2x \cos x}{x^3}$$

$$f'''(x) = \frac{(3x^2 - 6)\sin x + (6x - x^3)\cos x}{x^4}$$

$$f^{(4)}(x) = \frac{(x^4 - 12x^2 + 24)\sin x + (4x^3 - 24x)\cos x}{x^5}$$

$$f^{(5)}(x) = \frac{1}{x^6}\left[(-5x^4 + 60x^2 - 120)\sin x + (x^5 - 20x^3 + 120x)\cos x\right]$$

as can be verified with detailed calculations or with a CAS.

Using L'Hospital's rule when needed, we can determine the values of the derivatives of f evaluated at $x = 0$:

$$f'(0) = \lim_{x\to 0} \frac{f(x) - f(0)}{x - 0}$$

$$= \lim_{x\to 0}\left[\frac{\frac{\sin x}{x} - 1}{x}\right] = \lim_{x\to 0} \frac{\sin x - x}{x^2}$$

$$= \lim_{x\to 0} \frac{\cos x - 1}{2x} = \lim_{x\to 0} \frac{-\sin x}{2}$$

$f'(0) = 0$,

$$f''(0) = \lim_{x\to 0} \frac{f'(x) - f'(0)}{x - 0}$$

$$= \lim_{x\to 0} \frac{\frac{x \cos x - \sin x}{x^2} - 0}{x}$$

$$= \lim_{x\to 0} \frac{x \cos x - \sin x}{x^3}$$

$$= \lim_{x\to 0} \frac{-x \sin x}{3x^2}$$

$$= -\frac{1}{3} \lim_{x\to 0} \frac{\sin x}{x}$$

$$= -\frac{1}{3}$$

$$f'''(0) = \lim_{x\to 0} \frac{f''(x) - f''(0)}{x - 0}$$

$$= \lim_{x\to 0} \frac{\frac{(2 - x^2)\sin x - 2x \cos x}{x^3} - \left(-\frac{1}{3}\right)}{x}$$

$$= \lim_{x\to 0} \frac{(2 - x^2)\sin x - 2x \cos x + \frac{1}{3}x^3}{x^4}$$

$= \lim_{x\to 0} \frac{1}{4x^3}\left[(2-x^2)\cos x - 2x\sin x - 2(\cos x - x\sin x) + x^2\right]$

$= \lim_{x\to 0} \frac{-x^2\cos x + x^2}{4x^3}$

$= \lim_{x\to 0} \frac{1-\cos x}{4x}$

$= \lim_{x\to 0} \frac{\sin x}{4}$

$f'''(0) = 0$

$f^{(4)}(0) = \lim_{x\to 0} \frac{f'''(x) - f'''(0)}{x-0}$

$= \lim_{x\to 0} \frac{(3x^2-6)\sin x + (6x - x^3)\cos x}{x^5}$

$= \lim_{x\to 0} \frac{1}{5x^4}\left[(3x^2-6)\cos x + 6x\sin x + (6-3x^2)\cos x - (6x - x^3)\sin x\right]$

$= \lim_{x\to 0} \frac{x^3\sin x}{5x^4}$

$= \frac{1}{5}\lim_{x\to 0} \frac{\sin x}{x}$

$f^{(4)}(0) = \frac{1}{5}$

$f^{(5)}(0) = \lim_{x\to 0} \frac{f^{(4)}(x) - f^{(4)}(0)}{x-0}$

$= \lim_{x\to 0} \frac{1}{x^6}\left[(x^4 - 12x^2 + 24)\sin x + (4x^3 - 24x)\cos x - \frac{1}{5}x^5\right]$

$= \lim_{x\to 0} \frac{1}{6x^5}\left[\cos x(x^4 - 12x^2 + 24 + 12x^2 - 24) + \sin x(4x^3 - 24x - 4x^3 + 24x) - x^4\right]$

$= \frac{1}{6}\lim_{x\to 0} \frac{x^4\cos x - x^4}{x^5}$

$= \frac{1}{6}\lim_{x\to 0} \frac{\cos x - 1}{x}$

$= \frac{1}{6}\lim_{x\to 0}[-\sin x]$

$f^{(5)}(0) = 0$

$f^{(6)}(0) = \lim_{x\to 0} \frac{f^{(5)}(x) - f^{(5)}(0)}{x-0}$

$= \lim_{x\to 0} \frac{1}{x^7}\left[(-5x^4 + 60x^2 - 120)\sin x + (x^5 - 20x^3 + 120x)\cos x\right]$

$= \lim_{x\to 0} \frac{1}{7x^6}\left[\sin x(-20x^3 + 120x - x^5 + 20x^3 - 120x) + \cos x(-5x^4 + 60x^2 - 120 + 5x^4 - 60x^2 + 120)\right]$

$= \frac{1}{7}\lim_{x\to 0} \frac{-x^5\sin x}{x^6}$

$= -\frac{1}{7}\lim_{x\to 0} \frac{\sin x}{x}$

$f^{(6)}(0) = -\frac{1}{7}$

Thus, the first 4 non-zero terms of the Maclaurin series for f are

$1 - \frac{1}{3}\frac{x^2}{2!} + \frac{1}{5}\frac{x^4}{4!} - \frac{1}{7}\frac{x^6}{6!} + \cdots$

$= 1 - \frac{x^2}{3!} + \frac{x^4}{5!} - \frac{x^6}{7!} + \cdots$

The Maclaurin series for $\sin x$ is

$x - \frac{x^3}{3!} + \frac{x^5}{5!} - \frac{x^7}{7!} + \cdots$

which, when divided by x, is equal to the Maclaurin series for $f(x)$.

51.

$$f(t) = \sum_{k=0}^{3} \frac{f^{(0)}(0)}{k!}t^k = 10+10t+t^2-\frac{t^3}{3!}$$

$$f(2) = 10+10(2)+(2)^2-\frac{(2)^3}{6} = \frac{98}{3} \text{ miles}$$

53. $f^{(k)}(x) = e^x$, $f^{(k)}(c) = e^c$, so

$$e^x = \sum_{k=0}^{\infty} \frac{f^{(k)}(c)}{k!}(x-c)^k = \frac{e^c}{k!}(x-c)^k$$

55. To generate the series, differentiate $(1+x)^r$ repeatedly. The series terminates if r is a positive integer, since all derivatives after the rth derivative will be 0. Otherwise the series is infinite.

57. This is the case when $r = \frac{1}{2}$, and then, the Maclaurin series is

$$1+\sum_{k=1}^{\infty} \frac{(1/2)(-1/2)\ldots(1/2-k+1)}{k!}x^k$$

59.

$$\cosh x = 1+\frac{1}{2}x^2+\frac{1}{24}x^4+\frac{1}{720}x^6+\cdots$$

$$\sinh x = x+\frac{1}{6}x^3+\frac{1}{120}x^5+\frac{1}{5040}x^7+\cdots$$
These match the cosine and sine series except that here all signs are positive.

8.8 Applications of Taylor Series

1. Using the first three terms of the sine series around the center $\pi/2$ we get $\sin 1.61 \approx 0.999231634426433$.

3. Using the first five terms of the cosine series around the center 0 we get $\cos 0.34 \approx 0.94275466553403$.

5. Using the first nine terms of the exponential series around the center 0 we get $e^{0.2} \approx 0.818730753079365$.

7. $\lim_{k\to\infty} \frac{\cos x^2 - 1}{x^4}$
$$= \lim_{k\to\infty} \frac{\left(-\frac{x^4}{2!}+\frac{x^8}{4!}-\frac{x^{12}}{6!}\cdots\right)-1}{x^4}$$
$$= -\frac{1}{2}.$$

9. Using
$$\ln x \approx (x-1)-\frac{1}{2}(x-1)^2+\frac{1}{3}(x-1)^3$$
we get
$$\lim_{k\to\infty} \frac{\ln x-(x-1)}{(x-1)^2}$$
$$= \lim_{k\to\infty} \frac{-2(x-1)-\frac{1}{2}(x-1)^2+\frac{1}{3}(x-1)^3}{(x-1)^2}$$
$$= -\frac{1}{2}.$$

11. $\lim_{k\to\infty} \frac{x^x - 1}{x}$
$$= \lim_{k\to\infty} \frac{\left(1+x+\frac{x^2}{2}+\frac{x^3}{3!}x^3+\cdots\right)-1}{x}$$
$$= 1$$

13. $\int_{-1}^{1} \frac{\sin x}{x}\,dx$
$$\approx \int_{-1}^{1} \left(1-\frac{x^2}{6}+\frac{x^4}{120}\right)dx$$
$$= x-\frac{x^3}{18}+\frac{x^5}{600}\Big|_{-1}^{1}$$
$$= \frac{1703}{900} \approx 1.8922.$$

15. $\int_{-1}^{1} e^{-x^2}\,dx$
$$\approx \int_{-1}^{1} \left(1-x^2+\frac{x^4}{2}-\frac{x^6}{6}+\frac{x^8}{24}\right)dx$$
$$= x-\frac{x^3}{3}+\frac{x^5}{10}-\frac{x^7}{42}+\frac{x^9}{(24)(9)}\Big|_{-1}^{1}$$
$$= \frac{5651}{3780} \approx 1.495.$$

17. $\int_{1}^{2} \ln x\,dx$
$$\approx \int_{1}^{2} (x-1)-\frac{1}{2}(x-1)^2$$
$$+\frac{1}{3}(x-1)^3-\frac{1}{4}(x-1)^4+\frac{1}{5}(x-1)^5\,dx$$
$$= \left(\frac{(x-1)^2}{2}-\frac{(x-1)^3}{6}+\right.$$
$$\left.\frac{(x-1)^4}{12}-\frac{(x-1)^5}{20}+\frac{(x-1)^6}{30}\right)\Big|_{1}^{2}$$
$$= \frac{2}{5}$$

19.
$$\left|\frac{a_{n+1}}{a_n}\right| \le \frac{x^2}{2^2(n+1)(n+2)}.$$
Since this ratio tends to 0, the radius of convergence for $J_1(x)$ is infinite.

21. We need the first neglected term to be less than 0.04. The k-th term is bounded by
$$\frac{10^{2k+1}}{2^{2k+2}k\ (k+2)!},$$
which is equal to 0.0357 for $k = 12$. Thus we will need the terms up

through
$k = 11$, that is, the first 12 terms of the series.

23. Let $f(x) = \dfrac{1}{\sqrt{1-x}}$. Then

$$f(x) \approx f(0) + f'(0)x = 1 + \frac{1}{2}x$$

Now let $x = \dfrac{v^2}{c^2}$. Then

$$\frac{1}{\sqrt{1-\dfrac{v^2}{c^2}}} \approx 1 + \frac{v^2}{2c^2}.$$

Thus

$$m(v) = m_0/\sqrt{1 - v^2/c^2} \approx m_0\left(1 + \frac{v^2}{2c^2}\right).$$

To increase mass by 10%, we want

$$1 + \frac{v^2}{2c^2} = 1.1$$

Solving for v, we have $\dfrac{1}{2c^2}v^2 = .1$ so $v^2 = .2c^2$, thus

$$v = \sqrt{.2}c \approx 83{,}000 \text{ miles per second.}$$

25. $w(x) = \dfrac{mgR^2}{(R+x)^2}$, $w(0) = mg$

$w'(x) = \dfrac{-2mg}{(R+x)^3}$, $w'(0) = \dfrac{-2mg}{R^3}$, so

$$w(x) \approx \sum_{k=0}^{\infty} \frac{f^{(k)}(0)}{k!}x^k = mg\left(1 - \frac{2x}{R}\right)$$

To reduce weight by 10%, we want $1 - \dfrac{2x}{R} = 9$. Solving for x, we have $\dfrac{-2x}{R} = -.1$, so $x = \dfrac{R}{20} \approx 200$ miles.

27. No, since x is much smaller than R for high-altitude locations on Earth. You have to go out to an altitude of more than 100 miles before you weigh significantly less.

29. The first neglected term is negative, so this estimate is too large.

31. Use the first two terms of the series for tanh:

$$\tanh x \approx x - \frac{1}{3}x^3$$

Substitute $\sqrt{\dfrac{g}{40m}}t$ for x, multiply by $\sqrt{40mg}$ and simplify. The result is

$$gt - \frac{g^2}{120m}t^3$$

33. $\dfrac{1}{\sqrt{1-x}} = (1+(-x))^{-1/2}$

$$= 1 - \frac{1}{2}(-x) + \frac{\left(-\frac{1}{2}\right)\left(-\frac{3}{2}\right)}{2!}(-x)^2 + \frac{\left(-\frac{1}{2}\right)\left(-\frac{3}{2}\right)\left(-\frac{5}{2}\right)}{3!}(-x)^3 + \frac{\left(-\frac{1}{2}\right)\left(-\frac{3}{2}\right)\left(-\frac{5}{2}\right)\left(-\frac{7}{2}\right)}{4!}(-x)^4 + \cdots$$

$$= 1 + \frac{1}{2}x + \frac{3}{8}x^2 + \frac{5}{16}x^3 + \frac{35}{128}x^4 + \cdots$$

35. $\dfrac{6}{\sqrt[3]{1+3x}} = 6\,(1+(3x))^{-1/3}$

$$= 6\left[1 - \frac{1}{3}(3x) + \frac{\left(-\frac{1}{3}\right)\left(-\frac{4}{3}\right)}{2!}(3x)^2 + \frac{\left(-\frac{1}{3}\right)\left(-\frac{4}{3}\right)\left(-\frac{7}{3}\right)}{3!}(3x)^3 + \frac{\left(-\frac{1}{3}\right)\left(-\frac{4}{3}\right)\left(-\frac{7}{3}\right)\left(-\frac{10}{3}\right)}{4!}(3x)^4 + \cdots\right]$$

$$= 6\left[1 - x + 2x^2 - \frac{14}{3}x^3 + \frac{35}{3}x^4 - \cdots\right]$$

$$= 6 - 6x + 12x^2 - 28x^3 + 70x^4 - \cdots$$

37. (a)

$$\sqrt{26} = \sqrt{25\left(\frac{26}{25}\right)} = 5\sqrt{1+\frac{1}{25}}$$

Since $\frac{1}{25}$ is in the interval of convergence $-1 < x < 1$, the binomial series can be used to get

$$\sqrt{1+\frac{1}{25}}$$
$$= 1 + \frac{1}{2}\left(\frac{1}{25}\right) - \frac{1}{8}\left(\frac{1}{25}\right)^2 + \frac{1}{16}\left(\frac{1}{25}\right)^3 - \frac{5}{128}\left(\frac{1}{28}\right)^4 + \cdots$$

Using the first four terms of the series implies that the error will be less than $\frac{5}{128}\left(\frac{1}{25}\right)^4 = 10^{-7}$.

Thus, $\sqrt{26}$

$$\approx 5\left[1 + \frac{1}{2}\left(\frac{1}{25}\right) - \frac{1}{8}\left(\frac{1}{25}\right)^2 + \frac{1}{16}\left(\frac{1}{25}\right)^3\right]$$
$$\approx 5.0990200$$

(b) $\sqrt{24}$

$$\approx 5\left[1 - \frac{1}{2}\left(\frac{1}{25}\right) - \frac{1}{8}\left(\frac{1}{25}\right)^2 - \frac{1}{16}\left(\frac{1}{25}\right)^3\right]$$
$$\approx 4.8989800$$

39. $(4+x)^3 = \left(4\left[1+\frac{x}{4}\right]\right)^3$

$$= 4^3\left[1+\frac{x}{4}\right]^3$$
$$= 4^3\left[1 + 3\left(\frac{x}{4}\right) + 3\left(\frac{x}{4}\right)^2 + \left(\frac{x}{4}\right)^3\right]$$
$$= 4^3 + 3(4^2)x + 3(4)x^2 + x^3$$
$$= 64 + 48x + 12x^2 + x^3$$

$$(1-2x)^4 = [1+(-2x)]^4$$
$$= 1 + 4(-2x) + \frac{(4)(3)}{2}(-2x)^2 + \frac{(4)(3)(2)}{3!}(-2x)^3 + (-2x)^4$$
$$= 1 - 8x + 24x^2 - 32x^3 + 16x^4$$

For a positive integer n, there are $n+1$ non-zero terms in the binomial expansion.

41.

$$\frac{1}{\sqrt{1-x}} = 1+\frac{1}{2}x+\frac{3}{8}x^2+\frac{5}{16}x^3+\frac{35}{128}x^4+\cdots$$

Therefore,

$$\frac{1}{\sqrt{1-x^2}} = 1+\frac{1}{2}x^2+\frac{3}{8}x^4+\frac{5}{16}x^6+\frac{35}{128}x^8+\cdots$$

and so $\sin^{-1} x - \int \frac{1}{\sqrt{1-x^2}}dx$

$$= x + \frac{1}{6}x^3 + \frac{3}{40}x^5 + \frac{5}{112}x^7 + \frac{35}{1152}x^9 + \cdots + \text{const.}$$

Since $\sin^{-1}(0) = 0$, it follows that the constant in the previous equation is 0. Thus, the Maclaurin series for the inverse sine function is

$$\sin^{-1} x = x+\frac{x^3}{6}+\frac{3x^5}{40}+\frac{5x^7}{112}+\frac{35x^9}{1152}+\cdots$$

43. $\sin x = x - \frac{x^3}{3!} + \frac{x^5}{5!} - \frac{x^7}{7!} + \frac{x^9}{9!} - \frac{x^{11}}{11!} + \frac{x^{13}}{13!} - \cdots$

$$\frac{\sin x}{x} = 1 - \frac{x^2}{3!} + \frac{x^4}{5!} - \frac{x^6}{7!} + \frac{x^8}{9!} - \frac{x^{10}}{11!} + \frac{x^{12}}{13!} - \cdots$$

$$\int_0^{\pi} \frac{\sin x}{x}dx$$

$$= \int_0^\pi \left[1 - \frac{x^2}{3!} + \frac{x^4}{5!} - \frac{x^6}{7!} + \frac{x^8}{9!} - \frac{x^{10}}{11!} + \frac{x^{12}}{13!} - \cdots\right] dx$$

$$\approx 1.851937$$

45. Using $e^x = \sum_{n=0}^{\infty} \frac{x^n}{n!}$ and $\frac{hc}{k} \approx 0.014$, we get

$$g(\lambda) = \sum_{n=1}^{\infty} \frac{\lambda^5}{n!}\left(\frac{hc}{k\lambda T}\right)^n \approx \sum_{n=1}^{\infty} \frac{\lambda^5\, 0.014^n}{(\lambda T)^n n!}$$

Therefore,

$$\frac{dg}{d\lambda}(\lambda) \approx \sum_{n=1}^{5} (5-n)\frac{\lambda^4\, 0.014^n}{(\lambda T)^n n!}$$

and the above expression equals to 0 when

$$\lambda = -\frac{0.002985852303}{T}.$$

This doesn't agree with Wien's law, but the value of λ might be too large as the radius of convergence for the Maclaurin series is infinite.

47. Using the series for $(1+x)^{3/2}$ around the center $x = 0$, we get
$S(d)$

$$\approx \frac{8\pi c^2}{3}\left(\frac{3}{32}\cdot\frac{d^2}{c^2} + \frac{3}{2048}\cdot\frac{d^4}{c^4} - \frac{1}{65536}\cdot\frac{d^6}{c^6}\right)$$

If we ignore the d^4 and d^6 terms, this simplifies to $\frac{\pi d^2}{4}$.

8.9 Fourier Series

1. $a_0 = \frac{1}{\pi}\int_{-\pi}^{\pi} f(x)\,dx = \frac{1}{\pi}\int_{-\pi}^{\pi} x\,dx$

$$= \frac{1}{\pi}\left.\frac{x^2}{2}\right|_{-\pi}^{\pi} = 0$$

$$a_k = \frac{1}{\pi}\int_{-\pi}^{\pi} x\cos(kx)\,dx$$

$$= \frac{1}{\pi}\left.\frac{x}{k}\sin(kx) + \frac{1}{k^2}\cos(kx)\right|_{-\pi}^{\pi}$$

$$= 0$$

$$b_k = \frac{1}{\pi}\int_{-\pi}^{\pi} x\sin(kx)\,dx$$

$$= \frac{1}{\pi}\left.\frac{-x}{k}\cos kx + \frac{1}{k^2}\sin kx\right|_{-\pi}^{\pi}$$

$$= \frac{-2}{k}(-1)^k$$

Hence,

$$f(x) = \sum_{k=1}^{\infty}(-1)^{k+1}\frac{2}{k}\sin(kx)$$

for $-\pi < x < \pi$.

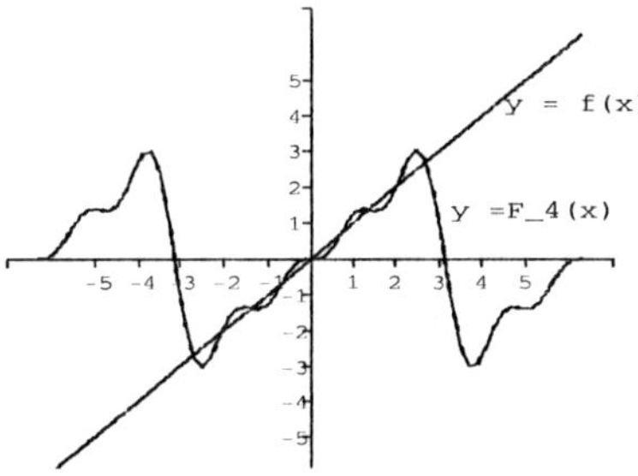

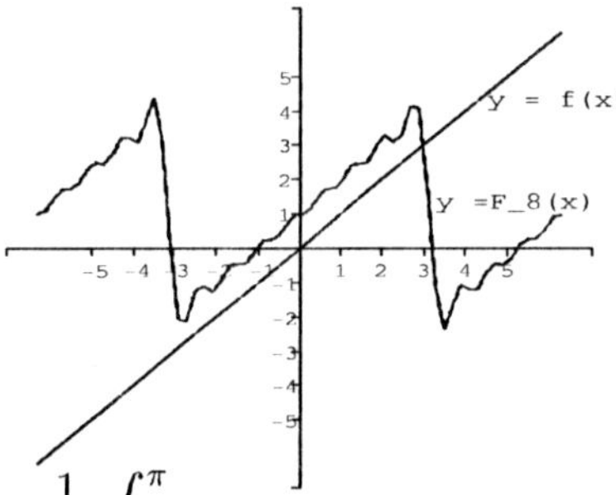

3. $a_0 = \frac{1}{\pi}\int_{-\pi}^{\pi} f(x)\,dx$

$$= \frac{1}{\pi}\int_{-\pi}^{0} (-2x)\,dx + \frac{1}{\pi}\int_0^{\pi} 2x\,dx$$

$$= \frac{1}{\pi}\left.-x^2\right|_{-\pi}^{0} + \frac{1}{\pi}\left.x^2\right|_0^{\pi} = 2\pi$$

$$a_k = \frac{1}{\pi}\int_{-\pi}^{\pi} f(x)\cos(kx)\,dx$$

$$= \frac{1}{\pi}\int_{-\pi}^{0} -2x\cos(kx)\,dx + \frac{1}{x}\int_0^{\pi} 3x\cos(kx)\,dx$$

$$= \frac{1}{\pi}\left.-\frac{2x}{k}\sin(kx) - \frac{2}{k^2}\cos(kx)\right|_{-\pi}^{0}$$

$$+\frac{1}{\pi}\left.\frac{2x}{k}\sin(kx)+\frac{2}{k^2}\cos(kx)\right|_0^{\pi}$$
$$=\frac{4}{k^2\pi}[(-1)^k-1]$$

So $a_{2k=0}$ and $a_{2(k-2)} = \dfrac{-8}{(2k-1)^2\pi}$.

$$b_k=\frac{1}{\pi}\int_{-\pi}^{\pi} f(x)\sin(kx)\,dx$$
$$=\frac{1}{\pi}\int_{-\pi}^{0} -2x(kx)\,dx$$
$$+\frac{1}{\pi}\int_{0}^{\pi} 2x\sin(kx)\,dx$$
$$=\frac{1}{\pi}\left.\frac{2x}{k}\cos(kx)-\frac{2}{k^2}\sin(kx)\right|_{-\pi}^{0}$$
$$+\frac{1}{\pi}\left.\frac{-2x}{k}\cos(kx)+\frac{2}{k^2}\sin(kx)\right|_0^{\pi}$$
$$=\frac{2}{k}\cos(k\pi)-\frac{2}{k}\cos k\pi$$
$$=0$$

So

$$f(x)=\frac{2\pi}{2}+\sum_{k=1}^{\infty}\frac{-8}{(2k-1)^2\pi}\cos(2k-1)x$$

$$f(x)=\pi-\sum_{k=1}^{\infty}\frac{8}{(2k-1)^2\pi}\cos(2k-1)x$$

for $-\pi < x < \pi$.

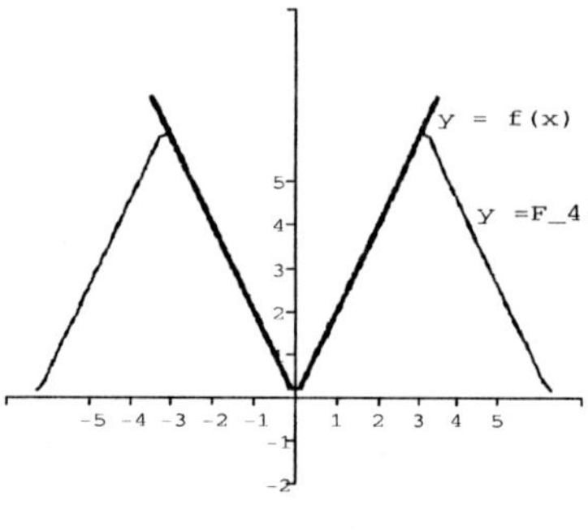

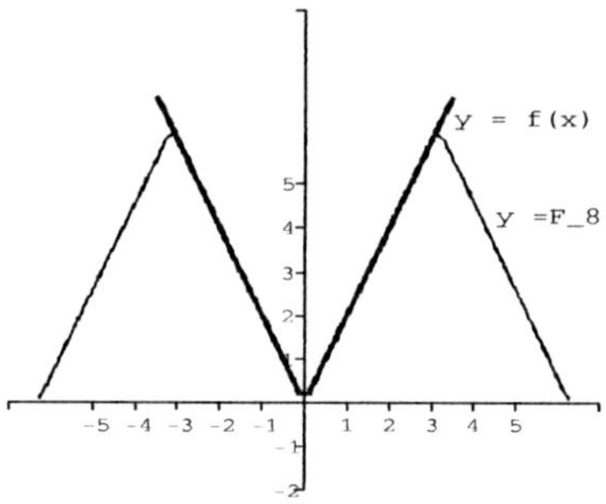

5. $a_0=\dfrac{1}{\pi}\displaystyle\int_{-\pi}^{0} 1\,dx+\frac{1}{\pi}\int_{0}^{\pi}(-1)\,dx=0$

$$a_k=\frac{1}{\pi}\int_{-\pi}^{0}\cos(kx)\,dx$$
$$+\frac{1}{\pi}\int_{0}^{\pi}[-\cos(kx)]\,dx$$
$$=\frac{1}{\pi}\frac{1}{k}\sin kx\Big|_{-\pi}^{0}-\frac{1}{\pi}\frac{1}{k}\sin kx\Big|_{0}^{\pi}$$
$$=0$$

$$b_k=\frac{1}{\pi}\int_{\pi}^{0}[\sin(kx)]\,dx$$
$$+\frac{1}{\pi}\int_{0}^{\pi}[-\cos(kx)]\,dx$$
$$=\frac{1}{\pi}\,-\frac{1}{k}\cos kx\Big|_{-\pi}^{0}-\frac{1}{\pi}\,-\frac{1}{k}\cos kx\Big|_{0}^{\pi}$$

So $b_{2k}=0$ and $b_{2k-1}=\dfrac{-4}{(2k-1)\pi}$.

Thus

$$f(x)=\sum_{k=1}^{\infty}\frac{-4}{(2k-1)\pi}\sin(2k-1)x,$$

for $-\pi < x < \pi$.

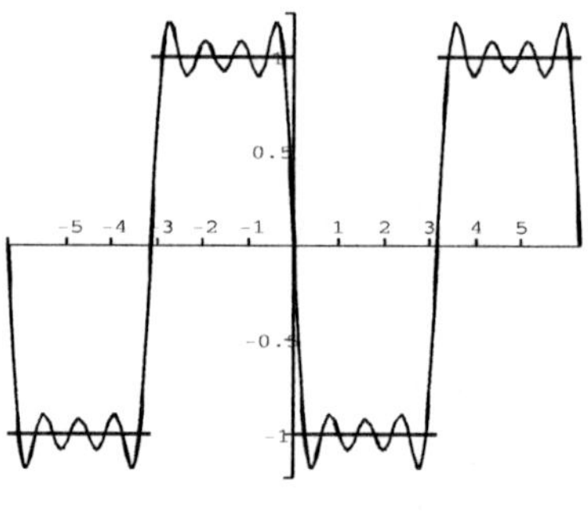

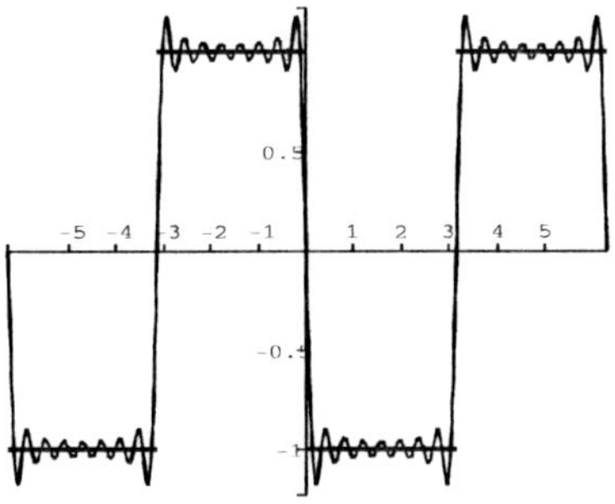

7. $f(x) = 3\sin 2x$ is already periodic on $[-2\pi, 2\pi]$.

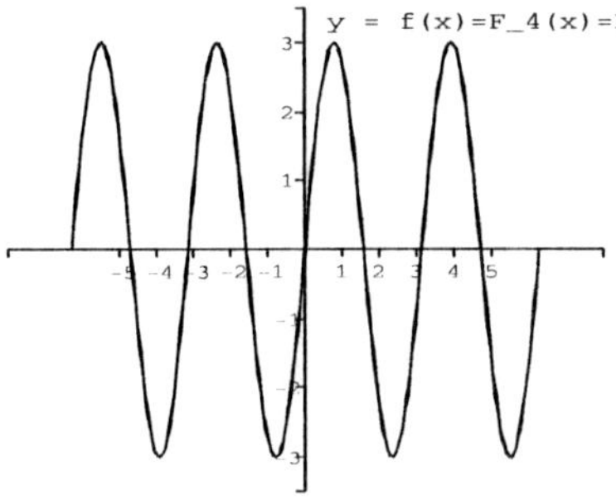

9. $$a_0 = \int_{-1}^{1} (-x)\,dx = \left.\frac{-x^2}{2}\right|_{-1}^{1} = 0$$

$$\begin{aligned} a_k &= \int_{-1}^{1} (-x)\cos(k\pi x)\,dx \\ &= \left.\frac{-x}{k\pi}\sin k\pi x - \frac{1}{(k\pi)^2}\cos k\pi x\right|_{-1}^{1} \\ &= \frac{-1}{k\pi}\sin k\pi - \frac{1}{(k\pi)^2}\cos k\pi \\ &\quad -\left[\frac{1}{k\pi}\sin(-k\pi) - \frac{1}{(k\pi)^2}\cos(-k\pi)\right] \\ &= 0 \end{aligned}$$

$$\begin{aligned} b_k &= \int_{-1}^{1} (-x)\sin(k\pi x)\,dx \\ &= \left.\frac{x}{k\pi}\cos k\pi x - \frac{1}{(k\pi)^2}\sin k\pi x\right|_{-1}^{1} \\ &= \frac{1}{k\pi}\cos k\pi - \frac{1}{(k\pi)^2}\sin k\pi \\ &\quad -\left[\frac{-1}{k\pi}\cos(-k\pi) - \frac{1}{(k\pi)^2}\sin(-k\pi)\right] \\ &= \frac{2}{k\pi}\cos k\pi \quad = \frac{2}{k\pi}(-1)^k \end{aligned}$$

So

$$f(x) = \sum_{k=1}^{\infty}(-1)^k\frac{2}{k\pi}\sin k\pi x$$

11. $$a_0 = \frac{1}{1}\int_{-1}^{1} x^2\,dx = \left.\frac{x^3}{3}\right|_{-1}^{1} = \frac{2}{3}$$

$$\begin{aligned} a_k &= \frac{1}{1}\int_{-1}^{1} x^2\cos k\pi x\,dx \\ &= \frac{x^2}{k\pi}\sin k\pi x + \frac{2x}{(k\pi)^2}\cos k\pi x \\ &\quad \left. -\frac{2}{(k\pi)^3}\sin k\pi x\right|_{-1}^{1} \\ &= \frac{4}{(k\pi)^2}(-1)^k \end{aligned}$$

$$\begin{aligned} b_k &= \frac{1}{1}\int_{-1}^{1} x^2\sin k\pi x\,dx \\ &= \frac{x^2}{k\pi}\cos k\pi x + \frac{2x}{(k\pi)^2}\sin k\pi x \\ &= 0 \end{aligned}$$

Hence,

$$f(x) = \frac{1}{3} + \sum_{k=1}^{\infty}\frac{(-1)^k 4}{(k\pi)^2}\cos k\pi x.$$

for $-1 < x < 1$.

13. $$\begin{aligned} a_0 &= \int_{-1}^{1} f(x)\,dx = \int_{-1}^{0} 0\,dx + \int_{0}^{1} x\,dx \\ &= \left.\frac{x^2}{2}\right|_{0}^{1} = \frac{1}{2} \end{aligned}$$

$$\begin{aligned} a_k &= \int_{-1}^{0} 0\,dx + \int_{0}^{1} x\cos(k\pi x)\,dx \\ &= \left.\frac{x}{k\pi}\sin(k\pi x) + \frac{1}{(k\pi)^2}\cos(k\pi x)\right|_{0}^{1} \end{aligned}$$

So $a_{2k} = 0$ and $a_{2k-1} = \dfrac{-2}{2(k-1)^2\pi^2}$.

$$\begin{aligned} b_k &= \int_{-1}^{0} 0\,dx + \int_{0}^{1} x\cos(k\pi x)\,dx \\ &= \left.\frac{-x}{k\pi}\cos(k\pi x) + \frac{1}{(k\pi)^2}\sin(k\pi x)\right|_{0}^{1} \\ &= \frac{-1}{k\pi}(-1)^k \end{aligned}$$

Hence,

$$f(x) = \frac{1}{4} + \sum_{k=1}^{\infty} \left[\frac{-2}{(2k-1)^2\pi^2} \cos(2k-1)\pi x + \frac{(-1)^{k+1}}{k\pi} \sin k\pi x \right]$$

for $-1 < x < 1$.

15.

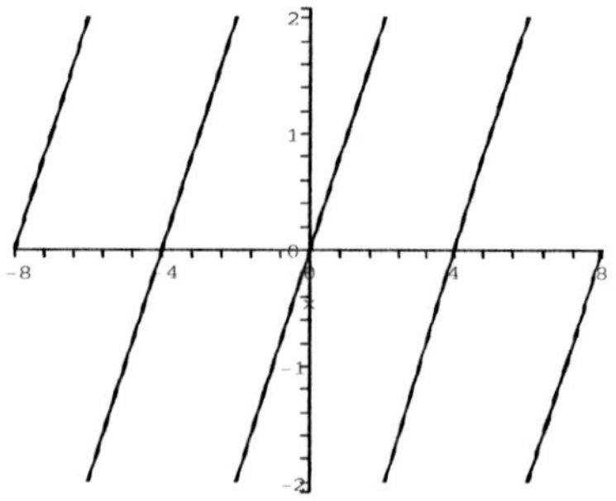

17.

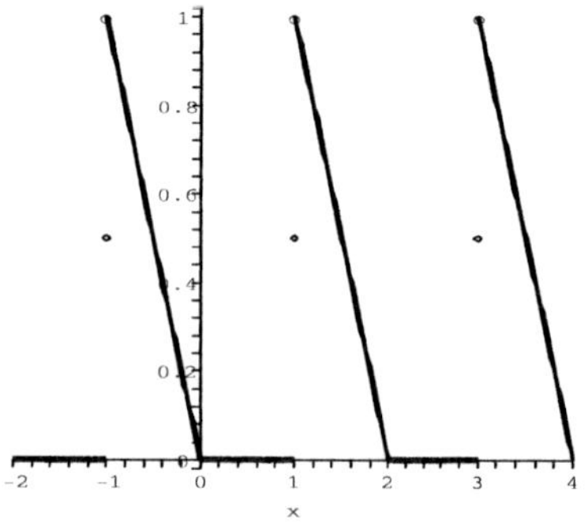

19.

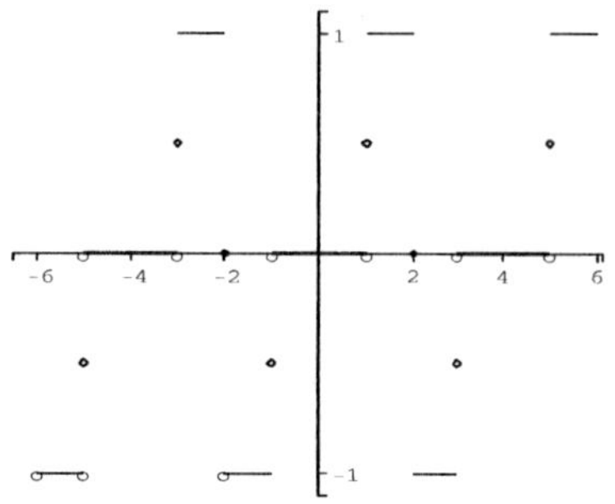

21. $f(x) = x^2 = \frac{1}{3} + \sum_{k=1}^{\infty} (-1)^k \frac{4}{k^2\pi^2} \cos k\pi x$

$$f(1) = 1 = \frac{1}{3} + \sum_{k=1}^{\infty} (-1)^k \frac{4}{k^2\pi^2} (-1)^k$$

$$\frac{2}{3} = \sum_{k=1}^{\infty} \frac{4}{k^2\pi^2} \text{ so } \frac{2}{3} = \frac{4}{\pi^2} \sum_{k=1}^{\infty} \frac{1}{k^2}, \text{ so}$$

$$\frac{\pi^2}{6} = \sum_{k=1}^{\infty} \frac{1}{k^2}$$

23.

$$f(x) = \frac{\pi}{2} - \sum_{k=1}^{\infty} \frac{4}{(2k-1)^2\pi} \cos(2k-1)x$$

$$f(0) = 0 = \frac{\pi}{2} - \sum_{k=1}^{\infty} \frac{4}{(2k-1)^2\pi} \cos 0$$

$$\frac{\pi}{2} = \sum_{k=1}^{\infty} \frac{4}{(2k-1)^2\pi} = \frac{4}{\pi} \sum_{k=1}^{\infty} \frac{1}{(2k-1)^2}$$

$$\frac{\pi^2}{8} = \sum_{k=1}^{\infty} \frac{1}{(2k-1)^2}$$

25. Note that f is continuous except at $x = k\pi$, where $k \in \mathbb{Z}$. Also $f'(x) = 0$ except at $x = k\pi$. Thus, f' is continuous for $-\pi < x < \pi$, except at $x = 0$. By the Fourier Convergence theorem (Theorem 9.1), the Fourier series for f converges to f for all x, except at $x = k\pi$, where the Fourier series converges to $\frac{1}{2}(-1+1) = 0$.

27. Note that f is continuous for all x except at $x = 2k+1$, where $k \in \mathbb{Z}$. Also note that $f'(x) = -1$ except at $x = 2k+1$, so that f' is continuous except at $x = 2k+1$. By the Fourier Convergence theorem (Theorem 9.1), the Fourier series for f converges to f for all x except at $x = 2k+1$, where the Fourier series converges to $\frac{1}{2}(-1+1) = 0$.

29. If $f(x) = \cos x$, then $f(-x) = \cos(-x) = \cos x = f(x)$ so $\cos x$ is even.

If $f(x) = \sin x$, then
$f(-x) = \sin(-x) = -\sin x = -f(x)$
so $\sin x$ is odd.

If $f(x) = \cos x + \sin x$, then
$f(-x) = \cos(-x) + \sin(-x)$
$= \cos x - \sin x \neq f(x)$ and
$\cos x - \sin x \neq -f(x)$
so $\cos x + \sin x$ is neither even nor odd.

31. If $f(x)$ is odd and $g(x) = f(x)\cos x$ then
$g(-x) = f(-x)\cos(-x)$
$= -f(x)\cos x = -g(x)$ is odd.

Also if $h(x) = f(x)\sin x$ then
$h(-x) = f(-x)\sin(-x)$
$= -f(x)(-\sin x)$
$= f(x)\sin x = h(x)$
so $h(x)$ is even.

If $f(x)$ and $g(x)$ are even, then
$f(-x)g(-x) = f(x)g(x)$,
so $f(x)g(x)$ is even.

33. We have the Fourier series expansion

$$f(x) = \frac{a_0}{2} + \sum_{k=1}^{\infty}\left[a_k \cos\left(\frac{k\pi x}{l}\right) + b_k \sin\left(\frac{n\pi x}{l}\right)\right].$$

Multiply both sides of this equation by $\cos\left(\frac{n\pi x}{l}\right)$ and integrate with respect to x on the interval $[-l, l]$ to get

$$\int_{-l}^{l} f(x)\cos\left(\frac{n\pi x}{l}\right) dx$$
$$= \int_{-\ell}^{\ell} \frac{a_0}{2}\cos\left(\frac{n\pi x}{l}\right) dx + \sum_{k=1}^{\infty}\left[a_k \int_{-l}^{l}\cos\left(\frac{k\pi x}{l}\right)\cos\left(\frac{n\pi x}{l}\right) dx + b_k \int_{-l}^{l}\sin\left(\frac{k\pi x}{l}\right)\cos\left(\frac{n\pi x}{l}\right) dx\right]$$

Well $\int_{-l}^{1}\cos\left(\frac{n\pi x}{l}\right) dx = 0$ for all n. And,

$$\int_{-l}^{l}\sin\left(\frac{k\pi x}{l}\right)\cos\left(\frac{n\pi x}{l}\right) dx$$
$$= \begin{cases} 0 & \text{if } n \neq k \\ l & \text{if } n = k. \end{cases}$$

So when $n = k$, we have

$$\int_{-l}^{l} f(x)\cos\left(\frac{k\pi x}{l}\right) dx = a_k l, \text{ so}$$
$$a_k = \frac{1}{l}\int_{-l}^{l} f(x)\cos\left(\frac{k\pi x}{l}\right) dx.$$

Now multiply both sides of the original equation by $\sin\left(\frac{n\pi x}{l}\right)$ and integrate on $[-l, l]$, we have

$$\int_{-l}^{l} f(x)\sin\left(\frac{n\pi x}{l}\right) dx$$
$$= \int_{-l}^{l}\frac{a_0}{2}\sin\left(\frac{n\pi x}{l}\right) dx + \sum_{k=1}^{\infty}\left[a_k \int_{-l}^{l}\cos\left(\frac{k\pi x}{l}\right)\sin\left(\frac{n\pi x}{l}\right) dx + b_k \int_{-l}^{l}\sin\left(\frac{k\pi x}{l}\right)\sin\left(\frac{n\pi x}{l}\right) dx\right]$$

Well $\int_{-l}^{l}\sin\left(\frac{n\pi x}{l}\right) dx = 0$. And

$$\int_{-l}^{l}\cos\left(\frac{k\pi x}{l}\right)\sin\left(\frac{n\pi x}{l}\right) dx = 0.$$

Also $\int_{-l}^{l}\sin\left(\frac{k\pi x}{l}\right)\sin\left(\frac{k\pi x}{l}\right) dx$
$$= \begin{cases} 0 & \text{if } n \neq k \\ l & \text{if } n = k \end{cases}.$$

So when $n = k$ we have

$$\int_{-l}^{l} f(x)\sin\left(\frac{k\pi x}{l}\right) dx = b_k l. \text{ So}$$
$$b_k = \frac{1}{l}\int_{-l}^{l} f(x)\sin\left(\frac{k\pi x}{l}\right) dx.$$

35. Since $f(x) = x^3$ is odd, the Fourier Series will contain only sine.

37. Since $f(x) = e^x$ is neither even nor odd, the Fourier Series will contain both.

39. Because $g(x)$ is an odd function, its series contains only sine terms, so

$f(x)$ contains sine terms and the constant 1.

41. The graph of the limiting function for the function in Exercise 9, with $k = 1000$:

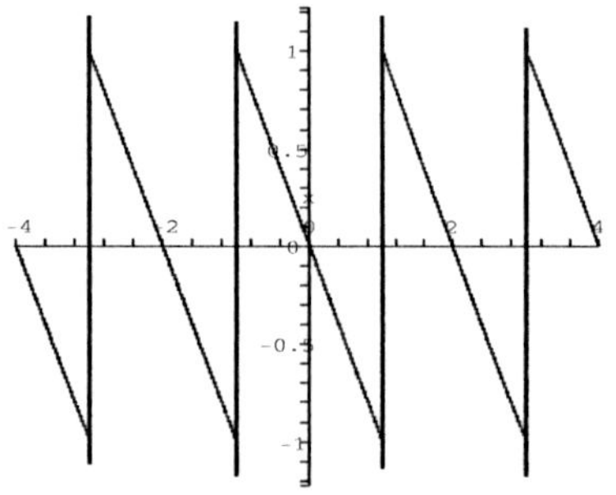

The graph of a pulse wave of width 1/3:

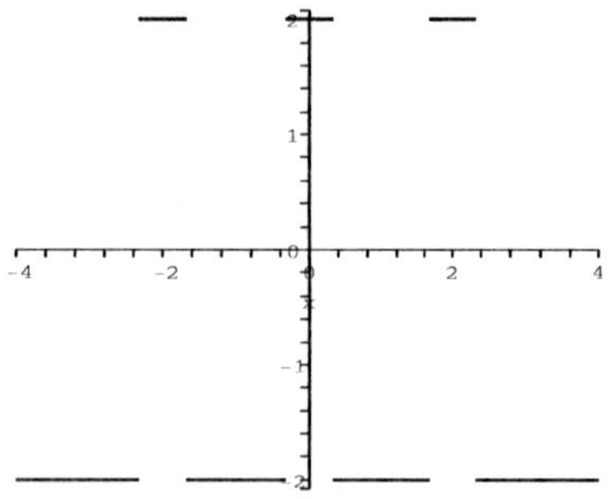

The graph of a pulse wave of width 1/4:

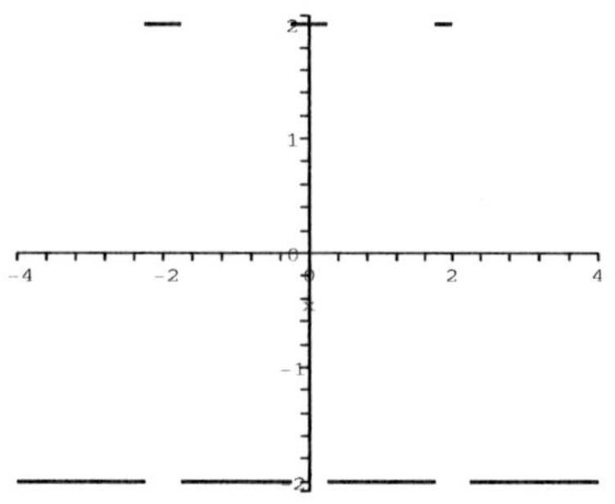

43. The cutoff frequency, n, corresponds to the partial sum $F_n(x)$ because in both, everything after the nth term is zero.

$f(x) = -x$ on $[-1, 1]$, so

$$f(x) = \sum_{k=1}^{\infty} \frac{(-1)^k 2}{k\pi} \sin k\pi x, \text{ thus}$$

$$F_2(x) = \frac{2}{\pi}\left[-\sin \pi x + \frac{1}{2}\sin 2\pi x\right]$$

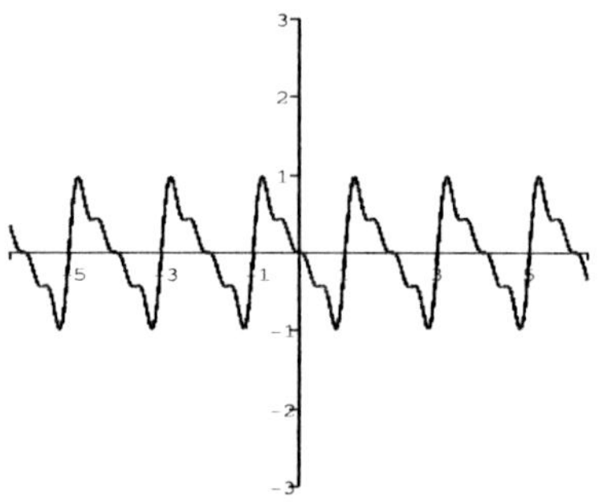

$$F_4(x) = \frac{2}{\pi}\left[-\sin \pi x + \frac{1}{2}\sin 2\pi -\frac{1}{3}\sin 3\pi x + \frac{1}{4}\sin 4\pi x\right]$$

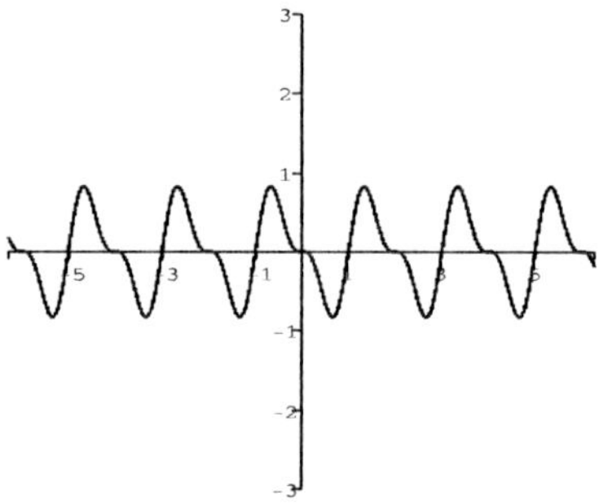

$$F_2(x) = \frac{8}{\pi}\left[\sin \pi x + \frac{1}{3}\sin 3\pi x\right]$$

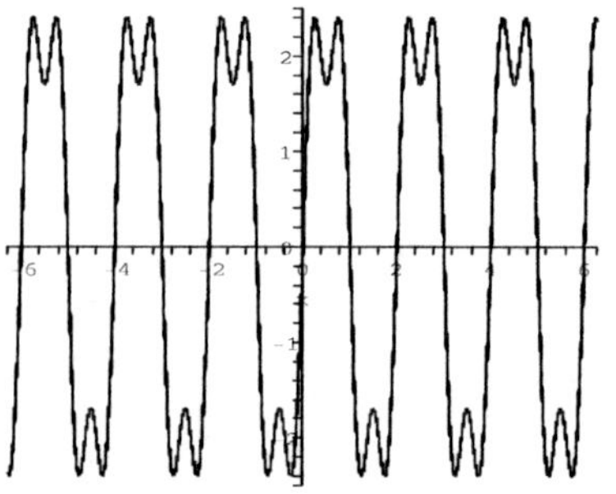

$$F_4(x) = \frac{8}{\pi}\left[\sin \pi x + \frac{1}{3}\sin 3\pi x +\frac{1}{5}\sin 5\pi x + \frac{1}{7}\sin 7\pi x\right]$$

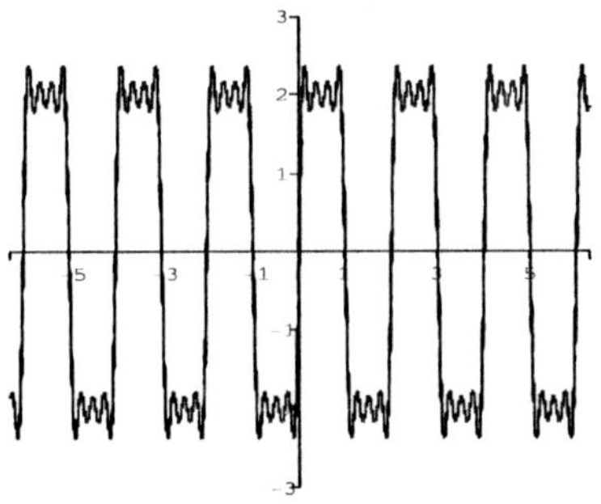

Because the wave is smoother.

45. The amplitude varies slowly because the frequency of $2\cos(0.2t)$ is small compared to the frequency of $\sin(8.1t)$. The variation of the amplitude explains why the volume varies, since the volume is proportional to the amplitude.

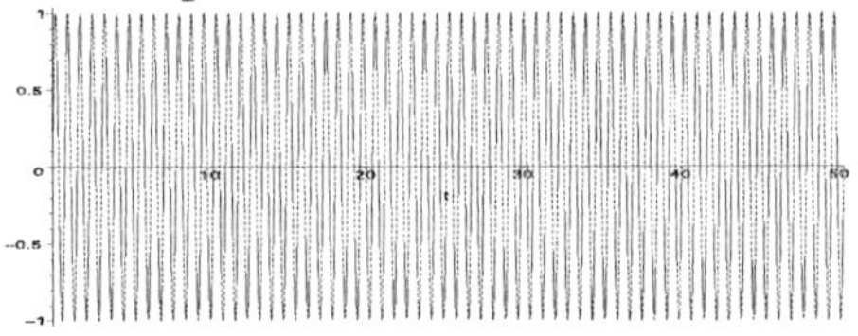

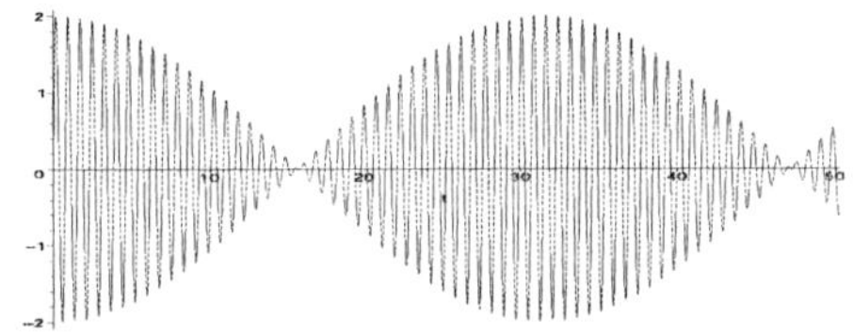

47. $f(x) = \dfrac{a_0}{2} + \sum_{k=1}^{\infty} [a_k \cos kx + b_k \sin kx]$

$$[f(x)]^2 = \frac{a_0^2}{4} + a_0 \sum_{k=1}^{\infty} [a_k \cos kx + b_k \sin kx] + \left[\sum_{k=1}^{\infty} (a_k \cos kx + b_k \sin kx)\right]^2$$

Now, note that

$$\frac{1}{\pi}\int_{-\pi}^{\pi} \frac{a_0^2}{4} dx = \frac{a_0^2}{4\pi}(2\pi) = \frac{a_0^2}{2} \text{ and}$$

$$\frac{1}{\pi}\int_{-\pi}^{\pi} a_0 \sum_{k=1}^{\infty} [a_k \cos kx + b_k \sin kx] dx$$

$$= \frac{a_0}{\pi} \sum_{k=1}^{\infty} \left[a_k \int_{-\pi}^{\pi} \cos kx dx + b_k \int_{-\pi}^{\pi} \sin kx dx\right]$$

$$= 0$$

Now consider the last term, which can be expanded as

$$\left[\sum_{k=1}^{\infty} (a_k \cos kx + b_k \sin kx)\right]^2$$

$$= \left[\sum_{k=1}^{\infty} (a_k \cos kx + b_k \sin kx)\right] \cdot \left[\sum_{m=1}^{\infty} (a_m \cos mx + b_m \sin mx)\right]$$

$$= \sum_{k=1}^{\infty} \sum_{m=1}^{\infty} [a_k a_m \cos kx \cos mx + a_k b_m \cos kx \sin mx + a_m b_k \cos mx \sin kx + b_k b_m \sin kx \sin mx]$$

Since $\int_{-\pi}^{\pi} \cos kx \sin mx dx = 0$ and

$$\int_{-\pi}^{\pi} \cos kx \cos mx dx = \int_{-\pi}^{\pi} \sin kx \sin mx dx = \begin{cases} 1 \text{ if } k = m \\ 0 \text{ otherwise,} \end{cases}$$

it follows that

$$\frac{1}{\pi}\int_{-\pi}^{\pi} \sum_{k=1}^{\infty} [f(x)]^2 dx$$

$$= \frac{a_0^2}{2} + \frac{1}{\pi} \sum_{k=1}^{\infty} \sum_{m=1}^{\infty} \left[a_k a_m \int_{-\pi}^{\pi} \cos kx \cos mx dx + b_k b_m \int_{-\pi}^{\pi} \sin kx \sin mx dx\right]$$

$$= \frac{a_0^2}{2} + \frac{1}{\pi} \sum_{k=1}^{\infty} [a_k^2 \pi + b_k^2 \pi]$$

$$= \frac{a_0^2}{2} + \sum_{k=1}^{\infty} [a_k^2 + b_k^2]$$

$$= \frac{a_0^2}{2} + \sum_{k=1}^{\infty} A_k^2$$

49. The plots from a CAS indeed show that the modified Fourier series

$$\frac{1}{2}+\sum_{k=1}^{n} \frac{2n}{[(2k-1)\pi]^2} \sin \frac{(2k-1)\pi}{n} \sin(2k-1)x$$

reduce the Gibbs phenomenon in this case.

The case $n = 4$

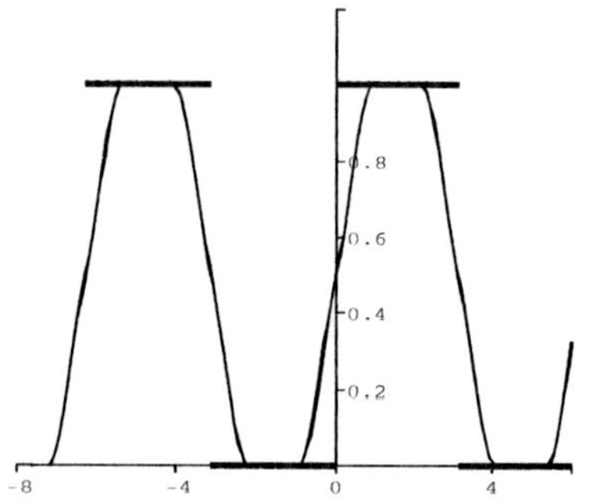

The case $n = 8$

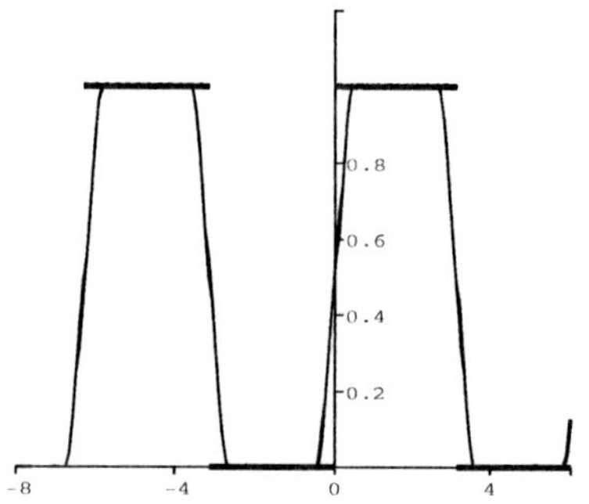

Ch. 8 Review Exercises

1. $\lim_{n\to\infty} \frac{4}{3+n} = 0$

3. $\lim_{n\to\infty} \frac{n}{n^2+4} = \lim_{n\to\infty} \frac{1/n}{1+4/n^2} = 0$

5.

$$0 < \frac{4\cdot 4\cdot 4\cdot \ldots 4}{1\cdot 2\cdot 3 \ldots n} < 4\cdot\frac{4}{2}\cdot\frac{4}{3}\cdot\frac{4}{4}\cdot\frac{4}{n} = \frac{128}{3n}$$

and

$$\lim_{n\to\infty} \frac{128}{3n} = 0,$$

so by the Squeeze Theorem, the original series converges to 0.

7. $\{\cos \pi n\}_{n=1}^{\infty} = \{-1, 1, -1, \ldots\}$ diverges

9. diverges

11. can't tell

13. diverges

15. converges

17. converges

19. $\sum_{k=0}^{\infty} 4\left(\frac{1}{2}\right)^k$ is a geometric series with $a = 4$ and $|r| = \frac{1}{2} < 1$ so the series converges to $\frac{a}{1-r} = \frac{4}{1-1/2} = 8$.

21. $\sum_{k=0}^{\infty} \left(\frac{1}{2}\right)^k$ is a geometric series with $a = 1$ and $|r| = \frac{1}{4} < 1$, so the series converges to $\frac{a}{1-r} = \frac{1}{1-1/4} = \frac{4}{3}$.

23. $|S - S_k| \le a_5 \le .01$ so $S \approx S_4 \approx -0.41$.

25. $\lim_{k\to\infty} \frac{2k}{k+3} = \lim_{k\to\infty} \frac{2}{1+\frac{3}{k}} = 2 \neq 0$ so by the k-th Term Test for Divergence, the original series diverges.

27. $\lim_{k\to\infty} |a_k| = \lim_{k\to\infty} \frac{4}{(k+1)^{1/2}} = 0$ and

$$\frac{a_{k+1}}{a_k} = \frac{4}{(k+2)^{1/2}} \cdot \frac{(k+1)^{1/2}}{4}$$

$= \dfrac{(k+1)^{1/2}}{(k+2)^{1/2}} < 1$ for all $k \geq 0$ so $a_{k+1} < a_k$ for all $k \geq 0$, so by the Alternating Series Test, the original series converges.

29. Diverges, by the p-test: $p = \dfrac{7}{8} < 1$.

31. Using the Limit Comparison Test, let $a_k = \dfrac{\sqrt{k}}{k^3+1}$ and $b_k = \dfrac{1}{k^{5/2}}$. Then

$$\lim_{k\to\infty} \frac{a_k}{b_k} = \lim_{k\to\infty} \frac{\sqrt{k}}{k^3+1} \cdot \frac{k^{5/2}}{1} = 1 > 0$$

so because $\displaystyle\sum_{k=1}^{\infty} \frac{1}{k^{5/2}}$ is a convergent p-series $\left(p = \dfrac{5}{2} > 1\right)$, the original series converges.

33. Using the Alternating Series Test, $\displaystyle\lim_{k\to\infty} \frac{4^k}{k!} = 0$ and

$$\frac{a_{k+1}}{a_k} = \frac{4k+1}{(k+1)!} \cdot \frac{k!}{4} \leq 1$$

for $k \geq 3$, so $a_{k+1} \leq a_k$ for $k \geq 3$, thus the original series converges.

35. Converges, using the Alternating Series Test, since

$$\lim_{k\to\infty} |a_k| = \lim_{k\to\infty} \ln\left(1 + \frac{1}{k}\right) = \ln 1 = 0$$

and

$$a_{k+1} = \ln\left(\frac{k+2}{k+1}\right) < \ln\left(\frac{k+1}{k}\right) = a_k$$

37. Using the Limit Comparison Test, let $a_k = \dfrac{2}{(k+3)^2}$ and $b_k = \dfrac{1}{k^2}$, so

$$\lim_{k\to\infty} \frac{a_k}{b_k} = \lim_{k\to\infty} \frac{2k^2}{(k+3)^2} = 2 > 0$$

so because $\displaystyle\sum_{k=1}^{\infty} \frac{1}{k^2}$ is a convergent p-series ($p = 2 > 1$), the original series converges.

39. Diverges, by the k-th Term Test.

41. Converges, by the Comparison Test, ($e^{1/k} < e$), and the p-series Test ($p = 2 > 1$).

43. Converges, by the Ratio Test:
$$\lim_{k\to\infty} \left|\frac{a_{k+1}}{a_k}\right| = \lim_{k\to\infty} \left|\frac{4^{k+1}k!^2}{4^k(k+1)!^2}\right|$$
$$= \lim_{k\to\infty} \frac{4}{(k+1)^2} = 0 < 1.$$

45. Converges, by the Alternating Series Test.

47.

$$\left|\frac{\sin k}{k^{3/2}}\right| = \frac{\sin k|}{k^{3/2}} \leq \frac{1}{k^{3/2}}$$

because $|\sin k| \leq 1$ for all k. And $\displaystyle\sum_{k=1}^{\infty} \frac{1}{k^{3/2}}$ is a convergent p-series $\left(p = \dfrac{3}{2} > 1\right)$. So by the Comparison Test, $\displaystyle\sum_{k=1}^{\infty} \left|\frac{\sin k}{k^{3/2}}\right|$ converges, so the original series converges absolutely.

49. Using the Limit Comparison Test, $a_k = \dfrac{2}{(3+k)^p}$ and $b_k = \dfrac{1}{k^p}$, so

$$\lim_{k\to\infty} \frac{a_k}{b_k} = \lim_{k\to\infty} \frac{2k^p}{(3+k)^p} = 2 > 0$$

so the series $\displaystyle\sum_{k=1}^{\infty} \frac{2}{(3+k)^p}$ converges if and only if the p-series $\displaystyle\sum_{k=1}^{\infty} \frac{1}{k^p}$ converges, which happens when $p > 1$.

51. $|S - S_k| \le a_{k+1} = \dfrac{3}{(k+1)^2} \le 10^{-6}$, so $1732.05 \le k+1$, so $1731.05 \le k$. Hence $k = 1732$ terms.

53. $f(x) = \dfrac{1}{4+x} = \displaystyle\sum_{k=0}^{\infty} \frac{(-1)^k x^k}{4^{k+1}}$
which is a geometric series that converges when $\left|-\dfrac{x}{4}\right| < 1$, or $-4 < x < 4$. Thus, $r = 4$.

55. $f(x) = \dfrac{3}{3+x^2} = \displaystyle\sum_{k=0}^{\infty} \frac{(-x^2)^k}{3^k}$
is a geometric series that converges for $\left|-\dfrac{x^2}{3}\right| < 1$ so $x^2 < 3$ or $|x| < \sqrt{3}$, so $-\sqrt{3} < x < \sqrt{3}$. Thus, $r = \sqrt{3}$.

57. $\dfrac{1}{4+x} = \displaystyle\sum_{k=0}^{\infty} \frac{(-1)^k x^k}{4^{k+1}}$ with $r = 4$. By integrating both sides, we get

$$\int \frac{1}{4+x}\,dx = \sum_{k=0}^{\infty} \frac{(-1)^k}{4^{k+1}} \int x^k\,dx$$

so

$$\ln(4+x) = \sum_{k=0}^{\infty} \frac{(-1)^k x^{k+1}}{(k+1)4^{k+1}} + \ln 4$$

with $r = 4$.

59. The Ratio Test gives $|x| < 1$, so $|r| = 1$. Since the series diverges for both $x = -1$ and $x = 1$, the interval of convergence is $(-1, 1)$.

61. The Ratio Test gives $|x| < 1$, so the radius of convergence is $|r| = 1$. The series diverges for $x = -1$ and converges for $x = 1$. Thus the interval of convergence is $(-1, 1]$.

63. The Ratio Test gives that the series converges absolutely for all $x \in (-\infty, \infty)$. Thus the interval of convergence is $(-\infty, \infty)$.

65. The Ratio Test gives that $3|x-2| < 1$, and this is when $-\dfrac{1}{3} < x - 2 < \dfrac{1}{3}$ or $\dfrac{5}{3} < x < \dfrac{7}{3}$. The series diverges for both $x = \dfrac{5}{3}$ and $x = \dfrac{7}{3}$. So the interval of convergence is $\left(\dfrac{5}{3}, \dfrac{7}{3}\right)$.

67. $\sin x = \displaystyle\sum_{k=0}^{\infty} \frac{(-1)^k x^{2k+2}}{(2k+1)!}$

69. $P_4(x) = (x-1) - \dfrac{1}{2}(x-1)^2 + \dfrac{1}{3}(x-1)^3 - \dfrac{1}{4}(x-1)^4$

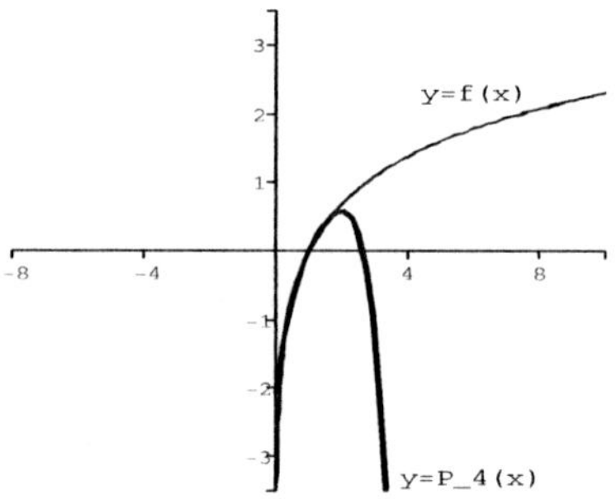

71. $|R_n(1.2)| = \left|\dfrac{f^{(n+1)}(z)(1.2-1)^{n+1}}{(n+1)!}\right|$ for some $z \in (1, 1.2)$.

Hence

$$|R_n(1.2)| < \frac{(n-1)!}{(n+1)!}(0.2)^{n+1}$$

and
$R_8(1.2) < 10^{-8}$. Consequently, $\ln(1.2) \approx P_8(1.2) \approx .1823215086$

73. Since $e^x = \displaystyle\sum_{k=0}^{\infty} \frac{x^k}{k!}$ for all x, we have

$$e^{-3x^2} = \sum_{k=0}^{\infty} \frac{(-3x^2)^k}{k!}$$

for all x, and so the radius of convergence is ∞.

75. $\int_0^1 \tan^{-1} x\, dx$

$= \int_0^1 x - \frac{1}{3}x^3 + \frac{1}{5}x^5 - \frac{1}{7}x^7 + \frac{1}{9}x^9\, dx$

$= \frac{x^2}{2} - \frac{x^4}{12} + \frac{x^6}{30} - \frac{x^8}{56} + \frac{x^{10}}{90}\Big|_0^1$

$= \frac{1117}{2520} \approx .4432539683$

Compare this estimation with the actual value

$$\int_0^1 \tan^{-1} x\, dx \approx .4388245732.$$

77. $a_0 = \frac{1}{2}\int_{-2}^{2} x\, dx = \frac{1}{2}\frac{x^2}{x}\Big|_{-2}^{2} = 0$

$a_k = \frac{1}{2}\int_{-2}^{2} x\cos\left(\frac{k\pi x}{2}\right) dx = 0$

$b_k = \frac{1}{2}\int_{-2}^{2} x\sin\left(\frac{k\pi x}{2}\right) dx$

$= -\frac{4}{k\pi}(-1)^k$

So

$$f(x) = \sum_{k=1}^{\infty} \frac{(-1)^{k+1}4}{k\pi} \sin\left(\frac{k\pi x}{2}\right).$$

for $-2 < x < 2$.

79.

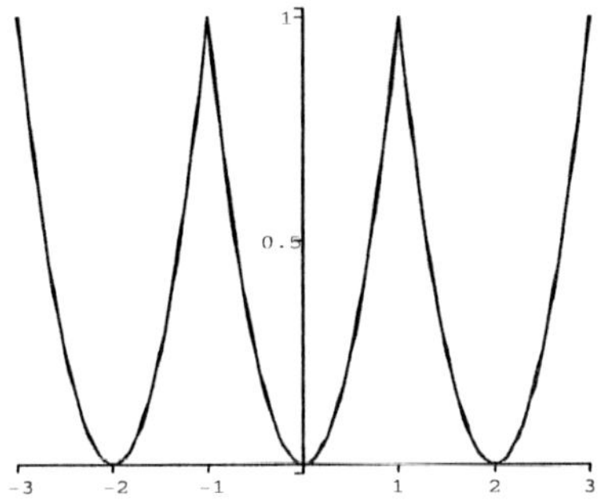

81.

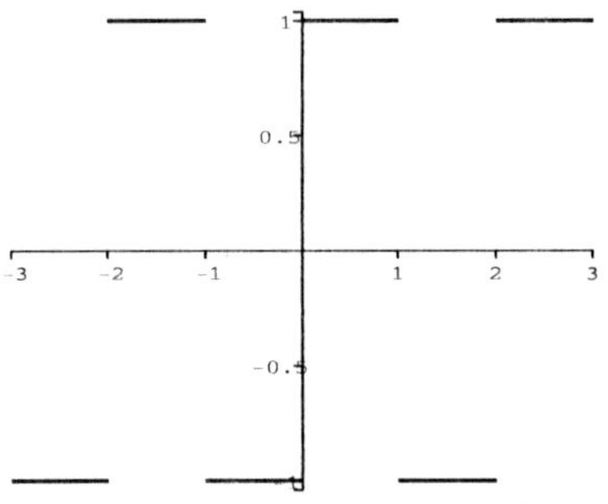

83.

$$\frac{1}{2} + \frac{1}{8} + \frac{1}{32} = \cdots = \sum_{k=0}^{\infty} \frac{1}{2}\left(\frac{1}{4}\right)^4$$

is a geometric series with $a = \frac{1}{2}$ and $|r| = \frac{1}{4} < 1$, so it converges to

$$\left(\frac{\frac{1}{2}}{1 - \frac{1}{4}}\right) = \left(\frac{\frac{1}{2}}{\frac{3}{4}}\right) = \frac{2}{3}.$$

85. $a_{n+1} = a_n + a_{n-1}$,

$\frac{a_{n+1}}{a_n} = \frac{a_n}{a_n} + \frac{a_{n-1}}{a_n} = 1 + \frac{a_{n-1}}{a_n}$

Let $r = \lim_{n\to\infty} \frac{a_{n+1}}{a_n}$, then take limit as $n \to \infty$ on both sides of the equation in above, we get

$$\lim_{n\to\infty} \frac{a_{n+1}}{a_n} = \lim_{n\to\infty} 1 + \lim_{n\to\infty} \frac{a_{n-1}}{a_n}$$

$r = 1 + \frac{1}{r}, r^2 - r - 1 = 0$

$r = \frac{1 \pm \sqrt{5}}{2}$

But r is the limit of the ratios of positive integers, so r cannot be negative, and this gives

$$r = \lim_{n\to\infty} \frac{a_{n+1}}{a_n} = \frac{1+\sqrt{5}}{2}.$$

87. Let $a_1 = 1 + \frac{1^2}{2}$ and so on,

n	a_n
1	1.5
2	1.153846154
3	1.381578947
4	1.141689373
5	1.344065167
6	1.218123800
7	1.325788792
8	1.229873907
9	1.314993721
10	1.237502842

Chapter 9

Parametric Equations and Polar Coordinates

9.1 Plane Curves and Parametric Equations

1. Equation: $\left(\frac{x}{2}\right)^2 + \left(\frac{y}{3}\right)^2 = 1$

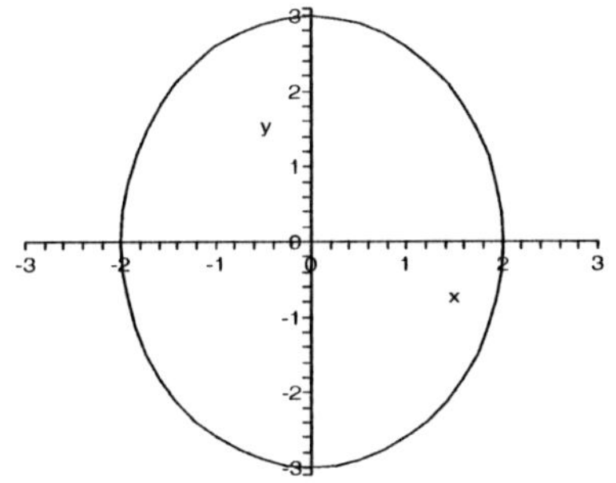

3. $t = \frac{y}{3}$

$x = -1 + 2\left(\frac{y}{3}\right)$

$\frac{2}{3}y = x + 1$

$y = \frac{3}{2}x + \frac{3}{2}$

This is a line, slope $3/2$, y-intercept $3/2$.

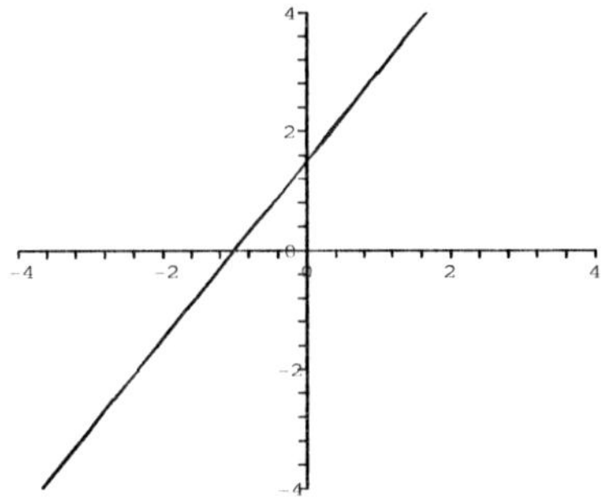

5. $t = x - 1$

$y = (x-1)^2 + 2$

$y = x^2 - 2x + 3$

This is a parabola, with vertex $(1, 2)$ opening up.

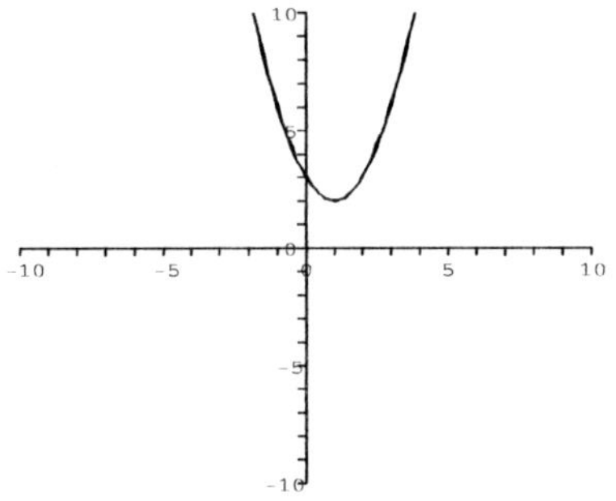

7. $t = \frac{y}{2}$

$x = \left(\frac{y}{2}\right)^2 - 1 = \frac{1}{4}y^2 - 1$

This is a parabola, vertex $(-1, 0)$ opening to the right.

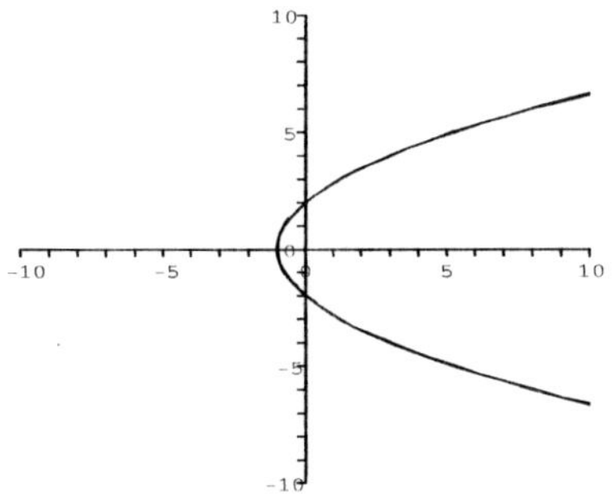

9. Equation: $y = 3x - 1$, $-1 \leq x \leq 1$

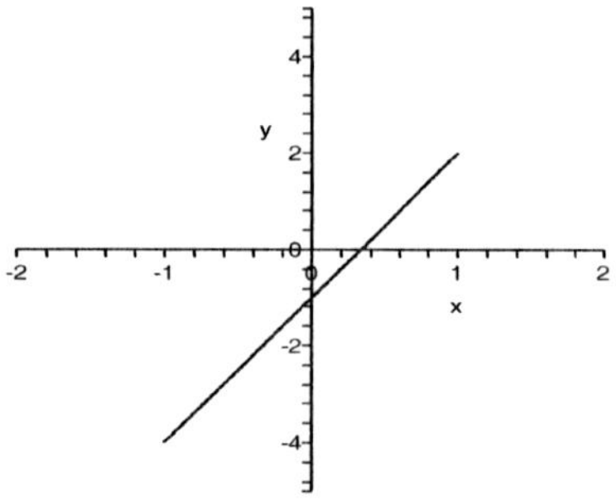

11. Graph is:

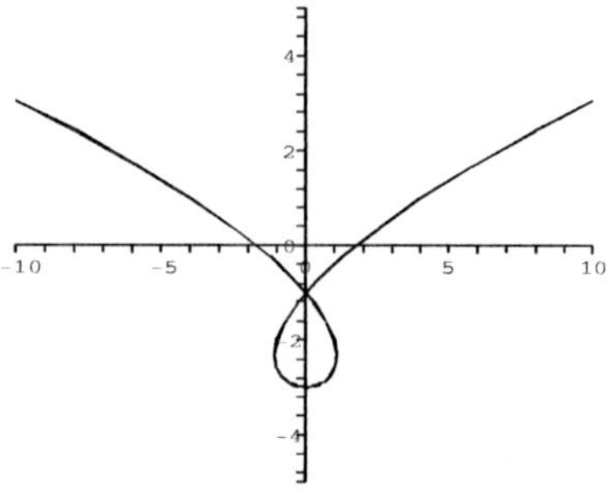

13. Graph is:

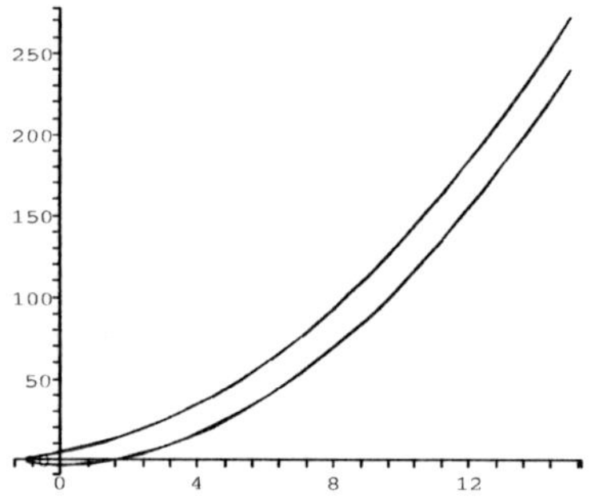

15. Graph is:

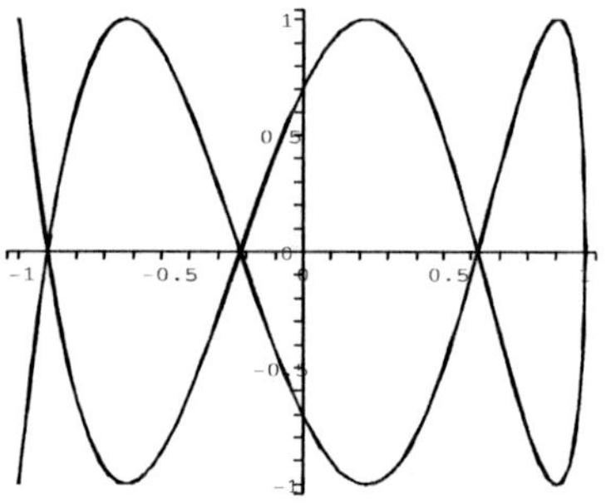

17. Graph is:

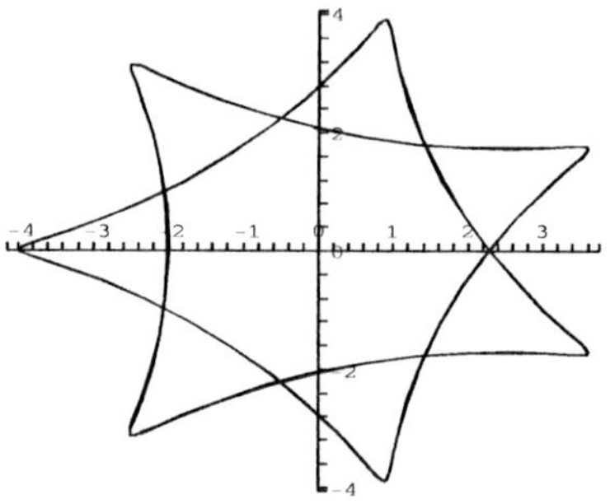

19. Graph is:

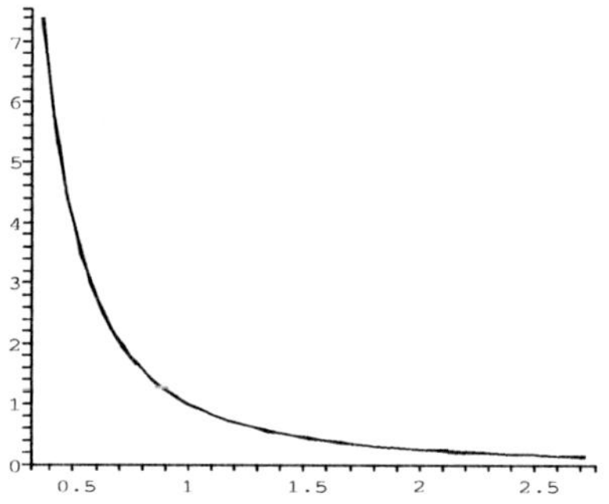

21. Integer values for k lead to closed curves, but irrational values for k do not.

23. Graph for $k = 2$:

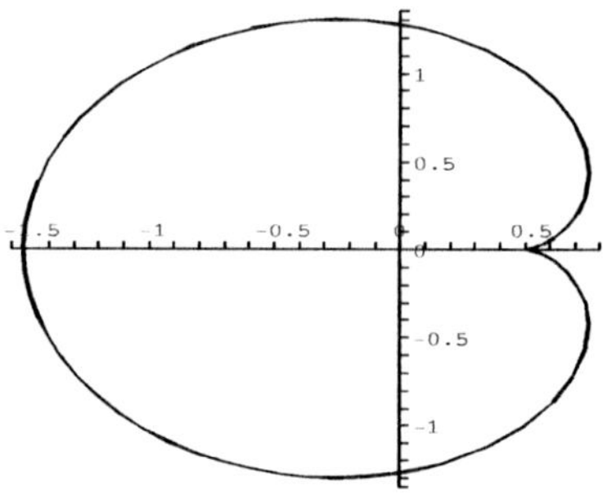

Graph for $k = 3$:

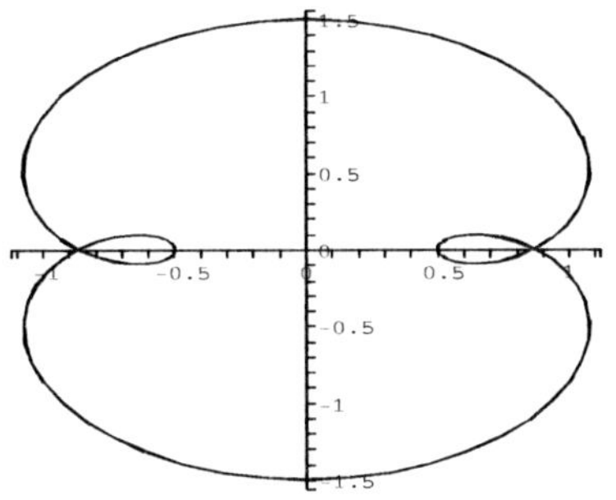

Graph for $k = 4$:

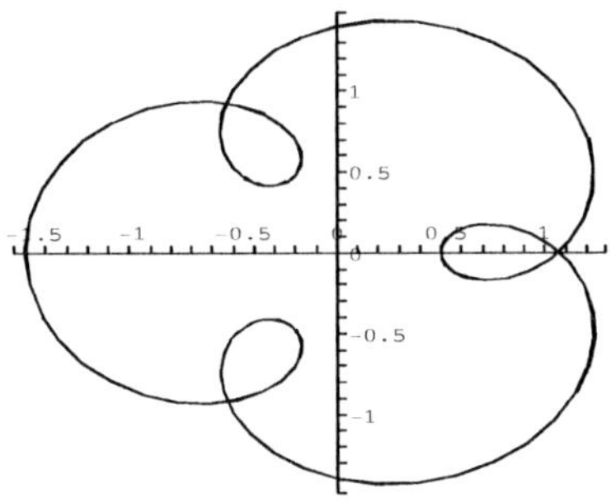

Graph for $k = 5$:

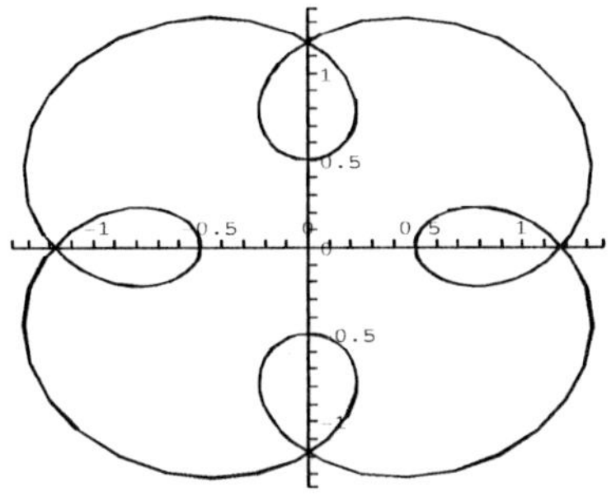

The graph has $k - 1$ "inner loops".

25. Since $(x + 1)^2 = y$, with $x \geq -1$, this is the right half of an upward-opening parabola. It has to be C.

27. x is bounded below by -1. y is bounded below by -1 and above by 1. From
$y = \sin(t) = \sin(\pm\sqrt{x+1})$
$= \pm\sin(\sqrt{x+1})$,
it has some of the features of a double sine curve, but the length of the cycles get longer as x increases. On the basis of this alone, it could be B or E, but the first y-intercept after $x = -1$ will be when
$\sqrt{x+1} = \pi$
$x = \pi^2 - 1 \approx 8.9$
and this is B.

29. x and y both oscillate between -1 and 1, but with different periods. This has to be A.

31. Use the model $x = a + tb$, $y = c + dt$, $(0 \leq t \leq 1)$, with $t = 0$ corresponding to $(0, 1)$ and $t = 1$ corresponding to $(3, 4)$.
$t = 0$: $0 = x = a$, $1 = y = c$
$t = 1$: $3 = x = 0 + b$, $4 = y = 1 + d$
So $b = 3$, $d = 3$, and the equations are
$x = 3t$, $y = 1 + 3t$.

33. Use the model $x = a + tb$, $y = c + dt$, $(0 \leq t \leq 1)$, with $t = 0$ corresponding to $(-2, 4)$ and $t = 1$ corresponding to $(6, 1)$.
$t = 0$: $-2 = x = a$, $4 = y = c$
$t = 1$: $6 = x = -2 + b$, $1 = y = 4 + d$
So $b = 8$, $d = -3$, and the equations are
$x = -2 + 8t$, $y = 4 - 3t$.

35. $x = t$ and $y = t^2 + 1$ from $t = 1$ to $t = 2$.

37. We set $x = -t + 2$ from $t = 0$ to $t = 2$ so that x travels from 2 to 0. So now we have $t = 2 - x$. Squaring this and taking the negative (to get the $-x^2$ in the formula for y) gives
$-t^2 = -(2 - x)^2 = -4 + 4x - x^2$.
There is no x term in the formula for y, so we will have to eliminate the x term above by adding $4t$. We now have
$-t^2 + 4t = -(2 - x)^2 + 4(2 - x)$
$= -4 + 4x - x^2 + 8 - 4x$
$= -x^2 + 4$

So now we see that we need to subtract 2 to get y. Our final equations are:
$x = -t + 2,\ y = -t^2 + 4t - 2$
from $t = 0$ to $t = 2$.

39. Use the model
$x = a + b\cos t,\ y = c + b\sin t$
from $t = 0$ to $t = 2\pi$
The center is at $(2, 1)$, therefore $a = 2$ and $c = 1$. The radius is 3, therefore $b = 3$, and the equations are

$$x = 2 + 3\cos t, y = 1 + 3\sin t.$$

41. We set the x-values equal to each other, and the y-values equal to each other, giving us a system of two equations with two unknowns:
$t = 1 + s$
$t^2 - 1 = 4 - s$
The first equation is already solved for t, so we plug this expression for t into the second equation and then solve for s:
$(1 + s)^2 - 1 = 4 - s$
$1 + 2s + s^2 - 1 = 4 - s$
$s^2 + 3s - 4 = 0$
$(s + 4)(s - 1) = 0$
So the two possible solutions for s are $s = -4$ and $s = 1$. Since $t = 1 + s$, the corresponding t-values are $t = -3$ and $t = 2$, respectively.

43. We must solve the system of equations:
$t + 3 = 1 + s$
$t^2 = 2 - s$
We plug $s = 2 - t^2$ into the first equation and solve for t:
$t + 3 = 1 + 2 - t^2$
$t^2 + t = 0$
$t(t + 1) = 0$
So $t = 0$ or $t = -1$. Since $s = 2 - t^2$, the corresponding s-values are $s = 2$ and $s = 1$, respectively.

45. The missile from example 1.9 follows the path
$$\begin{cases} x = 100t \\ y = 80t - 16t^2 \end{cases}$$
for $0 \le t \le 5$. Set the x-values equal:
$100t = 500 - 500(t - 2) = 1500 - 500t$
$600t = 1500$
$t = 5/2$
Now check the y-values at $t = 5/2$:

$$80(\tfrac{5}{2}) - 16(\tfrac{5}{2})^2 = 200 - 100 = 100$$
$$208(\tfrac{5}{2} - 2) - 16(\tfrac{5}{2} - 2)^2 = 104 - 4 = 100$$

Therefore the interceptor missile will hit its target at time $t = 5/2$.

47. The initial missile is launched at time $t = 0$ and the interceptor missile is launched 2 minutes later. Let T be the time variable for the interceptor missile. Then when $T = 0$, $t = 2$ and from then on $T = t - 2$.

49. At time t, the position of the sound wave, as described by these parametric equations, is a circle of radius t centered at the origin. This makes sense as the sound wave would propagate equally in all directions.

51. The graph will look as follows, where we have marked the position of the jet with a diamond:

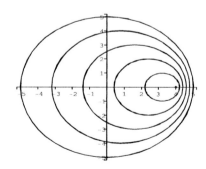

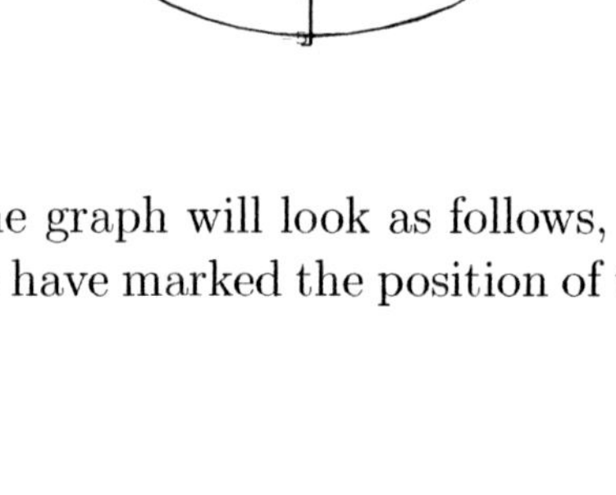

53. The graph will look as follows, where we have marked the position of the jet with a diamond:

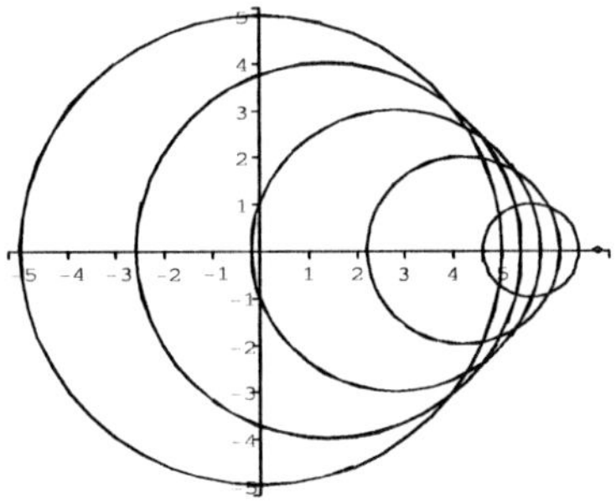

55. The shock waves are the lines formed by connecting the tops of the circles. In three dimensions, these lines will radiate out equally in all directions, i.e., will have circular cross sections. The three-dimensional figure formed by revolving the shock wave lines about the x-axis is a cone, which has circular cross sections, as expected.

57. Since distance = rate · time, the distance from the point $(0, D)$ to the position of the object at (x, y) is $v \cdot t$. We then find:

$$\begin{cases} x = (v \sin\theta)t \\ y = D - (v\cos\theta)t \end{cases}$$

59.

$$\begin{aligned} h(t) &= \lim_{dt\to 0} \frac{x(t+dt) - x(t)}{\Delta T} \\ &= \lim_{dt\to 0} \frac{v\sin\theta(t+dt) - v\sin\theta t}{\Delta T} \\ &= \lim_{dt\to 0} \frac{v\sin\theta dt}{\Delta T} \\ &= \lim_{dt\to 0} \frac{v\sin\theta}{\Delta T/dt} \\ &= \frac{v\sin\theta}{\lim_{dt\to 0}\frac{\Delta T}{dt}} \\ &= \frac{v\sin\theta}{T'(t)} \end{aligned}$$

Since $T = t + L(t)$, then $T'(t) = 1 + L'(t)$. Thus,

$$h(t) = \frac{v\sin\theta}{1 + L'(t)}.$$

61. From exercise 60,

$$h(0) = \frac{cv\sin\theta}{c - v\cos\theta}.$$

The maximum value of $h(0)$ occurs when $(d/d\theta)h(0) = 0$:

$$\begin{aligned} \frac{1}{cv}\frac{d}{d\theta}[h(0)] &= \\ &= \frac{(c - v\cos\theta)\cos\theta - \sin\theta(v\sin\theta)}{(c - v\cos\theta)^2} \\ &= \frac{c\cos\theta - v\cos^2\theta - v\sin^2\theta}{(c - v\cos\theta)^2} \\ &= \frac{c\cos\theta - v(\cos^2\theta + \sin^2\theta)}{(c - v\cos\theta)^2} \\ &= \frac{c\cos\theta - v}{(c - v\cos\theta)^2} \end{aligned}$$

Thus,

$$\frac{d}{d\theta}[h(0)] = 0 \Leftrightarrow 0 = c\cos\theta - v$$

$$\Leftrightarrow \cos\theta = \frac{v}{c}$$

Thus, the maximum value of $h(0)$ occurs when $\cos\theta = v/c$. When $\cos\theta = v/c$,

$$\begin{aligned} \sin\theta &= \sqrt{1 - \cos^2\theta} \\ &= \sqrt{1 - \tfrac{v^2}{c^2}} \\ &= \gamma^{-1} \end{aligned}$$

The maximum value of $h(0)$ is

$$\begin{aligned} h(0)_{\max} &= \frac{cv\gamma^{-1}}{c - v(v/c)} \\ &= \frac{v\gamma^{-1}}{1 - \frac{v^2}{c^2}} \\ &= \frac{v\gamma^{-1}}{\gamma^{-2}} \\ &= v\gamma \end{aligned}$$

63.

$$\begin{cases} x = \cos 2t \\ y = \sin t \end{cases}$$

We have the identity $\cos 2t = 1 - 2\sin^2 t$, so we have the x-y equation $x = 1 - 2y^2$.

This shows that the graph is part of a parabola, with vertex $(1, 0)$ and opening to the left. However the moving point never goes left of $x = -1$, forever going back and forth on this parabola, making an about-face every time x reaches -1.

$$\begin{cases} x = \cos t \\ y = \sin 2t \end{cases}$$

We have the identity $\sin 2t = 2\cos t \sin t$ so
$\sin^2 2t = 4\cos^2 t \sin^2 t$
$= 4\cos^2 t(1 - \cos^2 t)$
Therefore $y^2 = 4x^2(1 - x^2)$ or $y = \pm 2x\sqrt{1 - x^2}$.

This is a considerably more complicated graph, a figure-eight in which the moving point cycles smoothly around, starting out (when $t = 0$) by moving upward from $(1, 0)$ and completing the figure every time t passes an integral multiple of 2π. The moving point passes the origin twice during each cycle (first when $t = \pi/2$ and later when $t = 3\pi/2$).

65. The plot (including both pieces) is:

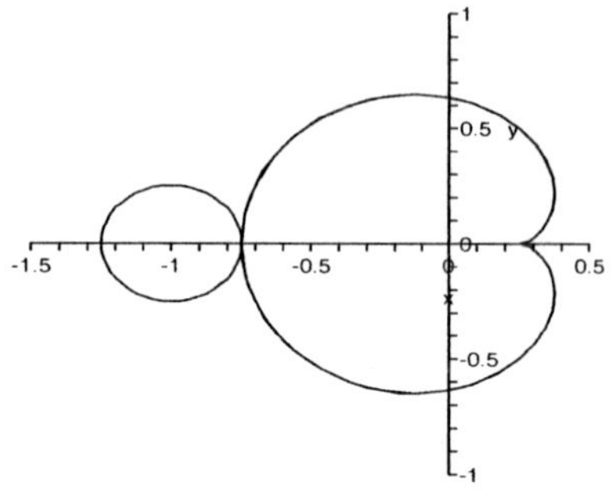

9.2 Calculus and Parametric Equations

1. $x'(t) = 2t$
$y'(t) = 3t^2 - t$

(a) $\dfrac{3(-1)^2 - 1}{2(-1)} = -1$

(b) $\dfrac{3(1)^2 - 1}{2(1)} = 1$

(c) We need to solve the system of equations:
$t^2 - 2 = -2$
$t^3 - t = 0$
The only solution to the first equation is $t = 0$. As this is also a solution to the second equation, this solution will work (and is the only solution). The slope at $(-2, 0)$ (i.e., the slope at $t = 0$) is:

$$\frac{3(0)^2 - 1}{2(0)} = \frac{-1}{0}$$

So we see that there is a vertical tangent line at this point.

3. $x'(t) = -2\sin t$
$y'(t) = 3\cos t$

(a) $\dfrac{3\cos(\pi/4)}{-2\sin(\pi/4)} = \dfrac{-3}{2}$

(b) $\dfrac{3\cos(\pi/2)}{-2\sin(\pi/2)} = \dfrac{0}{-2} = 0$

(c) We need to solve the system of equations:
$2\cos t = 0$
$3\sin t = 3$
The solutions to the first equation are $t = \pi/2 + k\pi$ for any

integer k. The solutions to the second equation are $t = \pi/2 + 2k\pi$ for any integer k. So the values of t that solve both are $t = \pi/2 + 2k\pi$. From (b), we see that these values all give slope 0.

5. $x'(t) = -2\sin 2t$
$y'(t) = 4\cos 4t$

(a) $\dfrac{4\cos(\pi)}{-2\sin(\pi/2)} = \dfrac{-4}{-2} = 2$

(b) $\dfrac{4\cos(2\pi)}{-2\sin(\pi)} = \dfrac{4}{0}$
Thus there is a vertical tangent line at this point.

(c) We need to solve the system of equations:
$\cos 2t = \sqrt{2}/2$
$\sin 4t = 1$
The first equation requires that $2t = \pm\pi/4 + 2\pi k$ or $t = \pm\pi/8 + \pi k$ for any integer k. The second equation requires that $4t = \pi/2 + 2\pi k$ or $t = \pi/8 + \pi k/2$ for any integer k. Thus we have $t = \pi/8 + \pi k$. At these values of t, we have slope:
$$\frac{4\cos(4(\pi/8 + \pi k))}{-2\sin(2(\pi/8 + \pi k))} = \frac{4\cos(\pi/2)}{-2\sin(\pi/4)} = \frac{0}{-\sqrt{2}} = 0$$

7. $x'(t) = 2t$
$y'(t) = 3t^2 - 1$
We must solve the system of equations:
$t^2 - 2 = -2$
$t^3 - t = 0$
The solutions to the first equation are $t = \pm 1$. The solutions to the second are $t = 0$ and $t = \pm 1$, so $t = \pm 1$ are the solutions to the system of equations. These are the values at which we must find slopes.

$$t = 1: \quad \frac{3-1}{2(1)} = 1$$
$$t = -1: \quad \frac{3-1}{2(-1)} = -1$$

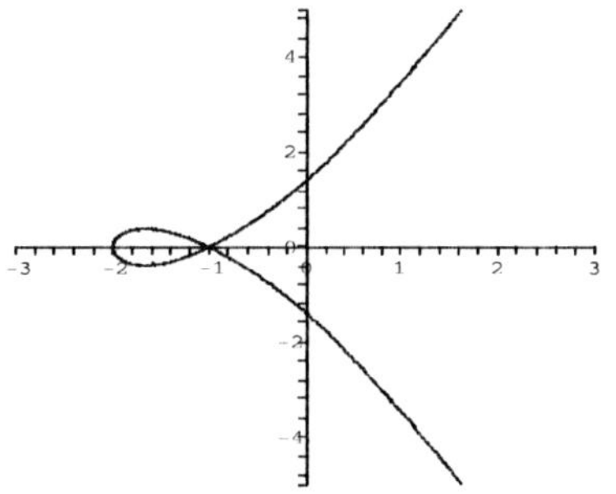

9. $x'(t) = -2\sin 2t$
$y'(t) = 4\cos 4t$

(a) We need $\cos 4t = 0$, so $4t = \pi/2 + k\pi$ or $t = \pi/8 + k\pi/4$ for any integer k. We need to check that the denominator is not 0 at these values.
$\sin(2(\pi/8 + k\pi/4))$
$= \sin(\pi/4 + k\pi/2)$.
This is nonzero for all k, so all of these t values give horizontal tangent lines. We plug these t values into the expressions for x and y to find the corresponding points. Depending on the value of k, we get the points $(\sqrt{2}/2, 1)$, $(-\sqrt{2}/2, -1)$, $(-\sqrt{2}/2, 1)$, and $(\sqrt{2}/2, -1)$.

(b) We need $\sin 2t = 0$, so $2t = k\pi$ or $t = k\pi/2$ for any integer k. We need to check that the numerator is not 0 at these values.
$$\cos(4(k\pi/2)) = \cos(2k\pi) \neq 0$$
The corresponding points are $(1, 0)$ and $(-1, 0)$.

11. $x'(t) = 2t$
$y'(t) = 4t^3 - 4$

(a) We want $4(t^3 - 1) = 0$ so the only solution is $t = 1$, which works since the denominator $(2t)$ is not zero for this t. This t-value corresponds to the x-y point $(0, -3)$.

(b) We want $2t = 0$ so $t = 0$, which does indeed give an undefined slope as the numerator will be -4. The corresponding point is $(-1, 0)$.

13. $x'(t) = -2\sin t + 2\cos 2t$
$y'(t) = 2\cos t - 2\sin 2t$

(a) We are looking for values of t for which $y'(t) = 0$, i.e., we want $\cos t = \sin 2t$.
Since $\sin 2t = 2\sin t \cos t$, we need $\cos t = 2\sin t \cos t$. This happens when $\cos t = 0$ (when $t = \pi/2 + k\pi$ for any integer k) or when $\sin t = 1/2$. This occurs when $t = \pi/5 + 2k\pi$ or $t = 5\pi/6 + 2k\pi$ for any integer k.

(b) We are looking for values of t for which $x'(t) = 0$, i.e., we want $\sin t = \cos 2t$.
Since $\cos 2t = 1 - 2\sin^2 t$, we need to solve
$2\sin^2 t + \sin t - 1 = 0$.
Using the quadratic formula (with $\sin t$ replacing the usual x), we see that $\sin t = 1/2$ or $\sin t = -1$. The former gives $t = \pi/5 + 2k\pi$ or $t = 5\pi/6 + 2k\pi$ for any integer k while the latter gives $t = 3\pi/2 + 2k\pi$ for any integer k.

15. $x' = -2\sin t$, $y' = 2\cos t$

(a) At $t = 0$, $x' = -2\sin 0 = 0$,
$y' = 2\cos 0 = 2$.
Speed $= \sqrt{0^2 + 2^2} = 2$.
Motion is up.

(b) At $t = \pi/2$, $x' = -2\sin(\pi/2) = -2$, $y' = 2\cos(\pi/2) = 0$.
Speed $= \sqrt{(-2)^2 + 0^2} = 2$.
Motion is to the left.

17. $x' = 20$, $y' = -2 - 32t$

(a) At $t = 0$, $x' = 20$, $y' = -2$.
Speed $= \sqrt{20^2 + (-2)^2} = 2\sqrt{101}$.
Motion is to the right and slightly down.

(b) At $t = 2$, $x' = 20$,
$y' = -2 - 64 = -66$.
Speed $= \sqrt{20^2 + (-66)^2}$
$= 2\sqrt{1189}$.
Motion is to the right and down.

19. $x' = -4\sin 2t + 5\cos 5t$,
$y' = 4\cos 2t - 5\sin 5t$

(a) At $t = 0$, $x' = -4\sin 0 + 5\cos 0 = 5$,
$y' = 4\cos 0 - 5\sin 0 = 4$.
Speed $= \sqrt{5^2 + 4^2} = \sqrt{41}$.
Motion is to the right and up.

(b) At $t = \pi/2$,
$x' = -4\sin \pi + 5\cos(5\pi/2) = 0$,
$y' = 4\cos \pi - 5\sin(5\pi/2) = -9$.
Speed $= \sqrt{0^2 + (-9)^2} = 9$.
Motion is down.

21. $x'(t) = -3\sin t$ and this curve is

traced out counterclockwise, so

$$\begin{aligned}A &= -\int_0^{2\pi} 2\sin t(-3\sin t)\,dt\\ &= 6\int_0^{2\pi} \sin^2 t\,dt\\ &= 6\left(\frac{1}{2}t - \frac{1}{2}\sin t\cos t\right)\Big|_0^{2\pi}\\ &= 6\pi\end{aligned}$$

23. $x'(t) = -\frac{1}{2}\sin t + \frac{1}{2}\sin 2t$ and this curve is traced out counterclockwise, so

$$\begin{aligned}A &= -\int_0^{2\pi} y(t)x'(t)\,dt\\ &\approx 1.178\end{aligned}$$

where $y(t) = -\frac{1}{2}\sin t + \frac{1}{2}\sin 2t$.

25. $x'(t) = -\sin t$ and this curve is traced out clockwise, so

$$\begin{aligned}A &= \int_{\pi/2}^{3\pi/2} \sin 2t(-\sin t)\,dt\\ &= -2\int_0^{2\pi} \sin^2 t\cos t\,dt\\ &= -\frac{2}{3}\sin^3 t\Big|_{\pi/2}^{3\pi/2}\\ &= -\frac{2}{3}((-1)^3 - 1^3) = \frac{4}{3}\end{aligned}$$

27. $x'(t) = 3t^2 - 4$ and this curve is traced out counterclockwise, so

$$\begin{aligned}A &= -\int_{-2}^{2} (t^2-3)(3t^2-4)\,dt\\ &= -\int_{-2}^{2} (3t^4 - 13t^2 + 12)\,dt\\ &= -\left[\frac{3t^5}{5} - \frac{13t^3}{3} + 12t\right]\Big|_{-2}^{2}\\ &= -\frac{256}{15}\end{aligned}$$

29. Crossing x − axis $\Rightarrow y = 0 \Rightarrow t = n\pi$
$x' = -4\cos t\sin t - 2\sin t$,
$y' = 2\sin^2 t + 2(1-\cos t)\cos t$

At $t = n\pi$, n even, $(x,y) = (3,0)$ and $x' = 0$, $y' = 0$, speed $= 0$.
At $t = n\pi$, n odd, $(x,y) = (-1,0)$ and $x' = 0$, $y' = -4$, speed $= 4$.

31. The $3t$ and $5t$ indicate a ratio of 5-to-3.

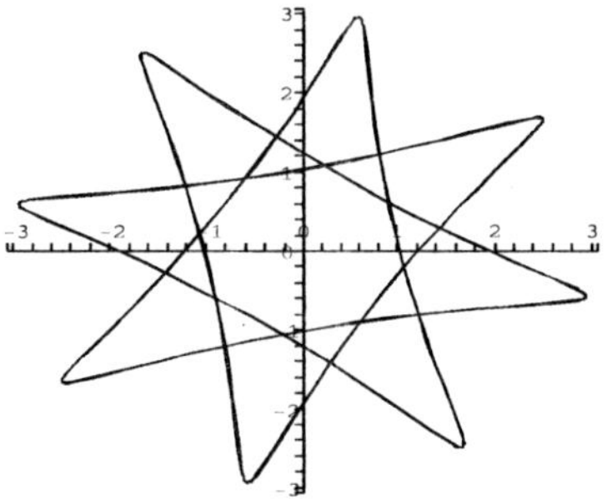

33. Use $x = 2\cos t + \sin 3t$,
$y = 2\sin t + \cos 3t$.

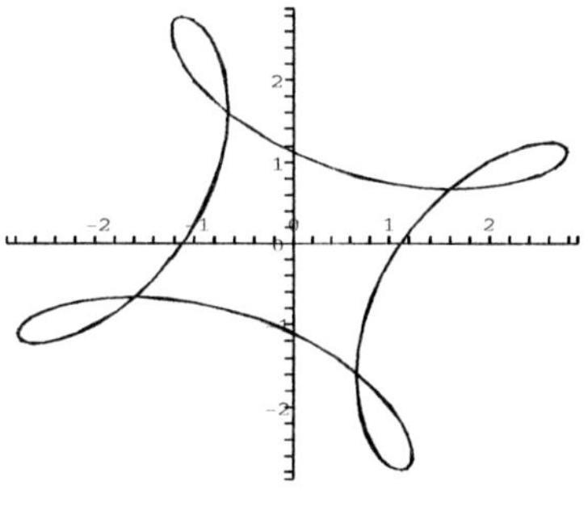

$x' = -2\sin t + 3\cos 3t$
$y' = 2\cos t - 3\sin 3t$
$s(t) =$
$\sqrt{(x'(t))^2 + (y'(t))^2}$
$= \sqrt{4 + 9 - 12(\sin t\cos 3t + \cos t\sin 3t)}$
$= \sqrt{13 - 12\sin 4t}$
Minimum speed $= \sqrt{13-12} = 1$;
maximum speed $= \sqrt{13+12} = 5$.

35. $x' = 4\cos 4t$, $y' = 4\sin 4t$

$$\begin{aligned}s(t) &= \sqrt{(4\cos 4t)^2 + (4\sin 4t)^2}\\ &= \sqrt{16(\sin^2 4t + \cos^2 4t)}\end{aligned}$$

$= 4$

The slope of the tangent line is $y'/x' = \tan 4t$, and the slope of the origin-to-object line is $y/x = -\cot 4t$, and the product of the two slopes is -1.

37. $\dfrac{y(t)}{x(t)} = \tan\theta$, where θ is the angle the object makes with the observer at the origin and the positive x axis. The derivative

$$\left[\frac{y(t)}{x(t)}\right]' = \sec^2\theta\frac{d\theta}{dt}$$

will be positive if $\dfrac{d\theta}{dt}$ is positive, and the object will be moving counter-clockwise. If the derivative is negative, then $\dfrac{d\theta}{dt}$ is negative, and the object will be moving clockwise.

39. We decide for convenience that we will have the circle rolling out on the positive x-axis with the center starting at $(0, r)$. The center moves out in proportion to time, i.e., $x_c = vt$, $y_c = r$ (constant).

Now, we have seen by example that if the circle were simply rotating in place, the motion of the given point relative to the center could be described by $x = r\cos(-2\pi t)$, $y = r\sin(-2\pi t)$. (The minus sign is there to accommodate *clockwise* rotation induced by the circle rolling *forward*, the factor of 2π is present to yield *one* revolution in the first unit of time.). In summary, the path (x_p, y_p) of the selected point on the rim satisfies

$$\begin{cases} x_p - x_c = r\cos(-2\pi t) \\ \qquad\quad\;\; = r\cos(2\pi t), \\ y_p - y_c = r\sin(-2\pi t) \\ \qquad\quad\;\; = -r\sin(2\pi t)). \end{cases}$$

The final instructions are to *add* these numbers to the coordinates of the center. We can see why. It produces

$$\begin{cases} x_p = x_c + (x_p - x_c) \\ \quad\; = vt + r\cos(2\pi t), \\ y_p = y_c + (y_p - y_c) \\ \quad\; = r - r\sin(2\pi t). \end{cases}$$

To find the speed, we'll need $x'(t) = v - 2\pi r\sin(2\pi t)$ and $y'(t) = -2\pi r\cos(2\pi t)$. Plugging these in and simplifying we find that the speed is given by

$$\begin{aligned} s(t) &= \sqrt{(x'(t))^2 + (y'(t))^2} \\ &= \sqrt{v^2 - 4\pi rv\sin(2\pi t) + 4\pi^2 r^2}. \end{aligned}$$

To find the extrema, we take the derivative:

$$s'(t) = \frac{-4\pi^2 rv\cos(2\pi t)}{\sqrt{v^2 - 4\pi rv\sin(2\pi t) + 4\pi^2 r^2}}.$$

This is equal to 0 when $\cos(2\pi t) = 0$, i.e., when $2\pi t = \pi/2 + k\pi$ or $t = 3k/4$ for any integer k. If k is even, $s(t) = \sqrt{v^2 + 4\pi^2 r^2}$. If k is odd, $s(t) = \sqrt{v^2 \pm 4\pi rv + 4\pi^2 r^2}$, with the sign ambiguity depending on k.

The graph for $v = 3$ and $r = 2$:

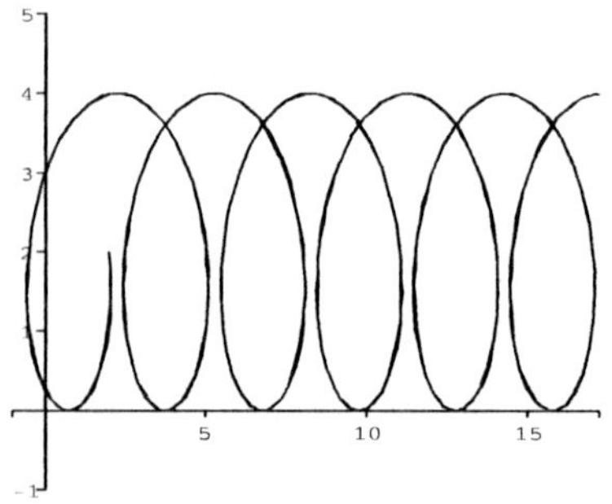

41. There are an infinite number of solutions, depending on the initial position of the small circle, and the direction of motion. Assume that the center of the small circle begins at $(a-b, 0)$ and moves clockwise. Then the path of the center is $((a-b)\cos t, (a-b)\sin t)$. The path of a point about the center of the small circle is $(b\cos\theta, -b\sin\theta)$. To determine the relationship between θ and t, observe that when the small circle rotates through an angle θ, the distance $b\theta$ that its circumference moves matches up with an arc of length at on the large circle. Thus, $b\theta = at$, and so $\theta = \frac{a}{b}t$.

Thus, the path traced out by a point on the small circle is

$$\begin{cases} x = (a-b)\cos t + b\cos\left(\frac{a}{b}t\right) \\ y = (a-b)\sin t - b\sin\left(\frac{a}{b}t\right) \end{cases}$$

If $a = 2b$, then the path of a point on the small circle is

$$(b\cos t + b\cos 2t, b\sin t - b\sin 2t)$$

The path now passes through the origin.

The graph for $a = 5$ and $b = 3$:

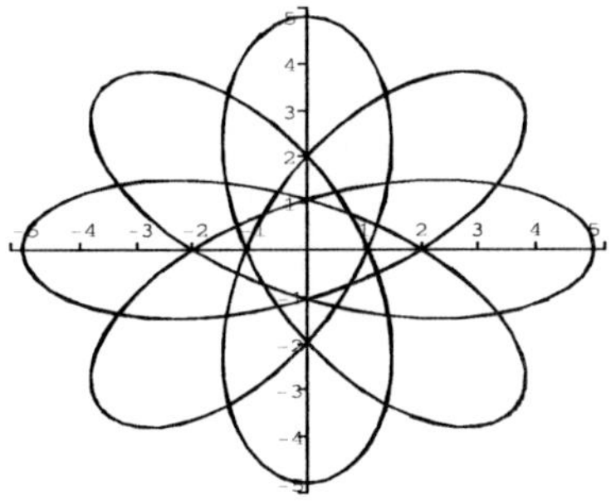

The equation for the slope of the tangent line at t is:

$$\left.\frac{dy}{dx}\right|_t = \frac{\frac{dy}{dt}(t)}{\frac{dx}{dt}(t)}$$
$$= \frac{(a-b)\cos t - a\cos(\frac{a}{b}t)}{-(a-b)\sin t - a\sin(\frac{a}{b}t)}.$$

One point at which the tangent line is vertical is when $t = 0$, i.e. at the point $(a, 0)$.

43. Let $x = 2\cos t$ and $y = 2\sin t$. Then,

$$\begin{aligned} \frac{dx}{dt} &= -2\sin t \\ \frac{d^2x}{dt^2} &= -2\cos t \\ \frac{d^2x}{dt^2}(\pi/6) &= -2\frac{\sqrt{3}}{2} = -\sqrt{3} \\ \frac{dy}{dt} &= 2\cos t \\ \frac{d^2y}{dt^2} &= -2\sin t \\ \frac{d^2y}{dt^2}(\pi/6) &= -2\frac{1}{2} = -1 \\ \frac{\frac{d^2y}{dt^2}(\pi/6)}{\frac{d^2x}{dt^2}(\pi/6)} &= \frac{-1}{-\sqrt{3}} \\ &= \frac{\sqrt{3}}{3} \end{aligned}$$

Now, eliminating the parameter t to obtain an equation involving only the variables x and y, we get

$$\begin{aligned} \left(\frac{x}{2}\right)^2 + \left(\frac{y}{2}\right)^2 &= \cos^2 t + \sin^2 t \\ \frac{x^2}{4} + \frac{y^2}{4} &= 1 \\ x^2 + y^2 &= 4 \end{aligned}$$

Differentiating implicitly with respect to x:

$$\begin{aligned} 2x + 2yy' &= 0 \\ x + yy' &= 0 \quad (*) \\ 1 + y'y' + yy'' &= 0 \quad (**) \end{aligned}$$

Solving equation (*) for y', we get $y' = -x/y$. Substituting the result into equation (**), we get

$$1+\left(\frac{-x}{y}\right)\left(\frac{-x}{y}\right)+yy''=0$$
$$1+\frac{x^2}{y^2}+yy''=0$$
$$y^2+x^2+y^3y''=0$$
$$4+y^3y''=0$$
$$y''=-\frac{4}{y^3}$$

Thus,

$$\frac{d^2y}{dx^2}\left(\sqrt{3},1\right)=-\frac{4}{1^3}=-4$$

9.3 Arc Length and Surface Area in Parametric Equations

1. $x'(t)=-2\sin t$
$y'(t)=4\cos t$

$$\begin{aligned} s &= \int_0^{2\pi}\sqrt{(-2\sin t)^2+(4\cos t)^2}\,dt \\ &= \int_0^{2\pi}\sqrt{4\sin^2 t+16\cos^2 t}\,dt \\ &\approx 19.3769 \end{aligned}$$

3. $x'(t)=3t^2-4$
$y'(t)=2t$

$$\begin{aligned} s &= \int_{-2}^{2}\sqrt{(3t^2-4)^2+(2t)^2}\,dt \\ &\approx 15.6940 \end{aligned}$$

5. $x'(t)=-4\sin 4t$
$y'(t)=4\cos 4t$

$$\begin{aligned} s &= \int_0^{\pi/2}\sqrt{(-4\sin 4t)^2+(4\cos 4t)^2}\,dt \\ &= \int_0^{\pi/2}\sqrt{16\sin^2 4t+16\cos^2 4t}\,dt \\ &= \int_0^{\pi/2}\sqrt{16}\,dt \\ &= \int_0^{\pi/2}4\,dt = 4\left(\frac{\pi}{2}\right)=2\pi \end{aligned}$$

7. $x'(t)=\cos t-t\sin t$
$y'(t)=t\cos t+\sin t$

$$\begin{aligned} s &= \int_{-1}^{1}\sqrt{(y'(t))^2+(x'(t))^2}\,dt \\ &= \int_{-1}^{1}\sqrt{t^2+1}\,dt \\ &= 2\int_0^1\sqrt{t^2+1}\,dt \\ &= \sqrt{2}+\ln(1+\sqrt{2})\approx 2.29559 \end{aligned}$$

9. $x'(t)=2\cos 2t\cos t-\sin 2t\sin t$
$y'(t)=2\cos 2t\sin t+\sin 2t\cos t$

Expanding and then combining like terms gives

$(x'(t))^2+(y'(t))^2=4\cos^2 2t+\sin^2 2t$

so

$$\begin{aligned} s &= \int_0^{\pi/2}((x'(t))^2+(y'(t))^2)^{1/2}\,dt \\ &= \int_0^{\pi/2}(4\cos^2 2t+\sin^2 2t)^{1/2}\,dt \\ &\approx 2.422 \end{aligned}$$

11. $x'(t)=\cos t$
$y'(t)=\pi\cos\pi t$

$$\begin{aligned} s &= \int_0^{\pi}(\cos^2 t+\pi^2\cos^2\pi t)^{1/2}\,dt \\ &\approx 6.914 \end{aligned}$$

13. $x(0) = 0$, $y(0) = 0$, $x(1) = \pi$, $y(1) = 2$
$x'(t) = \pi$
$y'(t) = t^{-1/2}$

$$\begin{aligned} T &= \int_0^1 k\sqrt{\frac{\pi^2 + 1/t}{2\sqrt{t}}}\,dt \\ &= \int_0^1 k\left(\frac{\pi^2}{2t^{1/2}} + \frac{t^{-3/2}}{2}\right)^{1/2} dt \\ &\approx k \cdot 4.486 \end{aligned}$$

15. $x(0) = -\frac{1}{2}\pi(\cos 0 - 1) = 0$,
$y(0) = 0 + \frac{7}{10}\sin 0 = 0$,
$x(1) = -\frac{1}{2}\pi(\cos\pi - 1) = \pi$,
$y(1) = 2 + \frac{7}{10}\sin\pi = 2$

$x'(t) = \frac{\pi^2}{2}\sin\pi t$
$y'(t) = 2 + \frac{7}{10}\pi\cos\pi t$

$$T = \int_0^1 k\sqrt{\frac{(x'(t))^2 + (y'(t))^2}{y(t)}}\,dt$$

Here,

$$\frac{(x'(t))^2 + (y'(t))^2}{y(t)} = \frac{\frac{\pi^4}{4}\sin^2\pi t + 4 + \frac{14}{5}\pi\cos\pi t}{2t + \frac{7}{10}\sin\pi t} + \frac{\frac{49}{100}\pi^2\cos^2\pi t}{2t + \frac{7}{10}\sin\pi t}$$

so $T \approx k \cdot 4.457$.

17. Slope is $\dfrac{t^{-1/2}}{\pi}$; so at $t = 0$ the slope is undefined.

$$s = \int_0^1 \left(\pi^2 + \frac{1}{t}\right)^{1/2} dt \approx 3.8897$$

The time for this path was approximately $k \cdot 4.486$ which is slightly slower than the time for the cycloid path, which was approximately $k \cdot 4.443$.

19. Slope is

$$\frac{2 + \frac{7}{10}\cos\pi t}{\frac{1}{2}\pi^2\sin\pi t};$$

at $t = 0$ the slope is undefined, so there is a vertical tangent line. We have

$$s = \int_0^1 \sqrt{f(t)}\,dt$$

where

$$f(t) = \frac{1}{4}\pi^4\sin^2\pi t + \left(2 + \frac{7}{10}\cos\pi t\right)^2$$

so $s \approx 3.87$.
The time for this path was approximately $k \cdot 4.457$ which is just slightly slower than the time for the cycloid path, which was approximately $k \cdot 4.443$.

21. $x'(t) = 2t$
$y'(t) = 3t^2 - 4$

$$\begin{aligned} s &= \int_{-2}^0 2\pi|t^3 - 4t|\sqrt{(2t)^2 + (3t^2 - 4)^2}\,dt \\ &= \int_{-2}^0 2\pi|t^3 - 4t|\sqrt{9t^4 - 20t^2 + 16}\,dt \\ &\approx 85.823 \end{aligned}$$

23. $x'(t) = 2t$
$y'(t) = 3t^2 - 4$

$$\begin{aligned} s &= \int_{-1}^1 2\pi|t^2 - 1|\sqrt{(2t)^2 + (3t^2 - 4)^2}\,dt \\ &\approx 29.696 \end{aligned}$$

25. $x'(t) = 3t^2 - 4$
$y'(t) = 2t$

$$\begin{aligned} s &= \int_0^2 2\pi|t^3 - 4t|\sqrt{(3t^2 - 4)^2 + (2t)^2}\,dt \\ &\approx 85.823 \end{aligned}$$

27. Parametric equations for the midpoint of the ladder are given by:

$$\begin{cases} x &= 4\sin\theta \\ y &= 4\cos\theta \end{cases}$$

for $0 \le \theta \le \pi/2$.

The distance, given by the formula for arc length is

$$\begin{aligned} s &= \int_0^{\pi/2} \sqrt{16\sin^2\theta + 16\cos^2\theta}\, d\theta \\ &= \int_0^{\pi/2} 4\, d\theta \\ &= 4\theta \Big|_0^{\pi/2} \\ &= 2\pi. \end{aligned}$$

29. $x'(t) = \cos(\pi t^2)$
$y'(t) = \sin(\pi t^2)$
Therefore, the arc length for $a \le t \le b$ is

$$\begin{aligned} &\int_a^b \sqrt{\left(\frac{dx}{dt}\right)^2 + \left(\frac{dy}{dt}\right)^2}\, dt \\ &= \int_a^b \sqrt{\cos^2(\pi t^2) + \sin^2(\pi t^2)} dt \\ &= \int_a^b dt \\ &= [t]_a^b \\ &= b - a \end{aligned}$$

Thus, the arc length of the curve is equal to the time interval; in other words, the parametrization being used is one for which the corresponding speed of a particle is 1. This means that as the size of the spiral decreases the spiralling rate increases.

In the special case that $a = -2\pi$ and $b = 2\pi$, the length of the curve is $b - a = 2\pi - (-2\pi) = 4\pi$.

9.4 Polar Coordinates

1. $r = 2$, $\theta = 0$
$x = 2\cos 0 = 2$,
$y = 2\sin 0 = 0$
Rectangular representation: $(2, 0)$

3. $r = -2$, $\theta = \pi$
$x = -2\cos\pi = 2$,
$y = -2\sin\pi = 0$
Rectangular representation: $(2, 0)$

5. $r = 3$, $\theta = -\pi$
$x = 3\cos(-\pi) = -3$,
$y = 3\sin(-\pi) = 0$
Rectangular representation: $(-3, 0)$

7. $r = \sqrt{2^2 + (-2)^2} = \pm\sqrt{8} = \pm 2\sqrt{2}$
$\tan\theta = \frac{-2}{2} = -1$
The point $(x, y) = (2, -2)$ is in Quadrant IV, so θ could be $-\pi/4$.

All polar representations:
$(2\sqrt{2}, -\frac{\pi}{4} + 2\pi n)$
or
$(-2\sqrt{2}, \frac{3\pi}{4} + 2\pi n)$
where n is any integer.

9. $r = \sqrt{0^2 + 3^2} = \pm 3$
Since the point $(0, 3)$ is on the positive y-axis, we can take $\theta = \pi/2$.

All polar representations:
$(3, \frac{\pi}{2} + 2\pi n)$
or
$(-3, \frac{3\pi}{2} + 2\pi n)$
where n is any integer.

11. $r = \sqrt{3^2 + 4^2} = \pm 5$
$\tan\theta = \frac{4}{3}$
The point $(x, y) = (3, 4)$ is in Quadrant I; $\theta = \tan^{-1}(\frac{4}{3}) \approx 0.9273$.

All polar representations:
$(5, \tan^{-1}(\frac{4}{3}) + 2\pi n)$
or
$(-5, \tan^{-1}(\frac{4}{3}) + \pi + 2\pi n)$
where n is any integer.

13. $r = 2$, $\theta = -\pi/3$
$x = 2\cos\left(-\frac{\pi}{3}\right) = 1$
$y = 2\sin\left(-\frac{\pi}{3}\right) = -\sqrt{3}$
Rectangular representation: $\left(1, -\sqrt{3}\right)$

15. $r = 0$, $\theta = 3$
$x = 0\cos(3) = 0$
$y = 0\sin(3) = 0$
Rectangular representation: $(0, 0)$

17. $r = 4$, $\theta = \pi/10$
$x = 4\cos\left(\frac{\pi}{10}\right) \approx 3.8042$
$y = 4\sin\left(\frac{\pi}{10}\right) \approx 1.2361$
Rectangular representation:
$(3.8042, 1.2361)$

19.

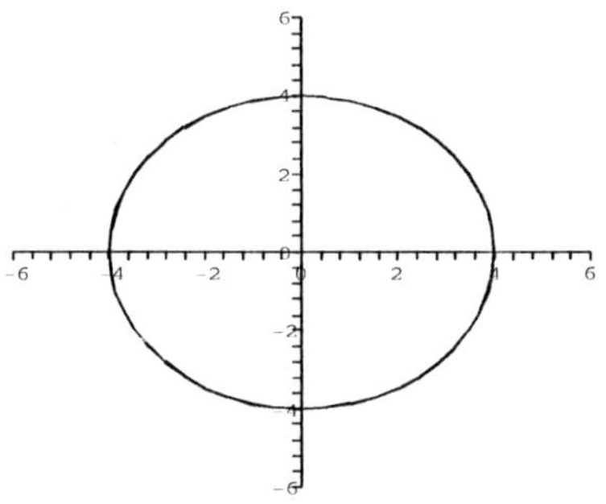

This is a circle centered at the origin with radius 4. The equation is

$$x^2 + y^2 = 4^2 = 16$$

21.

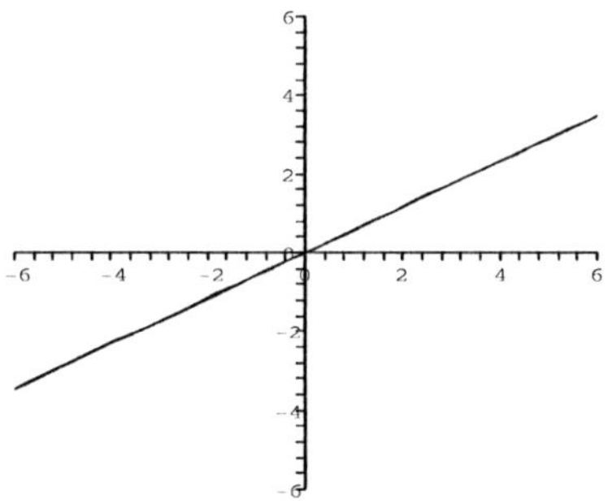

This is a line through the origin making an angle $\pi/6$ to the x-axis. We have

$$\frac{y}{x} = \tan\theta = \tan\frac{\pi}{6} = \frac{1}{\sqrt{3}}, \text{ so}$$

$$y = \frac{1}{\sqrt{3}}x$$

23.

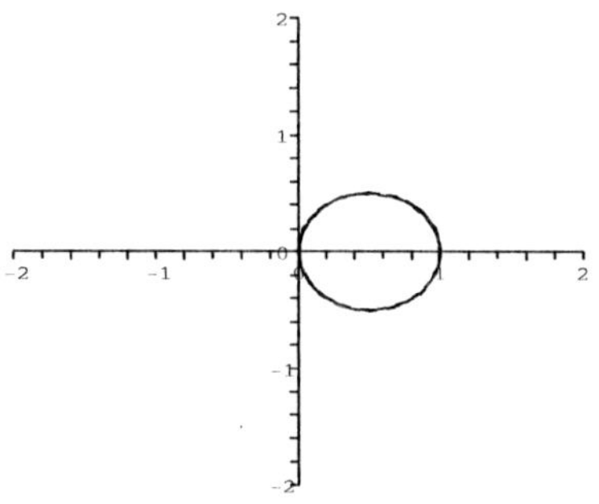

One knows from studying the examples that the figure is a circle centered at $(1/2, 0)$ with radius $1/2$. One could simply write down the equation as

$$\left(x - \frac{1}{2}\right)^2 + y^2 = \left(\frac{1}{2}\right)^2.$$

Failing the recognition, write

$$r = \cos\theta$$
$$r^2 = r\cos\theta$$
$$x^2 + y^2 = x$$

The two equations are the same.

25.

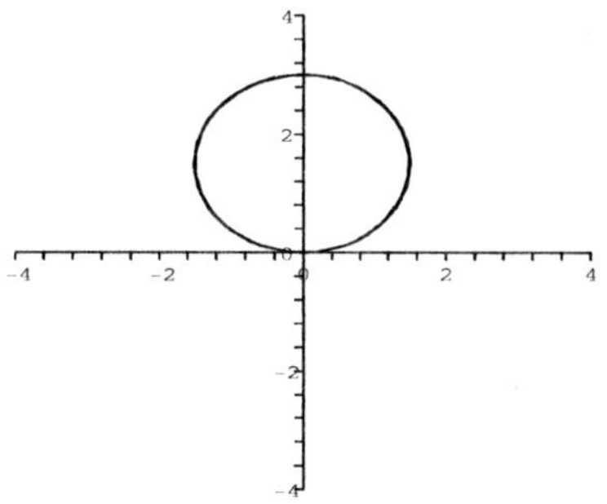

$$r = 3\sin\theta$$
$$r^2 = 3r\sin\theta$$
$$x^2 + y^2 = 3y$$

27.

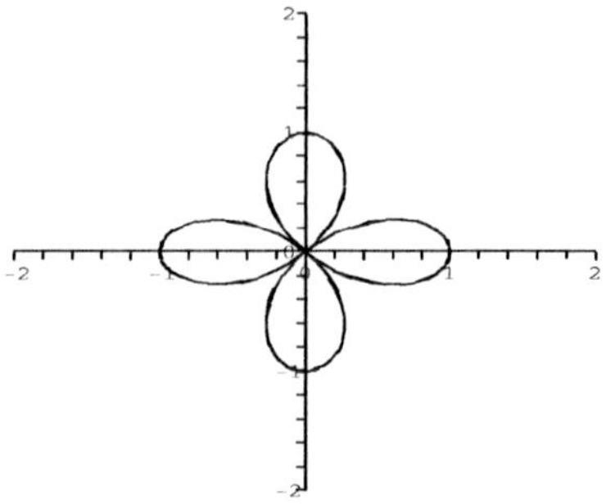

$r = \cos 2\theta = 0$ when $2\theta = \pi/2 + k\pi$ for any integer k, i.e., when $\theta = \pi/4 + k\pi/2$ for any integer k.

$0 \le \theta \le 2\pi$ produces one copy of the graph.

29. $r = \sin 3\theta = 0$ when $3\theta = k\pi$, i.e., $\theta = k\pi/3$ for any integer k.

$0 \le \theta \le \pi$ produces one copy of the graph.

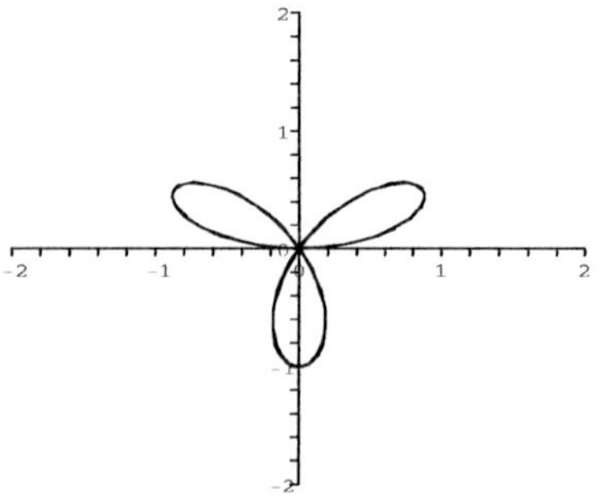

31. $r = 0$ when $3 + 2\sin\theta = 0$ or $\sin\theta = -3/2$. This never happens so r is never 0.

$0 \le \theta \le 2\pi$ produces one copy of the graph.

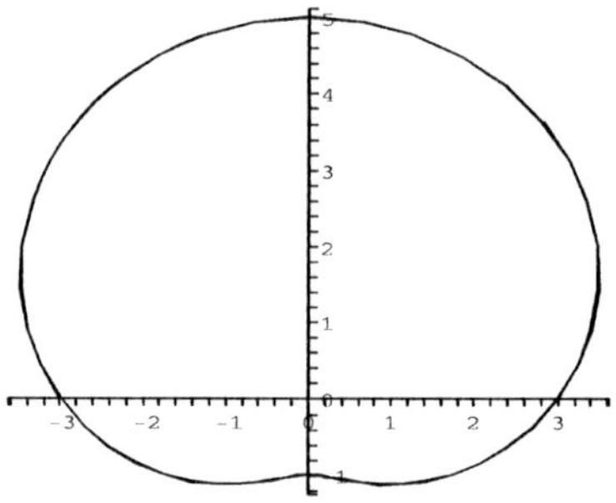

33. $r = 0$ when $2 - 4\sin(\theta) = 0$, hence $\sin\theta = 1/2$, so $\theta = \pi/6 + 2k\pi$ or $\theta = 5\pi/6 + 2k\pi$ for integers k.

$0 \le \theta \le 2\pi$ produces one copy of the graph.

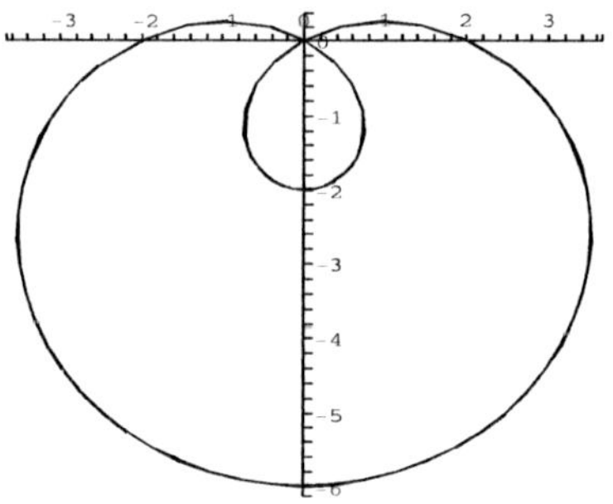

35. $r = 0$ when $\theta = 3\pi/2 + 2k\pi$ for integers k.

$0 \le \theta \le 2\pi$ produces one copy of the graph.

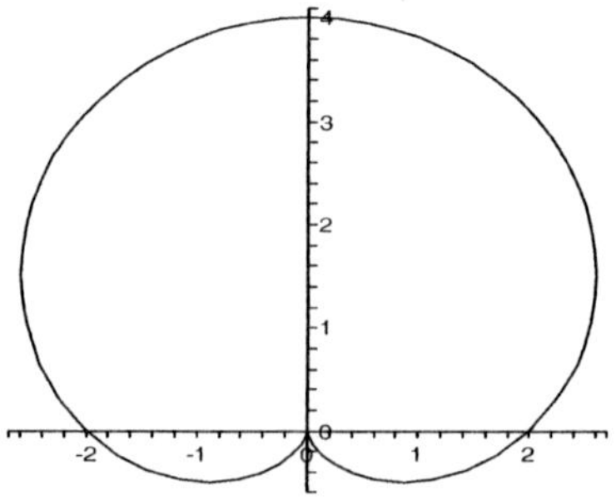

37. $r = \frac{1}{4}\theta = 0$ only when $\theta = 0$.

This graph does not repeat itself, i.e., for completion, one would have to graph for all real numbers θ.

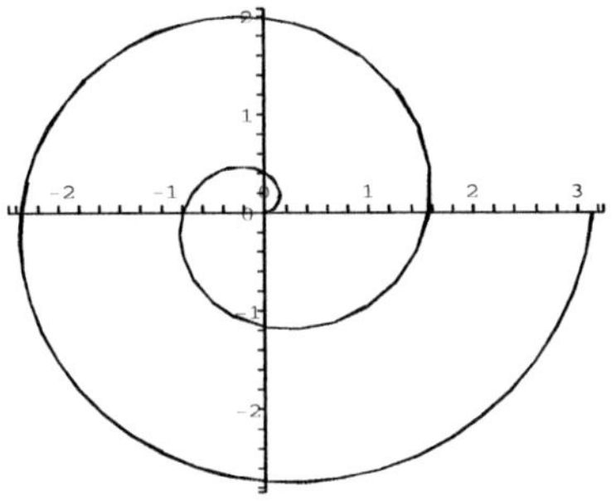

39. $r = 2\cos(\theta - \pi/4) = 0$ when $\theta = \pi/4 = \pi/2 + k\pi$, i.e., $\theta = 3\pi/4 + k\pi$ for integers k.

$0 \le \theta \le 2\pi$ produces one copy of the graph.

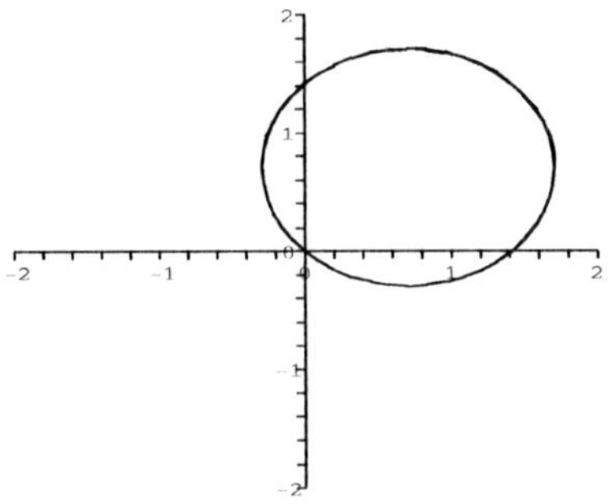

41. $r = \cos\theta + \sin\theta = 0$ when $\theta = 3\pi/4 + 2k\pi$ or $\theta = 7\pi/4 + 2k\pi$ for integers k.

$0 \le \theta \le 2\pi$ produces one copy of the graph.

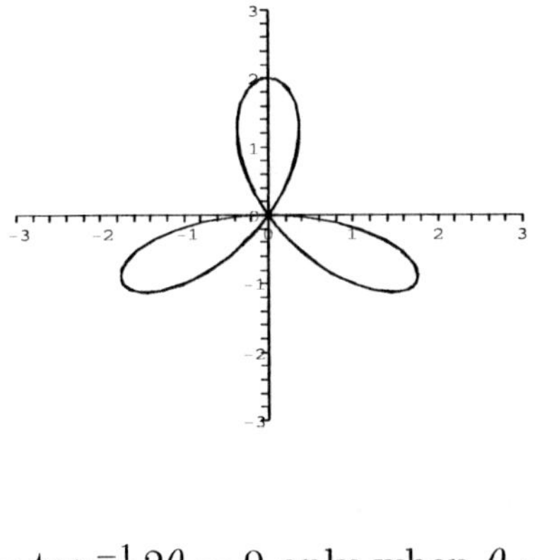

43. $r = \tan^{-1} 2\theta = 0$ only when $\theta = 0$.

This graph does not repeat itself, i.e., for completion, one would have to graph for all real numbers θ.

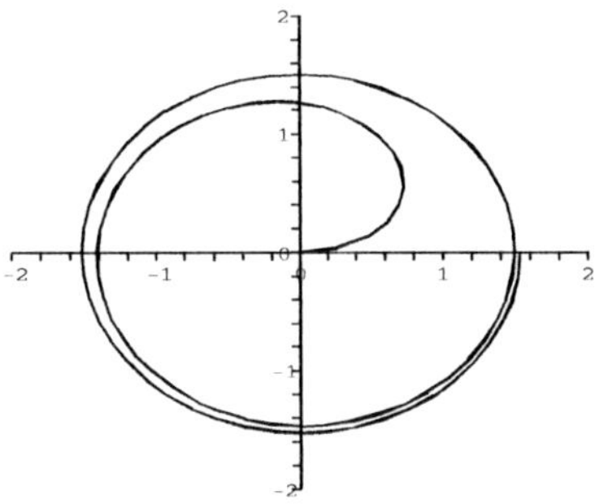

45. $r = 2 + 4\cos 3\theta = 0$ when $\cos 3\theta = -1/2$, i.e., when $\theta = 2\pi/9 + 2k\pi/3$ or $\theta = 4\pi/9 + 2k\pi/3$ for any integer k.

$0 \le \theta \le 2\pi$ produces one copy of the graph.

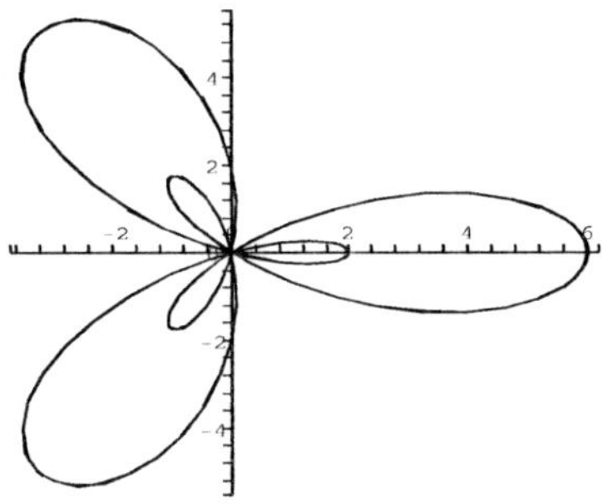

47. $r = \dfrac{2}{1 + \sin\theta} \ne 0$ for any θ. It is undefined when $\sin\theta = -1$, i.e., at $\theta = -\pi/2 + 2k\pi$ for integers k.

$-\pi/2 < \theta < 3\pi/2$ produces one copy of the graph.

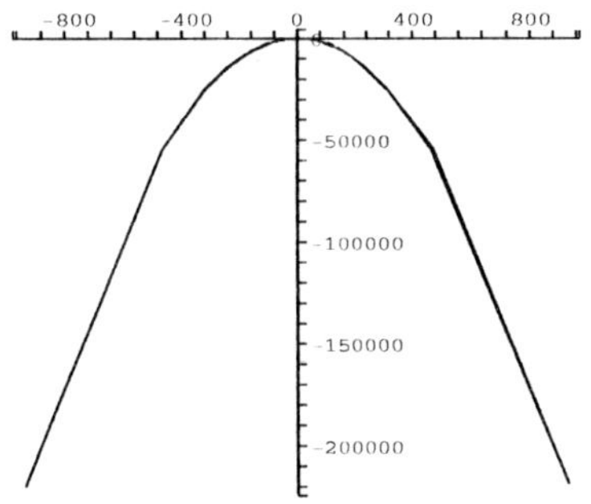

49. $r = \dfrac{2}{1+\cos\theta} \neq 0$ for any θ. It is undefined when $\cos\theta = -1$, i.e., at $\theta = (2k+1)\pi$ for integers k.

$\pi < \theta < 3\pi$ produces one copy of the graph.

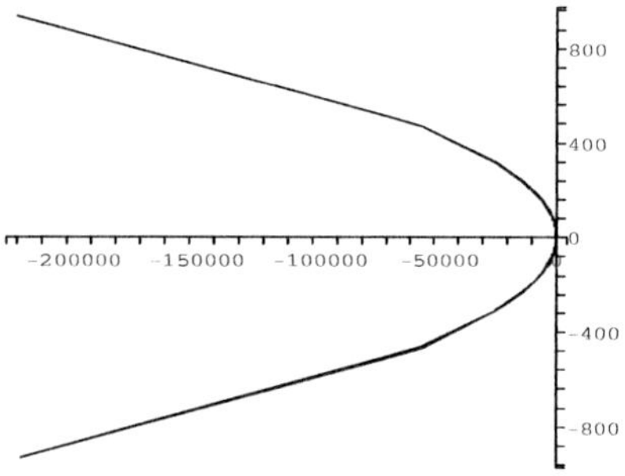

51.

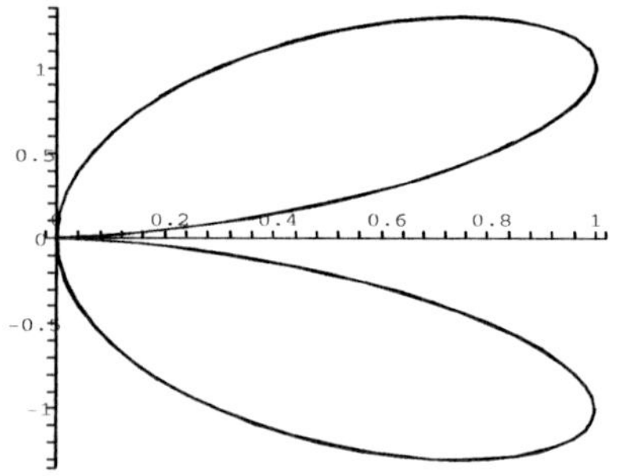

There is no graph to the left of the y-axis because the value of the x-coordinate of every point is positive, which we see as follows:

$$\begin{aligned} x &= r\cos\theta \\ &= 4\cos\theta\sin^2\theta\cos\theta \\ &= 4\cos^2\theta\sin^2\theta \end{aligned}$$

53. One guesses based on a number of examples in the text and the problems, that the figure is a circle of center $(a/2, 0)$ and radius $a/2$. To verify it:

$$r = a\cos\theta,\ r^2 = ar\cos\theta$$
$$x^2 + y^2 = ax,\ x^2 - ax + y^2 = 0$$
$$x^2 - ax + \frac{a^2}{4} + y^2 = \frac{a^2}{4}$$
$$\left(x - \frac{a}{2}\right)^2 + y^2 = \left(\frac{a}{2}\right)^2$$

55. $\cos n\theta$ is a rose with n petals if n is odd, and $2n$ petals if n is even.

57.

$$\begin{aligned} y^2 - x^2 &= 4, \\ r^2\sin^2\theta - r^2\cos^2\theta &= 4, \\ r^2(\cos^2\theta - \sin^2\theta) &= -4, \\ r^2 &= \frac{-4}{\cos(2\theta)} \end{aligned}$$

This is an acceptable answer. Before going any farther, one should note that this will require $\cos(2\theta) < 0$, which, as a subset of $[0, 2\pi)$, puts θ in one of the two open intervals

$$\left(\frac{\pi}{4}, \frac{3\pi}{4}\right) \cup \left(\frac{5\pi}{4}, \frac{7\pi}{4}\right).$$

Subject to that quantification, one could go

$$r = \frac{2}{\sqrt{-\cos(2\theta)}} = 2\sqrt{-\sec(2\theta)},$$

eschewing the possible minus sign since the use of the minus sign merely duplicates points already "present and accounted for."

All this confirms what we might already know about the curve: it is a hyperbola, opening up and down, with asymptotes formed by the two lines $y = x$ and $y = -x$ (which correspond to the endpoints of the stated domain-intervals for θ).

59. $r = 4$

61. $r\sin\theta = 3$

$$r = \frac{3}{\sin\theta} = 3\csc\theta$$

63. Graph for $0 \leq \theta \leq \pi$:

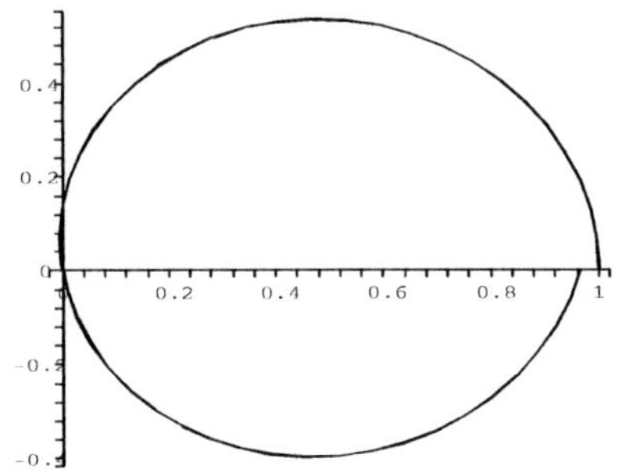

Graph for $0 \leq \theta \leq 2\pi$:

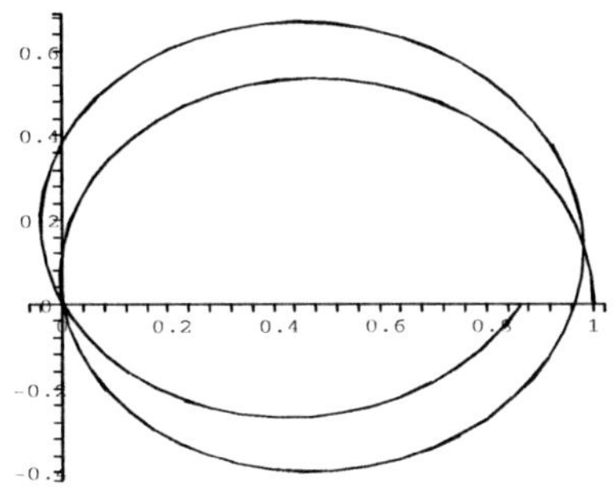

Graph for $0 \leq \theta \leq 3\pi$:

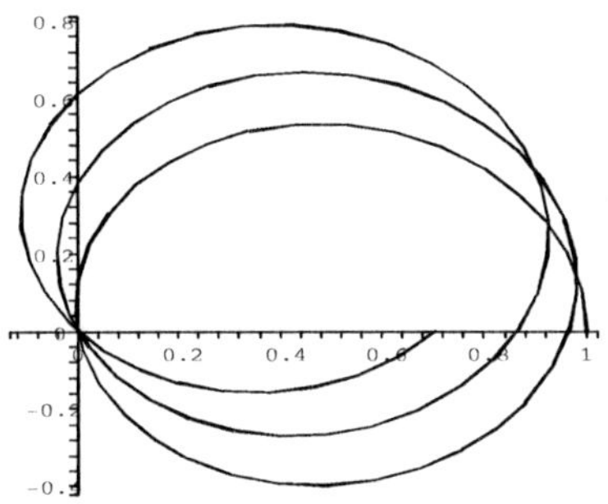

Graph for $0 \leq \theta \leq 6\pi$:

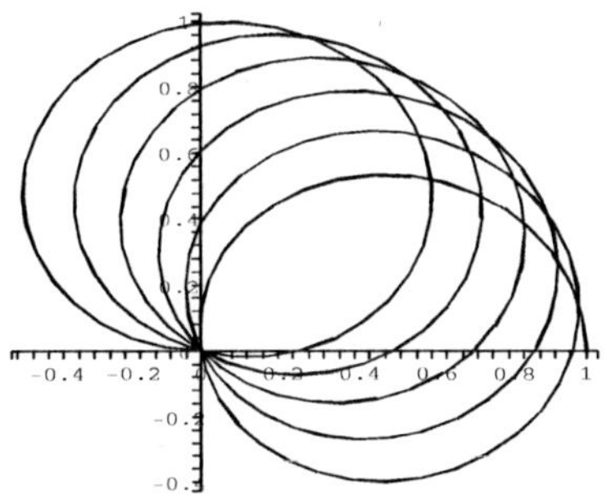

Graph for $0 \leq \theta \leq 12\pi$:

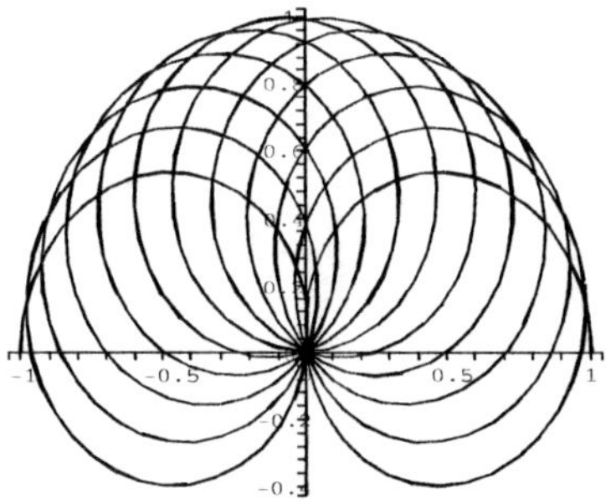

Graph for $0 \leq \theta \leq 23\pi$:

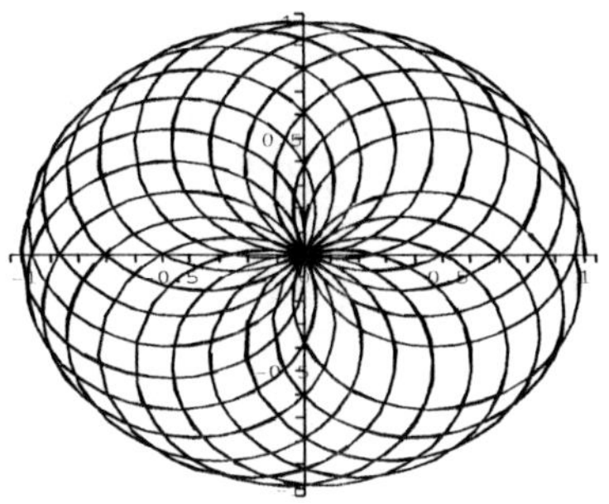

Graph for $0 \leq \theta \leq 24\pi$:

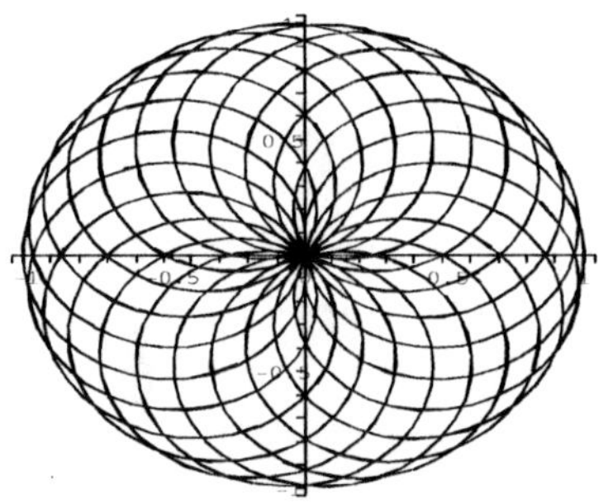

After 24π, the graph will repeat itself.

65. It is clear enough that there is a critical angle formed by the two lines from the ball, tangent to the hole. In one possible set-up, the ball is at the origin O, the hole is centered at $D = (d, 0)$ (on the x-axis, at distance d), and the critical angle A_0 is being measured from the tangent line to the center line of the hole (the x-axis). If the upper point of tangency is denoted by T, one must recall that in the triangle TOD, the right angle is at T rather than at D. This makes OD the hypotenuse,

$$\sin A_0 = \frac{\text{opposite}}{\text{hypotenuse}} = \frac{h}{d},$$

and $A_0 = \sin^{-1}(h/d)$. Any acceptable angle A has to satisfy $-A_0 < A < A_0$.

67. There are no more calculations to be done at this point. According to the conclusions of #65, it must be the case that $A_1 = -A_0$ while $A_2 = A_0$. From #66 one finds

$$r_1(A) = d\cos(A) - \sqrt{d^2\cos^2 A - (d^2 - h^2)}$$

and given in this problem is

$$\begin{aligned} r_2(A) &= d + b\left(1 - \left[\frac{A}{\sin^{-1}(h/d)}\right]^2\right) \\ &= d + b\left(1 - \left[\frac{A}{A_0}\right]^2\right). \end{aligned}$$

One can observe, if the ball is to just drop into the side of the hole ($A \approx A_0$), that r_1 is just about $\sqrt{d^2 - h^2}$ and r_2 is just about d. There is very little margin for error on the speed.

9.5 Calculus and Polar Coordinates

1. $f(\theta) = \sin 3\theta$
$f'(\theta) = 3\cos 3\theta$

$$\begin{aligned} \left.\frac{dy}{dx}\right|_{\theta=\frac{\pi}{3}} &= \frac{3\cos\pi\sin\frac{\pi}{3} + \sin\pi\cos\frac{\pi}{3}}{3\cos\pi\cos\frac{\pi}{3} - \sin\pi\sin\frac{\pi}{3}} \\ &= \frac{3(\frac{\sqrt{3}}{2})}{3(\frac{1}{2})} = \sqrt{3} \end{aligned}$$

3. $f(\theta) = \cos 2\theta$
$f'(\theta) = -2\sin 2\theta$

$$\begin{aligned} \left.\frac{dy}{dx}\right|_{\theta=0} &= \frac{-2\sin 0\sin 0 + \cos 0\cos 0}{-2\sin 0\cos 0 - \cos 0\sin 0} \\ &= \frac{1}{0} \text{ undefined} \end{aligned}$$

5. $f(\theta) = 3\sin\theta$
$f'(\theta) = 3\cos\theta$

$$\begin{aligned} \left.\frac{dy}{dx}\right|_{\theta=0} &= \frac{3\cos 0\sin 0 + 3\sin 0\cos 0}{3\cos 0\cos 0 - 3\sin 0\sin 0} \\ &= \frac{0}{3} = 0 \end{aligned}$$

7. $f(\theta) = \sin 4\theta$
$f'(\theta) = 4\cos 4\theta$

$$\begin{aligned} \left.\frac{dy}{dx}\right|_{\theta=\frac{\pi}{4}} &= \frac{4\cos\pi\sin\frac{\pi}{4} + \sin\pi\cos\frac{\pi}{4}}{4\cos\pi\cos\frac{\pi}{4} - \sin\pi\sin\frac{\pi}{4}} \\ &= \frac{-4(\frac{\sqrt{2}}{2})}{-4(\frac{\sqrt{2}}{2})} = 1 \end{aligned}$$

9. $f(\theta) = \cos 3\theta$
$f'(\theta) = -3\sin 3\theta$

$$\begin{aligned} \left.\frac{dy}{dx}\right|_{\theta=\frac{\pi}{6}} &= \frac{-3\sin\frac{\pi}{2}\sin\frac{\pi}{6} + \cos\frac{\pi}{2}\cos\frac{\pi}{6}}{-3\sin\frac{\pi}{2}\cos\frac{\pi}{6} - \cos\frac{\pi}{2}\sin\frac{\pi}{6}} \\ &= \frac{-3(\frac{1}{2})}{-3(\frac{\sqrt{3}}{2})} = \frac{1}{\sqrt{3}} \end{aligned}$$

11. $|r|$ is a maximum when $\sin 3\theta = \pm 1$. This occurs when $3\theta = \pi/2 + k\pi$ or $\theta = \pi/6 + k\pi/3$ for any integer k. We have $f(\theta) = \sin 3\theta$ so $f'(\theta) = 3\cos 3\theta$. Thus the slope of the tangent line is:

$$\begin{aligned} &\left.\frac{dy}{dx}\right|_{\theta=\frac{\pi}{6}+\frac{k\pi}{3}} \\ &= \frac{0 + \sin(\frac{\pi}{2} + k\pi)\cos(\frac{\pi}{6} + \frac{k\pi}{3})}{0 - \sin(\frac{\pi}{2} + k\pi)\sin(\frac{\pi}{6} + \frac{k\pi}{3})} \\ &= \frac{\sin(\frac{\pi}{2} + k\pi)\cos(\frac{\pi}{6} + \frac{k\pi}{3})}{-\sin(\frac{\pi}{2} + k\pi)\sin(\frac{\pi}{6} + \frac{k\pi}{3})} \\ &= -\frac{\cos(\frac{\pi}{6} + \frac{k\pi}{3})}{\sin(\frac{\pi}{6} + \frac{k\pi}{3})} \end{aligned}$$

Here, the first terms of the numerator and denominator are always 0 since both have a factor of $\cos(\frac{\pi}{2}+k\pi)$.
At $\theta = \frac{\pi}{6}+\frac{k\pi}{3}$, $r = \sin(\frac{\pi}{2}+k\pi)$ so the point in question is either

$$\left(\cos\left(\frac{\pi}{6}+\frac{k\pi}{3}\right), \sin\left(\frac{\pi}{6}+\frac{k\pi}{3}\right)\right)$$

or

$$\left(-\cos\left(\frac{\pi}{6}+\frac{k\pi}{3}\right), -\sin\left(\frac{\pi}{6}+\frac{k\pi}{3}\right)\right).$$

In either case, the slope of the radius connecting the point to the origin is

$$\frac{\sin(\frac{\pi}{6}+\frac{k\pi}{3})}{\cos(\frac{\pi}{6}+\frac{k\pi}{3})}$$

which is the negative recipricol of the slope of the tangent line found above. Therefore the tangent line is perpendicular to the radius connecting the point to the origin.

13. $|r|$ is a maximum when $\cos 2\theta = 0$. This occurs when $2\theta = \pi/2 + k\pi$ or $\theta = \pi/4 + k\pi/2$ for any integer k. We have $f(\theta) = 2 - 4\sin 2\theta$ so $f'(\theta) = -8\cos 2\theta$. Thus the slope of the tangent line is:

$$\left.\frac{dy}{dx}\right|_{\theta=\frac{\pi}{4}+\frac{k\pi}{2}}$$
$$= \frac{0+(2-4\sin(\frac{\pi}{2}+k\pi))\cos(\frac{\pi}{4}+\frac{k\pi}{2})}{0-(2-4\sin(\frac{\pi}{2}+k\pi))\sin(\frac{\pi}{4}+\frac{k\pi}{2})}$$
$$= \frac{(2-4\sin(\frac{\pi}{2}+k\pi))\cos(\frac{\pi}{4}+\frac{k\pi}{2})}{-(2-4\sin(\frac{\pi}{2}+k\pi))\sin(\frac{\pi}{4}+\frac{k\pi}{2})}$$
$$= -\frac{\cos(\frac{\pi}{4}+\frac{k\pi}{2})}{\sin(\frac{\pi}{4}+\frac{k\pi}{2})}$$

Here, the first terms of the numerator and denominator are always 0 since both have a factor of $\cos(\frac{\pi}{2}+k\pi)$.

At $\theta = \frac{\pi}{4}+\frac{k\pi}{2}$, $r = 2-4\sin(\frac{\pi}{2}+k\pi)$ so the point in question is either

$$\left(-2\cos\left(\frac{\pi}{4}+\frac{k\pi}{2}\right), -2\sin\left(\frac{\pi}{4}+\frac{k\pi}{2}\right)\right)$$

or

$$\left(6\cos\left(\frac{\pi}{4}+\frac{k\pi}{2}\right), 6\sin\left(\frac{\pi}{4}+\frac{k\pi}{2}\right)\right).$$

In either case, the slope of the radius connecting the point to the origin is

$$\frac{\sin(\frac{\pi}{4}+\frac{k\pi}{2})}{\cos(\frac{\pi}{4}+\frac{k\pi}{2})}$$

which is the negative recipricol of the slope of the tangent line found above. Therefore the tangent line is perpendicular to the radius connecting the point to the origin.

15. One leaf is traced out over the range $\pi/6 \le \theta \le \pi/2$, so the area A is

$$A = \frac{1}{2}\int_{\pi/6}^{\pi/2} \cos^2 3\theta \, d\theta$$
$$= \frac{1}{6}\left(\frac{1}{2}\cdot 3\theta + \frac{1}{2}\sin 3\theta\cos 3\theta\right)\Bigg|_{\pi/6}^{\pi/2}$$
$$= \frac{1}{6}\left(\frac{3\pi}{4}-\frac{\pi}{4}\right) = \frac{\pi}{12}$$

17. Endpoints for the inner loop are given by $\theta = \sin^{-1}(3/4) \approx 0.848$ and $\theta = \pi - \sin^{-1}(3/4) \approx 2.294$, so the area A is

$$A = \frac{1}{2}\int_{0.848}^{2.294} (3-4\sin\theta)^2 \, d\theta$$
$$= \frac{1}{2}\int_{0.848}^{2.294} (9-24\sin\theta+16\sin^2\theta) \, d\theta$$
$$= \frac{1}{2}\int_{0.848}^{2.294} (17-24\sin\theta-8\cos 2\theta) \, d\theta$$
$$= \frac{1}{2}(17\theta+24\cos\theta-4\sin 2\theta)\Bigg|_{0.848}^{2.294}$$
$$\approx 0.3806$$

19. The curve is traced out over the range $0 \le \theta \le \pi$, so the area A is

$$\begin{aligned} A &= \frac{1}{2}\int_0^{\pi} 4\cos^2\theta\, d\theta \\ &= 2\int_0^{\pi} \cos^2\theta\, d\theta \\ &= 2\left(\frac{1}{2}\theta + \frac{1}{2}\sin\theta\cos\theta\right)\Big|_0^{\pi} \\ &= 2\left(\frac{\pi}{2}\right) = \pi \end{aligned}$$

21. A small loop is traced out over the range $7\pi/12 \le \theta \le 11\pi/12$, so the area A is

$$\begin{aligned} A &= \frac{1}{2}\int_{7\pi/12}^{11\pi/12} (1+2\sin 2\theta)^2\, d\theta \\ &= \frac{1}{2}\int_{7\pi/12}^{11\pi/12} (1+4\sin 2\theta + 4\sin^2 2\theta)\, d\theta \\ &= \frac{1}{2}\int_{7\pi/12}^{11\pi/12} (1+4\sin 2\theta \\ &\qquad + 2(1-\cos 4\theta))\, d\theta \\ &= \frac{1}{2}\left(3\theta - 2\cos 2\theta - \frac{1}{2}\sin 4\theta\right)\Big|_{7\pi/12}^{11\pi/12} \\ &= \frac{\pi}{2} - \frac{3\sqrt{3}}{4} \approx 0.2718 \end{aligned}$$

23. Endpoints for the inner loop are given by

$$\theta = \frac{-\sin^{-1}(-2/3)}{3} \approx 1.2904$$

and

$$\theta = \frac{2\pi + \sin^{-1}(-2/3)}{3} \approx 1.8512,$$

so the area A is

$$\begin{aligned} A &= \frac{1}{2}\int_{1.2904}^{1.8512} (2+3\sin 3\theta)^2\, d\theta \\ &= \frac{1}{2}\int_{1.2904}^{1.8512} (4+12\sin 3\theta + 9\sin^2 3\theta)\, d\theta \\ &\approx \frac{1}{2}(-8.193) = -4.0965 \end{aligned}$$

25. This region is traced out over $-\pi/6 \le \theta \le 7\pi/6$, so the area A is given by

$$A = A_1 - A_2$$

where

$$A_1 = \frac{1}{2}\int_{-\pi/6}^{7\pi/6} (3+2\sin\theta)^2\, d\theta$$

and

$$A_2 = \frac{1}{2}\int_{-\pi/6}^{7\pi/6} 2^2\, d\theta.$$

This works out to

$$A \approx 24.187$$

27. This region is traced out over $5\pi/6 \le \theta \le 13\pi/6$, so the area A is given by

$$A = A_1 - A_2$$

where

$$A_1 = \frac{1}{2}\int_{5\pi/6}^{13\pi/6} 2^2\, d\theta.$$

and

$$A_2 = \frac{1}{2}\int_{5\pi/6}^{13\pi/6} (1+2\sin\theta)^2\, d\theta.$$

This works out to

$$A = \frac{2\pi}{3} + \frac{3\sqrt{3}}{2}.$$

29. Since the graph is symmetric through the x-axis, we can take the portions of the area we need above the x-axis and multiple these by 2. We then need the area of $r = 1$ on the region $0 \le \theta \le \pi/2$ (multiplied by 2) plus the area of $r = 1 + \cos\theta$ on the region

$\pi/2 \leq \theta \leq \pi$ (multiplied by 2). We find that the area A is

$$A = \frac{1}{2} \cdot 2 \int_0^{\pi/2} 1^2 \, d\theta + \frac{1}{2} \cdot 2 \int_{\pi/2}^{\pi} (1 + \cos\theta)^2 \, d\theta$$
$$= \int_0^{\pi/2} 1 \, d\theta + \int_{\pi/2}^{\pi} (1 + \cos\theta)^2 \, d\theta$$
$$= \frac{\pi}{2} + \frac{3\pi}{4} - 2$$
$$= \frac{5\pi}{4} - 2$$

31. To find the points of intersection, we graph the function

$$y = 1 - 2\sin x - 2\cos x$$

and look for the roots. We find $x \approx 1.9948$ and $x \approx 5.8592$. Note that the point $(0, 0)$ is not an intersection point since the two graphs pass through the origin at different values of θ.

33. To find the points of intersection in this case, we just set the two expressions for r equal to each other and solve for θ:

$$1 + \sin\theta = 1 + \cos\theta$$
$$\sin\theta = \cos\theta$$

This occurs when $\theta = \pi/4$ or $\theta = 5\pi/4$.

35. $f(\theta) = 2 - 2\sin\theta$
$f'(\theta) = -2\cos\theta$
The arc length is:

$$s = \int_0^{2\pi} \sqrt{4\cos^2\theta + (2 - 2\sin\theta)^2} \, d\theta$$
$$= \int_0^{2\pi} \sqrt{4\cos^2\theta + 4 - 8\sin\theta + 4\sin^2\theta} \, d\theta$$
$$= \int_0^{2\pi} \sqrt{8(1 - \sin\theta)} \, d\theta$$
$$= 16$$

37. $f(\theta) = \sin 3\theta$
$f'(\theta) = 3\cos 3\theta$
The arc length is:

$$s = \int_0^{\pi} \sqrt{\sin^2 3\theta + 9\cos^2 3\theta} \, d\theta$$
$$\approx 6.683$$

39. $f(\theta) = 1 + 2\sin 2\theta$
$f'(\theta) = 4\cos 2\theta$
The arc length is:

$$s = \int_0^{2\pi} \sqrt{16\cos^2 2\theta + (1 + 2\sin 2\theta)^2} \, d\theta$$
$$\approx 20.016$$

41. We use the same setup as in example 5.7, except here we use the line $y = -0.6$ which corresponds to $r = -0.6\csc\theta$. To find the limits of integration, we must solve the equation

$$2 = -0.6\csc\theta.$$

We find $\theta_1 \approx 3.4463$ and $\theta_2 \approx 5.9785$. With these limits of integration, we find the area A of the filled region:

$$A = \int_{\theta_1}^{\theta_2} 2 \, d\theta - \int_{\theta_1}^{\theta_2} \frac{1}{2}(-0.6\csc\theta)^2 \, d\theta$$
$$= (2\theta + 0.18\cot\theta)\Big|_{\theta_1}^{\theta_2}$$
$$\approx 3.9197$$

The fraction of oil remaining in the tank is then approximately $/4\pi \approx$ or about % of the total capacity of the tank.

43. We use the same setup as in example 5.7, except here we use the line $y = 0.4$ which corresponds to $r = 0.4\csc\theta$.

To find the limits of integration, we must solve the equation

$$2 = 0.4 \csc \theta.$$

We find $\theta_1 \approx 0.2014$ and $\theta_2 \approx 2.9402$. With these limits of integration, we find the area A of the filled region:

$$A = \int_{\theta_1}^{\theta_2} 2\,d\theta - \int_{\theta_1}^{\theta_2} \frac{1}{2}(0.4 \csc \theta)^2\,d\theta$$
$$= (2\theta + 0.08 \cot \theta)\Big|_{\theta_1}^{\theta_2}$$
$$\approx 4.694$$

The fraction of oil remaining in the tank is then approximately $/4\pi \approx$ or about % of the total capacity of the tank.

45. To show that the curve passes through the origin at each of these three values of θ, we simply show that $r = 0$ for each.

$\theta = 0$: $r = \sin 0 = 0$
$\theta = \pi/3$: $r = \sin \pi = 0$
$\theta = 2\pi/3$: $r = \sin 2\pi = 0$

We now find the slope at each of the three angles:

$$\left.\frac{dy}{dx}\right|_{\theta=0} = \frac{3\sin 0 + 0}{3 - 0} = 0$$
$$\left.\frac{dy}{dx}\right|_{\theta=\pi/3} = \frac{-3(\frac{\sqrt{3}}{2}) + 0}{-3(\frac{1}{2}) - 0} = \sqrt{3}$$
$$\left.\frac{dy}{dx}\right|_{\theta=\pi/3} = \frac{3(\frac{\sqrt{3}}{2}) + 0}{3(-\frac{1}{2}) - 0} = -\sqrt{3}$$

As the graph passes through the origin at each of these three angles, it does so at a different slope.

47. If $r = cf(\theta)$ then $r' = cf'(\theta)$. The arc length is:

$$s = \int_a^b \sqrt{(cf(\theta))^2 + (cf'(\theta))^2}\,d\theta$$
$$= \int_a^b \sqrt{c^2((f(\theta))^2 + (f'(\theta))^2)}\,d\theta$$
$$= |c| \int_a^b \sqrt{(f(\theta))^2 + (f'(\theta))^2}\,d\theta$$
$$= |c| L$$

49. Let s represent the arc length from $\theta = d$ to $\theta = c$. Also note that since $f(\theta) = ae^{b\theta}$, it follows that $f'(\theta) = abe^{b\theta}$. Then,

$$s = \int_d^c \sqrt{[f'(\theta)]^2 + [f(\theta)]^2}d\theta$$
$$= \int_d^c \sqrt{a^2b^2e^{2b\theta} + a^2e^{2b\theta}}d\theta$$
$$= \int_d^c ae^{b\theta}\sqrt{b^2+1}d\theta$$
$$= a\sqrt{b^2+1} \int_d^c e^{b\theta}\,d\theta$$
$$= a\sqrt{b^2+1}\left(\frac{1}{b}\right)\left[e^{b\theta}\right]_d^c$$
$$= \frac{a\sqrt{b^2+1}}{b}\left[e^{bc} - e^{bd}\right]$$

Therefore, the total arc length is

$$\lim_{d\to-\infty} s = \lim_{d\to-\infty} \frac{a\sqrt{b^2+1}}{b}\left[e^{bc} - e^{bd}\right]$$
$$= \frac{a\sqrt{b^2+1}}{b}\left[e^{bc} - \lim_{d\to-\infty} e^{bd}\right]$$
$$= \frac{a\sqrt{b^2+1}}{b}e^{bc}$$
$$= \frac{\sqrt{b^2+1}}{b}R$$

since the distance from the origin to the starting point is $R = ae^{bc}$.

9.6 Conic Sections

1. Since the focus is $(0, -1)$ and the directrix is $y = 1$, we see that the vertex must be $(0,0)$. Thus, $b=0$, $c=0$ and $a=-\frac{1}{4}$. The equation for the parabola is:

$$y=-\frac{1}{4}x^2.$$

3. Since the focus is $(3,0)$ and the directrix is $x=1$, we see that the vertex must be $(2,0)$. Thus, $b=0$, $c=2$ and $a=\frac{1}{4}$. The equation for the parabola is:

$$x=\frac{1}{4}y^2+2.$$

5. From the given foci and vertices we find that $x_0=0$, $y_0=3$, $c=2$, $a=4$ and $b=\sqrt{12}$. The equation for the ellipse is:

$$\frac{x^2}{12}+\frac{(y-3)^2}{16}=1.$$

7. From the given foci and vertices we find that $y_0=1$, $x_0=4$, $c=2$, $a=4$ and $b=\sqrt{12}$. The equation for the ellipse is:

$$\frac{(x-4)^2}{16}+\frac{(y-1)^2}{12}=1.$$

9. From the given foci and vertices we find that $y_0=0$, $x_0=2$, $c=2$, $a=1$ and $b=\sqrt{3}$. The equation for the hyperbola is:

$$\frac{(x-2)^2}{1}-\frac{y^2}{3}=1.$$

11. From the given foci and vertices we find that $x_0=2$, $y_0=4$, $c=2$, $a=1$ and $b=\sqrt{3}$. The equation for the hyperbola is:

$$\frac{(y-4)^2}{1}-\frac{(x-2)^2}{3}=1.$$

13. This is a parabola with $a=2$, $b=-1$ and $c=-1$.
vertex: $(-1,-1)$
focus: $(-1,-1+\frac{1}{8})$
directrix: $y=-1-\frac{1}{8}$

15. This is an ellipse with $x_0=1$, $y_0=2$, $a=3$, $b=2$ and $c=\sqrt{5}$.
foci: $(1,2-\sqrt{5})$, $(1,2+\sqrt{5})$
vertices: $(1,5)$, $(1,-1)$

17. This is a hyperbola with $x_0=1$, $y_0=0$, $a=3$, $b=2$ and $c=\sqrt{13}$.
foci: $(1-\sqrt{13},0)$, $(1+\sqrt{13},0)$
vertices: $(4,0)$, $(-2,0)$

19. This is a hyperbola with $y_0=-1$, $x_0=-2$, $a=4$, $b=2$ and $c=\sqrt{20}$.
foci: $(-2,-1+\sqrt{20})$, $(-2,-1-\sqrt{20})$
vertices: $(-2,3)$, $(-2,-5)$

21. Dividing both sides by 9 gives:

$$\frac{(x-2)^2}{9}+y^2=1$$

This is an ellipse with $x_0=2$, $y_0=0$, $a=3$, $b=1$ and $c=\sqrt{8}$.
foci: $(2-\sqrt{8},0)$, $(2+\sqrt{8},0)$
vertices: $(5,0)$, $(-1,0)$

23. Solving for y gives:

$$y=\frac{1}{4}(x+1)^2-2$$

This is a parabola with $a=1/4$, $b=-1$ and $c=-2$.
vertex: $(-1,-2)$
focus: $(-1,-1)$
directrix: $y=-3$

25. This is a parabola with focus $(2,1)$ and directrix $y=-3$. Thus the vertex is $(2,-1)$ and $b=2$, $c=-1$ and $a=\frac{1}{8}$. The equation is

$$y=\frac{1}{8}(x-2)^2-1.$$

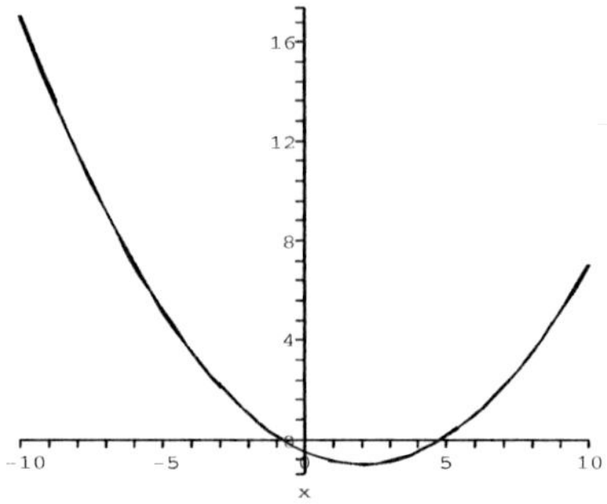

27. This is an ellipse with foci $(0, 2)$ and $(4, 2)$ and center $(2, 2)$. Since the sum of the distances to the foci must be 8, the vertices are $(-2, 2)$ and $(6, 2)$. Thus $y_0 = 2$, $x_0 = 2$, $c = 2$, $a = 4$ and $b = \sqrt{12}$. The equation is

$$\frac{(x-2)^2}{16} + \frac{(y-2)^2}{12} = 1.$$

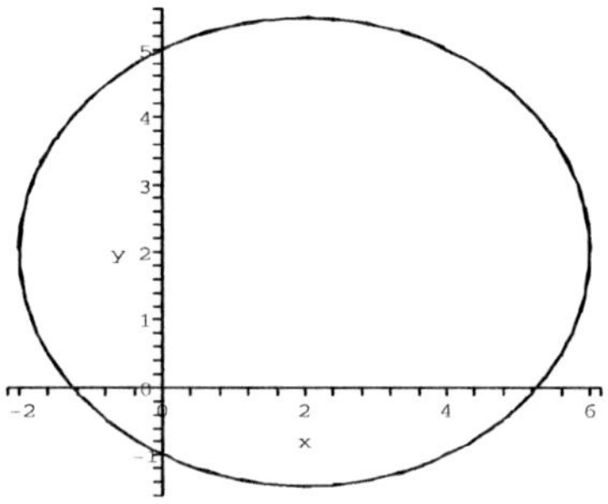

29. This is a hyperbola with foci $(0, 4)$ and $(0, -2)$ and center $(0, 1)$. Since the difference of the distances to the foci must be 4, the vertices are $(0, 3)$ and $(0, -1)$. Thus $x_0 = 0$, $y_0 = 1$, $c = 3$, $a = 2$ and $b = \sqrt{5}$. The equation is

$$\frac{(y-1)^2}{4} - \frac{x^2}{5} = 1.$$

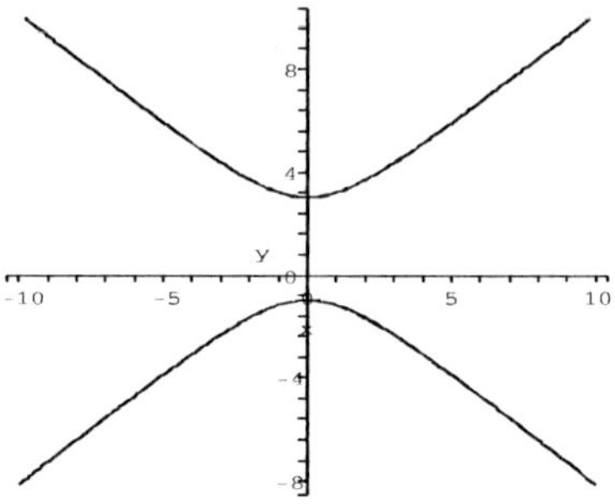

31. Since $x = 4y^2$, we have $a = 4$, $b = 0$ and $c = 0$. The lightbulb should be placed at the focus, i.e., $(\frac{1}{16}, 0)$.

33. Since $y = 2x^2$, we have $a = 2$, $b = 0$ and $c = 0$. The microphone should be placed at the focus, i.e., $(0, \frac{1}{8})$.

35. We have $c = \sqrt{a^2 - b^2} = \sqrt{124 - 24} = 10$, so that the foci are 20 inches apart. The transducer should be located 20 inches away from the kidney stone and aligned so that the line segment from the kidney stone to the transducer lies along the major axis of the elliptical reflector.

37. From the equation, we have $y_0 = -4$, $x_0 = 0$, $a = 1$, $b = \sqrt{15}$ and $c = 4$. Thus the foci are $(0, -8)$ and $(0, 0)$. Light rays following the path $y = cx$ would go through the focus $(0, 0)$, so they are reflected toward the focus at $(0, -8)$.

39. From the equation, we have $x_0 = 0$, $y_0 = 0$, $a = \sqrt{3}$, $b = 1$ and $c = 2$. Thus the foci are $(-2, 0)$ and $(2, 0)$. Light rays following the path $y = c(x-2)$ would go through the focus $(2, 0)$, so they are reflected toward the focus at $(-2, 0)$.

41. We have $x_0 = 0$, $y_0 = 0$ and $c = \sqrt{400 - 100} = 10\sqrt{3}$. The foci are

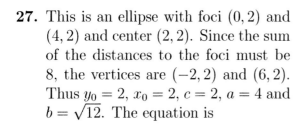

$(\pm 10\sqrt{3}, 0)$ so the desks should be placed along the major axis of the ellipse $10\sqrt{3}$ units from the center (at $(0,0)$).

43. Let $f(t)$ be the object's distance from the spectator at time t. We know this is a quadratic function, so it must be of the form $f(t) = at^2 + bt + c$ for some constants a, b and c. At time $t = 0$, the spectator has the object, so the distance from the spectator is 0. Thus $f(0) = a(0)^2 + b(0) + c = c = 0$. So now our function is of the form $f(t) = at^2 + bt$. Plugging in $t = 2$ and $t = 4$ (and their respective values of $f(t)$) gives the following two equations:

$$4a + 2b = 28$$
$$16a + 4b = 48.$$

Multiplying the first equation by 2 and then subtracting the result from the second equation gives $8a = -8$ or $a = -1$. Plugging $a = -1$ into either of the above equations gives $b = 16$. Thus the equation for the distance of the object from the spectator is $f(t) = -t^2 + 16t$. Graphing this function gives:

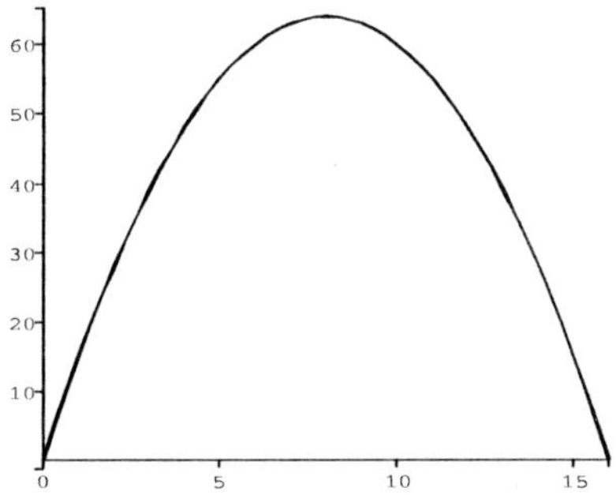

Since this object returns to the spectator (at time $t = 16$ seconds the object is again at 0 meters from the spectator) we guess that the object is a boomerang.

9.7 Conic Sections in Polar Coordinates

1. From Theorem 7.2 part (i) we have

$$r = \frac{0.6(2)}{0.6\cos\theta + 1} = \frac{1.2}{0.6\cos\theta + 1}.$$

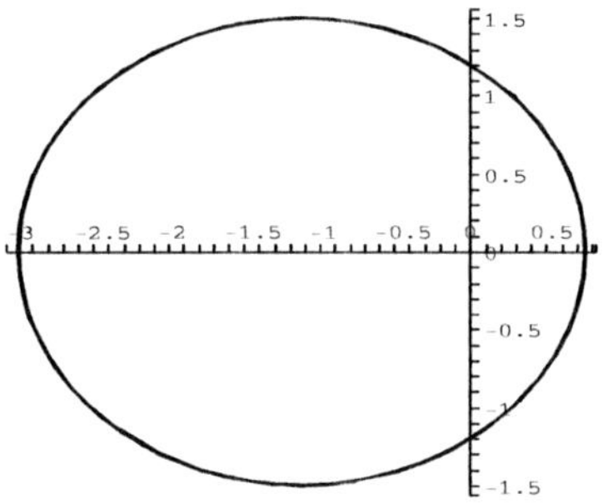

3. From Theorem 7.2 part (i) we have

$$r = \frac{2}{\cos\theta + 1}.$$

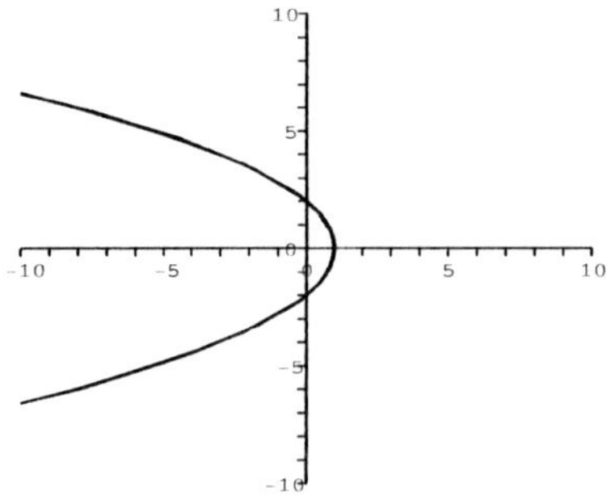

5. From Theorem 7.2 part (iii) we have

$$r = \frac{1.2}{0.6\sin\theta + 1}.$$

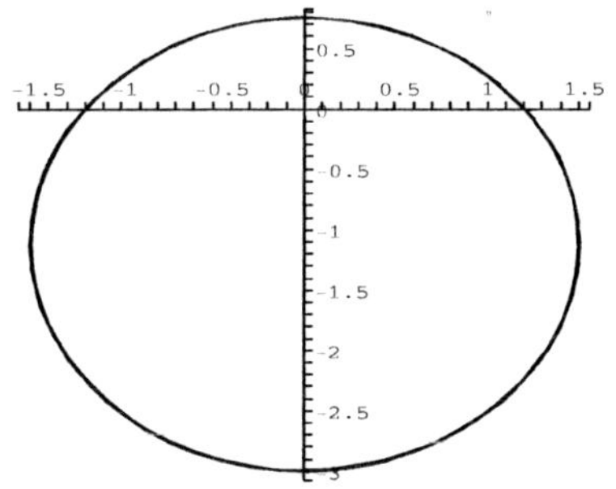

7. From Theorem 7.2 part (iii) we have

$$r = \frac{2}{\sin\theta + 1}.$$

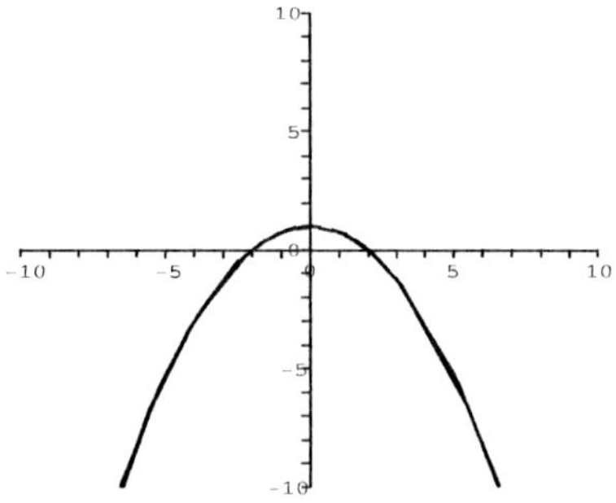

9. From Theorem 7.2 part (ii) we have

$$r = \frac{-0.8}{0.4\cos\theta - 1}.$$

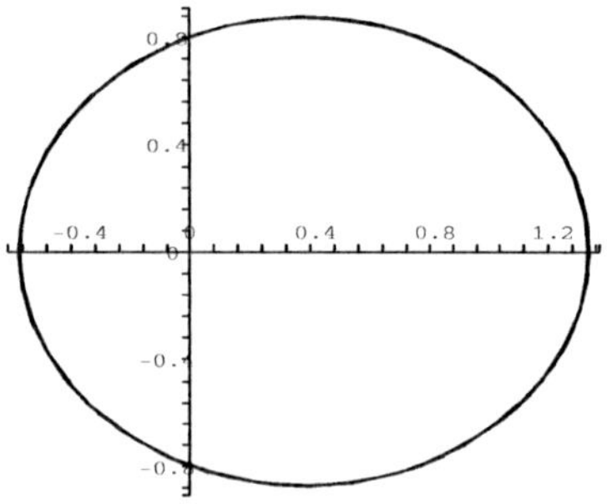

11. From Theorem 7.2 part (ii) we have

$$r = \frac{-4}{2\cos\theta - 1}.$$

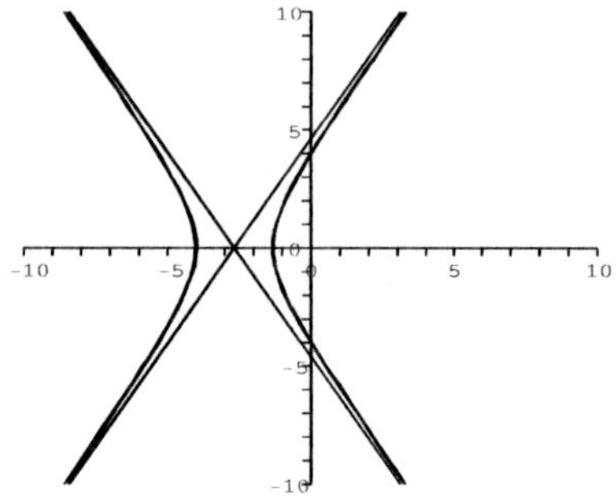

13. From Theorem 7.2 part (iv) we have

$$r = \frac{-0.8}{0.4\sin\theta - 1}.$$

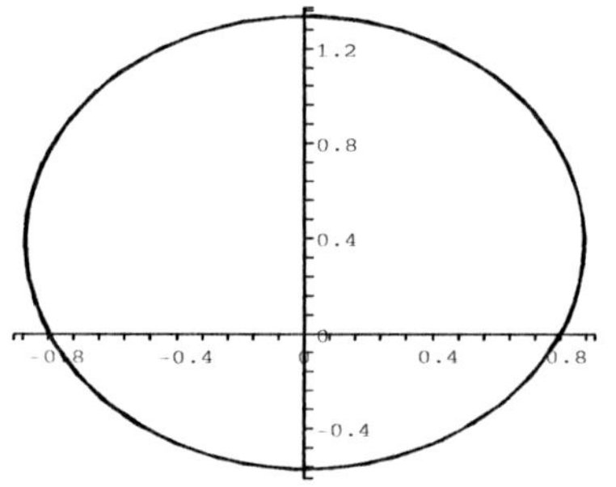

15. From Theorem 7.2 part (iv) we have

$$r = \frac{-2}{\sin\theta - 1}.$$

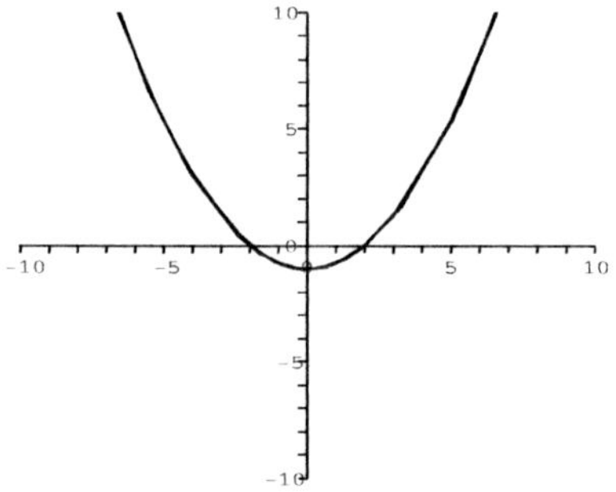

17. This is a hyperbola:

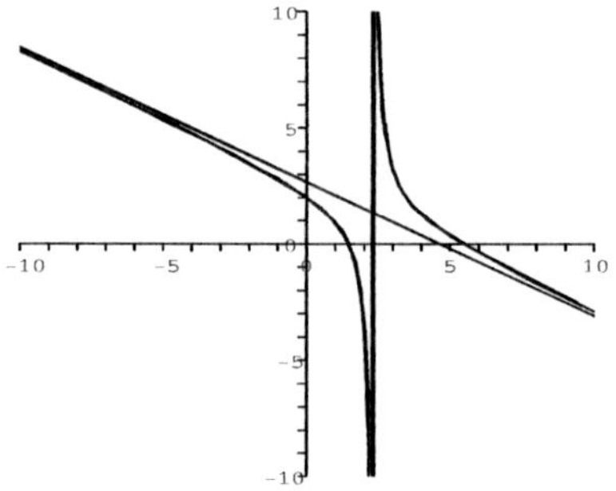

19. This is an ellipse:

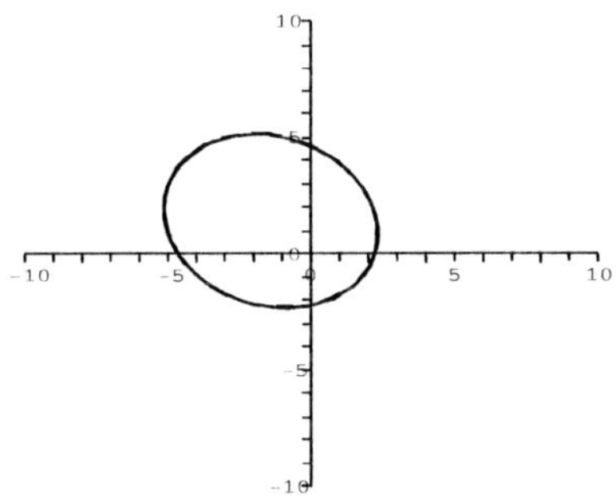

21. This is a parabola:

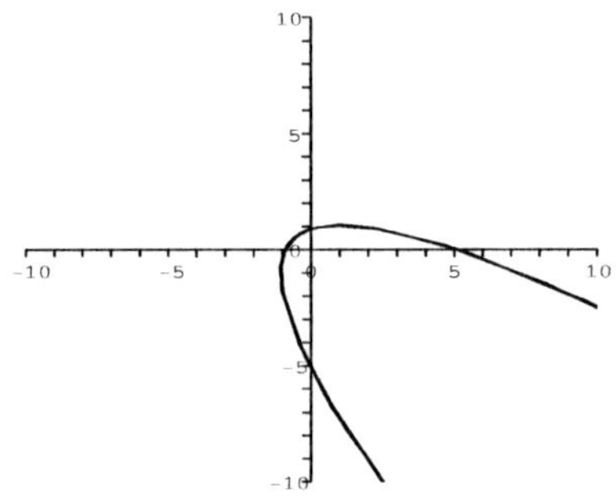

23. This is an ellipse with center $(-1, 1)$ and major axis parallel to the x-axis. The parametric equations (for $0 \leq t \leq 2\pi$) are:

$$\begin{cases} x = 3\cos t - 1 \\ y = 2\sin t + 1 \end{cases}$$

25. The right half of this hyperbola is given by

$$\begin{cases} x = 4\cosh t - 1 \\ y = 3\sinh t \end{cases}$$

while the left half is given by

$$\begin{cases} x = -4\cosh t - 1 \\ y = 3\sinh t \end{cases}$$

27. We solve for y to get:

$$y = -\frac{x^2}{4} + 1.$$

The parametric equations are then given by:

$$\begin{cases} x = t \\ y = -t^2/4 + 1 \end{cases}$$

29. We have

$$\begin{aligned} A_1 &= \frac{1}{2}\int_0^{\pi/2} \left(\frac{2}{\sin\theta + 2}\right)^2 d\theta \\ &= \frac{1}{2}\left(\frac{8}{27}\sqrt{3} - \frac{2}{3}\right) \\ &\approx 0.473 \end{aligned}$$

$$\begin{aligned} A_2 &= \frac{1}{2}\int_{3\pi/2}^{4.953} \left(\frac{2}{\sin\theta + 2}\right)^2 d\theta \\ &\approx 0.472 \end{aligned}$$

and

$$s_1 = \int_0^{\pi/2} g(\theta)\, d\theta \approx 1.266$$

$$s_2 = \int_{3\pi/2}^{4.953} g(\theta)\, d\theta \approx 0.481$$

where

$$g(\theta) = \sqrt{\frac{4\cos^2\theta}{(\sin\theta + 2)^4} + \frac{4}{(\sin\theta + 2)^2}}.$$

Therefore the two areas are approximately the same, while the average speed on the portion of the orbit from $\theta = 0$ to $\theta = \pi/2$ is a little more than two-and-a-half times the average speed on the portion of the orbit from $\theta = 3\pi/2$ to $\theta = 4.953$.

31. For any point (x, y) on the curve, the distance to the focus is $\sqrt{x^2 + y^2}$ and the distance to the directrix is $x - d$. We then have

$$\begin{aligned} \sqrt{x^2 + y^2} &= e(x - d) \\ r &= e(r\cos\theta - d) \\ r - er\cos\theta &= -ed \\ r(1 - e\cos\theta) &= -ed \\ r &= \frac{-ed}{1 - e\cos\theta} \\ r &= \frac{ed}{e\cos\theta - 1}. \end{aligned}$$

33. For any point (x, y) on the curve, the distance to the focus is $\sqrt{x^2+y^2}$ and the distance to the directrix is $y - d$. We then have

$$\begin{aligned} \sqrt{x^2+y^2} &= e(y-d) \\ r &= e(r\sin\theta - d) \\ r(1 - e\sin\theta) &= -ed \\ r &= \frac{ed}{e\sin\theta - 1}. \end{aligned}$$

Ch. 9 Review Exercises

1. The given parametric equations for x and y satisfy the following x-y equation:

$$\left(\frac{x+1}{3}\right)^2 + \left(\frac{y-2}{9}\right)^2 = 1$$

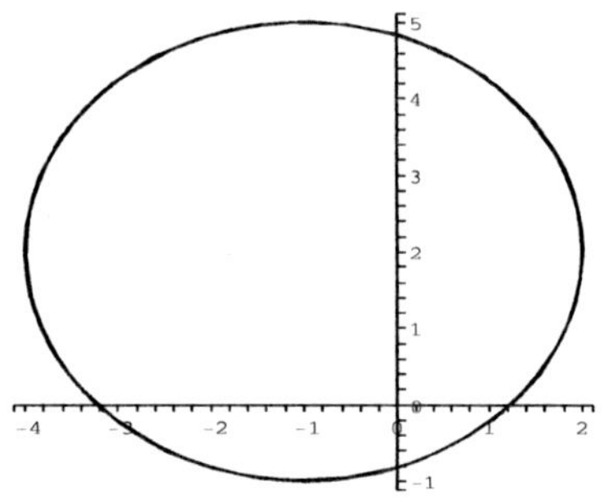

3. From the first equation we have $t^2 = x - 1$. We plug this into the second equation to get

$$y = (t^2)^2 = (x-1)^2$$

which is valid for $x \geq 1$.

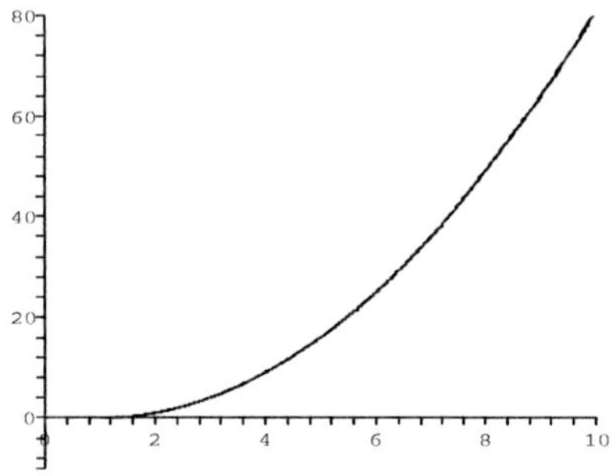

5.

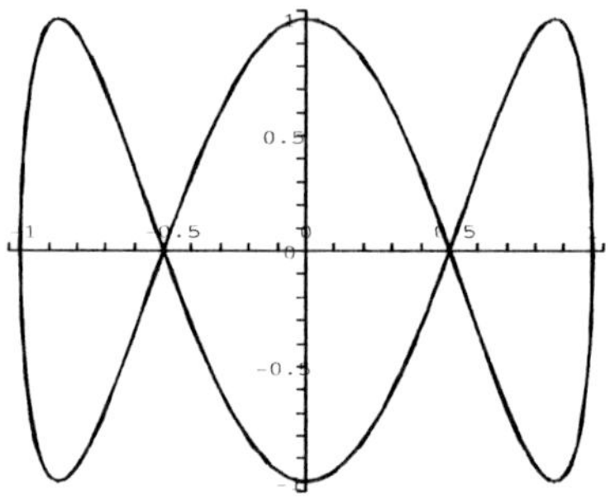

7.

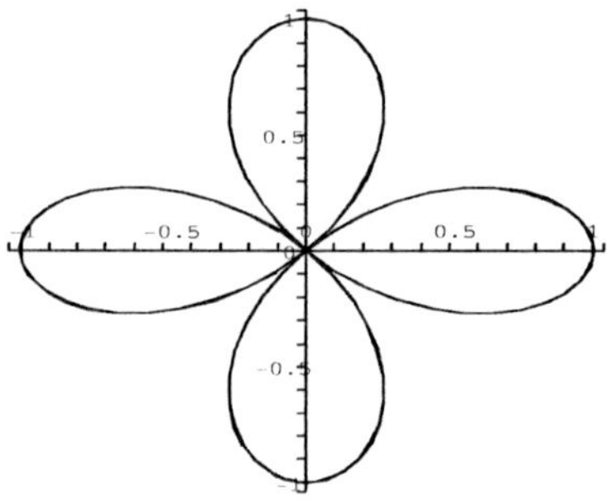

9. Solving the first equation for t gives $t = \sqrt{x+1}$, so $x \geq -1$. Plugging into the second equation gives $y = (x+1)^{3/2}$. This matches Figure C.

11. Sketching the graph, we find that this corresponds to Figure B.

13. We want equations of the form

$$\begin{cases} x &= at + b \\ y &= ct + d \end{cases}$$

on the interval $0 \leq t \leq 1$. At $t = 0$, we need $x = b = 2$ and $y = d = 1$. At $t = 1$, we then have $x = a + 2 = 4$ so $a = 2$ and $y = c + 1 = 7$ so $c = 6$. Therefore our parametric equations for $0 \leq t \leq 1$ are:

$$\begin{cases} x &= 2t + 2 \\ y &= 6t + 1 \end{cases}$$

15. $x'(t) = t^3 - 3t$
$y'(t) = 2t - 1$

(a)

$$\left.\frac{dy}{dx}\right|_{t=0} = \frac{y'(0)}{x'(0)} = \frac{-1}{-3} = \frac{1}{3}$$

(b)

$$\left.\frac{dy}{dx}\right|_{t=1} = \frac{y'(1)}{x'(1)} = \frac{1}{0} \text{ undefined}$$

(c) We need to find a value for t such that $x = t^3 - 3t = 2$ and $y = t^2 - t + 1 = 3$. The latter gives $t^2 - t - 2 = 0$. Solving for t gives $t = 2$ or $t = -1$. Both solve the former equation, so we check the slope at both values of t:

$$\left.\frac{dy}{dx}\right|_{t=-1} = \frac{y'(-1)}{x'(-1)} = \frac{-3}{0} \text{ undefined}$$

and

$$\left.\frac{dy}{dx}\right|_{t=2} = \frac{y'(2)}{x'(2)} = \frac{3}{9} = \frac{1}{3}$$

17. $x'(t) = 3t^2 - 3$
$y'(t) = 2t + 2$
$x'(0) = -3$; $y'(0) = 2$, so the motion is up and to the left.
speed $= \sqrt{(-3)^2 + 2^2} = \sqrt{13}$

19. $x'(t) = 3\cos t$
With $0 \le t \le 2\pi$, the curve is traced out clockwise, so the area A is

$$\begin{aligned} A &= -\int_0^{2\pi} 2\cos t \cdot 3\cos t\,dt \\ &= -6\int_0^{2\pi} \cos^2 t\,dt \\ &= -6\left(\frac{1}{2}t + \frac{1}{2}\sin t\cos t\right)\Big|_0^{2\pi} \\ &= -6\pi \end{aligned}$$

21. $x'(t) = -2\sin 2t$

$$\begin{aligned} A &= \int_{-1}^{1} \sin(\pi t)(-2\sin(2t))\,dt \\ &= -2\int_{-1}^{1} \sin(\pi t)\sin(2t)\,dt \\ &= -2\left(\frac{2\sin(2)\pi}{(\pi-2)(\pi+2)}\right) \\ &\approx -1.947 \end{aligned}$$

23. $x'(t) = -2\sin 2t$
$y'(t) = \sin \pi t$

$$\begin{aligned} s &= \int_{-1}^{1} \sqrt{4\sin^2 2t + \pi^2\cos^2 \pi t}\,dt \\ &\approx 5.2495 \end{aligned}$$

25. $x'(t) = -4\sin 4t$
$y'(t) = 5\cos 5t$

$$\begin{aligned} s &= \int_0^{2\pi} \sqrt{16\sin^2 4t + 25\cos^2 5t}\,dt \\ &\approx 27.185 \end{aligned}$$

27. $x'(t) = 3t^2 - 4$
$y'(t) = 4t^3 - 4$
Let $f(t) = (3t^2-4)^2 + (4t^2-4)^2$. Then

$$\begin{aligned} SA &= \int_{-1}^{1} 2\pi|t^4 - 4t|\sqrt{f(t)}\,dt \\ &\approx 81.247 \end{aligned}$$

29. Multiplying both sides by r gives $r^2 = 3r\cos\theta$. Substitution then gives
$x^2 + y^2 = 3x$.
We then complete the square (as follows) to obtain an equation:

$$x^2 - 3x + y^2 = 0$$

$$\left(x^2 - 3x + \frac{9}{4}\right) + y^2 - \frac{9}{4} = 0$$

$$\left(x - \frac{3}{2}\right)^2 + y^2 = \left(\frac{3}{2}\right)^2$$

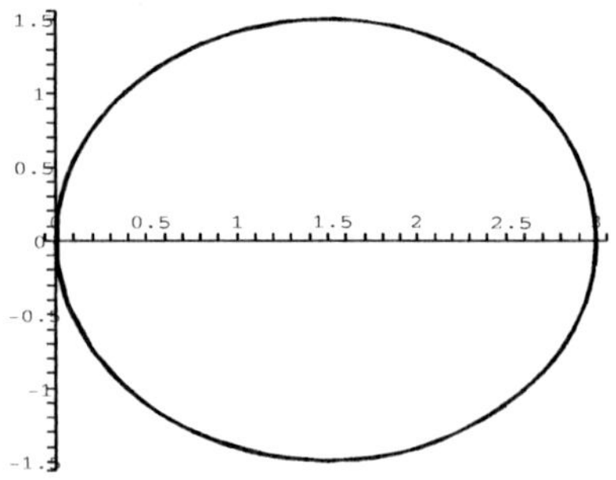

31. $2\sin\theta = 0$ when $\theta = k\pi$ for any integer k. One copy of the graph will be produced by the range $0 \le \theta \le 2\pi$.

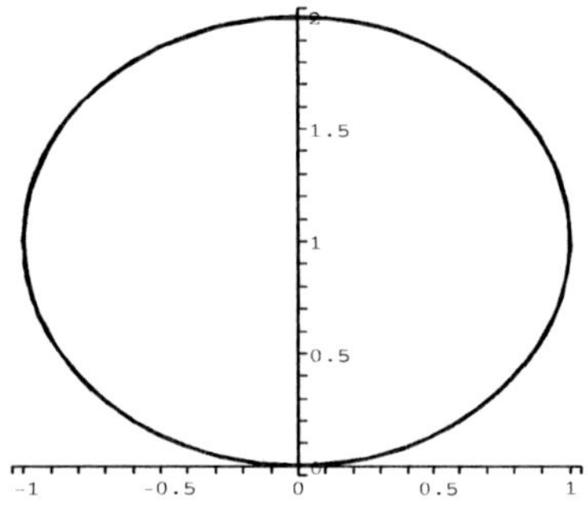

33. $2 - 3\sin\theta = 0$ when $\sin\theta = 2/3$ or $\theta = \sin^{-1}(2/3)$. One copy of the graph will be produced by the range $0 \le \theta \le 2\pi$.

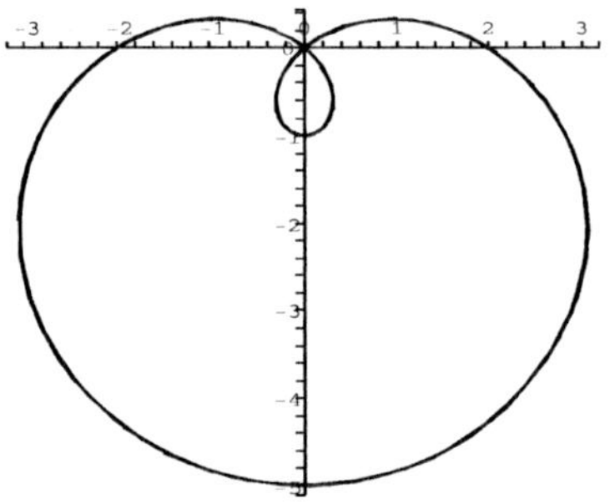

35. $r = 0$ when $\sin 2\theta = 0$, i.e., when $2\theta = k\pi$ or $\theta = k\pi/2$ for any integer k. One copy of the graph will be produced by the range $0 \le \theta \le \pi$.

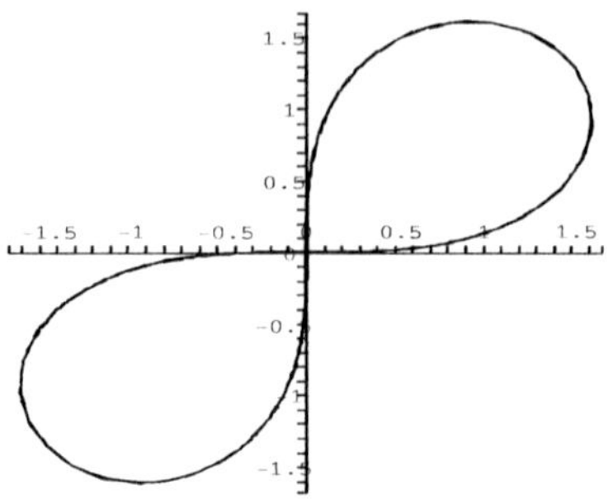

37. $\dfrac{2}{1+2\sin\theta}$ is not equal to 0 for any value of θ. One copy of the graph will be produced by the range $0 \le \theta \le 2\pi$.

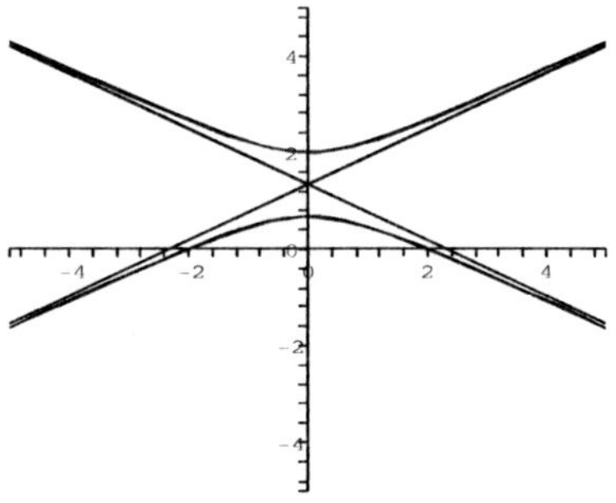

39. This is a circle of radius 3 centered at the origin, so a polar equation is $r = 3$ for $0 \le \theta \le 2\pi$.

41. If $f(\theta) = \cos 3\theta$ then $f'(\theta) = -3\sin 3\theta$. The slope of the tangent line is:

$$\left.\frac{dy}{dx}\right|_{\theta=\frac{\pi}{6}} = \frac{-3\sin\frac{\pi}{2}\sin\frac{\pi}{6} + \cos\frac{\pi}{2}\cos\frac{\pi}{6}}{-3\sin\frac{\pi}{2}\cos\frac{\pi}{6} - \cos\frac{\pi}{2}\sin\frac{\pi}{6}} = \frac{-3\sin\frac{\pi}{2}\sin\frac{\pi}{6}}{-3\sin\frac{\pi}{2}\cos\frac{\pi}{6}} = \frac{\sin\frac{\pi}{6}}{\cos\frac{\pi}{6}} = \frac{\frac{1}{2}}{\frac{\sqrt{3}}{2}} = \frac{1}{\sqrt{3}}$$

43. One leaf of $r = \sin 5\theta$ is traced out by

$0 \le \theta \le \pi/5$ so the area A is:

$$A = \int_0^{\pi/5} \frac{1}{2}(\sin^2 5\theta)\,d\theta$$
$$= \frac{1}{2}\int_0^{\pi/5} (1 - \cos 10\theta)\,d\theta$$
$$= \frac{1}{2}\left(\theta - \frac{\sin 10\theta}{10}\right)\Big|_0^{\pi/5}$$
$$= \frac{1}{2}\left(\frac{\pi}{5} - 0\right) = \frac{\pi}{10}$$

45. Endpoints for the inner loop are given by $\theta = \pi/6$ and $\theta = 5\pi/6$ so the area A is

$$A = \frac{1}{2}\int_{\pi/6}^{5\pi/6} (1 - 2\sin\theta)^2\,d\theta$$
$$= \frac{1}{2}\left(-3\sqrt{3} + 2\pi\right)$$
$$= \frac{-3\sqrt{3} + 2\pi}{2}$$

47. $1 = \sin\theta = 1 + \cos\theta$ when $\theta = \pi/4$ and $\theta = 5\pi/4$. Since we want the region from $\pi/4$ to $5\pi/4$, we find that the area A is

$$A = \frac{1}{2}\int_{\pi/4}^{5\pi/4} (1 + \sin\theta)^2\,d\theta - \frac{1}{2}\int_{\pi/4}^{5\pi/4} (1 + \cos\theta)^2\,d\theta$$
$$= \frac{1}{2}\left(\frac{3\pi}{2} + 2\sqrt{2}\right) - \frac{1}{2}\left(\frac{3\pi}{2} - 2\sqrt{2}\right)$$
$$= 2\sqrt{2}$$

49. $r = f(\theta) = 3 - 4\sin\theta$

$$f'(\theta) = -4\cos\theta$$
$$s = \int_0^{2\pi} \sqrt{f'(\theta)^2 + f(\theta)^2}\,d\theta$$
$$= \int_0^{2\pi} \sqrt{16 + 9 - 24\sin\theta}\,d\theta$$
$$= \int_0^{2\pi} \sqrt{25 - 24\sin\theta}\,d\theta$$
$$\approx 28.814$$

51. Since the focus is $(1, 2)$ and the directrix is $y = 0$, the vertex must be $(1, 1)$. Then $b = 1$, $c = 1$ and $a = 1/4$. The equation is

$$y = \frac{1}{4}(x - 1)^2 + 1.$$

53. Since the foci are $(2, 0)$ and $(2, 4)$ and the vertices are $(2, 1)$ and $(2, 3)$, we see that the center is $(2, 2)$. We then have $x_0 = 2$, $y_0 = 2$, $c = 2$, $a = 1$ and $b = \sqrt{3}$. The equation is

$$\frac{(y-2)^2}{1} - \frac{(x-2)^2}{3} = 1.$$

55. This is an ellipse with $x_0 = -1$, $y_0 = 3$, $a = 5$, $b = 3$ and $c = 4$.
foci: $(-1, 7)$ and $(-1, -1)$
vertices: $(-1, 8)$ and $(-1, -2)$

57. Solving for y gives $y = (x - 1)^2 - 4$. This is a parabola with $a = 1$, $b = 1$ and $c = -4$.
vertex: $(1, -4)$
focus: $(1, -4 + \frac{1}{4})$
directrix: $y = -4 - \frac{1}{4}$

59. The microphone should be placed at the focus, i.e., at $(0, \frac{1}{2})$.

61. Theorem 7.2 part (i) gives

$$r = \frac{(0.8)(3)}{0.8\cos\theta + 1} = \frac{2.4}{0.8\cos\theta + 1}.$$

63. Theorem 7.2 part (iii) gives

$$r = \frac{2.8}{1.4\sin\theta + 1}.$$

65. This is an ellipse with center $(-1, 3)$. Parametric equations are

$$\begin{cases} x &= 3\cos t - 1 \\ y &= 5\sin t + 3 \end{cases}$$

with $0 \le t \le 2\pi$.

67. We provide here a few of the graphs. From these, you can detect the patterns.

Graph of $r = 1 + (-2)\cos\theta$:

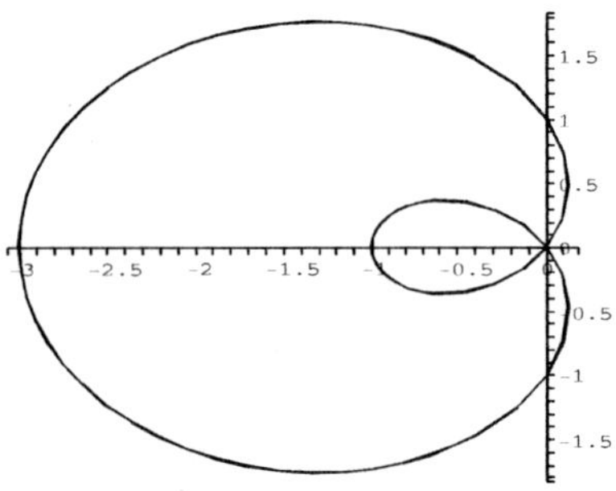

Graph of $r = 1 + (-0.5)\cos\theta$:

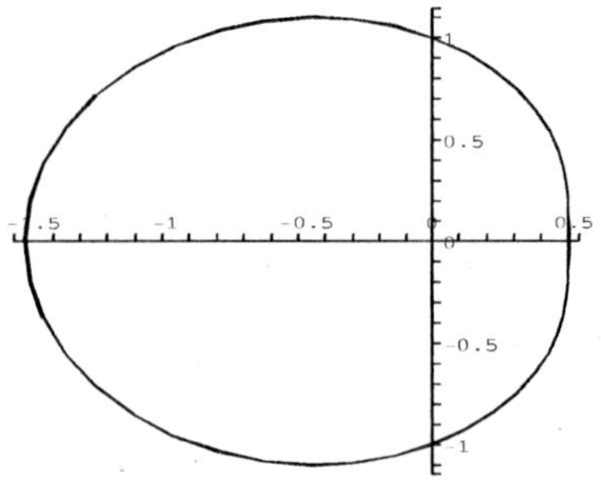

Graph of $r = 1 + (1)\cos\theta$:

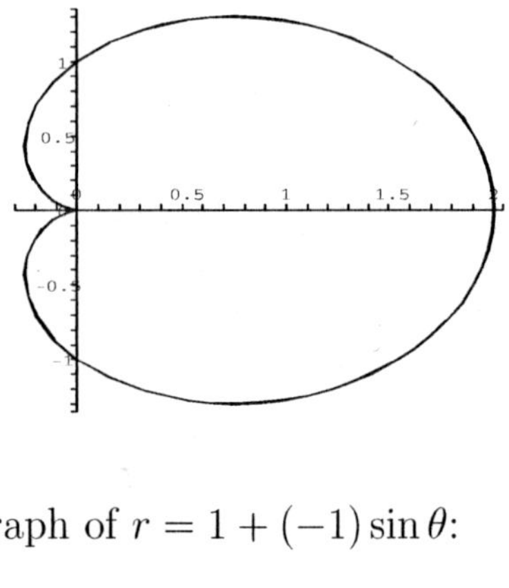

Graph of $r = 1 + (-1)\sin\theta$:

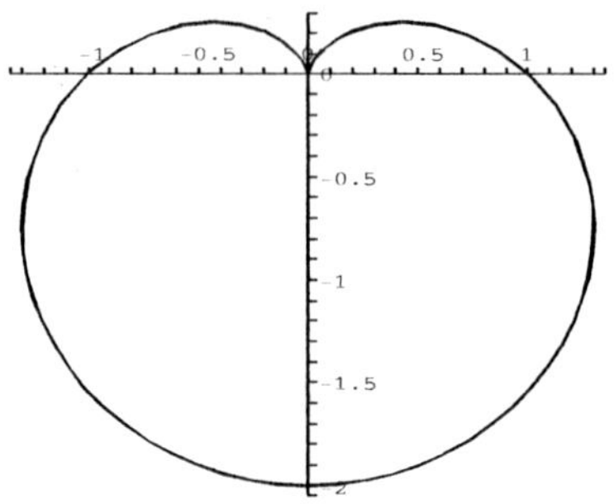

Graph of $r = 1 + (0.5)\sin\theta$:

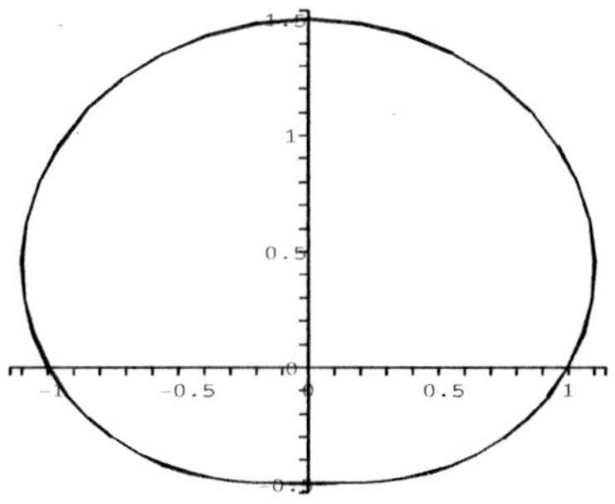

Graph of $r = 1 + (2)\sin\theta$:

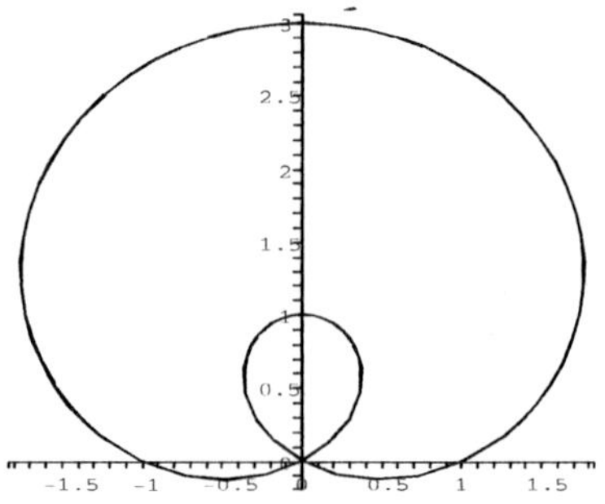